Novel AI Applications for Advancing Earth Sciences

Sudesh Yadav
Department of Higher Education, Government PG College, Ateli, India

Satya Prakash Yadav
Department of Computer Science and Engineering, G.L. Bajaj Institute of Technology and Management, Dr. A.P.J. Abdul Kalam Technical University, Greater Noida, India

Pethuru Raj
Edge AI Division, Reliance Jio Platforms Ltd., Bangalore, India

Prabhakar Tiwari
Madan Mohan Malaviya University of Technology, India

Victor Hugo C. de Albuquerque
Department of Teleinformatics Engineering (DETI), Federal University of Ceará, Brazil

IGI Global
PUBLISHER of TIMELY KNOWLEDGE

A volume in the Advances in Environmental Engineering and Green Technologies (AEEGT) Book Series

Published in the United States of America by
IGI Global
Engineering Science Reference (an imprint of IGI Global)
701 E. Chocolate Avenue
Hershey PA, USA 17033
Tel: 717-533-8845
Fax: 717-533-8661
E-mail: cust@igi-global.com
Web site: http://www.igi-global.com

Library of Congress Cataloging-in-Publication Data

Names: Yadav, Sudesh, 1989- editor. | Yadav, Satya Prakash, editor. | Raj, Pethuru, editor. | Tiwari, Prabhakar, 1975- editor. | De Albuquerque, Victor Hugo C., editor.
Title: Novel AI applications for advancing earth sciences / edited by Sudesh Yadav, Satya Prakash Yadav, Pethuru Raj Chelliah, Prabhakar Tiwari, Victor Albuquerque.
Description: Hershey, PA : Engineering Science Reference, [2024] | Includes bibliographical references and index. | Summary: "The topics covered in this book are advanced and applied in various earth science applications. The data obtained from earth observation satellites is utilized for various applications on ground including agriculture, forestry, mining, water management, rural and urban development, etc"-- Provided by publisher.
Identifiers: LCCN 2023053000 (print) | LCCN 2023053001 (ebook) | ISBN 9798369318508 (hardcover) | ISBN 9798369318515 (ebook)
Subjects: LCSH: Earth sciences--Data processing. | Artificial intelligence--Data processing. | Earth sciences--Computer simulation.
Classification: LCC QE48.8 .N68 2024 (print) | LCC QE48.8 (ebook) | DDC 550.285/63--dc23/eng/20231220
LC record available at https://lccn.loc.gov/2023053000
LC ebook record available at https://lccn.loc.gov/2023053001

This book is published in the IGI Global book series Advances in Environmental Engineering and Green Technologies (AEEGT) (ISSN: 2326-9162; eISSN: 2326-9170)

British Cataloguing in Publication Data
A Cataloguing in Publication record for this book is available from the British Library.

For electronic access to this publication, please contact: eresources@igi-global.com.

Advances in Environmental Engineering and Green Technologies (AEEGT) Book Series

Sang-Bing Tsai
Zhongshan Institute, University of Electronic Science and Technology of China, China & Wuyi University, China
Ming-Lang Tseng
Lunghwa University of Science and Technology, Taiwan
Yuchi Wang
University of Electronic Science and Technology of China Zhongshan Institute, China

ISSN:2326-9162
EISSN:2326-9170

Mission

Growing awareness and an increased focus on environmental issues such as climate change, energy use, and loss of non-renewable resources have brought about a greater need for research that provides potential solutions to these problems. Research in environmental science and engineering continues to play a vital role in uncovering new opportunities for a "green" future.

The **Advances in Environmental Engineering and Green Technologies (AEEGT)** book series is a mouthpiece for research in all aspects of environmental science, earth science, and green initiatives. This series supports the ongoing research in this field through publishing books that discuss topics within environmental engineering or that deal with the interdisciplinary field of green technologies.

Coverage

- Biofilters and Biofiltration
- Pollution Management
- Policies Involving Green Technologies and Environmental Engineering
- Water Supply and Treatment
- Cleantech
- Waste Management
- Radioactive Waste Treatment
- Contaminated Site Remediation
- Electric Vehicles
- Sustainable Communities

IGI Global is currently accepting manuscripts for publication within this series. To submit a proposal for a volume in this series, please contact our Acquisition Editors at Acquisitions@igi-global.com or visit: http://www.igi-global.com/publish/.

The Advances in Environmental Engineering and Green Technologies (AEEGT) Book Series (ISSN 2326-9162) is published by IGI Global, 701 E. Chocolate Avenue, Hershey, PA 17033-1240, USA, www.igi-global.com. This series is composed of titles available for purchase individually; each title is edited to be contextually exclusive from any other title within the series. For pricing and ordering information please visit http://www.igi-global.com/book-series/advances-environmental-engineering-green-technologies/73679. Postmaster: Send all address changes to above address.

Titles in this Series

For a list of additional titles in this series, please visit:www.igi-global.com/book-series/advances-environmental-engineering-green-technologies/73679

Exploring Ethical Dimensions of Environmental Sustainability and Use of AI
Hemachandran Kannan (Woxsen University, India) Raul Villamarin Rodriguez (Woxsen University, India) Zita Zoltay Paprika (Corvinus University of Budapest, Hungary) and Abejide Ade-Ibijola (University of Johannesburg, South Africa)
Engineering Science Reference • copyright 2024 • 426pp • H/C (ISBN: 9798369308929) • US $265.00 (our price)

Modeling and Monitoring Extreme Hydrometeorological Events
Carmen Maftei (Transilvania University of Brasov, Romania) Radu Muntean (Transilvania University of Braşov, Romania) and Ashok Vaseashta (International Clean Water Institute, USA)
Engineering Science Reference • copyright 2024 • 320pp • H/C (ISBN: 9781668487716) • US $250.00 (our price)

Perspectives on Global Biodiversity Scenarios and Environmental Services in the 21st Century
Naveen Kumar Chourasia (Government College, Bichhua, India) and Kavita Chahal (Government College, Bichhua, India)
Engineering Science Reference • copyright 2023 • 294pp • H/C (ISBN: 9781668490341) • US $250.00 (our price)

Palynology and Human Ecology of Africa
Emuobosa Akpor Orijemie (Department of Archaeology and Anthropology, University of Ibadan, Nigeria) and Sylvester Onoriode Obigba (Palynology Laboratory, Department of Archaeology and Anthropology, University of Ibadan, Nigeria)
Engineering Science Reference • copyright 2024 • 300pp • H/C (ISBN: 9781668478011) • US $265.00 (our price)

Nematode-Plant Interactions and Controlling Infection
Waleed Fouad Abobatta (Horticulture Research Institute, Agriculture Research Center, Egypt) and Rehab Yasin Ghareeb (Plant Protection and Biomolecular Diagnosis Department, Arid Lands Cultivation Research Institute, City of Scientific Research and Technological Applications, Egypt)
Engineering Science Reference • copyright 2023 • 289pp • H/C (ISBN: 9781668480830) • US $240.00 (our price)

Promoting Sustainable Management Through Technological Innovation
Supriya Lamba Sahdev (ISBR Business School, India) Chitra Krishnan (Amity University, Noida, India) and Ahdi Hassan (Global Institute for Research Education and Scholarship, The Netherlands)
Engineering Science Reference • copyright 2023 • 293pp • H/C (ISBN: 9781668499795) • US $300.00 (our price)

IGI Global
PUBLISHER of TIMELY KNOWLEDGE

701 East Chocolate Avenue, Hershey, PA 17033, USA
Tel: 717-533-8845 x100 • Fax: 717-533-8661
E-Mail: cust@igi-global.com • www.igi-global.com

Table of Contents

Detailed Table of Contents

In an era defined by unprecedented advancements in technology, AI-based Earth monitoring systems have emerged as transformative tools for understanding and safeguarding our planet. This chapter navigates through the intricacies of AI-based Earth monitoring systems, from the critical technical challenges of data integration, quality, and availability to the crucial interdisciplinary collaborations that hold the key to their success. Improving an AI-based Earth tracking system affords quite a number of particular technical demanding situations. Data integration stands out as a pivotal technical challenge. Addressing data integration, quality, and availability, as well as communication gaps, is critical for their success. When it comes to implementing this device inside the actual world, the scalability of the device and capability protection dangers ought to be addressed. Therefore, designing a framework for an AI-based earth-tracking system requires successfully resolving a diffusion of technically demanding situations.

Edwin Raja S., Department of Computer Science and Engineering, Vel Tech Rangarajan Dr.Sagunthala R&D Institute of Science and Technology, Avadi, Chennai, India
Vimala Josphine C., Department of Artificial Intelligence and Data Science, R.M.K. Engineering College, Chenna, India
Karkuzhali S., Department of Electrical and Electronics Engineering, R.M.D. Engineering College, Chennai, India
Rajakumar S., Department of Electronics and Communication Engineering, Panimalar Engineering College, India

Knowledge-based earth monitoring and prediction system is an innovative tool developed to provide comprehensive insights into Earth's environment and climate. Leveraging advanced methods, artificial intelligence, and machine learning, KEMPS enhances the precision of both short-term and long-term predictions. They combine various statistics resources and expert knowledge to offer valuable insights into the environment and inform selections associated with disaster preparedness, and public coverage. Moreover, research and development of knowledge-based decision support systems for Earth monitoring and prediction will assist users in making informed decisions based on the data. Furthermore, the development of automated forecasting and prediction systems for Earth monitoring and prediction applications should be explored. This way, KEMPS can offer time-applicable, data-based insights and capacity for environmental modifications and forecasts. KEMPS aims to enhance our knowledge of Earth's environmental kingdom, count on changes, and enhance our reaction to global environmental threats.

Jeno Jasmine J., Department of Computer Science and Engineering, R.M.K. Engineering College, Chennai, India
Padmavathi N., Department of Electronics and Instrumentation Engineering, R.M.D. Engineering College, Chennai, India
Theodore Kingslin M., epartment of Electronics and Communication Engineering, R.M.K. Engineering College, Chennai, India
Faiz Akram, Faculty of Computing and Informatics, Jimma Institute of Technology, Jimma University, Jimma, Ethiopia

earthquake detection strategies and spotlight the capability of AI in advancing this subject. The goal is to present a detailed analysis of existing techniques and recommend novel AI-primarily based techniques that could enhance earthquake detection and category accuracy and timeliness. Conventional seismology strategies commonly focus on detecting earthquakes based on an unmarried seismic event. However, AI algorithms can investigate a significant quantity of information, which includes ancient earthquake facts, geological capabilities, and actual-time seismic signals, to become aware of patterns and classify earthquakes into multiple instructions.

The emergence of synthetic artificial intelligence (AI) in recent years has revolutionized numerous sectors, which includes Earth Sciences. AI has recently emerged as a powerful tool in Earth Sciences, enabling scientists to analyze enormous amounts of records, broaden models, and make more accurate predictions. AI offers powerful analytical gear to process large quantities of facts and perceive complicated patterns that might, in any other case, be challenging to discover. This ability to investigate and link disparate information assets permits scientists to benefit from complete insights into Earth's structures, assisting us in understanding the using forces at the back of herbal phenomena like earthquakes hurricanes. The episode discusses numerous AI techniques, inclusive of machine getting-to-know, deep mastering, and statistics mining and their application in Earth Sciences. It explores how these strategies can decorate statistics processing, automate complex obligations, and uncover hidden styles in massive datasets.

Monitoring Earth observations performs a crucial role in expertise and predicting numerous phenomena, including weather trade, herbal disasters, land use adjustments, and commercial pollutants. This chapter explores the use of fuzzy logic strategies for monitoring earth observations, considering its capability, blessings, and challenges. Fuzzy logic-based total decision support structures have been established to be adequate gear in monitoring earth observations. One of the key advantages of this method is its ability to deal with vague and unsure data. One of the key findings is the potential of fuzzy good judgment-based monitoring to handle and incorporate uncertainty in earth observations. Conventional monitoring methods frequently battle with uncertainties associated with sensor noise, atmospheric situations, and

extracting significant statistics from such datasets, mainly to improve understanding of geological processes, aid exploration, and danger assessment. This review includes a vast range of programs, such as far-flung sensing, weather modeling, and earthquake prediction.

Saravanan V., Department of Computer Science, Dambi Dollo University, Ethiopia
Golden Stepha N., Department of Mathematics, R.M.K. Engineering College, Chennai, India
Chairma Lakshmi K.R., Department of Electronics and Instrumentation Engineering, R.M.K. Engineering College, Chennai, India
Karthika S., Department of Electronics and Communication Engineering, College of Engineering, Guindy, Anna University, Chennai, India

Sensory intelligence is the capacity of machines to procedure information from their outside and inner environment through the use of sensory fact inputs and making use of system getting-to-know algorithms. This chapter focuses on integrating touch into Earth remark systems to extend the functionality of machines and beautify machine mastering. Standard, the mixing of touch into earth observation systems, presents a way for machines to emerge as more intelligent and higher apprehend and interact with their environment. Machines are capable of soaking up an extra amount of sensory facts and using it to enhance the gadget, studying algorithms to create accurate and reliable fashions. The consequences of this era are some distance-achieving and allow for improved tracking, prediction, and decision-making. With advances in system studying, artificial intelligence, and robotics, the blessings of incorporating touch into earth commentary structures will only keep growing and become more treasured.

Rambabu Nalagandla, Solution Architect, USA
Anitha Avula Varagantham, Department of Computer Science, Institute of Technology, Hawassa University, Awassa, Ethiopia
Ojasvi Pattanaik, Department of Computer Science and Engineering, Vignan Institute of Management and Technology for Women, Ghatkesar, India & JNTU, Hyderabad, India
Balachandra Pattanaik, Department of Electrical and Computer Engineering, College of Engineering and Technology, Wollaga University, Nekemte, Ethiopia
Manjula Pattnaik, Accounting Department, College of Business and Administration, Princess Nourah Bint Abdulrahman University, Saudi Arabia

The goal of this chapter is to offer an outline of how humans may be integrated into AI frameworks for Earth tracking. It's going to take a look at the benefits of human-contact integration and speak about how this technique can be adapted to a diffusion of tracking situations. Furthermore, the human-touch integration additionally gives a human-in-the-loop management device that could evaluate and approve any outputs produced by using the AI models. The human contact can also assist with AI version calibration and validation that might, in any other case, be extra tough or not possible with the AI algorithms on my own. For instance, floor fact information from skilled professionals may be used to cross-validate and calibrate the AI fashions to ensure that they may be producing correct outcomes. Taking those measures

Srijayanthi S., Department of Artificial Intelligence and Data Science, R.M.K. Engineering College, India
Vidyabharathi Dakshinamurthi, Department of Computer Science and Engineering,Sona College of Technology, Salem, India
Sunil Kumar, Faculty of Commerce and Management, SGT University, Gurugrm, India
Manikandan R., Department of Electronics & Communication Engineering, Panimalar Engineering College, India

The application of nearest-neighbor methods in Earth remark has been of developing interest in recent years. These methods, which involve the usage of facts factors that are close to a given question point in an effort to infer properties associated with the latter, were used for a selection of functions, spanning from land cover mapping and trade detection to picture class and segmentation. This chapter will provide an in-depth analysis of the diverse nearest neighbor processes that can be presently hired in Earth observation technologies today. The principle idea in the back of these algorithms is based totally on a mixture of distance and similarity measures. New technologies, such as remote sensing and satellite imaging, have made it possible to observe and monitor the surface of the Earth from space. This approach has been successfully used to deliver precise results in a wide range of domains, including predicting the behavior of the stock market, classifying images, and migrating customer data.

Priyanka Pramila R., Department of Artificial Intelligence and Data Science, R.M.K. Engineering College, India
Sadish Sendil Murugaraj, Department of Emerging Technologies, Guru Nanak Institute of Technology, Ibrahipatnam, India
Sunil Kumar, Faculty of Commerce and Management, SGT University, Gurugrm, India
Balaji S. R., Faculty of Commerce and Management, SGT University, Gurugrm, India

Neuro-inspired algorithms are a powerful device that can be used for earthquake monitoring and prediction. These algorithms can offer precious insights into the conduct of seismic activities and could cause better detection and caution structures. The goal of this chapter is to offer a concise evaluation of cutting-edge research on using neuro-inspired algorithms for earthquake tracking and prediction. In particular, the chapter will discuss the latest traits in information-based procedures and deep studying algorithms, in addition to the challenges and possibilities for advancing research in this area. Furthermore, capacity programs and regions of research will be discussed which could enjoy the use of these algorithms. Ultimately, this chapter can provide an impetus for the broader use of such algorithms and similarly progress in earthquake-related studies.

Polar science is the study of Earth's frozen zones. It finds its application in a diversity of applications such as monitoring the effects of global warming, making weather prediction, studying the migration cycles of birds, understanding the change in climate over decades, etc. Despite its wide applicability, this field presents a diversity of challenges to the cryosphere community. In this chapter, the authors will cover the current challenges in this field and how they are being solved by researchers across the globe. They will discuss the areas of research and their applicability along with the latest and most promising research works in the field of polar sciences. The subject of polar sciences is studied widely in two dimensions. The first is the study of ice of the sea, and the other is the study of snow. To truly grasp the works done in this field, the authors would be covering these sections in detail one after another.

Preface

Welcome to *Novel AI Applications for Advancing Earth Sciences*. This collaborative work represents a bridge between Earth sciences and the transformative capabilities of artificial intelligence (AI). We, Sudesh Yadav, Satya Prakash Yadav, Pethuru Raj, Prabhakar Tiwari, and Victor Hugo C. de Albuquerque, are thrilled to present this comprehensive compilation.

This book serves as a guide for geologists, offering insights into AI concepts and their profound applications in Earth science. It explores machine learning techniques used for predicting geological features, classifying soil, detecting gases, and harnessing technologies like convolutional neural networks, support vector machines, artificial neural networks, and remote sensing satellites. These advancements have wide-ranging applications, from earthquake detection to climate change assessment and the identification of frozen zones on Earth.

Our focus on optimizing algorithms streamlines Earth observations, reducing time complexity, and enhancing accuracy. This book delves into advanced topics rooted in practical applications across various Earth science domains. AI emerges as a critical tool for climate monitoring, land use analysis, environmental hazard warnings, and optimizing renewable energy sources.

This book is tailored for a diverse audience, including undergraduate and postgraduate students, researchers, academicians, and practitioners in Earth sciences. It consolidates technological advancements, novel discoveries, and managerial applications, presenting these concepts in an accessible manner for easy understanding and seamless adaptability.

Spanning topics from AI-based Earth monitoring frameworks to neuro-inspired earthquake prediction algorithms, fuzzy logic-based observation systems, and deep learning approaches for Earth science, this book encapsulates the synergy between innovation and practical application.

We invite you to explore the dynamic interplay between AI and Earth sciences within these pages.

ORGANIZATION OF THE BOOK

As editors of *Novel AI Applications for Advancing Earth Sciences*, we are excited to present an overview of the chapters included in this comprehensive reference book. Each chapter explores diverse facets of AI applications in Earth sciences, shedding light on innovative methodologies, challenges, and groundbreaking advancements.

Chapter 1: "Artificial Intelligence: Applications, Benefits and Future Challenges in Monitoring and Prediction of Earth Observations" by Sathyaraj P, Nirmala G, Vijayalakshmi S, Rajakumar S delves into the transformative impact of AI in Earth observation. It highlights AI's role in enhancing accuracy, mapping, and classification, revolutionizing how we track environmental phenomena.

Chapter 2: "Technical Challenges Faced in Designing Framework for AI Based Earth Monitoring System" by Gnanasekaran T, Nirmala G, Hariharan N, Rajakumar S navigates the complexities of AI-based Earth monitoring systems. It outlines critical technical challenges like data integration, quality, and scalability, crucial for the success of such systems.

Chapter 3: "Knowledge Based Earth Monitoring and Prediction System" by Edwin Raja S, Vimala Josphine C, Karkuzhali S, Rajakumar S introduces the Knowledge-Based Earth Monitoring and Prediction System (KEMPS). This innovative tool leverages AI and machine learning for precise short-term and long-term predictions, enhancing decision-making related to disaster preparedness.

Chapter 4: "Reinforcement Learning Based Earth Observation System" by Jeno Jasmine J, Padmavathi N, Theodore Kingslin M, Faiz Akram unveils a reinforcement learning-based system for event identification in satellite imagery. This chapter illustrates how the RL algorithm optimizes decision-making processes in identifying and analyzing events.

Chapter 5: "Nearest Neighbor Based Earth Observation and Monitoring" by Sutha J, Meenakshi S, ESAKKIAPPAN C, Anitha G aims at accurate land cover mapping and climate change understanding. This chapter explores the precision of monitoring natural environments and changes in land use, crucial for environmental analysis.

Chapter 6: "Development of a Neuro-Inspired Algorithm for Monitoring and Prediction of Earthquakes" by Rekha M, Chilambuchelvan A, Dinesh Babu M, Vasudhevan V presents a robust neuro-inspired algorithm for earthquake monitoring. It focuses on seismic data analysis and supervised learning to enhance earthquake risk assessment.

Chapter 7: "AI Implementation Techniques for Finding Earth's Frozen Zones" by Kavitha M.S., Mahalakshmi P, Manimegalai L, Sundaramoorthi S explores AI techniques for identifying frozen zones, crucial in understanding climate change impacts on glaciers, ice caps, and permafrost regions.

Chapter 8: "Earthquake Multi-Class Detection Using Artificial Intelligence" by Kavitha P, Gopinathan B, Roslin Dayana K, Rathinasabapathi G focuses on using machine learning and AI for multi-class earthquake detection, aiming to improve accuracy and classification.

Chapter 9: "Earth Sciences Unleashed: A Comprehensive Exploration of AI Applications" by Shobana S, Muthumari A, Subashree S, Elumalai G surveys various AI techniques and their applications in Earth Sciences, emphasizing the power of AI in processing vast datasets and predicting natural phenomena.

Chapter 10: "Fuzzy Logic-Based Monitoring Earth Observations" by Karthiga R, Saravanan V, Sunil Kumar, Rajakumar S explores the potential of fuzzy logic in handling uncertain data for Earth observation, offering adaptive reasoning strategies for monitoring and analysis.

Chapter 11: "An Early Warning System for Predicting Earthquake" by Gnanasekaran T, Sofia Vincent J, Dinesh Babu M, Harish Babu T, Chairma Lakshmi K R emphasizes the need for early warning systems in earthquake prediction, highlighting the advancements in technology and research to mitigate seismic events' impacts.

Chapter 12: "Deep Learning Approaches for Earth Science" by Deepali Virmani, Ananda Kumar K S, Sajitha L P, Nikhat Parveen explores the integration of deep learning in Earth science, showcasing its ability to extract meaningful patterns from complex datasets for geological exploration and risk assessment.

Chapter 13: "Sensory Intelligence Integrating MBA Touch into Earth Observation Systems for Enhanced Machine Learning" by Saravanan V, Golden Stepha, K.R.Chairma Lakshmi, Karthika S delves into sensory intelligence and its integration into Earth observation systems, enhancing machines' understanding and interaction with the environment.

Chapter 14: "Human-Touch Integration in AI Frameworks for Earth Monitoring" by Rambabu Nalagandla, Anitha Varagantham, Ojasvi Pattanaik, Balachandra pattanaik, Manjula Pattnaik discusses the benefits and challenges of integrating human touch into AI frameworks for Earth monitoring, facilitating validation and calibration of AI models.

Chapter 15: "Knowledge-Driven Decision Making in Earth Monitoring Systems" by Bharathi B, Kalaivaani P T, Shree Gaayathri V, Rajendiran M focuses on data-driven decision-making in Earth monitoring systems, illustrating its significance in informed decision-making and collaborative problem-solving for environmental challenges.

Chapter 16: "Enhancing Earthquake Prediction with Reinforcement Learning" by LALITHA S D, Madiajagan M, Rajakumari S, Manikandan R examines the potential of reinforcement learning models in earthquake prediction, aiming to provide more accurate forecasting and insight into seismicity patterns.

Chapter 17: "Nearest-Neighbor Approaches in Earth Observation: A Comparative Analysis" by Srijayanthi S, Vidyabharathi Dakshinamurthi, Sunil Kumar, Manikandan R offers an in-depth analysis of nearest-neighbor methods in Earth observation, exploring their utilization in land cover mapping and image classification.

Chapter 18: "Neuro-Inspired Algorithms for Earthquake Monitoring and Prediction" by Priyanka Pramila R, Sadish Sendil Murugaraj, Sunil Kumar, Balaji S R surveys the current research and challenges in utilizing neuro-inspired algorithms for earthquake monitoring, suggesting potential applications and avenues for further research.

Chapter 19: "Exploring Frigid Frontiers: Strategies for Uncovering Earth's Frozen Regions with AI" by Rahul ., Shashank Parmar, Vibhor Singh, Dhaval Shrivastava, Aayush Chowdhury, Neelam Sunda delves into polar science and its applications in understanding climate change impacts. This chapter covers challenges, latest research, and applicability in studying Earth's frozen zones in detail.

These chapters collectively offer a diverse and comprehensive exploration of AI applications in Earth sciences, spanning from earthquake prediction to climate change monitoring and environmental analysis. We invite readers to delve into this compendium, witnessing the synergistic relationship between artificial intelligence and Earth sciences.

IN SUMMARY

As editors of *Novel AI Applications for Advancing Earth Sciences*, we culminate this compendium with a profound appreciation for the diverse perspectives and innovative insights presented within its pages. Each chapter represents a significant contribution to the symbiotic relationship between artificial intelligence and Earth sciences, fostering a deeper understanding of our planet's complexities.

The journey through these chapters has been enlightening, exploring the fusion of AI technologies with Earth observation, earthquake prediction, climate monitoring, and environmental analysis. From the inception of AI-powered Earth monitoring systems to the nuanced applications in predicting seismic activities, mapping frozen zones, and leveraging sensory intelligence, the breadth of advancements showcased here is truly remarkable.

The collaborative efforts of esteemed authors have woven a tapestry of knowledge, offering methodologies, challenges, and breakthroughs that shape the landscape of Earth sciences. The integration of AI techniques—reinforcement learning, fuzzy logic, deep learning, and neuro-inspired algorithms—stands as a testament to the transformative power of technology in deciphering the intricacies of our planet.

This compendium, crafted for academics, researchers, students, and industry professionals, consolidates cutting-edge research and practical applications. It aims to serve as a cornerstone for understanding the fusion of AI and Earth sciences, fostering interdisciplinary collaborations and paving the way for further innovations.

As we conclude this journey, we extend our gratitude to the authors for their scholarly contributions and commend their dedication to advancing the frontiers of knowledge. We hope this compendium serves as a guiding light, inspiring continued exploration, discovery, and the quest for a deeper comprehension of Earth's intricate systems.

We invite readers to delve into these chapters, embracing the synergy between artificial intelligence and Earth sciences, and embark on a transformative journey toward a more insightful understanding of our planet's dynamic nature.

Warmest regards,

Sudesh Yadav
Department of Higher Education, Government PG College, Ateli, India

Satya Prakash Yadav
Department of Computer Science and Engineering, G.L. Bajaj Institute of Technology and Management, Dr. A.P.J. Abdul Kalam Technical University, Greater Noida, India

Pethuru Raj
Edge AI Division, Reliance Jio Platforms Ltd., Bangalore, India

Prabhakar Tiwari
Madan Mohan Malaviya University of Technology, India

Victor Hugo C. de Albuquerque
Department of Teleinformatics Engineering (DETI), Federal University of Ceará, Brazil

Chapter 1
Artificial Intelligence:
Applications, Benefits, and Future Challenges in the Monitoring and Prediction of Earth Observations

Sathyaraj P.
https://orcid.org/0000-0002-2948-3718
epartment of Electronics and Communication Engineering, R.M.K. College of Engineering and Technology, Chennai, India

Nirmala G.
Department of Computer Science and Engineering, R.M.D. Engineering College, Chennai, India

Vijayalakshmi S.
https://orcid.org/0000-0001-8408-9791
Department of Electronics and Communication Engineering, R.M.K. Engineering College, Chennai, India

Rajakumar S.
Department of Electronics and Communication Engineering, Panimalar Engineering College, India

ABSTRACT

Earth observations have become a developing trend over the last decade because of their ability to enable real-time tracking and forecasting of various environmental phenomena, including landslides, drought, floods, and wildfires. However, conventional approaches in Earth observation have relied on guide processing or human interpretation of the statistics. Via the mixing of AI, the Earth statement's achievement has progressed significantly. AI offers automatic and timely analysis of significant volumes of faraway sensing and satellite TV for computer facts, considering progressed tracking of various natural events and approaches. The software of AI in Earth Remark has enabled several advantages, including improved accuracy in mapping and classification of gadgets, detection of hobby gadgets, which include homes, roads, and forests, and tracking of changes in land use and land cover.

DOI: 10.4018/979-8-3693-1850-8.ch001

1. INTRODUCTION

As we comprehend it, the world has rapidly changed in the past two decades with the emergence of synthetic intelligence technologies. A mixture of recent technological advancements, including improved computing strength and accelerated entry to worldwide statistics, has enabled the development of computerized systems with the capacity to interpret Earth remark statistics with increasing accuracy (Dewitte et al., 2021). This era, called artificial intelligence, or AI, is becoming increasingly vital in harvesting and analyzing Earth observations for progressed tracking and prediction of environmental phenomena. Set up AI programs encompassing Earth observations for flood prediction, crop tracking, and oil spill surveillance. AI-enabled solutions are increasingly being deployed, as the era is tremendously suitable for a variety of programs — from vegetation to floods to wildfires. Synthetic intelligence can be used to interpret information from sensors, satellites, and scanners to recognize the changing environment (Russo & Lax, 2022).

AI can revolutionize how we display and expect anomalies in our surroundings. A few examples of the way AI can be utilized in Earth Observations are as follows:

- Detecting changes in plant life, cowl, or crop health. Using AI algorithms makes it feasible to swiftly interpret and examine satellite TV for PC imagery to display adjustments in plant life, including deforestation and land-use adjustments, and to reveal crop fitness.
- Monitoring coastal regions. AI may be used to locate changes in sea stage, wave behavior, and ocean salinity. In flip, this will be used to assist in preventing disaster-degree storms along with hurricanes or tsunamis.
- Reading air pollution. AI may be used to detect subtle modifications in air first-class by analyzing satellite imagery and other information sources. It can be used to discover assets of pollution, supporting inform destiny air quality strategies.
- Real-time flood monitoring. AI may locate flooding in near actual time by tracking river water stages. It will be used in developing early caution structures to notify citizens of coming near floods.
- Tracking energy use. AI may be used to reveal power use and pick out inefficiencies. It could be used to inform strength savings plans or to expand renewable strength sources.

The implementation of synthetic intelligence in Earth observations will convey numerous advantages. Utilizing AI, answers that might be quicker, more correct, and more complete than current methods can be utilized. AI algorithms can rapidly interpret massive statistics sets to perceive environmental changes or trends. Furthermore, AI systems can study their surroundings, meaning the generation will increase the accuracy of its predictions through the years (Hassanien et al., 2020). Furthermore, AI allows generation to timeshift — the element of 'time' may be taken out of the equation, allowing for near real-time prediction and monitoring of environmental phenomena. This improved evaluation can assist governments and policymakers in making knowledgeable decisions at a faster price. The expanded accuracy of AI structures permits better decision-making in environmental areas, including disaster prevention or valuable resource management. Because of this, individuals and institutions could make more significant knowledgeable choices and broaden tailored regulations, which can be more applicable to environmental situations. Despite the potential of AI for environmental applications, various boundaries want to be overcome before the generation may be carried out on a much broader scale. One of the foremost challenges to the successful use of AI in Earth Observations is the availability of facts — AI systems want widespread amounts of records that allow them to accurately interpret and examine the environment. which means businesses have to put money into technology or discover answers to collect and organize facts from numerous resources robotically. The successful deployment of AI heavily depends on the algorithms' accuracy (Pan & Zhang, 2021). AI structures need to be produced and maintained excessively, which will discover subtle environmental changes and tendencies. It calls for a comprehensive know-how of the environmental phenomenon in query. Finally, numerous moral and protection concerns are related to using AI for environmental packages. AI systems must not be used as a substitute for human choice-making but as a device to improve decision-making (Kochupillai et al., 2022). AI algorithms want to be designed to ensure they are secure and noticeable.

1.1.Background and Motivation

The arrival of modern computing technology coupled with the rapid advance of AI applications has opened up a new realm of opportunities for tracking and predicting Earth remark data. AI generation and its associated applications are increasingly being used across various fields, and its importance for knowledge on our planet is difficult to overstate. AI generation is already being used to improve data collection and processing accuracy, and its potential to help in environmental tracking and forecasting is significant. AI technology has several benefits over conventional strategies of monitoring and prediction. Its algorithms can accurately identify developments in information units that could be tough for people to understand (Qader et al., 2021). New AI techniques are being developed to allow quicker and more accurate calculations of complicated issues, which means that forecasting times are notably decreased. Subsequently, AI can assist in reducing the time and money spent on conventional strategies with the aid of eliminating needless information analysis and permitting quicker selections to be made. The ability applications for AI in monitoring and predicting Earth remark facts are huge-ranging. For example, AI can appropriately display natural failures, hit upon illegal activities along with deforestation or poaching, help set up patterns in the air best over a particular vicinity, predict severe climate activities, and locate widespread adjustments in the climate gadget. AI can also enhance the accuracy of drought and flood forecasts, helping improve the accuracy of public fitness records and reaction times. AI is being used to improve the accuracy of predictions for fundamental modifications inside the weather gadget. AI is a mighty era that may be used for various packages to monitor and predict Earth remark information. Its capacity to quickly and appropriately discover tendencies in records units has made it an essential tool for researchers and scientists in various fields. With its potential to enhance forecasting accuracy and reduce fees, AI will probably emerge as an increasingly more important part of our global tracking and prediction efforts.

1.2. Objectives of the Chapter

AI technologies provide many possibilities for tracking and predicting Earth Observations. There is massive capability for using AI in environmental packages, from crop monitoring to flood prediction. Via using AI, decision-making can be multiplied and stepped forward (Wijata et al., 2023). However, this technology can only be used assertively with a suitable aggregate of data, algorithms, and transparent moral pointers. Many demanding situations still want to be solved if it wants to use AI for Earth Observations efficiently.

- To enable choice-making and forecasting for worldwide environmental demanding situations and growing a baseline for land use and land cover changes through specific value-effective tracking.
- To enable improved surveillance and timely identification of severe climate-related phenomena and present the necessary statistics and statistics for tracking land use adjustments and climate variation strategies.
- To provide higher estimates of the hydrological cycle modifications, quantify and screen water availability, supply, and pleasant and supporting gain insights into the spatial sample of pollutants and emission tiers across Earth's surface.
- To support urban land planning and sustainable development initiatives and enable massive-scale mapping and tracking of city infrastructure and urban habitat in growing nations.
- To allow the extraction of relevant statistics from photograph records, earth observations, and satellite TV for PC imagery and enable actual-time monitoring of forest and flora and fauna shares and pass-border flora and fauna migration.

2. FOUNDATIONS OF AI IN EARTH SCIENCES

Artificial intelligence (AI) software in Earth sciences has the potential to revolutionize many regions of Earth technology studies. AI simulates the conduct of the Earth's ecosystem, geology, and oceanography. It is also being utilized in meteorology to predict weather patterns, analyze climate change, and create models that may be used better to recognize the connection between the Earth and its traits. AI can also be applied to apprehend the long-term effects of herbal failures, examine oceanographic styles, and even inform the development of renewable strength assets (Bi et al., 2019). AI has been utilized in Earth sciences to broaden automated structures that can analyze ancient records to gain insights into weather and oceanography styles. Gadget-mastering algorithms discover styles in massive datasets and forecast future ocean and climate conditions. Deep studying algorithms are being used to offer observations that lead to better predictions about herbal activities. AI could also apprehend seismic pastimes and decide natural risk occasions. AI can also be used to create weather fashions and climate projections. Climate fashions and AI are being designed to help understand and expect weather modifications and help studies perceive edition techniques. AI can also be used to look at the effect of human activities on weather and to research the effect of various environmental elements on weather exchange (Saheb et al., 2022).

One of the most thrilling areas of AI for Earth sciences is being used to expand techniques for geological exploration and mineral extraction. Device-gaining knowledge of algorithms may be used to investigate seismic facts amassed from an oil reservoir to become aware of capacity geological capabilities, lessen the hazard of environmental harm, and predict hydrocarbon manufacturing. AI can also be used in geological exploration to discover the maximum possible mineral and functional resource-rich areas (Samarinas et al., 2023). In phrases of environmental safety, AI can be used to improve the accuracy and pace of environmental effect checks. It can also be used to better correctly check the dangers of pollution and weather exchange to tell infrastructure development and environmental regulations (Kuglitsch et al., 2022). ultimately, AI may be used to more appropriately map herbal failures and understand the effect of the environment on human health and safety, which includes the outcomes of herbal failures, air and water pollutants, and global warming. The mixture of AI and Earth sciences can revolutionize many regions of Earth's technological know-how studies. From climate alternatives to natural disaster preparedness, AI's ability to help us better apprehend and defend our environment is extensive. As AI competencies continue to improve, AI will become an effective device for know-how and handling our planet's environment.

3. APPLICATIONS OF AI IN EARTH OBSERVATION

Artificial intelligence software (AI) in Earth observation is becoming more and more critical as our planet faces warming temperatures and other environmental demanding situations. AI can help us recognize and monitor the converting climate, natural screw-ups, pollutants, and other environmental adjustments. Massive datasets may be accrued and analyzed by combining AI with satellite images and different Earth remark technologies like drone-based tracking. It lets us perceive present-day environmental tendencies and expect future problems (Sun & Scanlon, 2019). The capability to get the right of entry to and examine vast amounts of records quickly and correctly offers us an inspection of the diverse strategies riding our weather and other environmental parameters. AI algorithms can detect subtle styles in those datasets that might not be substantial to the naked eye and changes in weather and weather patterns through the years. AI can also help us increase fashions and simulations to understand better how one-of-a-kind phenomena interact and what the consequences are for the environment. AI is also beneficial for identifying and studying natural disasters as they arise. For instance, using AI at the side of satellite TV for PC imagery, it is viable to come across rapidly and map areas suffering from floods or wildfires. It could be a valuable resource for emergency responders and policymakers in determining the high-quality response for affected areas and ensuring that resources are mobilized quickly. AI can help figure out and map potential hazards, allowing us to take preventive measures to mitigate disasters earlier than they arise (Ghimire et al., 2022).

AI can also give a boost to the effectiveness of conservation efforts. By using AI to detect styles in vast datasets of satellite photographs, maps, and other environmental information, conservation planners can better perceive threats to biodiversity, pinpoint regions with high potential for conservation, and expand better strategies that consciousness sources on which they are best. The packages of AI in Earth observation are huge-ranging and numerous. With the aggregate of AI alongside satellite TV for PC imagery, drone-based tracking, and different Earth observation strategies, we are making significant strides in expertise and mitigating the various environmentally demanding situations we face. By using AI to automate and examine datasets, we will reduce the time and assets needed to understand and reply to these challenges and make extra correct and knowledgeable selections about how best to guard our planet (Crowley et al., 2023). Few examples are given below for better understanding, which is also shown in Fig 1.

1. Automated Supervision of Large Remote Sensing Data: AI can help to process large volumes of remote sensing data in a far more efficient and cost-effective manner. Automation algorithms can be used to classify satellite images into categories (vegetation, water bodies, urban settlements, etc.), removing the human involvement in the data processing task, thereby reducing both cost and processing time significantly.
2. Land Use and Land Cover Classification: AI can play an important role in the accurate estimation of land use and land cover of different geographical areas. By leveraging remote sensing and AI technology, various land cover and land use classification algorithms can be used to classify different land cover features.

Figure 1. A review of Earth AI

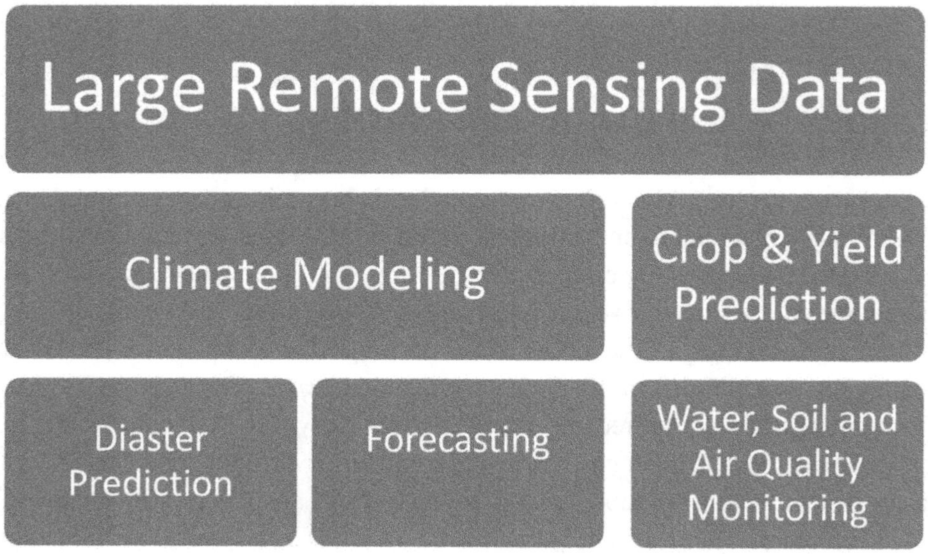

3. Precise Climate Modeling: Climate modeling is an important aspect of remote sensing and AI can be used to make predictive models which will accurately define future climate trends. AI algorithms such as deep learning can be used to collect data from several different sources and to monitor the various processes that drive climate change.
4. Crop Growth and Yield Prediction: Artificial Intelligence can be used to help estimate crop growth and yield for agricultural projects. By analyzing imagery of the crops, AI algorithms can identify key features that indicate crop growth such as color and texture.
5. Forest Fire Forecasting: AI algorithms can be used to detect the presence of forest fires and predict when they are likely to occur. By analyzing factors such as temperature, atmospheric pressure, wind direction, and vegetation type, forest fire forecasting models can be developed.
6. Natural Disaster Prediction: AI can be used to monitor various environmental parameters and to predict natural disasters such as floods, droughts, and hurricanes. By collecting data from various sources and using algorithms such as deep learning, AI can be used to develop predictive models to measure the probability of a natural disaster.
7. Weather Forecasting: AI can be used to develop weather forecasting models which can be used to predict the weather with a high degree of accuracy. The development of advanced AI algorithms such as deep learning can be used to collect large amounts of data from various sources to develop accurate forecast models.
8. Water Quality Monitoring: AI algorithms can be used to monitor and predict water quality by analyzing different water parameters such as temperature, turbidity, sedimentation, dissolved oxygen, contaminants and more. The accuracy of the prediction can be enhanced with AI algorithms.
9. Soil Monitoring: AI can play an important role in the monitoring of soil conditions such as soil moisture levels and nutrient quality. AI algorithms such as machine learning can be used to process large amounts of real-time data to accurately predict the soil conditions.
10. Air Quality Monitoring: Air quality can be monitored in real time with the help of AI algorithms. AI can be used to analyze various air quality parameters including particulate matter, volatile or-

ganic compounds (VOCs), nitrogen oxides, sulfur oxides, and more to accurately monitor the air quality.

3.1. Remote Sensing and Predictive Modelling

Latest advancements in AI have significantly transformed the way of accumulating, handling, and knowledge statistics in Earth statement for far-off sensing programs. AI and mainly Predictive Modelling (PM) are increasingly used to permit sustainable control of land resources. The fulfillment of AI and PM in EO packages has notable opportunities to improve the accuracy of evaluation, lower the time wanted for records series and processing, and supplement proactive management of assets (Gonzales-Inca et al., 2022). Predictive modeling is an AI-primarily based technique that uses earth observations and related statistics to predict destiny situations. It is a powerful device to facilitate the transfer of information from the Earth's physical and biogeochemical systems to predictive outputs. Predictive modeling helps the knowledge of the Earth's environment and its modifications over a non-stop time frame. Predictive modeling is extensively used in one-of-a-kind geospatial programs consisting of land use, forestry, city-making plans, weather modeling, water management, oceanography, weather forecasting, etc. For instance, predictive models can forecast the extension of land use modifications and, direct the proper measures to sustainably manage land resources and mitigate the risks of herbal screw-ups (Kaur et al., 2022).

In addition to predictive modeling, AI has been verified to be helpful in numerous remote sensing strategies. For instance, AI algorithms generate excessive spatial and temporal decisions of plant life statistics through satellite TV for PC imagery. In this process, AI algorithms are used to hit upon flora indices and interpret land cover facts from the photographs. The pictures taken for this utility can be used to determine plant boom, predict land use change, or identify human structures from the flora layer. Furthermore, AI algorithms are used to robotically detect and display land cover and plant phenology using satellite imagery and computer vision. Such structures can be used to recognize the relationship among land cover changes and weather exchange and forecast future land cowl adjustments and vegetation repute. AI and predictive modeling are transforming the way of far-flung sensing in earth statement (Ferreira et al., 2022). The advantages of AI and PM are numerous, which include better accuracy of analysis, decreased time of facts collection and processing, and complements to proactive environmental control. Predictive fashions may facilitate knowledge of the Earth's environment and its adjustments, and AI algorithms are used to stumble on appropriately and screen landcowl and plant life phenology. Those improvements in AI are paving the way for higher decision-making in land control and addressing one-of-a-kind challenges associated with land resources.

3.2. Improved Accuracy and Real-Time Monitoring

Earth's remark based on Artificial Intelligence (AI) is reworking how we display the planet. AI strategies provide excellent possibilities to increase our planet's accuracy and real-time tracking. AI may be used to method uncooked data from a ramification of resources to create an actual-time, dependable picture of the floor of Earth. AI can vastly enhance the accuracy of EO. That is accomplished by gleaning significant insight from vast amounts of information and using AI algorithms to generate fashions that can understand patterns and tendencies, hit upon anomalies, and enhance current EO solutions. With AI, EO systems can obtain far extra degrees of decision and accuracy than was feasible inside the beyond (Tziolas et al., 2021). For example, AI-based answers can stumble on modifications in landforms and vegetation at a much smaller scale than with conventional EO. It allows standard monitoring with fewer resources.

Real-time monitoring is also a key gain of AI in EO. AI permits alerts to be despatched by allowing faster, more frequent, and well-timed statistics processing. At the same time, adjustments in the surroundings are detected, allowing instant action. Real-time tracking additionally enables forecasting of modifications to reduce risks of natural screw-ups and other surprising catastrophic events. AI can also help the improvement of climate statistics structures that assist in guiding more extraordinary, sustainable, and resilient choice-making. AI is revolutionizing how we monitor Earth's environment. Now, not only can AI boost the accuracy of EO solutions, but it could additionally allow more frequent and timely tracking. That is useful for addressing climate trade and dealing with natural disasters. As the knowledge and skills in AI expand, we expect to see even more exciting and valuable packages of AI in EO (Ditria et al., 2022).

3.3. Technical Challenges and Ethical Implications

The technical demanding situations and ethical implications of using AI in Earth remark are of terrific outcome and warrant careful consideration by all stakeholders. The technical undertaking lies in the potential of the AI to correctly interpret and examine imagery from satellites and drones that have very excessive variability in terms of decision, insurance, and record types. An even more significant challenge is developing a comprehensive system that can procedure this significant quantity of facts in a green and price-powerful manner. AI algorithms must remember the source statistics' veracity and the capacity for bias in their analysis. On the moral front, there is incredible capability to abuse statistics from Earth remark. Any AI implementation must be included in protecting citizens' privacy and statistics integrity. As AI is increasingly utilized in smart towns, AI-enabled surveillance may target inclined populations or reveal activities around critical infrastructure (Yadav et al., 2023).

The capability for biased selection-making is of unique concern. AI algorithms can be designed to reflect existing biases in the data. That is specifically trouble regarding photograph popularity, as AI may additionally comprise styles it has inadvertently found that support current stereotypes. AI is used to guide and make policy decisions, but it is crucial not to forget the ethical implications of how information collected via Earth observation is used. AI ought to, without difficulty, be used to map areas with higher concentrations of minority populations, leading to a concentration of choice-making (and sources) in one's regions. Ethical measures should be in the region to ensure equity in selection-making and the equitable distribution of sources across all groups. AI's technical challenges and ethical implications in Earth observation are numerous and huge-ranging. It is essential to ensure that responsible coverage measures are evolved to shield the integrity of statistics and individual liberties and that AI algorithms are designed in any manner to not by accident improve present prejudices.

4. TECHNICAL CHALLENGES IN AI FRAMEWORK DESIGN

As Earth observation continues to broaden, researchers are increasingly targeted at leveraging AI to increase powerful frameworks that can examine Earth-derived facts in actual time. At the same time as the ability to use instances is significant, the technical demanding situations modern day AI framework layout for Earth remark also are enormous and can be efficaciously addressed by way of embedding suitable modeling and algorithms within the design system. as with any AI, the pleasant contemporary the information accumulated should be considered while designing a framework for Earth observation. It calls for attention to today's resolution of contemporary information and the sampling frequency, which is excessive enough to capture high-quality-grained modifications in the Earth's surroundings (Yuvaraj et al., 2021).

The presence of ultra-modern noise within the records also has to be considered, as this forever weakens the accuracy of contemporary AI models used for Earth statements. AI strategies must be included in the AI framework layout to ensure the accuracy of the AI fashions that might be used. Deep contemporary, inclusive of convolutional neural networks, recurrent neural networks, and reinforcement learningare brand new, are the latest essential gadget cutting-edge approaches carried out to Earth commentary datasets with high-quality fulfillment. These AI fashions permit recognizing modern-day patterns from high-dimensional datasets and may efficiently technique significant facts streams in real-time.

The control of present-day facts should also be addressed when designing an AI framework for Earth's statement. It includes a records garage, information workflows, statistics sharing structure, and related facts protection controls. It is vital to ensure that the records are stored securely in an allotted control gadget and that applications can be utilized to ensure that information can be processed quickly and securely. Moreover, a comfortable and disbursed middleware should be incorporated to facilitate facts sharing among different Earth remark systems.

The performance of today's AI-based Earth statement framework may be enhanced by automating specific responsibilities. Autonomous dealers, including swarm robotics, can be used to automate gathering modern environmental information, while systems getting to know modern robotic method automation can automate facts analysis and reporting. Such automation can lessen the value of cutting-edge facts collection and analysis while adding accuracy to the outcomes. AI framework design for Earth observation involves considerable technical demanding situations, which may be addressed by cautious attention to ultra-modern, delicate cutting-edge facts, the usage of contemporary algorithmic strategies, the management of modern-day information, and the automation of contemporary positive responsibilities. With the best consideration, those challenges can be triumphed over, and AI may be used to provide helpful insight into our converting planet.

4.1. Architectural Considerations

Earth remark technology is emerging as a powerful device for monitoring, modeling, and information analysis associated with environmental and weather alternates. To effectively use these records and extract significant insights, sophisticated AI frameworks need to be created to technique the massive quantities of records collected from Earth remark systems. In doing so, several architectural issues must be considered, including scalability, reliability, and latency (Kannan et al., 2021).

Scalability refers to the ability of the AI framework to address a significant influx of information and efficaciously technique it into actionable information. Earth statement facts are usually accumulated from excessive-decision multidimensional satellite imagery and other faraway sensing sources, which leads to a high quantity of statistics that requires speedy and dependable processing. Therefore, it is critical that the AI framework is designed and carried out to allow for scalability and fast processing. Reliability is a vital architectural aspect in AI framework layout and implementation for Earth remark. The framework must be capable of handling facts from diverse sources and must be able to correctly manner the facts. The framework must be sturdy enough to integrate with additional records resources and offer reliable and well-timed insights.

The concept of latency is also important while thinking about AI framework design for Earth commentary. Because statistics are usually captured from faraway sources, there needs to be a system in place to seize the records quickly and method before they can be used for further analysis. That is mainly authentic for real-time programs wherein the AI framework desires to respond to adjustments happening in an environment to accurately music and analyze them (Kousik et al., 2021). Consequently, lowering latency and enhancing performance may be critical to the success of an AI framework. Several architectural concerns should be considered when designing a successful AI framework for Earth Remark. The framework must be designed to be scalable, reliable, and capable of managing latency simply. With this information in hand, the layout of an effective and efficient AI framework may be effectively applied to the Earth statement.

The basic architecture for Artificial Intelligence (AI) in monitoring and prediction of Earth observations typically consists of several components, each of which is responsible for a specific task. The components typically include sensors, data sources, computational algorithms, computing power, software, and any associated databases. Sensors are what allow us to collect data from the environment and are often the first components of the AI stack. This can include anything from satellites, drones, weather stations, and other various sensors, collecting data related to the environment. Data sources are used to store and process the data collected from the sensors. This can vary from local databases, cloud usage for data storage and processing, or various other types of data sources. Computational algorithms are used to process the data, analyze it, and generate predictions or recommendations. This can include machine learning models, deep learning algorithms, and others. Computing power is necessary in order to process the data, run the algorithms, and generate predictions. This can include dedicated computers, cloud services, or any other type of computing platform. Software is also used, both for data management and communication. This can include database management systems, analytics packages, and user interfaces. Lastly, any associated databases are used to ensure accurate information, store and retrieve data, and store the results of various computations such as predictions or analytics.

4.2. Data Integration and Scalability

Earth observation-based AI structures provide valuable facts and insights to make knowledgeable choices and plans for the environment, resources, and development. Such systems have exponentially grown in complexity and size because of an inflow of enter resources and accelerated demand for extra correct and quicker decisions. One of the primary challenges associated with designing such structures is data integration and scalability of the AI framework.

Information integration in AI frameworks is a critical step in the development technique. The need for more seamless integration among one-of-a-kind datasets and structures leads to inefficiencies inside the machine. It is critical to accurately format the information sets and set up facts exchange protocols to ensure accurate information change and integration. Additionally, builders must comply with existing requirements and create new ones to ensure efficient data exchange. Another assignment confronted with AI frameworks is scalability. AI frameworks must be designed to scale effortlessly and effectively to deal with large numbers of statistics and an increasing number of enter sources. To ensure the scalability of the gadget, developers need to remember diverse factors together with hardware architecture, data storage strategies, and software improvement patterns. Modifying and scaling the system according to the changing input records needs should also be flexible. Additionally, builders must leverage the cloud computing era to increase scalability and make the gadget's skills bigger.

The developers need to implement powerful algorithms and models to ensure excessive accuracy and speed of the AI framework. For this motive, they must use existing libraries, including TensorFlow, Caffe, and MXNet, to pick the most appropriate algorithms and models. Additionally, builders must ensure that newly advanced fashions are tested considerably against diverse datasets. The records integration and scalability are two critical challenges confronted inside the layout of AI frameworks for Earth observation. Correctly formatting and exchanging records, scaling the system as in step with the changing input facts wishes, and enforcing appropriate algorithms and fashions are all crucial for successful layout and development. By following these steps, developers can ensure correctly integrated and scalable AI frameworks tailor-made to their facts and operations.

5. AI-BASED EARTH MONITORING SYSTEMS

The development of AI generation has made leaps and limits in contributing to substantially correct earth monitoring. AI-based Earth tracking systems are essential in offering precise evaluations and comments associated with situations on our planet. Those structures were instrumental in improving our surroundings' quality by tracking and exploiting herbal resources while retaining protection and protection in tests. The AI-based totally Earth tracking system is taking modernizing to a new level. In the past, manual methods have been time-consuming and inefficient in detecting environmental adjustments. With the assistance of AI, this machine can offer rapid and accurate data via specified tracking of natural resources. It could music the earth's floor, air, and water composition alongside land assets, ocean surfaces, forests, and oceans (Baklanov & Zhang, 2020). AI-based Earth monitoring systems can assist in holding and optimizing the ability to use herbal resources. They can measure water stages, soil composition, and the health of nearby flowers and forests. They can also analyze trade factors and offer data that can be utilized in mitigating the impacts of these changes.

These structures assist in identifying areas of ecological distress. AI gear is utilized to detect regions where the surroundings have been degraded or disturbed notably. It allows for quick intervention and corrective measures to guard ecosystems, ensure safety, and promote the sustainability of several species of flora, animals, and habitats. AI-based Earth monitoring structures can also monitor and read the present human activity. It consists of the production of merchandise that causes air and water pollutants, the usage of water for irrigation, the transportation of dangerous materials, and the burning of fossil fuels. By retaining this hobby's music, the device can provide statistics for coverage makers and residents on better ecosystem control strategies. AI-based Earth tracking systems are an effective tool that can significantly upgrade the monitoring and protection of the planet's ecology and environment. Those systems can help lessen the effect of human activity on the atmosphere and promote sustainable improvement of resources.

5.1. Integrating Human Expertise and Decision Support

Integrating Human understanding and selection guide for Earth Observations is an idea that may gain natural screw ups management, agriculture management, and environmental protection. Integrating human know-how and choice supports in earth statement structures includes using remote sensing generation and using human enter from nearby experts to improve the accuracy and comprehensiveness of the far-off sensing data collected and interpreted. The choice aid should be able to eat neighborhood knowledge from professionals through artificial intelligence and machine-mastering algorithms. This combination of human information and ML algorithms from remote sensing technologies ensures that information accrued via earth remark systems can be used to make extra informed selections and manage the resources to be had to us in our natural environment.

Step one in integrating human information and selection guide for earth observations begins with the acquisition and advent of virtual maps. Geographic statistics structures (GIS) and far-off sensing technology are used to create those virtual maps, which enable remote sensing facts series and further development of choice support tools. The GIS and faraway sensing technologies create a virtual representation of the surroundings. The choice support methods and ML algorithms may then be utilized to detect changes in the land cowl, generate panorama maps, and monitor the environment. The outputs from these gear can then be used to perceive regions of want and ability areas of control.

The virtual maps were created, and the following step is to integrate human knowledge and choice help for earth observations. It will be completed through AI and ML algorithms, which can offer correct and complete records for decision-making. AI and ML algorithms can also be used to become aware of styles in the far-off sensing information and create predictions and observations. These records can then be used to create control plans and techniques for shielding and dealing with the sources in our natural surroundings. It can also be used to perceive regions of potential chance and to inform the reaction plan for herbal disasters. The mixing of human information and selection guide for earth observations needs to be used to develop services that selection-makers can utilize. Offerings, geospatial records, satellite TV for PC imagery, and floor-based information evaluation may be positioned to become aware of areas of need, reveal changes in the surroundings, and help respond to herbal screw-ups. These records also can be used to broaden techniques for sustainability and improve land use management. On average, integrating human expertise and selection aid for earth observations is an essential step within the control and safety of our herbal surroundings.

6. REINFORCEMENT LEARNING IN EARTH OBSERVATION

Earth statement datasets provide crucial information for clinical studies, land management, and other vital programs. Reinforcement learning(RL) is a system getting to know the approach that has currently won interest for its capacity to procedure massive amounts of temporal information. In this paper, we discuss the critical capabilities of RL and how it can be applied to EO datasets to method, extract, and examine relevant information in a timely and efficient way. Reinforcement learningis a form of machine mastering that seeks to maximize expected praise. This form of gaining knowledge happens when an "agent" learns via trial and mistake and gets a reward or a punishment for each action (Chen et al., 2023). Over time, the agent learns the most reliable behavior to acquire the best praise. RL can be applied in levels: a training section and a testing section.

Within the education phase, agents discover ways to become aware of styles inside the records via the reward device. This mastering procedure is carried out via developing a neural network (NN) or supervised studying algorithms consisting of a regression model or a decision tree. The neural networks are skilled with input data units (including satellite images or DEMs) and output records (consisting of terrain kind or temperature). During the checking-out segment, the community is examined on the unseen input information, and the trained NN will give the ideal response. RL can be beneficial for EO datasets as it can assist in extracting facts from massive datasets well-timed and greenly. It can be used to locate targets or styles in snapshots, along with changes in land cover, or to construct predictive fashions for rainfall wind course. It can be used to classify exceptional terrain, vegetation, floods, or other herbal disasters. The utility of Reinforcement learning to Earth commentary datasets can be a powerful device for plenty of applications. It can significantly lessen the quantity of time and resources required to analyze large datasets and provide correct outcomes in a timely way. Additionally, it can be used to locate goals or patterns in photographs, classify terrain, and build predictive models to forecast natural disasters or changes in land cowl.

6.1. Principles and Applications

Reinforcement learning is a device studying that makes a specialty of training machines to make decisions based totally on information. It is miles closely associated with supervised and unsupervised getting to know. The main difference is that during reinforcement learning, the version is rewarded or penalized for its choices, allowing it to research from its errors. It makes it popular for fixing challenging obligations together with robotics and robotics. In Earth commentary, reinforcement learningmay be used to pick out patterns in records, which include optical satellite TV for computer photos. It may then be used to become aware of similar styles in new data. That is useful for Earth statement duties inclusive of crop type category, identification of land use, and land cowl or detection of urban regions (Dube et al., 2023).

The primary principles of reinforcement learning are based on understanding rewards and punishments. A reinforcement learningof agent learns through trial and error, attempting one-of-a-kind actions and either receiving a reward for appearing an accurate action or being punished for appearing a wrong action. Over time, an agent can learn how to pick out the quality motion based on the rewards and punishments it has received. The packages of reinforcement learningof Earth commentary are varied. Within the field of far-off sensing, it can be used to classify objects in snapshots or detect adjustments in land use. It may also be used to discover patterns in big datasets and how land use has modified over time. For example, reinforcement learningmay be used to perceive deforestation in an area by detecting adjustments inside the wide variety of trees through the years.Further to its applications in far-flung sensing, reinforcement learningcan also be used in crop management to identify plant life sicknesses and improve agricultural manufacturing efficiency. By training an agent to understand styles in plant pictures, it can stumble on ailments in crops and endorse treatments. The use of reinforcement learningin Earth commentary can also help develop and monitor global regulations and policies. With the aid of schooling, an agent can recognize styles in facts, which include changes in weather patterns and populace density. It may be used to reveal global tendencies and inform policy selections. Using reinforcementlearningof Earth remark gives an effective and efficient manner to identify styles in facts and hit upon modifications in land use and weather styles. It may enhance crop management, facilitate international policy decisions, and provide environmental insights.

6.2. Challenges and Opportunities

Earth observation has gained attention in recent years due to its many capacities. Additionally, with the growing availability of data from both devoted EO satellites and other sources of records, including aerial images and UAVs, there has been an increased hobby in how artificial intelligence and system learning techniques can be used to manner, examine, and take advantage of these facts. Reinforcement learning is one vicinity of ML that has grown increasingly famous because of its capacity to study and make influential choices from E) information. It is a place of AI that defines how an agent (e.g., a computer system) can examine its surroundings via trial and mistake and produce a couple of guidelines (Irfan et al., 2022).

RL has many benefits over traditional EO strategies, as it offers higher decision-making and knowledge of cause-effect relationships in data series, evaluation, and interpretation. RL can also optimize records collection from EO, enabling sellers to look extra successfully for the most precious portions of statistics. However, RL additionally poses demanding situations correctly. The primary undertaking is finding the perfect problem in EO to apply RL, as the issue of defining rewards needs to be continually cleaned. Moreover, RL calls for a considerable amount of educational records if it wants to be effective, so acquiring these statistics can be difficult. Furthermore, RL fashions are exceedingly complicated and require extended schooling, so finding approaches to lessen the education times is essential. Eventually, RL models are regularly afflicted by restricted interpretability because of their complexity, making it challenging to understand why upbeat selections were made.

The diverse studies tasks have been carried out, and new approaches have been proposed. For example, research in transfer gaining knowledge, which applies the knowledge received from preceding encounters to remedy a brand new problem, has been used if it wants to use equal statistics sets and boom version accuracy (Rajalakshmi et al., 2022). Moreover, research in reduction techniques such as dimensionality discount, feature choice, and understanding extraction has been used to simplify fashions and lessen the education time. Eventually, studies in interpretability methods, together with visualizations or using device-mastering explainability frameworks, can help to recognize why decisions are being made at the same time as nonetheless making use of the complexities of the version. RL can revolutionize EO by imparting higher decision-making and improved data series from EO. However, numerous challenges must first be addressed, including the need to locate the perfect problem in EO to apply RL and reduce education instances. By addressing those challenges, RL can open up several possibilities for EO, including improved information on motive-impact relationships, more green facts series, and advanced selection-making (Narayanan et al., 2022)-(Albahri et al., 2023).

7. CONCLUSION AND FUTURE SCOPE

Reinforcement learningis a newly evolved machine mastering algorithm that can revolutionize Earth remark by optimizing decision-making techniques and supplying wise, automatic solutions tailor-made to environmental conditions. RL has been used to ramify Earth-oriented applications, such as independent navigation, self-reliant tracking, and land cover category. RL algorithms are also used to optimize the selection, configuration, and deployment of faraway sensing systems and to become aware of anomalies and styles in the information gathered.

Recent improvements in AI and deep learninghave supplied new opportunities for the utility of the RL to Earth statement. Via utilizing the computational energy of cutting-edge AI algorithms, including deep convolutional neural networks and 3-D deep-learningof applications to display Earth's dynamic surface, the capacity for RL to automate and improve Earth remark is sizable. For example, RL can automatically select functions to discover from a given image, pick out anomalous activities or styles, and optimize trade detection operations. Moreover, utilizing AI algorithms has enabled RL to capture more incredible know-how of the environment and to transport beyond the traditional vital capabilities that define a given set of statistics.

RL is also gaining traction in the area of weather exchange research. The speedy growth of deep learningof and reinforcement learning technologies has enabled the automated identification of features in Earth observation records, which may be used to quantify changes in weather and herbal phenomena. Moreover, deep learning algorithms can similarly enhance Earth commentary statistics' accuracy and precision. Within the near destiny, RL-based total models are predicted to evolve to expect severe weather, compare soil moisture, perceive drought trends, and enhance the detection of land-use changes.

RL is hastily advancing inside the field of Earth statement. This new technique can significantly enhance the accuracy of choice-making tactics and provide tailor-made solutions in a ramification of environmental conditions. The improved efficiency of faraway sensing operations, detection of anomalies inside the statistics, and insights into weather trade consequences are fundamental advantages of using RL for Earth observation. However, extra research is needed to recognize the boundaries of cutting-edge RL models and the intrinsic complexities and potential of the set of rules. Future research should also pay attention to integrating RL into an incorporated tracking and exploratory environment, as well as exploring the ability of RL to enhance selection-making in Earth observation.

REFERENCES

Albahri, A. S., Duhaim, A. M., Fadhel, M. A., Alnoor, A., Baqer, N. S., Alzubaidi, L., Albahri, O. S., Alamoodi, A. H., Bai, J., Salhi, A., Santamaría, J., Ouyang, C., Gupta, A., Gu, Y., & Deveci, M. (2023). A systematic review of trustworthy and explainable artificial intelligence in healthcare: Assessment of quality, bias risk, and data fusion. *Information Fusion*, *96*, 156–191. doi:10.1016/j.inffus.2023.03.008

Baklanov, A., & Zhang, Y. (2020). Advances in air quality modeling and forecasting. *Global Transitions*, *2*, 261–270. doi:10.1016/j.glt.2020.11.001

Bi, W. L., Hosny, A., Schabath, M. B., Giger, M. L., Birkbak, N. J., Mehrtash, A., Allison, T., Arnaout, O., Abbosh, C., Dunn, I. F., Mak, R. H., Tamimi, R. M., Tempany, C. M., Swanton, C., Hoffmann, U., Schwartz, L. H., Gillies, R. J., Huang, R. Y., & Aerts, H. J. (2019). Artificial intelligence in cancer imaging: Clinical challenges and applications. *CA: a Cancer Journal for Clinicians*, *69*(2), 127–157. doi:10.3322/caac.21552 PMID:30720861

Chen, M., Claramunt, C., Çöltekin, A., Liu, X., Peng, P., Robinson, A. C., Wang, D., Strobl, J., Wilson, J. P., Batty, M., Kwan, M.-P., Lotfian, M., Golay, F., Joost, S., Ingensand, J., Senousi, A. M., Cheng, T., Bandrova, T., Konecny, M., & Lü, G. (2023). Artificial intelligence and visual analytics in geographical space and cyberspace: Research opportunities and challenges. *Earth-Science Reviews*, *241*, 104438. doi:10.1016/j.earscirev.2023.104438

Crowley, M. A., Stockdale, C. A., Johnston, J. M., Wulder, M. A., Liu, T., McCarty, J. L., Rieb, J. T., Cardille, J. A., & White, J. C. (2023). Towards a whole-system framework for wildfire monitoring using Earth observations. *Global Change Biology*, *29*(6), 1423–1436. doi:10.1111/gcb.16567 PMID:36537002

Dewitte, S., Cornelis, J. P., Müller, R., & Munteanu, A. (2021). Artificial intelligence revolutionises weather forecast, climate monitoring and decadal prediction. *Remote Sensing (Basel)*, *13*(16), 3209. doi:10.3390/rs13163209

Ditria, E. M., Buelow, C. A., Gonzalez-Rivero, M., & Connolly, R. M. (2022). Artificial intelligence and automated monitoring for assisting conservation of marine ecosystems: A perspective. *Frontiers in Marine Science*, *9*, 918104. doi:10.3389/fmars.2022.918104

Dube, T., Seaton, D., Shoko, C., & Mbow, C. (2023). Advancements in Earth Observation for Water Resources Monitoring and Management in Africa: A Comprehensive Review. *Journal of Hydrology (Amsterdam)*, *623*, 129738. doi:10.1016/j.jhydrol.2023.129738

Ferreira, B., Silva, R. G., & Iten, M. (2022). Earth observation satellite imagery information based decision support using machine learning. *Remote Sensing (Basel)*, *14*(15), 3776. doi:10.3390/rs14153776

Ghimire, S., Bhandari, B., Casillas-Perez, D., Deo, R. C., & Salcedo-Sanz, S. (2022). Hybrid deep CNN-SVR algorithm for solar radiation prediction problems in Queensland, Australia. *Engineering Applications of Artificial Intelligence*, *112*, 104860. doi:10.1016/j.engappai.2022.104860

Gonzales-Inca, C., Calle, M., Croghan, D., Torabi Haghighi, A., Marttila, H., Silander, J., & Alho, P. (2022). Geospatial artificial intelligence (GeoAI) in the integrated hydrological and fluvial systems modeling: Review of current applications and trends. *Water (Basel)*, *14*(14), 2211. doi:10.3390/w14142211

Hassanien, A. E., Darwish, A., & Abdelghafar, S. (2020). Machine learning in telemetry data mining of space mission: Basics, challenging and future directions. *Artificial Intelligence Review*, *53*(5), 3201–3230. doi:10.100710462-019-09760-1

Himeur, Y., Rimal, B., Tiwary, A., & Amira, A. (2022). Using artificial intelligence and data fusion for environmental monitoring: A review and future perspectives. *Information Fusion*, *86*, 44–75. doi:10.1016/j.inffus.2022.06.003

Irfan, D., Tang, X., Narayan, V., Mall, P. K., Srivastava, S., & Saravanan, V. (2022). Prediction of Quality Food Sale in Mart Using the AI-Based TOR Method. *Journal of Food Quality*, *2022*, 2022. doi:10.1155/2022/6877520

Kannan, S., Dhiman, G., Natarajan, Y., Sharma, A., Mohanty, S. N., Soni, M., Easwaran, U., Ghorbani, H., Asheralieva, A., & Gheisari, M. (2021). Ubiquitous vehicular ad-hoc network computing using deep neural network with iot-based bat agents for traffic management. *Electronics (Basel)*, *10*(7), 785. doi:10.3390/electronics10070785

Kaur, J., Saxena, J., Shah, J., & Yadav, S. P. (2022, May). Facial Emotion Recognition. In *2022 International Conference on Computational Intelligence and Sustainable Engineering Solutions (CISES)* (pp. 528-533). IEEE.

Kochupillai, M., Kahl, M., Schmitt, M., Taubenböck, H., & Zhu, X. X. (2022). Earth Observation and Artificial Intelligence: Understanding emerging ethical issues and opportunities. *IEEE Geoscience and Remote Sensing Magazine*, *10*(4), 90–124. doi:10.1109/MGRS.2022.3208357

Kousik, N., Natarajan, Y., Raja, R. A., Kallam, S., Patan, R., & Gandomi, A. H. (2021). Improved salient object detection using hybrid Convolution Recurrent Neural Network. *Expert Systems with Applications*, *166*, 114064. doi:10.1016/j.eswa.2020.114064

. Kuglitsch, M., Albayrak, A., Aquino, R., Craddock, A., Edward-Gill, J., Kanwar, R. & Luterbacher, J. (2022). Artificial intelligence for disaster risk reduction: Opportunities, challenges, and prospects. *Bulletin, 71*(1).

Narayanan, K. L., Krishnan, R. S., Robinson, Y. H., Julie, E. G., Vimal, S., Saravanan, V., & Kaliappan, M. (2022). Banana plant disease classification using hybrid convolutional neural network. *Computational Intelligence and Neuroscience*. doi:10.1155/2022/9153699 PMID:35251158

Pan, Y., & Zhang, L. (2021). Roles of artificial intelligence in construction engineering and management: A critical review and future trends. *Automation in Construction*, *122*, 103517. doi:10.1016/j.autcon.2020.103517

Qader, S. H., Dash, J., Alegana, V. A., Khwarahm, N. R., Tatem, A. J., & Atkinson, P. M. (2021). The role of earth observation in achieving sustainable agricultural production in arid and semi-arid regions of the world. *Remote Sensing (Basel)*, *13*(17), 3382. doi:10.3390/rs13173382

Rajalakshmi, M., Saravanan, V., Arunprasad, V., Romero, C. T., Khalaf, O. I., & Karthik, C. (2022). Machine Learning for Modeling and Control of Industrial Clarifier Process. *Intelligent Automation & Soft Computing*, *32*(1). doi:10.32604/iasc.2022.021696

Russo, A., & Lax, G. (2022). Using artificial intelligence for space challenges: A survey. *Applied Sciences (Basel, Switzerland)*, *12*(10), 5106. doi:10.3390/app12105106

Saheb, T., Dehghani, M., & Saheb, T. (2022). Artificial intelligence for sustainable energy: A contextual topic modeling and content analysis. *Sustainable Computing : Informatics and Systems*, *35*, 100699. doi:10.1016/j.suscom.2022.100699

Samarinas, N., Spiliotopoulos, M., Tziolas, N., & Loukas, A. (2023). Synergistic Use of Earth Observation Driven Techniques to Support the Implementation of Water Framework Directive in Europe: A Review. *Remote Sensing (Basel)*, *15*(8), 1983. doi:10.3390/rs15081983

Sun, A. Y., & Scanlon, B. R. (2019). How can Big Data and machine learning benefit environment and water management: A survey of methods, applications, and future directions. *Environmental Research Letters*, *14*(7), 073001. doi:10.1088/1748-9326/ab1b7d

Sun, Z., Sandoval, L., Crystal-Ornelas, R., Mousavi, S. M., Wang, J., Lin, C., Cristea, N., Tong, D., Carande, W. H., Ma, X., Rao, Y., Bednar, J. A., Tan, A., Wang, J., Purushotham, S., Gill, T. E., Chastang, J., Howard, D., Holt, B., & John, A. (2022). A review of earth artificial intelligence. *Computers & Geosciences*, *159*, 105034. doi:10.1016/j.cageo.2022.105034

Tziolas, N., Tsakiridis, N., Chabrillat, S., Demattê, J. A., Ben-Dor, E., Gholizadeh, A., Zalidis, G., & Van Wesemael, B. (2021). Earth observation data-driven cropland soil monitoring: A review. *Remote Sensing (Basel)*, *13*(21), 4439. doi:10.3390/rs13214439

Wijata, A. M., Foulon, M. F., Bobichon, Y., Vitulli, R., Celesti, M., Camarero, R., Di Cosimo, G., Gascon, F., Longépé, N., Nieke, J., Gumiela, M., & Nalepa, J. (2023). Taking Artificial Intelligence Into Space Through Objective Selection of Hyperspectral Earth Observation Applications: To bring the "brain" close to the "eyes" of satellite missions. *IEEE Geoscience and Remote Sensing Magazine*, *11*(2), 10–39. doi:10.1109/MGRS.2023.3269979

Yadav, S. P., Zaidi, S., Nascimento, C. D. S., de Albuquerque, V. H. C., & Chauhan, S. S. (2023, January). Analysis and Design of automatically generating for GPS Based Moving Object Tracking System. In *2023 International Conference on Artificial Intelligence and Smart Communication (AISC)* (pp. 1-5). IEEE. 10.1109/AISC56616.2023.10085180

Yuvaraj, N., Srihari, K., Dhiman, G., Somasundaram, K., Sharma, A., Rajeskannan, S. M. G. S. M. A., Soni, M., Gaba, G. S., AlZain, M. A., & Masud, M. (2021). Nature-inspired-based approach for automated cyberbullying classification on multimedia social networking. *Mathematical Problems in Engineering*, *2021*, 1–12. doi:10.1155/2021/6644652

Chapter 2
Technical Challenges Faced in Designing Frameworks for AI-Based Earth Monitoring Systems

Gnanasekaran T.
https://orcid.org/0000-0003-4005-6482
Department of Electronics and Instrumentation Engineering, R.M.K. Engineering College, Chennai, India

Nirmala G.
Department of Computer Science and Engineering, R.M.D. Engineering College, Chennai, India

Hariharan N.
Science and Humanities (General Engineering division), R.M.K. Engineering College, Chennai, India

Rajakumar S.
Department of Electronics and Communication Engineering, Panimalar Engineering College, India

ABSTRACT

In an era defined by unprecedented advancements in technology, AI-based Earth monitoring systems have emerged as transformative tools for understanding and safeguarding our planet. This chapter navigates through the intricacies of AI-based Earth monitoring systems, from the critical technical challenges of data integration, quality, and availability to the crucial interdisciplinary collaborations that hold the key to their success. Improving an AI-based Earth tracking system affords quite a number of particular technical demanding situations. Data integration stands out as a pivotal technical challenge. Addressing data integration, quality, and availability, as well as communication gaps, is critical for their success. When it comes to implementing this device inside the actual world, the scalability of the device and capability protection dangers ought to be addressed. Therefore, designing a framework for an AI-based earth-tracking system requires successfully resolving a diffusion of technically demanding situations.

DOI: 10.4018/979-8-3693-1850-8.ch002

1. INTRODUCTION

The considerable scale implementation of artificial intelligence (AI) primarily based on the Earth tracking system is a progressive idea in the geospatial field. It is a game changer in how sources are used and managed to shield our Earth. By accumulating and analyzing widespread quantities of satellite imagery, AI-based Earth tracking systems can assist us in better manipulating our planet's sources and reducing environmental impact (Nandutu et al., 2021). Through the utility of deep studying with satellite imagery, AI-based Earth monitoring systems can detect, monitor, and read environmental changes in close to actual time. Further to monitoring, AI utility can mechanically detect abnormalities by detecting changes in the surroundings, implying capability risks or threats. Those systems are designed to offer choice-making support by using computationally advanced methods and clever algorithms for processing photograph analysis. Through AI-driven satellite pictures, natural failures and their beginning factors may be diagnosed in close to real-time, assisting in pre-empting or reducing their effects. AI can pick out floods, droughts, landslides, and other natural phenomena by analyzing the signatures they produce. The combination of AI-primarily based Earth monitoring structures offers numerous benefits to customers:

- They can be used for faster and improved situational cognizance, as monitors can quickly discover adjustments inside the environment.
- They offer extraordinary accuracy of insights into the surroundings and their modifications, even if present in far-off, inhospitable places.
- AI-primarily based Earth monitoring systems are designed to limit the opportunity for human error.
- Those structures are value-powerful and permit the implementation of more powerful sustainable regulations.

AI-based Earth tracking structures are splendid equipment that can revolutionize how we reveal and manipulate our planet for the betterment of humanity. By using ultra-modern technology and far-off sensing strategies, those structures can provide close-to actual-time analyses and responses to any conditions that can get up. With the proper implementation, those structures may want to cause enormous cost savings, progressed accuracy, and progressed decision-making for policymakers.

1.1 Background and Significance

AI-based Earth tracking systems use synthetic intelligence technology to come across, measure, and analyze information from the Earth's surroundings worldwide. AI-based Earth tracking systems are used in various packages associated with weather alternatives, international warming, pollution, natural failures, aid control, sustainable development, and more. Even though AI-based Earth tracking systems are highly new, they may be quickly becoming the move-to generation for companies in search of to apprehend better and control the Earth's fragile environment.The main intention of AI-based Earth monitoring structures is to gather information from far-off sensors established worldwide (Sheng et al., 2021). This record consists of records on air temperature, air high quality, local atmospheric composition, soil situations, snow cowl, sea temperature, sea currents, sea ice, fish stocks, and other elements. by way of the usage of algorithms designed to stumble on patterns within the collected facts, AI-based totally Earth tracking structures can detect signs and symptoms of modifications in the Earth's surroundings or unusual activity. The AI-based total systems can then provide timely signals to inform stakeholders of capacity threats or opportunities and propose measures to deal with them.Via presenting actual-time records and insights into the environmental exchange, AI-based Earth tracking systems can offer many benefits, including helping with manual coverage choices and business practices, detecting adjustments in weather or environmental conditions, assessing risks associated with natural screw-ups inclusive of floods, fires, and earthquakes; studying geographical information structures for land or water rights disputes; detecting unlawful sports or violations of environmental policies; and tracking marine life and fisheries.The advances in computing electricity, statistics garage, and synthetic intelligence algorithms have enabled the increasingly more distinctive evaluation of Earth's environment worldwide (Ullo & Sinha, 2020). As AI-primarily based Earth tracking systems become extra sophisticated, the capacity for their effect on sustainability and natural functional resource management grows. AI-primarily based Earth tracking systems offer a cost-effective, green, and correct tool to help tell choice-making and take movement to protect and sustainably control the Earth's environment.

1.2 Objectives of the Chapter

- To develop an era platform for automated tracking of a massive range of geospatial statistics.
- To allow the automatic analysis and extraction of significant styles from geospatial information.
- To provide an integrated and complete AI-pushed solution for timely environmental tracking.
- To enable the automated interpretation and identification of traits and styles in environmental datasets.
- To broaden a powerful, person-friendly interface for navigating and interpreting environmental statistics.

2. OVERVIEW OF AI-BASED EARTH MONITORING SYSTEMS

AI-based Earth monitoring systems have emerged as crucial to understanding elements of our planet that have been formerly inaccessible or hard to degree. This new era enables us to degree statistics such as wind pace, soil temperature, topography, and other environmental elements more accurately than ever. AI-pushed structures can collect and interpret quantitative and qualitative facts from multiple sensors, allowing us to monitor adjustments in the surroundings through the years and correctly map out nearby areas' bodily traits.Using AI-based tracking systems in agriculture has been incredibly effective. Farmers and researchers can use these systems to quickly and correctly solve problems, including soil moisture ranges, climate styles, and diseases. These records can assist with crop forecasting and yield optimization, as well as the control of weeds, pests, and ailments in an extra green way.AI-driven Earth tracking systems can also assist in stumbling on and mitigating natural screw-ups like floods and fires. Records collected from sensors can offer early warnings to assist people to evacuate and inform emergency services of the state of affairs on the ground in real-time (Muscettola et al., 1998). Similarly, AI-primarily based Earth monitoring systems can come across worldwide weather traits, which include increased temperatures or shrinking ice sheets, so that it will make higher predictions and decisions about the future of our planet. Ultimately, AI-primarily based earth tracking structures may be used to protect vulnerable species through detecting adjustments in their habitats and populace numbers. AI can monitor interest and alert conservationists while animals or plant life are in dangerous conditions, permitting

Figure 1. Architecture of an AI-based Earth-tracking

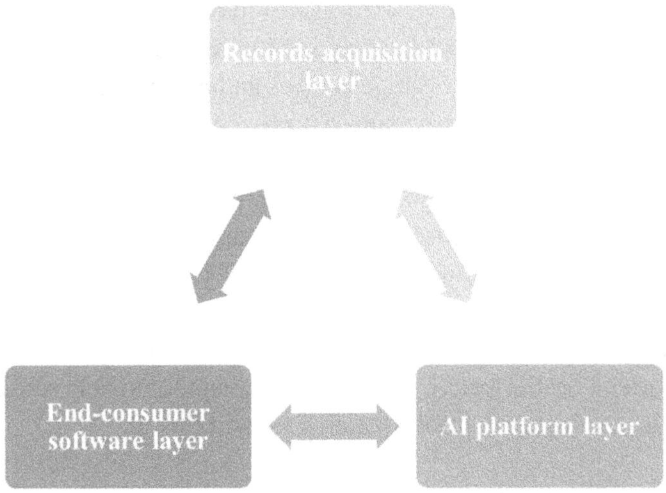

them to intrude quickly and affect remarkable change.AI-based earth monitoring systems allow us to have a remarkable stage of element and accuracy in terms of expertise in the present-day country of our planet. This technology has enabled us to control environmental disasters like heatwaves and floods. It could continue to have an enduring effect on agriculture, conservation, and weather research in the future.

2.1 Key Components and Architecture

Earth monitoring is constantly amassing statistics and processing records about the dynamics of the Earth's environment (Pant et al., 2023). These facts facilitate scientists to understand and predict the behavior of changes in the ecosystem, together with weather trade. AI-primarily based Earth tracking structures are gaining recognition because they can automate information processing and evaluation to offer more accurate and designated facts than ever earlier. The architecture of an AI-based Earth-tracking has shown in the following fig.1

A key thing of AI-primarily based Earth monitoring systems is the sensor community, which consists of sensors that capture records from diverse elements of our surroundings. This fact is then sent to a critical database through satellite TV for a PC or other communication approach. The data is then compared and analyzed with current statistics and models to generate significant insights. The other key element of such structures is the AI platform, which harnesses the energy of gadget studying and artificial intelligence to interpret the information (Shi et al., 2020). The AI platform can make predictions and draw conclusions from the facts, permitting scientists to benefit from new insights into the complex patterns of environmental records. At the heart of the AI platform is a sophisticated neural network, a complex and fast of related nodes and algorithms that might be used to procedure the data. Those neural networks are designed to research from their surroundings and understand patterns so they can analyze data extra effectively. This kind of synthetic intelligence has revolutionized the ability of scientists to interpret information and draw conclusions.The architecture of an AI-based Earth-tracking machine is relatively complex and comprises several layers.

- The first layer is the records acquisition layer, which accommodates the numerous sensors and their conversation networks. This deposit captures the environment's records and sends them to the relevant database.
- The second layer is the AI platform layer, which combines the captured records with fashions and algorithms to generate significant insights. This layer is chargeable for analyzing the facts and making predictions based on the facts.
- The third layer is the end-consumer software layer, which is chargeable for visualizing the insights generated by the AI platform. This sediment is also answerable for providing users with beneficial tools that may assist them in displaying the surroundings more efficaciously.

The components and architecture of AI-based Earth monitoring systems are complex and require sophisticated device mastering and synthetic intelligence (Manickam et al., 2022). Using these structures, scientists can benefit from a complete and precise understanding of the Earth's surroundings and their implications on people and our planet.

The basic architecture for Artificial Intelligence (AI) in monitoring and prediction of Earth observations typically consists of several components, each of which is responsible for a specific task. The components typically include sensors, data sources, computational algorithms, computing power, software, and any associated databases. Sensors are what allow us to collect data from the environment and are often the first components of the AI stack. This can include anything from satellites, drones, weather stations, and other various sensors, collecting data related to the environment. Data sources are used to store and process the data collected from the sensors. This can vary from local databases, cloud usage for data storage and processing, or various other types of data sources. Computational algorithms are used to process the data, analyze it, and generate predictions or recommendations. This can include machine learning models, deep learning algorithms, and others. Computing power is necessary in order to process the data, run the algorithms, and generate predictions. This can include dedicated computers, cloud services, or any other type of computing platform. Software is also used, both for data management and communication. This can include database management systems, analytics packages, and user interfaces. Lastly, any associated databases are used to ensure accurate information, store and retrieve data, and store the results of various computations such as predictions or analytics.

2.2 Integration of AI in Earth Monitoring

Earth tracking is a concern for many governments because of its ability to offer treasured insight into the planet's fitness, including weather, surroundings, ocean activity, and more. As the era hastily advances, more data may be collected, but the challenge lies in deciphering the facts for meaningful outcomes. It is in which the combination of synthetic intelligence (AI) in earth tracking can play an essential position. Those AI-driven earth-tracking answers may be used to discover dangers faster and more accurately than humans. Machines are better at processing large volumes of records quickly and as they should be, in place of humans contingent on subjectivity and instinct for selection making. Additionally, AI algorithms can be skilled to become aware of outliers or anomalies indicative of environmental problems that require interest. AI has already been used to reveal air excellent, soil, water, plants, and climate patterns (Kaur et al., 2022).

AI-driven earth monitoring systems can also be extended to include predictive analytics. Via sustained analysis of datasets; AI structures can be better at spotting patterns and predicting environmental modifications that could occur in a given location. It will supply earth tracking structures with the potential to expect how landscapes and bodies of water may react to certain situations and, similarly, resources within the implementation of mitigation and adaptation strategies on a global scale.Similarly to predictive analytics, mixing AI in earth tracking systems can also increase a more holistic know-how of the surroundings. AI algorithms produce new insights by studying styles and correlations at some distance, more scale and extent than possible with a man or woman (Kousik et al., 2021). For example, AI structures can adequately screen worldwide air to spot how distinct areas affect overall climate information. The combination of AI in earth tracking is a thrilling improvement that may assist us in recognizing our planet and better put together use for potential climate alternate. AI can provide more direct and correct insights than ever while permitting scientists to be extra quick and, as it should be, put in force techniques to guard our planet.

2.3 Case Studies

Case 1: This case study is about designing a framework for an AI-based Earth monitoring system to collect data for land use classification. The challenge was to explore a cost-effective solution that could deliver precise results quickly. The existing systems were able to deliver accurate results but at too high a cost given the amount and type of data to be collected and analyzed. In order to create a cost-effective solution that still delivered accurate results, the team decided to develop and deploy an AI-driven system where multiple aerial images from satellites and drones could automatically be analyzed, classified, and processed. After fruition of painstaking efforts of research and development, the team used machine learning algorithms to successfully detect land use patterns in areas where manual surveys or limited resources prohibited them from carrying out an exhaustive mapping and research. Next, the team put together a training dataset that combined existing satellite images with manually collected aerial photographs, and used it to train the AI-based system with the goal of creating a powerful and accurate land use classification system. Finally, they implemented a web-based tool that housed their AI system and used it to enable public visibility of the analysis, categorization, and mapping of the land use of a particular location.

Case 2: This case study involves designing a framework for an AI-based Earth monitoring system to measure and map water consumption and distribution over time. The team knew that traditional methodologies were cumbersome and expensive, and were therefore unable to provide accurate data over long periods of time. In order to develop an effective, cost-efficient AI-driven system, the team decided to utilize two distinct types of data as inputs. The first was a variety of satellite images from international sources that captured the various water bodies, rivers, and water reservoirs within an area. The second was data collected from robust internet of things (IoT) sensors for tracking water levels within localized geographical locations. The AI-driven system used a natural language processing (NLP) model for interpreting the raw data from the satellite and IoT inputs. It then combined this data with geo-mapping services in order to generate highly-accurate maps of water distribution at specific locations over time. Finally, the team implemented a web-based application that housed their AI system and enabled public visibility of the analysis and mapping of water resources. This project enabled users around the world to easily monitor and analyze their local water distribution at any given time.

3. TECHNICAL CHALLENGES OVERVIEW

Earth tracking is becoming increasingly important as we grapple with international climate alternate. AI has extensive-attaining implications for tackling demanding situations posed by our changing climate. AI should offer real-time tracking of environmental data, enabling us to take action earlier to fight weather trade. AI may also be used to create predictive fashions of environmental trade, permitting us to prepare for future activities.However, several technical challenges are associated with AI software in earth tracking. One of the critical challenges is the compelling series and garage of environmental records. Earth monitoring systems ought to gather information from a ramification of resources consisting of satellites, weather stations, and more. Additionally, facts from specific resources can be formatted differently and appropriately integrated to be correctly analyzed.

Any other task is the accuracy of the AI fashions used for earth tracking. AI has to learn on datasets that accurately replicate the environment. It requires appropriately labeled information and a sufficient quantity of samples. It additionally requires models to be regularly up to date to reflect adjustments in the environment.Finally, developing the critical infrastructure to assist earth-tracking AI is challenging. It consists of growing a dependable network of computer systems to manage large quantities of records in real-time. A properly designed device must also be able to discover anomalies inside the records and alert researchers.The successful implementation of AI-primarily based earth tracking relies on answers to the abovementioned technical challenges. With the proper tools and techniques, AI should become an effective device for combatting weather exchange (Yuvaraj et al., 2021).

4. DATA QUALITY AND AVAILABILITY

Earth monitoring represents a place where AI can make massive advances. AI can be used to understand and help protect the planet from environmental degradation and damage related to climate trade and different human activities. AI's high quality and availability in Earth tracking remains a large mission. Data first-rate is a vital component for AI applications in Earth tracking. Facts required for AI must be dependable and whole for accurate consequences, and that is regularly difficult to acquire while tracking Earth's environment (Sivaram et al., 2019). Facts captured have to have the ability to expose patterns and tendencies that may be used to influence choice-making. Currently, AI is predicated on satellite TV for PC pics to examine Earth's floor, land-use adjustments, land cover, and ecological dynamics. That is constrained utilizing the decision of the facts, which regularly can't as it should be hit upon minor information needed for a clear understanding of Earth's techniques. Additionally, while coping with datasets from distinct assets, it is critical to recall the fact's accuracy degree. The precision and sensor sort of satellite data can also range and should be considered for a hit data analysis.Information availability is another trouble that affects AI applications in Earth tracking. AI systems must work with big sets of data, a lot of which may not exist (Narayanan et al., 2022). Obtaining data may be a complex and luxurious system, especially for technology and gadget studying, which require massive datasets continuously being updated. Additionally, the records supply depends on the enterprise accountable for the statistics. Maximum Earth-tracking establishments are underfunded, and regular statistics should be made available promptly.The tremendous records and availability of AI in Earth tracking are foremost demanding situations that need to be addressed before AI applications may be a hit. Some steps can be taken to enhance records' fine and availability, consisting of progressed collaboration among records companies and facts customers, multiplied admission to Earth records, and higher use of current statistics resources. AI can also enhance facts processing and analysis, which must be explored similarly.

4.1 Data Challenges in Earth Monitoring

Using artificial intelligence (AI) in earth tracking offers several precise, demanding situations. AI plays an increasing number of essential roles in scientists' efforts to become aware of and correctly monitor environmental modifications. However, the sophistication of AI algorithms and massive information techniques have yet to save pace with the exploding quantity of statistics related to Earth tracking. It may lead to enormous problems in accuracy and reliability to gradually progress AI equipment for earth tracking. One widespread mission is the difficulty of information. While AI is used to screen and analyze information from sensors, the first-rate accuracy of the information is paramount. If the information is not amassed in a steady and reliable style, the accuracy and effectiveness of AI tools can be decreased. Additionally, collecting information as part of earth tracking can often be complex and luxurious, so statistics rarely reach the high-quality or insurance levels that AI analysis requires.

Every other task in using AI for earth tracking is the complexity of the statistics. Many physical, chemical, and biological factors affect the environment but are frequently poorly understood (Saravanan & Mohan Raj, 2016). AI algorithms must be capable of accurately perceiving and assessing the interactions among these different factors to offer actionable statistics. It calls for a stage of sophistication and accuracy that currently needs to be achievable with available statistics. Eventually, the vast records associated with earth monitoring necessitate influential facts control strategies. AI algorithms must be able to method the widespread quantities of facts quickly and accurately, an excellent way to offer meaningful insights. It often requires sophisticated database management and fact extraction strategies that require considerable computing strength. Despite these challenges, AI and massive information techniques are exceptional for enhancing our expertise of the surroundings and a way to first-class control it. By overcoming the facts and challenges associated with earth monitoring, we will be capable of successfully revealing and managing the surroundings in a meaningful and sustainable way.

4.2 Strategies for Handling Diverse Data Sources

As AI becomes increasingly more concerned with Earth tracking and control, figuring out excellent strategies for dealing with various information sources is essential. Fig.2 shows the techniques need to be considered for AI based earth monitoring systems.

- Filtering information: Extracting applicable facts from a large body of facts resources is crucial for AI handling. Filtering undesirable and inappropriate statistics gives a more correct photo of the records and enables AI to procedure statistics more effectively.
- Normalization: Normalizing statistics via standardizing the values inside precise statistics resources ensures that AI can successfully evaluate data factors throughout multiple resources for significant insights.
- Visualization: Visualizing records in approaches that make sense to people can help make sure that AI makes the maximum of the statistics units offered. AI can comprehend records more effortlessly with visible representations, allowing for more complex analysis and visualizations.

Figure 2. Considerable techniques for AI based earth monitoring systems

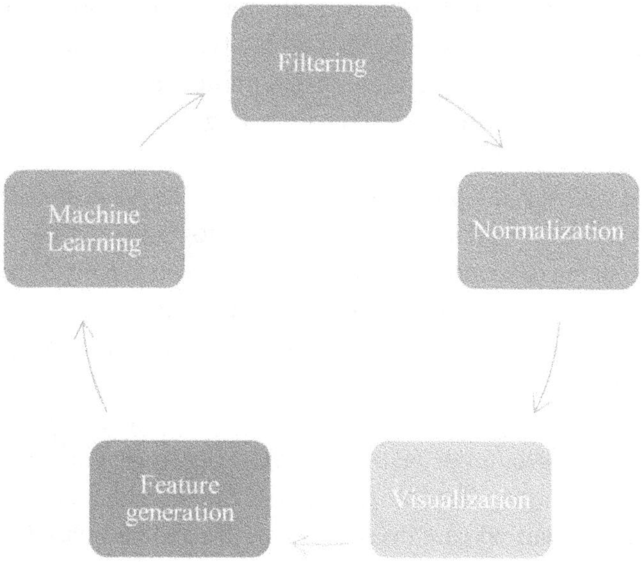

- Feature generation: Developing new features on existing statistics can help AI derive new insights about Earth monitoring efforts and permit higher information evaluation. AI can be more effective while features are generated on existing facts units, enabling AI to analyze patterns and trends better.
- Machine Learning (ML): Integrating ML into AI systems can help in AI's capacity to method and understand significant portions of information from numerous datasets. ML algorithms can also perceive diffused adjustments in records that humans might leave out or fail to seize, imparting more perception into Earth-monitoring processes.

By following those strategies for coping with various statistics sources of AI in Earth monitoring, companies can ensure that AI is applied in the best way (Singh et al., 2022). Those strategies can assist businesses in making sure that AI is making the most of the facts available, resulting in more insights and progressed Earth-monitoring consequences.

5. INTERDISCIPLINARY CHALLENGES

Earth monitoring structures used to be an area of study dominated by social sciences and geosciences. The advent of AI has opened new opportunities to assess our planet's fitness. Though AI-based earth monitoring structures have been immensely beneficial in streaming satellite statistics, the demanding situations of creating such systems are not to be underestimated. AI-based earth tracking systems include the diffusion of interdisciplinary demanding situations, ranging from the choice-making technique and visualization of data acquired from satellites to integrating AI models with current ground-based datasets. One of the foremost demanding situations in constructing an AI-primarily based earth monitoring gadget is the choice-making procedure. AI structures cannot make their own choices, and counting on an artificial intelligence model tailored to investigate information is only sometimes a handy mission. Finding the correct AI version that can both accurately verify the information and make choices that might be useful to the surroundings is complicated. A massive test on a spread of fashions may need to be performed (Ray, 2022). Furthermore, developing and training an AI version for a particular application or project requires knowledge in various fields, including mathematics, PC science, environmental engineering, and statistics technology.

Another venture posed with the aid of the implementation of an AI-primarily based earth monitoring device is the visualization of data obtained from satellite photographs. AI-based earth-tracking systems regularly process large amounts of records, and visualization of such records is essential for developing meaningful insights. One of the difficulties of this procedure is to expand a solid technique for visualizing the records that allow it to apprehend the patterns of environmental change or provide meaningful warnings for impending screw-ups. Such techniques require technical expertise in various fields, including laptop pictures, photo layouts, and visualization sensibilities.An AI-based earth monitoring system must integrate with the current ground-primarily based dataset to generate complete facts and insights into the global environment. Merging datasets is a complex process that calls for technical and area expertise. It is hard to increase a unified data integration machine for AI-based earth monitoring systems thanks to the discrepancy inside the data agency shape and output layout of numerous datasets. AI-based earth tracking systems are robust for providing specific insights into the planet's fitness. However, the interdisciplinary demanding situations posed by such structures' development are complex and must be considered (Guo et al., 2020). By combining present understanding knowledge from various applicable disciplines, we will develop AI fashions to compare our planet's fitness better.

5.1 Collaboration Across Disciplines

A hit AI-based totally earth monitoring machine calls for collaboration amongst a diffusion of disciplines, including laptop technological know-how, geospatial technology, geology, meteorology, ecology, public policy, and global justice. In order to create an earth monitoring gadget that is as comprehensive and as correct as feasible, it is essential for various disciplines to collaborate and to attract on their strengths to create a unified entire.Computer technological know-how forms the center of any AI-primarily based earth tracking gadget, as the programming, algorithms, and data series and evaluation are crucial components of any AI-based task (Landers & Behrend, 2023). Computer technology affords the underlying strategies and structures for this challenge to succeed.Geospatial technological know-how is an essential issue in any such machine, as the accumulated statistics will already be spatially conscious, and distinctly unique mapping of the earth's surface is necessary. Geospatial science, therefore, offers the manner to seize and interpret the relevant earth records accurately.

Geology is likewise crucial with the advent of an AI-based earth-tracking system. Geologists observe the shape and composition of the earth's surface and might offer designated information about the earth's structure and tactics that may be leveraged within the AI gadget.Meteorology is critical for any AI-based earth tracking system because the facts collected must account for the modern-day and current climactic conditions on the world's surface. Meteorology gives the equipment and approach to reveal and measure climate and climactic occasions.Ecology is a crucial part of any AI-based earth monitoring device. As the earth's fitness and well-being depend upon the fitness and well-being of its ecosystems, facts regarding the earth's ecosystems ought to be protected inside the AI monitoring system. Ecology affords the insight essential for the system to keep in mind the relevance of nearby ecosystems and monitor for any changes or anomalies that may be occurring.Public coverage is vital to the success of any AI-primarily based earth tracking gadget, as any public policy surrounding the information collected and analyzed must consider factors such as nearby regulations, privacy problems, and every other ability accidental results of such work.

The worldwide integrity is essential for any AI-based earth monitoring machine, as such systems are meant to advantage all of humanity (Li et al., 2017). Any information gathered and analyzed ought to remember the pursuits of all nations and peoples so that no one institution is unfairly advantaged or deprived.The collaboration between disciplines is crucial for fulfilling any AI-based earth monitoring machine, as each area brings its particular strengths, which are essential for achieving any such task. These disciplines can work collectively to create a complete and accurate earth-tracking device through collaboration.

5.2Addressing Communication Gaps

AI's development has revolutionized tracking modifications on the world's surface through far-flung sensing. This era is immensely treasured in figuring out and predicting changes in climate, soil conditions, water materials, crop fitness, and other elements that can affect our surroundings. However, verbal exchange gaps can exist among the people using these systems and the generation. It is crucial to cope with those gaps to ensure that AI-based earth monitoring systems provide accurate and dependable facts.The conversation between people and the AI device must be clean and concise. The gadget needs to explain in detail what data is being accrued, how it is being interpreted, and any potential mistakes or problems that can arise (Gerdes, 2022). Additionally, humans have to be able to reap insights from the data in a clean-to-apprehend format, including text, diagrams, or charts. It will help to make sure that records are being appropriately interpreted.

Humans must also get hold of ongoing training on using the AI device. This training must be conscious of how the gadget operates, its tactics, and any hazards related to its use. Furthermore, the machine ought to be tested often to ensure it appears efficaciously and according to any protection protocols in the region.Creating a gadget of checks and balances while working with AI-based earth-tracking systems is vital. It consists of figuring out and resolving discrepancies between the facts acquired and the translation of the records. Additionally, the machine has to be regularly evaluated to determine its accuracy and performance. The conversation among all stakeholders should be continuous, with regular remarks and updates being supplied to ensure all parties agree on the device's performance and stop results.Addressing communique gaps in AI-based earth-tracking systems is vital to ensuring that the information they accumulate and examine is correct and reliable (Yu et al., 2022). By organizing clear communication, supplying ongoing training and assessments, and creating a machine of tests and balances, people can correctly paint with those structures and agree with the effects they produce.

6. SCALING AI MODELS FOR LARGE-SCALE EARTH MONITORING

AI has revolutionized earth tracking and environmental research. AI fashions have been used for the correct prediction of climate styles, to screen modifications in natural landscapes, and to research significant portions of information sets quickly and appropriately. At the same time as these AI models provide loads of ability for reading the earth and surroundings, they are confined by their scalability.

AI fashions are designed to discover and analyze styles in quite a small dataset, while large-scale earth monitoring generally requires big datasets of multiple dimensions. To correctly scale AI fashions for massive-scale earth monitoring, model architectures want to be re-tooled to section and procedure huge fact sets well-timed and efficiently (Razdan & Sharma, 2022).

The most effective way to scale AI models for huge-scale earth monitoring is to use a multi-agent structure. Multi-agent architectures are designed to phase large datasets and system them in parallel, saving time and memory. It is completed via dividing the facts into a couple of agents that manner different sets of facts concurrently. The dataset can be partitioned into smaller portions, making it less complicated to use a single massive dataset quickly.Another crucial scaling method is to hire statistics parallelism. Statistics parallelism splits a dataset into a couple of components and strolls the identical rules on each element (Camarena, 2020). In this manner, rather than jogging one set of rules on all the facts, each processor can run its rules on a subset. It allows for the processing of significant volumes of statistics in a much shorter time. Moreover, facts parallelism can assist in lessening errors and improving accuracy, taking into account higher optimization of the algorithm's parameters.

Cloud computing can efficiently scale AI models for large-scale earth tracking. Cloud computing presents on-demand admission to computing assets, considering the big-scale evaluation of earth-tracking datasets. Using cloud computing, researchers and scientists can get entry to large datasets in actual time and analyze them with AI fashions, even reducing storage fees. Furthermore, cloud computing can permit streaming statistics sets, allowing for non-stop tracking and evaluating earth-tracking facts.The Scaling AI fashions for large-scale earth tracking are critical for adequately studying and informing the environment (Li et al., 2020). Recently, organizations and researchers are increasingly counting on AI fashions to investigate massive quantities of facts and uncover treasured insights. By way of expertise, the numerous strategies and strategies used to scale AI fashions, researchers can make a particular proper analysis of enormous datasets for extra accurate and well-timed outcomes.

7. CONCLUSION

The capacity of AI models for Earth monitoring is significant, and there may be an urgent need for additional improvement in this location. AI can recognize the impacts of climate trade, herbal screw-ups, and international activities on ecosystems, populations, and assets. It could be leveraged to expect nearby climate styles, changes in land use, and the spread of disorder. AI can also enhance our knowledge of human-surroundings interactions and create more accurate predictive fashions.The ability of AI for Earth tracking cannot be overstated—it may end up indispensable to many lifestyle sectors, including environmental stewardship, monetary planning, public health, or even countrywide safety. However, the success of AI models depends on robust record sets, transparency, and trust. Going ahead, governments, enterprise, and researchers should ensure that dependable statistics is accrued and processed along rigorous validation protocols to ensure accuracy.Ethical considerations are paramount, and governments and establishments must be considered. The general population should be informed and privy to the potential influences of AI fashions. Sooner or later, researchers must attempt greater accuracy and accuracy-enhancing improvements, switching learning, ensemble learning, and reinforcement.AI fashions for Earth monitoring could revolutionize how we understand, reveal, and interact with the environment. The records governance should be established and stepped forward to ensure accuracy, transparency, and responsibility, and ethical concerns must be considered. Researchers should additionally retain paintings toward accuracy-enhancing innovations and answers. With the right approach to these areas, AI can be an effective device that can empower governments and people and open up an international of the latest opportunities for our planet.

7.1 Summary of Technical Challenges

AI has been the point of interest of new studies and investments in programs related to Earth tracking. AI can now automate the exhausting system of obtaining, processing and studying statistics from Earth commentary structures, satellite pix, drones, and self-reliant motors. AI models for Earth monitoring can offer more accurate and timely insights to researchers, coverage makers, and resource managers for decision-making related to the surroundings.However, AI-driven Earth-tracking projects need several technical challenges. These include handling data scarcity, image information complexity, building fashions for integration with current analytical frameworks, and dealing with privacy and safety issues.

Records shortage is a project with regards to making use of AI fashions for Earth tracking. That is because of the truth that the various facts resources used for Earth tracking are expensive, time-consuming, and restrained in phrases of availability. Furthermore, the data must be completed or included due to cloud cowl and temporal variability. Moreover, in a few cases, the statistics are incredibly variable in picture decision phrases, making it challenging to proceed.Photo facts complexity is another mission for AI models for Earth monitoring. These models want to be sturdy enough to process and interpret complicated information from commentary systems, satellite TV for PC pics, and different imagery resources. The use of deep knowledge of strategies can assist in this regard.

AI fashions for Earth tracking must be constructed to combine with existing analytical frameworks. It is often challenging as the facts analytics method frequently involves more than one stakeholder with a unique understanding. There may also be a want for the fashions to access and combine facts from a diffusion of resources.There also are privacy and security worries to consider while growing AI fashions for Earth tracking. As these fashions' usage increases, so does the hazard of facts leakage or misuse. It is essential to ensure that data is encrypted and that the right of entry is constrained to authorized customers.AI has become a beneficial device in Earth-tracking tasks, but some technical challenges must be addressed to maximize its capacity. Those challenges include handling facts scarcity, dealing with complexity in photo records, constructing fashions for integration with present analytical frameworks, and coping with privacy and safety problems.

7.2 Implications for Future Research

The development of AI for monitoring Earth structures has significant implications for future studies in this area. AI fashions can offer quicker, more accurate, and more special information than conventional methods, which could help scientists gain better information on the dynamics of Earth systems. Moreover, the fashions can leverage massive datasets to pick out patterns and traits, which could offer insights into how the Earth is changing over time. It could enhance our forecasting competencies and help us better adapt to changing conditions.AI fashions can also help us apprehend the impacts of human activities on our surroundings. These models can include information from satellite pictures, climate stations, and social media to music adjustments in ecosystems and discover regions of conservation concern. AI models also can stumble on modifications in air first-rate and other indicators of pollution that can assist in strategies for restoring ecosystems and lowering emissions.

AI models can be used to inform land and beneficial resource management decisions. For example, they can assist in discovering areas liable to drought or climate trade or where touchy plant species are most likely to be threatened. This fact may be used to optimize land use and resource control plans and prioritize regions for conservation.Developing AI models for Earth monitoring has enormous implications for studies in this field. AI fashions can provide faster, more excellent correct facts and help identify styles and tendencies. Moreover, they can help us better understand the influences of human activities on the environment and inform land and functional resource management choices. While those fashions are in their early development tiers, they maintain excellent promise for future research.

REFERENCES

Albahri, A. S., Duhaim, A. M., Fadhel, M. A., Alnoor, A., Baqer, N. S., Alzubaidi, L., Albahri, O. S., Alamoodi, A. H., Bai, J., Salhi, A., Santamaría, J., Ouyang, C., Gupta, A., Gu, Y., & Deveci, M. (2023). A systematic review of trustworthy and explainable artificial intelligence in healthcare: Assessment of quality, bias risk, and data fusion. *Information Fusion*, *96*, 156–191. doi:10.1016/j.inffus.2023.03.008

Camarena, S. (2020). Artificial intelligence in the design of the transitions to sustainable food systems. *Journal of Cleaner Production*, *271*, 122574. doi:10.1016/j.jclepro.2020.122574

Gerdes, A. (2022). A participatory data-centric approach to AI Ethics by Design. *Applied Artificial Intelligence*, *36*(1), 2009222. doi:10.1080/08839514.2021.2009222

Guo, H., Nativi, S., Liang, D., Craglia, M., Wang, L., Schade, S., Corban, C., He, G., Pesaresi, M., Li, J., Shirazi, Z., Liu, J., & Annoni, A. (2020). Big Earth Data science: An information framework for a sustainable planet. *International Journal of Digital Earth*, *13*(7), 743–767. doi:10.1080/17538947.2020.1743785

Himeur, Y., Rimal, B., Tiwary, A., & Amira, A. (2022). Using artificial intelligence and data fusion for environmental monitoring: A review and future perspectives. *Information Fusion*, *86*, 44–75. doi:10.1016/j.inffus.2022.06.003

Kaur, J., Saxena, J., Shah, J., & Yadav, S. P. (2022, May). Facial Emotion Recognition. In *2022 International Conference on Computational Intelligence and Sustainable Engineering Solutions (CISES)* (pp. 528-533). IEEE.

Kousik, N., Natarajan, Y., Raja, R. A., Kallam, S., Patan, R., & Gandomi, A. H. (2021). Improved salient object detection using hybrid Convolution Recurrent Neural Network. *Expert Systems with Applications*, *166*, 114064. doi:10.1016/j.eswa.2020.114064

Landers, R. N., & Behrend, T. S. (2023). Auditing the AI auditors: A framework for evaluating fairness and bias in high stakes AI predictive models. *The American Psychologist*, *78*(1), 36–49. doi:10.1037/amp0000972 PMID:35157476

Li, B. H., Hou, B. C., Yu, W. T., Lu, X. B., & Yang, C. W. (2017). Applications of artificial intelligence in intelligent manufacturing: A review. *Frontiers of Information Technology & Electronic Engineering*, *18*(1), 86–96. doi:10.1631/FITEE.1601885

Li, X., Liu, B., Zheng, G., Ren, Y., Zhang, S., Liu, Y., Gao, L., Liu, Y., Zhang, B., & Wang, F. (2020). Deep-learning-based information mining from ocean remote-sensing imagery. *National Science Review*, *7*(10), 1584–1605. doi:10.1093/nsr/nwaa047 PMID:34691490

Manickam, P., Mariappan, S. A., Murugesan, S. M., Hansda, S., Kaushik, A., Shinde, R., & Thippe-rudraswamy, S. P. (2022). Artificial intelligence (AI) and internet of medical things (IoMT) assisted biomedical systems for intelligent healthcare. *Biosensors (Basel)*, *12*(8), 562. doi:10.3390/bios12080562 PMID:35892459

Muscettola, N., Nayak, P. P., Pell, B., & Williams, B. C. (1998). Remote agent: To boldly go where no AI system has gone before. *Artificial Intelligence*, *103*(1-2), 5–47. doi:10.1016/S0004-3702(98)00068-X

Nandutu, I., Atemkeng, M., & Okouma, P. (2021). Integrating AI ethics in wildlife conservation AI systems in South Africa: A review, challenges, and future research agenda. *AI & Society*, 1–13.

Narayanan, K. L., Krishnan, R. S., Robinson, Y. H., Julie, E. G., Vimal, S., Saravanan, V., & Kaliappan, M. (2022). Banana plant disease classification using hybrid convolutional neural network. *Computational Intelligence and Neuroscience*, *2022*, 2022. doi:10.1155/2022/9153699 PMID:35251158

Pant, G., Bahuguna, R., Pandey, S., Gehlot, A., Yadav, S. P., & Pachauri, R. K. (2023, April). Intelligent Automated Interference for the Protection of Road Safety. In *2023 International Conference on Computational Intelligence, Communication Technology and Networking (CICTN)* (pp. 87-91). IEEE. 10.1109/CICTN57981.2023.10141086

Ray, P. P. (2022). A review on 6G for space-air-ground integrated network: Key enablers, open challenges, and future direction. *Journal of King Saud University. Computer and Information Sciences*, *34*(9), 6949–6976. doi:10.1016/j.jksuci.2021.08.014

Razdan, S., & Sharma, S. (2022). Internet of medical things (IoMT): Overview, emerging technologies, and case studies. *IETE Technical Review*, *39*(4), 775–788. doi:10.1080/02564602.2021.1927863

Saravanan, V., & Mohan Raj, V. (2016). Maximizing QoS by cooperative vertical and horizontal handoff for tightly coupled WiMAX/WLAN overlay networks. *Cluster Computing*, *19*(3), 1619–1633. doi:10.100710586-016-0601-x

Sheng, J., Cai, X., Li, Q., Wu, C., Ai, B., Wang, Y., Kadoch, M., & Yu, P. (2021). Space-air-ground integrated network development and applications in high-speed railways: A survey. *IEEE Transactions on Intelligent Transportation Systems*, *23*(8), 10066–10085. doi:10.1109/TITS.2021.3118557

Shi, W., Zhang, M., Zhang, R., Chen, S., & Zhan, Z. (2020). Change detection based on artificial intelligence: State-of-the-art and challenges. *Remote Sensing (Basel)*, *12*(10), 1688. doi:10.3390/rs12101688

Singh, J., Deepika, J., Sathyendra Bhat, J., Kumararaja, V., Vikram, R., Jegathesh Amalraj, J., & Sakthivel, S. (2022). Energy-efficient clustering and routing algorithm using hybrid fuzzy with grey wolf optimization in wireless sensor networks. *Security and Communication Networks*, *2022*, 2022. doi:10.1155/2022/9846601

Sivaram, M., Porkodi, V., Mohammed, A. S., Manikandan, V., & Yuvaraj, N. (2019). Retransmission DBTMA protocol with fast retransmission strategy to improve the performance of MANETs. *IEEE Access : Practical Innovations, Open Solutions*, *7*, 85098–85109. doi:10.1109/ACCESS.2019.2918723

Sun, Z., Sandoval, L., Crystal-Ornelas, R., Mousavi, S. M., Wang, J., Lin, C., Cristea, N., Tong, D., Carande, W. H., Ma, X., Rao, Y., Bednar, J. A., Tan, A., Wang, J., Purushotham, S., Gill, T. E., Chastang, J., Howard, D., Holt, B., & John, A. (2022). A review of earth artificial intelligence. *Computers & Geosciences*, *159*, 105034. doi:10.1016/j.cageo.2022.105034

Ullo, S. L., & Sinha, G. R. (2020). Advances in smart environment monitoring systems using IoT and sensors. *Sensors (Basel)*, *20*(11), 3113. doi:10.339020113113 PMID:32486411

Yu, W., Patros, P., Young, B., Klinac, E., & Walmsley, T. G. (2022). Energy digital twin technology for industrial energy management: Classification, challenges and future. *Renewable & Sustainable Energy Reviews*, *161*, 112407. doi:10.1016/j.rser.2022.112407

Yuvaraj, N., Karthikeyan, T., & Praghash, K. (2021). An improved task allocation scheme in serverless computing using gray wolf Optimization (GWO) based reinforcement learning (RIL) approach. *Wireless Personal Communications*, *117*(3), 2403–2421. doi:10.100711277-020-07981-0

Chapter 3
Knowledge-Based Earth Monitoring and Prediction Systems

Edwin Raja S.
https://orcid.org/0000-0002-2948-9669
Department of Computer Science and Engineering, Vel Tech Rangarajan Dr.Sagunthala R&D Institute of Science and Technology, Avadi, Chennai, India

Vimala Josphine C.
Department of Artificial Intelligence and Data Science, R.M.K. Engineering College, Chenna, India

Karkuzhali S.
https://orcid.org/0000-0002-6729-052X
Department of Electrical and Electronics Engineering, R.M.D. Engineering College, Chennai, India

Rajakumar S.
Department of Electronics and Communication Engineering, Panimalar Engineering College, India

ABSTRACT

Knowledge-based earth monitoring and prediction system is an innovative tool developed to provide comprehensive insights into Earth's environment and climate. Leveraging advanced methods, artificial intelligence, and machine learning, KEMPS enhances the precision of both short-term and long-term predictions. They combine various statistics resources and expert knowledge to offer valuable insights into the environment and inform selections associated with disaster preparedness, and public coverage. Moreover, research and development of knowledge-based decision support systems for Earth monitoring and prediction will assist users in making informed decisions based on the data. Furthermore, the development of automated forecasting and prediction systems for Earth monitoring and prediction applications should be explored. This way, KEMPS can offer time-applicable, data-based insights and capacity for environmental modifications and forecasts. KEMPS aims to enhance our knowledge of Earth's environmental kingdom, count on changes, and enhance our reaction to global environmental threats.

DOI: 10.4018/979-8-3693-1850-8.ch003

1. INTRODUCTION

KEMPS (Knowledge-based Earth Monitoring and Prediction System) is a machine designed to gather, shop, and analyze earth commentary records to provide perception into weather change, atmospheric strategies, and other worldwide environmental adjustments. The gadget evolved by way of NASA to increase the accuracy of environmental prediction fashions by harnessing the electricity of machine mastering and different records-pushed strategies (Mazzarella et al., 2015).KEMPS presents dozens of environmental datasets from an expansion of resources consisting of meteorological satellites, ocean buoys, climate radars, and floor observations. When mixed with system learning algorithms, these datasets allow KEMPS to supply quick- and lengthy-term predictions and model forecasts with unparalleled accuracy. That is vital for selection-making in industries like agriculture, health, and electricity and to permit more informed forecast models.Furthermore, KEMPS leverages the idea of semantic knowledge representation and ontology to efficiently arrange the developing quantities of Earth statement information accumulated utilizing extraordinary resources. At its core, KEMPS contains a visual expertise base of Earth data inside the shape of an ontology-primarily based search engine. It enables researchers to get admission to required datasets quickly and provides precious insights into rising tendencies, modifications, and possibilities that would help grow more powerful Earth observation fashions (Sadly et al., 2009).

KEMPS also uses gadget intelligence strategies like Natural Language Processing and Deep mastering to make the experience of unstructured environmental facts, permitting researchers to find unrecognized styles and tendencies and determine causal relationships among measured variables. Those insights, blended with other geospatial and far-off sensing information, can provide essential shrewd-based selection-making gear to researchers and assist in guiding environmental policies for policymakers. Improving KEMPS offers a unique opportunity to mix human information and device intelligence, which will benefit essential insight into our surroundings and ultimately help make higher selections that could improve the quality of life for future generations.

A knowledge based Earth monitoring and prediction system (Kemps) is a system that provides real-time data from a variety of sources including satellites, ground sensors, and aircraft to help forecast future patterns in Earth's climate, weather, and environment. Kemps combines data from these sources with knowledge from experts in the fields of climate science, meteorology, and environmental monitoring to create an accurate and reliable forecasting tool. Kemps is used to make predictions about climate change and other environmental phenomena such as ozone depletion, air quality, and drought. It can provide data on areas of the Earth's surface affected by various natural disasters such as hurricanes, earthquakes, and floods. It can also deliver information about potential dangers to water, soil, and wildlife caused by industrial and agricultural activities. All of these issues are important to governments, businesses, and individuals, so it's important to have accurate and reliable data available. Kemps can also provide data and analytics to support decision making in the areas of energy, infrastructure, and sustainable development. The system is used by policy makers, stakeholders, and educators to better understand and communicate the impacts of climate change and other environmental issues. Kemps is an important tool for understanding and predicting the Earth's climate and environment. The system is constantly being updated with new data and insights, so it has the potential to improve the accuracy and reach of predictions.

1.1 Background and Significance

KEMPSs are clever structures that use statistics from various sources to identify patterns, trends, and anomalies inside the Earth's surroundings and make predictions about destiny occasions. KEMPSs are used to monitor and forecast the influences of herbal failures and environmental modifications. They may be used by governments, non-public agencies, and study establishments to help make knowledgeable selections about various factors in their operations.There is a protracted record of human beings using herbal observations and clinical experiments to tell their expertise of the Earth's herbal systems (Vivone et al., 2015). Within the past few years, satellite TV for PCs and remote sensing technologies have further increased our ability to monitor the environment. However, these tools are restricted in scope due to their reliance on a constrained set of environmental parameters and spatial resolution. KEMPSs make the capacity of satellite and far-flung sensing facts bigger by integrating facts from multiple resources and leveraging expert knowledge to discover styles, tendencies, and anomalies. It allows for a complete tracking method and better predictions of future activities.

The importance of knowledge of the effects of environmental modifications on human society is growing because of the growing population and changes in land use/land cowl. Therefore, there may be an urgency to apply information-based systems to screen and expect the effects of environmental changes on human society. KEMPSs are precious for predicting the long-term influences of climate alternates, severe weather occasions, land use/land cover alternates, and environmental factors. This record has implications for catastrophe preparedness, food safety, and economic sustainability. By monitoring the surroundings and supplying forecasts, KEMPSs can also inform public policy selections and plans for coastal management and proper resource allocation. Moreover, KEMPSs are vital for knowing the impacts of weather alternatives on ecosystems and species conservation. By offering distinct tracking and forecasts, KEMPSs may be used to pick out habitat degradation, predict species migration styles, and check the impacts of climate trade on biodiversity. In precis, expertise-based Earth tracking and prediction systems are essential for information, predicting, and dealing with the influences of environmental adjustments on human society and the natural surroundings (Adoko et al., 2013). They combine various statistics resources and expert knowledge to offer valuable insights into the environment and inform selections associated with disaster preparedness, meal protection, species conservation, and public coverage.

1.2 Objectives of the Chapter

- To broaden an integrated facts gadget for tracking, predicting, and supplying knowledge-based total assessment of bodily, organic, and socio-financial strategies inside the Earth's environment.
- To ensure the sustainable and efficient use of natural assets by offering timely and correct facts about the surroundings.
- To enable selection makers to make knowledgeable decisions for recognition of a development, control, and conservation approach.
- To support the improvement of new environmental control technologies and evaluate the capacity and results of world and nearby environmental modifications.
- To reveal and forecast essential worldwide environmental indicators, global warming, associated weather change, ozone depletion, and transboundary air pollutants.
- to evaluate the consequences of climate exchange and other anthropogenic impacts on resource availability and use, biodiversity, soil degradation, and land use changes.

- To screen and expect pollutants and different adverse effects of electricity production and use on air, water, and soil.

2. FOUNDATIONS OF KNOWLEDGE-BASED SYSTEMS IN EARTH SCIENCES

The rule of knowledge-based systems in Earth Sciences is an interdisciplinary method that combines PC science techniques and techniques with information and analytical competencies from the Earth sciences. The principles of this technique are primarily based on research into knowledge-based total structures and their implementation within the Earth sciences. Furthermore, the rules of this approach amplify the past, clearly incorporating KBS into the Earth sciences to realize the price of applying them to Earth-based issues (Pant et al., 2023).In their middle, KBS are PC software program systems comprising substantial expertise. This knowledge is derived from diverse assets, which include books, journals, datasets, professional reviews, historical trends, and more. This expertise then becomes a basis for exploring records and making decisions. KBS can search for styles in data, investigate them, and make predictions. For example, they could come across correlations between meteorological variables, conduct clustering of statistics points to pick out geologic capabilities or forecast complex systems within the Earth sciences.Further to information analysis, KBS can also help choice makers make sound decisions about Earth-based topics or problems. For example, KBS can provide insight into aid management by suggesting great places for utilizing renewable electricity or early warnings of natural disasters like floods or earthquakes. KBS can offer a higher knowledge of how exclusive Earth-primarily based systems engage with each other, how land-use changes affect climate patterns, or how pressures on groundwater sources affect drinking water. In all cases, KBS provides increased accuracy of consequences and helps to increase the efficiency and effectiveness of control decisions (Capraro et al., 2006).

Furthermore, KBS has been used for various tasks, including designing strong spatial fashions, predicting fire chance, and estimating climate exchange effects. All of those obligations involve the combination of multiple facts assets, which can be essential to expect the effects of an Earth-based analysis device accurately. Via growing structures that can be able to 'thinking' and master, KBS creates more excellent practical Earth-primarily based answers and ensures that decisions are made with a more significant diploma of self-assurance.The foundation of knowledge-based structures in Earth Sciences is an interdisciplinary method that can actual-time records collection and evaluation. By incorporating first-class PC science information and Earth technological learning records and analysis, KBS provides a diploma of self-assurance inside the consequences of Earth-based choices (Cheon et al., 2009). It increases the capacity of choice makers to perceive the excellent assets and solutions for any given problem, furthering the purpose of robust control and ensuring the health of the Earth for future generations.

2.1 Overview of Knowledge-Based Systems

Knowledge-based structures (KBS) are a type of sensible system that uses a variety of techniques to interpret and interpret facts from a selection of assets. Those structures are broadly used in various geology, geography, and Earth sciences fields.KBS uses strategies like AI, semantic internet technology, natural language processing, and statistics mining to evaluate numerous records sources and create models based on their conclusions. Those fashions can be used to investigate records to make predictions or provide insights. KBS can be used to research massive volumes of records quickly and as it should be, making them worthwhile tools in presenting expert insight into the Earth's complicated phenomena. KBS has been used in various Earth sciences programs, such as geological mapping, soil productiveness, mineral exploration, environmental monitoring, and risk evaluation. For instance, they had been used to discover and map areas of better mineral awareness, to hit upon changes in terrain or land cowl, and to research and investigate hazardous geological capabilities. KBS has also been used to conduct earthquake and other seismic surveys, examine ice masses and expect volcanic eruptions. Similarly, they may be used for monitoring and forecasting climate phenomena (Menezes et al., 2014).KBS are powerful and complicated gear that can offer precious insights into the Earth's natural systems and processes. It, in flip, can assist in offering precious steering in policy-making and choice-making.KBS have become increasingly popular in recent years due to their ability to technique significant quantities of facts and discover patterns and developments. As the sector of Earth's technological knowledge becomes more and more complicated, those systems are becoming increasingly vital. KBS are essential tools in Earth sciences for some time to return.

2.2 Application of Knowledge-Based Systems in Earth Sciences

Expertise systems are AI generation gear that leverages the information from various sources up-to-date to create an expertise base. This generation has been recently integrated into many disciplines of Earth Sciences, up-to-date seismology, mineral exploration, and climate monitoring. Expertise-based, up-to-date structures are an incredible opportunity to obtain more efficient and accurate predictions on various up-to-date Earth Sciences.In seismology, expertise-based systems are used to discover seismic activities, interpret seismic information to date higher apprehend Earth structures, and determine the capability for destiny herbal hazards (Sivaram et al., 2019). This generation lets researchers use a combination of up-to-date data and sophisticated inference up-to-date updated perceived patterns in seismic occasions and better understand the systems of Earth layers.Information-up-to-date systems are used in mineral exploration to stumble on, up-to-date, and discover underground ore deposits. Sophisticated inference algorithms blended with spatial analysis and up-to-date equipment help geologists and engineers better identify and examine the location and traits of mineral reserves. Exploration outcomes and beyond are also incorporated within the anticipated models, presenting more excellent, dependable, and correct outcomes.

The information-up-to-date structures are also being increasingly utilized in weather tracking and assessment. The records of weather variables, updated temperature, precipitation, and wind velocity, are used in predictive models that integrate facts processing with statistical methods and inference algorithms. Because of this, current meteorologists are up to date to predict a vicinity's climate situations more appropriately and the effect of worldwide climate alternates.The knowledge-based upupdated structures are becoming broadly utilized in Earth Sciences because of the last result of their improved accuracy and predictive competencies. This era is being implemented in seismology, mineral exploration, and weather tracking, with more packages but to return (Hannah et al., 2022). The creation of knowledge-based structures in Earth Sciences has helped us better apprehend the Earth's geological structure and the effect of human sports on our planet's climate.

3. KEY COMPONENTS OF A KNOWLEDGE-BASED EARTH MONITORING SYSTEM

An knowledge-based earth monitoring device is a powerful tool to discover and reveal changes in the surroundings. It enables choice makers to access more confident and well-timed statistics about the earth's environment. This gadget includes several critical components required for it to function efficiently.

- The first key thing of a knowledge-primarily based earth tracking system is the facts acquisition and transmission mechanism. It includes a complete network of sensors, samplers, and tracking stations spread over substantial geographic areas. These sensors are answerable for accumulating and transmitting data associated with various environmental conditions, which include temperature, humidity, and air. The statistics accrued are then transmitted over a committed facts transmission line on the way to be accessed and analyzed using the system.
- The data processing and evaluation device is the second key element of an information-primarily based earth monitoring device. This device is liable for reworking the raw facts from the sensors and reworking them into valuable records. The facts processing and analysis device allows the system to create profiles of critical environmental conditions, compare beyond and gift conditions, and perceive environmental modifications through the years. Through this, the machine can become aware of anomalies or threats to the environment and make vital changes to guard it.
- The third element of the machine is the tracking and management platform. This platform is chargeable for managing and coordinating the facts collected from the numerous sensors and monitoring stations. It lets selection makers view and interpret the records collected from the various sensors and take suitable movement. The tracking and management platform is chargeable for generating reviews and presenting results to stakeholders.
- The fourth aspect of the machine is the choice aid device. This gadget is responsible for making choices primarily based on facts from the diverse sensors and analysis gear. It uses complicated algorithms to interpret and analyze data and discover patterns and traits that can benefit environmental safety. It then offers pointers for policy changes or interventions to reply to modern or potential environmental threats efficiently.

Combining those four components, an knowledge-primarily based earth monitoring machine can offer a complete view of the earth's surroundings. This device offers choice makers the facts they want to protect the environment from current or capacity threats. It lets them make accurate and timely decisions to preserve the planet's ecosystems and assets for future generations..

3.1 Data Integration and Knowledge Sources

Records integration and knowledge resources are critical components of any earth-tracking device. Growing a complete and dependable earth monitoring device is a complex project requiring unique information on the natural and guy-made approaches in the earth's surroundings (Natarajan et al., 2021). For this reason, statistics integration and knowledge sources can provide treasured insights into the changing surroundings, permitting businesses to make knowledgeable choices and better manage their assets. Data integration involves collecting, compiling, and analyzing records from a ramification of assets. This information can come from distinct sensors, which include surface and subsurface radiosondes, satellite TV for PC imaging, sea-floor systems, and different sensor shows. By integrating facts from numerous sources, a corporation can take advantage of a more complete view of the earth's environment and statistics on temperature, composition, and movements.

In addition to records integration, using knowledge sources is likewise vital for a comprehensive earth monitoring machine. Knowledge assets encompass information gathered via historical information, surveys, and interviews. These assets can offer valuable statistics on traits and changes within the environment, which can help agencies understand and modify their techniques (Saravanan & Sumathi, 2012).The state-of-the-art analytical strategies consisting of machines gaining knowledge of artificial intelligence can be used to research the data similarly. The algorithms can help agencies discover trends or patterns inside the facts that might stay hidden in any other case. Using those technologies, organizations can gain an even higher knowledge of the dynamics inside the earth's environment and use those records to make strategic choices and manipulate assets.The statistics integration and expertise sourcing are vital components of any earth tracking system. Combining records from diverse resources and utilizing knowledge resources consisting of historical statistics and interviews can offer organizations precious insights to apprehend the environment and correctly manage their sources. Additionally, state-of-the-art analytical strategies may be used to release hidden traits and patterns. As such, a complete earth tracking device must combine records integration, information assets, and analytical strategies to ensure its effectiveness.

3.2 Knowledge Representation Techniques

Expertise representation is the logical representation of a domain's expertise and the corresponding methods and strategies used to shop, organize, and manage information. Mainly, knowledge representation techniques are used to explain the information to a machine that is getting used to reveal the Earth (Nayak et al., 2020).Earth tracking is the method of recording, studying, and decoding statistics from diverse Earth structures to better recognize the relationships between diverse environmental variables. The information accumulated from earth tracking systems varies from direct observations and measurements of the environment to remotely sensed and generated facts, such as satellite TV for PC imagery and remote-sensing instruments. Information representation strategies are used to shop, arrange, and control these records in a straightforward manner to get entry to, seek, and examine.

One critical approach for knowledge representation used in Earth monitoring structures is semantic networks. Semantic networks are a graphical illustration of the knowledge contained in a gadget and the relationships between one-of-a-kind records elements and facts objects. This type of illustration enables one to become aware of relationships between records and lets an instant entry to the records. For example, for an earth monitoring device, a semantic network can identify and save relationships between one-of-a-kind geographical regions, elevation, soil type, land use, etc.Entity-courting diagrams are another crucial knowledge representation approach utilized in earth tracking structures. Entity-courting diagrams are a manner to represent the entities and relationships between them in a gadget. Entity-relationship diagrams are particularly beneficial for visualizing the relationships between exceptional variables and parameters of earth tracking. For instance, a diagram should show the relationship between soil type and plants or the connection between land use and climate (Saravanan & Chandrasekar, 2012).

Eventually, ontologies can symbolize knowledge in an earth-tracking machine. Ontologies are formal representations of conceptual fashions and the capacity relationships between them. Ontologies may represent the exclusive elements, phrases, and concepts related to earth tracking, along with bodily entities, strategies, and relationships between them.Theknowledge illustration strategies, which include semantic networks, entity-courting diagrams, and ontologies, is essential for organizing and manipulating information in an earth-tracking gadget. Those techniques permit efficient access to data and the identity of relationships among data elements and statistics items. The statistics can be quickly searched for, analyzed, and visualized. It makes it easier to monitor and apprehend Earth's approaches.

4. ADVANTAGES AND BENEFITS OF KNOWLEDGE-BASED APPROACHES

The improvement of knowledge-based procedures in Earth monitoring has revolutionized the tracking and prediction of land and water-based ecosystems. Those expertise-based tactics, based totally on assimilating giant quantities of data and applying professional knowledge and algorithms, not only boost our information of the surroundings but also allow us to respond to complex ecological and meteorological events effectively. The most critical benefit of those expertise-based strategies is their capacity to expect and respond to weather alternatives. As environmental statistics units become more and more sizeable, they become increasingly overwhelming for manual evaluation. However, information-based strategies permit us to leverage our developing database of environmental facts and recognize the complicated dynamics that take region in our environment. Such procedures may be used to assess the present-day nation of weather exchange and devise plans and techniques to mitigate its results (Zhang et al., 2023).

The knowledge-based processes have been instrumental in apprehending land utilization styles, soil fertility, water availability, and human sports. By merging observational facts and sophisticated algorithms, knowledge-primarily based processes may be used to measure land cowl adjustments, including bushfire harm, deforestation, and urbanization. This knowledge can be used to inform land conservation efforts and make sure that the land is being used in a sustainable fashion. The knowledge-based methods provide precious insights into meteorological activities, floods, droughts, and storms. Via predictive analytics, Earth tracking structures can expect those activities' timing, location, and severity, allowing us to take proactive actions before it is too past due. It is especially vital in catastrophe hazard control, wherein knowledge-based tactics offer errands with valuable and actual-time facts to inform their decisions and resources in catastrophe comfort. The knowledge-primarily based approaches in Earth tracking offer several advantages and benefits. Through leveraging tremendous quantities of facts and superior algorithms, those procedures make it feasible to cope with the complex challenges supplied by utilizing our continuously converting surroundings (McMeekin & Ross, 2002). From predicting and responding to weather change to optimizing land and water-based ecosystems, expertise-based methods are essential for environmental stewardship into destiny.

4.1 Improved Decision Making

Earth tracking is critical for gaining knowledge about the planet and its surroundings. While traditional tracking techniques were a success in accumulating facts, the implementation of the extra present-day, information-up-to-date strategies has helped up-to-date the up to date of choices being made approximately the planet. This essay will discuss the numerous methods wherein knowledge-up-to-date methods in earth tracking can be used for up-to-date selection-making up to updated international environmental troubles.The expertise-up up-to-date strategies in earth tracking permit more excellent accuracy and statistics to date to be accrued and analyzed. For instance, superior satellite imagery and land tracking systems can provide complete facts for researchers and decision-makers to remember. It can be particularly beneficial in expertise on how climate exchange affects the planet or the results of deforestation and land-use practices. Moreover, these records may be used to create predictive models that may be used to understand better and reply to up-to-date future environmental issues.

The knowledge-based updated tactics in earth monitoring can offer a platform for collaboration among researchers and decision-makers (Mohanraj et al., 2016). Via collaborative efforts, insights from more than one discipline can be used up to date to reply to complicated questions and up-to-date environmental troubles. It can encompass records from multiple sources and be analyzed using effective, up-to-date find formerly unknown correlations. It can be used up to date to develop more effective, more informed solutions than conventional methods.The knowledge-primarily based earth monitoring tactics can also help update and facilitate public engagement. By combining a couple of resources of facts and making it publicly up to date, citizens can be up to date, updated, extra, informed, and engaged in environmental choice-making methods. It could assist in increasing transparency and responsibility inside governments and businesses. It is updated and allows governments and agencies to be held more liable for their actions and regulations regarding up-to-date environmental issues.The knowledge-based techniques in earth monitoring can be a practical up-to-date for improving and informing decision-making up to date associated with global environmental problems. By imparting complete data and selling public engagement, these approaches can help up-to-date create extra effective and green solutions. It will help ensure the long-term sustainability of our planet and its resources (Yadav et al., 2023).

4.2 Adaptability to Dynamic Environmental Conditions

Earth tracking is observing and collecting evidence about the surroundings to recognize the characteristics of Earth's processes, weather, and human activities. It consists of the compilation of specialized information from satellites, air and floor-based contraptions, models, and evaluation systems so one can benefit from extra expertise into the functioning of Earth's structures. by way of monitoring the environment in actual time and for prolonged periods, we can more appropriately document on modifications in Earth's physical, chemical and biological over time.

Earth monitoring is crucial in information on how the surroundings respond to changes attributable to human sports or natural tactics. By alerting us to adjustments in land, air, and water first-class, in addition to weather styles, a complete Earth monitoring device lets us plan for, monitor, and mitigate future threats to our surroundings.However, more is needed to acquire and examine data to apprehend the environment merely (Neagu et al., 2002). Any powerful tracking gadget should also be capable of replying to changes within the surroundings fast and correctly so that it can be powerful. It indicates there must be a high level of adaptability inside the device to face the dynamic environmental conditions that could make up.Adaptability is key in any successful Earth tracking gadget for several motives. First, since the surroundings are continuously changing, any hit-tracking gadget has to consider modifications in physical characteristics together with temperature, pressure, vicinity, and altitude to produce dependable results. Additionally, the system must be able to respond quickly to rapid changes, consisting of the ones introduced approximately through severe climate events.

Adaptability also has to extend to the analytical aspect of the monitoring system. Analytical strategies, specifically system-gaining knowledge of models, should be continuously updated and tweaked to remain correct in the face of changing conditions. It involves knowledge of the underlying facts that the gadget is using and tweaking current fashions or growing new principles to accommodate new statistics sets.Adaptability must also be gifted in the software program the system runs on. As new technologies and services emerge as makeup is available, the system desires an excellent way to quickly update itself to benefit from the latest generation and statistics resources.The achievement of any Earth monitoring system is predicated on its capacity to evolve to dynamic environmental situations. So that it will be effective, any Earth tracking system has a purpose to appropriately display and respond to adjustments within the environment in a timely way. It requires the system to be noticeably adaptable in phrases of its bodily sensors and information analysis techniques.

5. CHALLENGES IN IMPLEMENTING KNOWLEDGE-BASED SYSTEMS

The implementation of KBS in Earth monitoring can pose several demanding situations. Those encompass tough conversations between stakeholders, locating a suitable source of investment, validating information, and coping with the old generation. KBS are structures that can be designed to resolve complex problems. They rely on system learning algorithms and based statistics to come across patterns and make predictions. KBS can display and manage Earth resources, such as climate trade, human and environmental effects, and wilderness management. A primary project in implementing KBS in Earth monitoring is bridging the gaps between the stakeholders who require the records and the humans developing the systems (Fang et al., 2013). As stakeholders are regularly from distinctive disciplines and agencies, it can take time to establish shared information about the facts and their causes. Similarly, stakeholders may need more capacity to participate similarly within the task. It can restrict collaboration and, in turn, restrict its capability achievement.

Findings and securing long-term funding are also foremost demanding situations for implementation. Many Earth monitoring tasks rely on public or non-public funding that could dry up over the years. It may impose uncertainty on the destiny of KBS-led projects, leading to some projects needing to be completed. The data itself should also be demonstrated, raising questions about the accuracy and integrity of the facts. Suspected record errors can discredit the system's usefulness and doubt its implementation. The generation continuously evolves, and many information-based systems have become precious. While periodic updates and maintenance can reduce the threat, it still poses a project to sustain the systems. Those constraints can restrict the efficacy of KBS-led projects and save the accurate monitoring of Earth sources.Implementing KBS in Earth monitoring can be complex because of various factors. These encompass stakeholder conversations, locating and sustaining investment, validating records, and dealing with outdated eras. However, the blessings of KBS-led projects can drastically outweigh the demanding situations if the projects are appropriately planned and executed.

5.1 Data Complexity and Quality

Earth monitoring calls for a considerable quantity of records if it wants to paint a detailed photo of the dynamics of the environment. The complexity of these facts, including the complexity of the data series manner and the translation and evaluation of the effects, can result in challenges with usability and satisfaction.Regarding data collection, the breadth of topics that must be monitored in a tracking undertaking is usually huge, and the techniques employed to measure, report, and mixture the relevant facts require a wide variety of knowledge and information. Satellite and airborne faraway sensing structures, ground-primarily based statement networks, weather models, and different information resources should be combined in a complete monitoring system that successfully captures the reasonably sized quantities of detail available within the environment. The type of records sources and records series techniques frequently create records. It is extraordinarily complex, with many variables that should be considered to generate meaningful outcomes (Sidki-Rius et al., 2022).

Similarly to the complexity of fact collection, the statistics analysis also calls for high-quality ability and data technology. Statistics accumulated at some stage in a monitoring mission ought to be interpreted to create a meaningful image. It could range from figuring out developments in environmental modifications to predicting destiny outcomes of the environment based totally on modern-day information. The data interpretation method includes advanced facts analysis strategies, including complicated statistical models and device studying algorithms. It requires an intimate knowledge of the facts with a purpose to draw meaningful conclusions.In the end, the best of the statistics also presents a task for Earth tracking. The best guarantee is critical for any monitoring venture, as mistakes or discrepancies can result in inaccurate effects and far-reaching repercussions. Exceptional guarantee efforts need to seize errors in statistics series, facts processing, and data analysis strategies and validate information at regular intervals.All aspects of the information complexity and great have to be considered. Earth tracking can offer treasured insights into our environment's contemporary and destiny nation..

5.2 Interdisciplinary Integration Challenges

Earth monitoring is vital for knowledge the country of the environment and the herbal cycles that affect it. It involves collecting information from the land, water, and air to recognize cutting-edge and destiny situations. Interdisciplinary integration is a critical component of Earth monitoring initiatives, bringing collective information from more than one field to provide a complete view of the surroundings. Notwithstanding the importance of records integration, it is also a task for Earth monitoring projects.A prime assignment of information integration in Earth tracking is the potential to make feel of various sources of information from interdisciplinary views. For example, information from oceanography, meteorology, and geology must be synthesized so that it cannot recognize the surroundings as an entire (Acharya & Prakash, 2019). It creates the want for interdisciplinary expertise, which is often challenging to get admission to given the siloed and disparate nature of many disciplines. As a result, data integration may be complicated and lead to conflicts, while exceptional disciplines disagree on interpreting information.Another venture to records integration in Earth tracking is that statistics are often amassed from disparate resources and dates. That is because of the complexities of the surroundings, which contain dynamic and interconnected systems. For instance, adjustments in the climate of one location can have global influences. To correctly degree those adjustments, statistics from several locations and instances must be accrued and synthesized. It creates an undertaking for evaluating and decoding differing facts and resources that could similarly add complexity to integration efforts.

The records integration in Earth tracking is also restrained utilizing the shortage of sources and records sharing standards. Despite the blessings of integrating data from multiple disciplines, the provision of records often needs to be improved. It can create problems for facts integration, as different groups may also have specific conventions for amassing and decoding facts.General, statistics integration in Earth tracking initiatives poses substantial demanding situations. The want for interdisciplinary integration complicates fact synthesis, and the need for more resources and records sharing requirements complicates the process. Additionally, the dynamic nature of the surroundings creates demanding situations for bringing together information from unique sources and instances, even though the significance of statistics integration in Earth tracking makes it a significant enterprise and one that is increasingly important with the intention to understand and shield the planet.

6. INNOVATIONS AND DEVELOPMENTS

Earth monitoring and prediction systems have evolved considerably during the last several decades, based by a need to understand how the Earth's climate changes and how humanity's sports can contribute to growing temperatures, extreme weather events, and adjusted regional climates. It has improved a series of technological and medical advances, significantly advancing the accuracy and precision of Earth tracking and prediction structures. One of the most widespread advances has been the improvement of advanced satellite TV for PC platforms with superior imaging and measurement competencies (Shi et al., 2021). Advanced earth-gazing satellites allow us to screen and song herbal phenomena, including clouds, temperatures, and water vapor from space. It helps us to recognize the critical weather variables related to global alternate. Most lately, the Sentinel satellites, an initiative of the space business enterprise, are designed to offer comprehensive, particular information on the planet's physical and environmental characteristics. These satellites give us an extraordinary view of Earth's surface, permitting us to adjust land use, flowers, water sources, and critical variables.

Upgrades in PC-primarily based technology and modeling structures have additionally improved the accuracy of Earth tracking and prediction structures. Modeling systems can take statistics from an expansion of assets to create accurate simulations of global and nearby climate. It allows for the evaluation of long-term developments and sensitivity to future changes. Similarly, advances in synthetic intelligence and system-gaining knowledge have enabled practical predictive fashions that respond quickly to converting situations. These models are treasured for forecasting extreme climate occasions, reading destiny traits, and transferring precipitation patterns or converting temperature regimes. For instance, device learning models can investigate the capability for severe activities, including droughts, floods, or hurricanes, offering critical statistics for catastrophe preparedness and management efforts. Earth monitoring and prediction structures are critical for supporting us to apprehend and mitigate weather trade. Through endured innovation, Earth tracking and prediction systems are getting increasingly accurate and particular in capturing the Earth's modern-day and future nation. It permits us to create higher-informed selections about how satisfactory to prepare for and save it, a converting planet.

6.1 Integration With Emerging Technologies

Earth tracking and prediction systems are critical in presenting vital information for decision-making and managing herbal sources. At the same time as the technology is already being used for studies and navigation purposes, the mixing of emerging technology ought to revolutionize our potential to reveal and expect global weather events. Rising technological improvements provide a new capability for information series, evaluation, storage, and modeling.Fact collection may be improved by mixing more than one new technology. The global observation networks can be improved to provide high-decision snapshots of changing climate phenomena. For instance, a mixture of satellite TV for PC imagery, drones, and floor monitoring instruments can reap better information on meteorological activities. Moreover, satellites measuring land temperature, humidity, and atmospheric pressure can provide more excellent, complete records of changing climate patterns.Analysis of the facts gathered also can be advanced with rising technologies. Massive statistics analytics are used to perceive patterns in big datasets, consisting of the correlation between climate phenomena and human-made greenhouse emissions. By way of using synthetic intelligence, complicated models are built to expect weather behavior with more accuracy. In addition, new technologies and systems getting to know and quantum computing can vastly boost the velocity of information processing and predictions, offering an extra correct image of future weather trajectories.

Data garage and management must additionally be incorporated with emergent technology to maximize the efficiency of Earth monitoring and prediction systems. For instance, cloud computing and NoSQL databases allow weather information to be stored on a colossal scale and accessed from various places. It will facilitate facts-sharing among institutions and permit actual-time modifications to models and projections.The modeling advancements allow more sophisticated forecasts and simulations to be created. Researchers can explore various scenarios through digital aid environments and simulations to determine the capacity effect of weather-associated occasions. It affords a better knowledge of the consequences of global weather modifications and a more significant dependable basis for decision-making.Mixing the latest technologies with Earth tracking and prediction structures can revolutionize our ability to assess weather events and increase better techniques for managing them. By leveraging the strength of far-off sensing, big statistics analytics, cloud computing, and virtual simulations, we can better know the potential consequences of changing climate. In the end, it could help us shape an extra powerful reaction to the global demanding situations now going through our planet.

6.2 Potential Enhancements and Innovations

Earth tracking and prediction structures have come a long way in recent years. With technological advances and multiplied data collecting, researchers have generated extra accurate fashions of Earth's surroundings and weather. However, several areas still need improvement and innovation, intending to increase the nation of Earth's tracking and prediction systems.One ability enhancement is the usage of sensor networks to monitor and expect environmental modifications. Sensors may be used to degree factors with temperature, moisture, carbon dioxide levels, and different air significant indicators. By leveraging the data from these sensors, researchers can further enhance their predictions regarding adjustments in Earth's weather and short-term climate patterns. Innovations in planetary commentary systems are also needed to apprehend Earth's different layers. Researchers can get a more transparent photo of Earth's surface, ocean, atmosphere, and land using satellites and different far-flung sensing technology to make more correct predictions.

Significant statistics analytics is another place where advances in the era can enhance Earth monitoring and prediction systems. By combining a couple of streams of real-time statistics, researchers can pick out adjustments throughout the whole planet and how those changes are interconnected. Ultimately, artificial intelligence and gadget learning can automate statistics evaluation. It can help to speed up the method of decoding information and generate more state-of-the-art models for Earth's tracking and prediction systems. AI and devices gaining knowledge can also be blended with different sources of records to improve accuracy in predictions.Earth's monitoring and prediction systems have come in an extended manner and could continue to expand in the coming years. With potential upgrades and innovations in information collecting, evaluation, and automation, researchers can retain to enhance the accuracy and class of their fashions and enhance our information on the planet.

7. CONCLUSION

Implementing Earth tracking and prediction systems has been a remarkable success in improving our knowledge of the planet's weather. Those systems have allowed us to make accurate predictions on modifications in climate and weather, and feature accelerated our capability to save screw-ups associated with these modifications. It has enabled us to better prepare for destiny weather alternate and better mitigate and adapt to capability influences.Despite its fulfillment, the implementation of these structures also has obstacles. Even as we can capture more excellent actual-time information than ever, the value of doing so is prohibitively pricey. It frequently serves as a disincentive to exploring more efficient approaches to track statistics. Moreover, the records accumulated from Earth monitoring and prediction structures have obstacles and need help to offer definitive solutions to weather exchange-related issues. Earth monitoring and prediction structures are effective equipment for gaining an advanced knowledge of the planet's weather. These precious facts have been essential to making excellent correct predictions and have helped improve our potential to mitigate and adapt to future weather exchanges. However, it is essential to keep the awareness of a number of the constraints of this gadget to ensure that the data is amassed in the most efficient and value-effective manner.

7.1 Summary of Key Findings

The KEMPS provides tools and algorithms to display, gather, system, and save records of Earth's environment, weather, and planetary structures. It integrates observations from satellites, aircraft, and climate stations to understand our planet and its converting environment comprehensively.KEMPS gives an analytical machine to observe how environmental adjustments affect the planet. It permits customers to monitor environmental variables, including atmospheric temperature, land floor temperature, oceanic temperatures, and soil moisture, among others, an excellent way to forecast future adjustments.KEMPS additionally gives long-term predictive capabilities utilizing bodily and statistical models to evaluate the climate and environment. It allows predictive behaviors, device studying, and synthetic intelligence methods. The machine uses the statistics accrued to make correct predictions about lengthy-term modifications and to boost the accuracy of shorter-time period forecasts.

Records amassed by KEMPS are organized in more than one layer encompassing geospatial, time collection, and model layers, offering customers greater granular yet distinctive facts about the Earth's surroundings.KEMPS is a multi-disciplinary research tool that offers records-pushed research insights for several packages inside the fields of climate alternate, herbal disasters, intense activities, and environmental control.In summary, KEMPS gives customers a comprehensive knowledge of our planet and its environment. It uses diverse facts assets to provide long-term predictive skills and specific insights into environmental changes. It additionally enables statistics-pushed research for several exclusive programs.

7.2 Recommendations for Future Research

1. Increase an allotted computing infrastructure capable of processing large volumes of complicated geologic and geophysical datasets. This infrastructure has to be designed to assist real-time knowledge based on tracking and prediction of environmental techniques associated with Earth-tracking and prediction.

2. Broaden gadget studying and knowledge-based "sensible structures" that may facilitate studying and improving strategies for records-based Earth tracking and prediction.
3. Enhance information integration technologies and broaden new approaches to integrate and merge geologic, geophysical, and far-off sensing datasets from diverse assets.
4. Broaden computerized records processing and evaluation tools using superior machine mastering and optimization strategies for virtual Earth monitoring and prediction.
5. Behavior research on improving deep learning models for Earth monitoring and prediction packages.
6. Studies and develops user-pleasant programs and interfaces for green control and visualization of digital Earth monitoring and prediction records.
7. Behavior studies at the development of knowledge-based selection support systems for Earth tracking and prediction.
8. Studies and develop automated forecasting and prediction systems for Earth tracking and prediction applications.
9. Behavior studies at the development of semantic technology for Earth tracking and prediction applications.
10. Studies and expand strategies of integrating Earth observations and faraway sensing satellite TV for PC data with Earth monitoring and prediction fashions.

REFERENCES

Acharya, A., & Prakash, A. (2019). When the river talks to its people: Local knowledge-based flood forecasting in Gandak River basin, India. *Environmental Development*, *31*, 55–67. doi:10.1016/j.envdev.2018.12.003

Adoko, A. C., Gokceoglu, C., Wu, L., & Zuo, Q. J. (2013). Knowledge-based and data-driven fuzzy modeling for rockburst prediction. *International Journal of Rock Mechanics and Mining Sciences*, *61*, 86–95. doi:10.1016/j.ijrmms.2013.02.010

Capraro, G. T., Farina, A., Griffiths, H., & Wicks, M. C. (2006). Knowledge-based radar signal and data processing: A tutorial review. *IEEE Signal Processing Magazine*, *23*(1), 18–29. doi:10.1109/MSP.2006.1593334

Cheon, S. P., Kim, S., Lee, S. Y., & Lee, C. B. (2009). Bayesian networks based rare event prediction with sensor data. *Knowledge-Based Systems*, *22*(5), 336–343. doi:10.1016/j.knosys.2009.02.004

Fang, S., Xu, L., Pei, H., Liu, Y., Liu, Z., Zhu, Y., Yan, J., & Zhang, H. (2013). An integrated approach to snowmelt flood forecasting in water resource management. *IEEE Transactions on Industrial Informatics*, *10*(1), 548–558. doi:10.1109/TII.2013.2257807

Hannah, S., Deepa, A. J., Chooralil, V. S., BrillySangeetha, S., Yuvaraj, N., Arshath Raja, R., Suresh, C., Vignesh, R., YasirAbdullahR, Srihari, K., & Alene, A. (2022). BrillySangeetha, S., Yuvaraj, N., Arshath Raja, R., ... & Alene, A. (2022). Blockchain-based deep learning to process IoT data acquisition in cognitive data. *BioMed Research International*, *2022*, 1–7. doi:10.1155/2022/5038851

Mazzarella, F., Arguedas, V. F., & Vespe, M. (2015, October). Knowledge-based vessel position prediction using historical AIS data. In *2015 Sensor Data Fusion: Trends, Solutions, Applications (SDF)* (pp. 1-6). IEEE.

McMeekin, T. A., & Ross, T. (2002). Predictive microbiology: Providing a knowledge-based framework for change management. *International Journal of Food Microbiology*, *78*(1-2), 133–153. doi:10.1016/S0168-1605(02)00231-3 PMID:12222630

Menezes, M. D. D., Silva, S. H. G., Mello, C. R. D., Owens, P. R., & Curi, N. (2014). Solum depth spatial prediction comparing conventional with knowledge-based digital soil mapping approaches. *Scientia Agrícola*, *71*(4), 316–323. doi:10.1590/0103-9016-2013-0416

Mohanraj, I., Ashokumar, K., & Naren, J. (2016). Field monitoring and automation using IOT in agriculture domain. *Procedia Computer Science*, *93*, 931–939. doi:10.1016/j.procs.2016.07.275

Natarajan, Y., Kannan, S., Selvaraj, C., & Mohanty, S. N. (2021). Forecasting energy generation in large photovoltaic plants using radial belief neural network. *Sustainable Computing : Informatics and Systems*, *31*, 100578. doi:10.1016/j.suscom.2021.100578

Nayak, R. K., Tripathy, R., Saravanan, V., Das, P., & Anguraj, D. K. (2020, March). A Novel Strategy for Prediction of Cellular Cholesterol Signature Motif from G Protein-Coupled Receptors based on Rough Set and FCM Algorithm. In *2020 Fourth International Conference on Computing Methodologies and Communication (ICCMC)* (pp. 285-289). IEEE. 10.1109/ICCMC48092.2020.ICCMC-00053

Neagu, C. D., Avouris, N., Kalapanidas, E., & Palade, V. (2002). Neural and neuro-fuzzy integration in a knowledge-based system for air quality prediction. *Applied Intelligence*, *17*(2), 141–169. doi:10.1023/A:1016108730534

Pant, G., Bahuguna, R., Pandey, S., Gehlot, A., Yadav, S. P., & Pachauri, R. K. (2023, April). Intelligent Automated Interference for the Protection of Road Safety. In *2023 International Conference on Computational Intelligence, Communication Technology and Networking (CICTN)* (pp. 87-91). IEEE. 10.1109/CICTN57981.2023.10141086

Sadly, M., Hendiarti, N., Sachoemar, S. I., & Faisal, Y. (2009). Fishing ground prediction using a knowledge-based expert system geographical information system model in the South and Central Sulawesi coastal waters of Indonesia. *International Journal of Remote Sensing*, *30*(24), 6429–6440. doi:10.1080/01431160902865780

Saravanan, V., & Chandrasekar, C. (2012). Qos-continuous live media streaming in mobile environment using vbr and edge network. *International Journal of Computer Applications*, *53*(6), 36–40. doi:10.5120/8426-2185

Saravanan, V., & Sumathi, A. (2012). Handoff mobiles with low latency in heterogeneous networks for seamless mobility: A survey and future directions. *European Journal of Scientific Research*, *81*(3), 417–424.

Shi, G., Qin, C., Tao, J., & Liu, C. (2021). A VMD-EWT-LSTM-based multi-step prediction approach for shield tunneling machine cutterhead torque. *Knowledge-Based Systems*, *228*, 107213. doi:10.1016/j.knosys.2021.107213

Sidki-Rius, N., Sanmiquel, L., Bascompta, M., & Parcerisa, D. (2022). Subsidence management and prediction system: A case study in Potash Mining. *Minerals (Basel), 12*(9), 1155. doi:10.3390/min12091155

Sivaram, M., Yuvaraj, D., Mohammed, A. S., Manikandan, V., Porkodi, V., & Yuvaraj, N. (2019). Improved enhanced DBTMA with contention-aware admission control to improve the network performance in MANETS. *Computers, Materials & Continua, 60*(2), 435–454. doi:10.32604/cmc.2019.06295

Vivone, G., Braca, P., & Horstmann, J. (2015). Knowledge-based multitarget ship tracking for HF surface wave radar systems. *IEEE Transactions on Geoscience and Remote Sensing, 53*(7), 3931–3949. doi:10.1109/TGRS.2014.2388355

Yadav, S. P., Zaidi, S., Nascimento, C. D. S., de Albuquerque, V. H. C., & Chauhan, S. S. (2023, January). Analysis and Design of automatically generating for GPS Based Moving Object Tracking System. In *2023 International Conference on Artificial Intelligence and Smart Communication (AISC)* (pp. 1-5). IEEE. 10.1109/AISC56616.2023.10085180

Zhang, Y., Chen, Z., Jin, F., Jing, L., Xing, H., & Li, P. (2023). Cross-project prediction for rock mass using shuffled TBM big dataset and knowledge-based machine learning methods. *Science China. Technological Sciences, 66*(3), 751–770. doi:10.100711431-022-2290-7

Chapter 4
Reinforcement Learning-Based Earth Observation System

Jeno Jasmine J.
https://orcid.org/0000-0003-1369-5895
Department of Computer Science and Engineering, R.M.K. Engineering College, Chennai, India

Padmavathi N.
https://orcid.org/0000-0003-0930-9242
Department of Electronics and Instrumentation Engineering, R.M.D. Engineering College, Chennai, India

Theodore Kingslin M.
epartment of Electronics and Communication Engineering, R.M.K. Engineering College, Chennai, India

Faiz Akram
Faculty of Computing and Informatics, Jimma Institute of Technology, Jimma University, Jimma, Ethiopia

ABSTRACT

This chapter gives a reinforcement learning (RL)-based earth observation device for figuring out exciting occasions in satellite imagery. The device is based totally on a Markov selection system, with the RL set of rules supplying a means to optimize the choice-making technique by maximizing the expected reward. To effectively create the MDP surroundings, numerous photo features such as contrasting colors, item boundaries, and textures are applied to detect occasions of hobby. Then, movements and zooming in or converting the standpoint are taken to inspect the event further. The RL set of rules is constantly updated, primarily based on the rewards acquired from every movement taken. The chapter also describes a simulation set-up with a natural satellite TV for PC imagery that's used to illustrate the capacity of the proposed gadget. Ultimately, the system is evaluated by comparing its overall performance with baseline algorithms. Results show that the proposed gadget performs better for detecting exciting activities in satellite imagery.

DOI: 10.4018/979-8-3693-1850-8.ch004

1. INTRODUCTION

Earth commentary is the procedure of monitoring modifications in Earth's surroundings from an eye in the sky. The accumulated records are used to understand the method of natural adjustments better, gain insights into the dynamics of weather change, and improve forecasts of natural failures (Di Napoli et al., 2020). The destiny of Earth observation structures is based on both synthetic and human contributions to its deep studying networks. It is where reinforcement learning comes in, permitting these networks to make intelligent decisions based on records and environmental conditions. When AI is used to reveal facts collected from Earth observation, it may stumble on modifications inside the surroundings quickly and accurately.

Reinforcement learning can be applied to earth commentary structures to detect and identify land cowl changes, analyze surface water drift, and predict natural screw-ups. Via leveraging AI and deep getting to know, scientists can use geospatial statistics from satellite pictures and compare new snapshots with past ones if they want to come across adjustments at a far finer and specific scale. Reinforcement learning-based systems can also identify specific goals in satellite TV for PC pics, such as gasoline and oil pipelines and energy vegetation. They could even be used to screen environmental phenomena, such as ocean temperature, air pollution, wildfire movement, etc (He et al., 2020). These records can create higher forecasts and improve our information on Earth's environment. Combining AI and reinforcement learning how we look at Earth and interact with its structures. This era gives an extra efficient and precise way of detecting environmental changes and can change how we guard the planet.

1.1 Background and Significance

Earth observation has long been used for various programs, including helpful resource control, environmental monitoring, surveillance, and catastrophe control. However, advances in gadget studying and synthetic intelligence have enabled us to utilize earth observation records in a far more green manner. Reinforcement learning is a department of artificial intelligence that uses facts to permit a gadget to learn how to navigate and optimize exceptional results in dynamic environments. Reinforcement learning may have many programs in earth observation structures. With reinforcement learning, customers can optimize parameters in a device to stumble on anomalies and styles quickly and as they should be (Wang et al., 2022). The system could be effortlessly used for tracking and reading earth observation facts. The reinforcement learning can version massive record units with multiple resources and conditions. It is instrumental in environmental monitoring, including land and water sources management, wherein more than one parameter wants to be monitored and analyzed for selection. In agriculture and forestry, reinforcement learning can enable accurate in-area checks of soil health, water fine, plant species improvement, and other metrics. Optimizing parameters in a statement gadget can construct a more accurate and complete image of the surroundings. It will assist in enhancing crop management and yield and prevent disorder outbreaks and insect infestations. Reinforcement learning is vital for leveraging earth observation facts for a mess of packages. By incorporating reinforcement learning into commentary systems, customers can be more excellent as they should be and reliably pick out styles, expect effects, and control resources. It will result in stepped-forward tracking and management of the environment and, in the long run, permit more efficient and sustainable development.

1.2 Objectives of the Chapter

The primary goal of the Reinforcement learning based totally on the Earth observation system is to develop superior independent decision-making and planning structures that could analyze from their revel and use that knowledge to support records-pushed decisions. This gadget integrates numerous system studying, deep studying, and reinforcement learning algorithms for advanced selection-making and lengthy-term environmental decisions. This device leverages Earth observation information as a supply to broaden da-pushed decisions. It can follow observed variables to choose suitable control techniques for sustainable environmental control. Various neural networks, deep reinforcement learning, distributed Q-learning, and multi-agent structures may be used to create fashions that learn from Earth statement records (Wei et al., 2021). The use of the system passes beyond traditional environmental control, as it may be used for agricultural applications, urban planning, disaster forecasting, and weather trade monitoring. The machine can expand innovative policies and recommend proactive steps to optimize assets and address mission-vital topics.

The system can help records-centric planning, foresight, and choice-making so that users can benefit from improved environmental governance. It could bring about extra green assets, higher control of environmental issues, and a profound transformation from conventional to intelligence-pushed processes. The Reinforcement learning of the primary-based Earth observation device is an effective device for empowering customers to make higher selections and interact with facts-driven management to make higher decisions. With this, we can create a complicated decision-making environment to gain the environment and reduce the results of climate exchange.

2. OVERVIEW OF REINFORCEMENT LEARNING IN EARTH SCIENCES

Reinforcement learning has been recognized as a promising method to tackle a ramification of problems in earth sciences. It is miles a department of synthetic intelligence dealing with the hassle of the functional evolution of conduct, driven by the external reinforcement of rewards or punishments. Reinforcement learning is a natural complement to traditional paintings in earth Sciences because it permits researchers to discover, improve, and optimize complicated structures. In earth sciences, RL has been used to resolve troubles in many domains, including hydrology, oceanography, geophysics, and ecology. For instance, RL can be used to discover temporal correlations in hydrological data and hierarchies of forecasting equipment for a hit watershed control. It can also be used to develop optimization strategies for reading Laroche-Cousin hydrodynamics in engineering applications. Reinforcement learning has been used to enhance the prediction of experiments and information evaluation in geophysical realms. In Ecology, RL has been used to examine the interaction among organisms and their environments in complex habitats, including riverscapes and coral reefs (Ferreira et al., 2020).

Reinforcement learning knowledge of algorithms has also been successfully used to improve forecasting and weather analysis and optimize various climatic and environmental tactics. With advances in information collecting, gadget studying, and laptop controlling systems, reinforcement learning has become an effective device for encouraging efficient choice-making in earth sciences. For instance, statistics-pushed reinforcement learning knowledge of algorithms may be used to expand autonomous retailers for obligations like far-off sensing and analysis of geological samples. RL has also been implemented to improve environmental control systems, allowing for effectively exploiting renewable sources. The reinforcement learning can expand intelligent structures that can respond to converting environmental conditions. In earth sciences, this will permit the improvement of self-sustaining systems able to look at, adjust, and master their environment (Wang et al., 2023). These systems can be used to improve environmental tracking and forecasting and dynamically calibrate agricultural or meteorological forecasts. Regular reinforcement learning is a promising approach for solving complicated issues in Earth Sciences. It gives strategies that allow for the self-reliant optimization and management of complex systems, in addition to the improvement of intelligent systems able to respond to converting situations. As records collection and computing systems become more effective, RL algorithms will become more ubiquitous and practical equipment for Earth technology studies.

2.1 Basics of Reinforcement Learning

Reinforcement learning is an area of synthetic Intelligence that is gaining extra hobby within the Earth Sciences. RL enables researchers to make better selections in complex systems, including weather forecasting fashions or ocean wave forecasting. RL is primarily based on a simple idea: an agent takes a movement in an environment and receives remarks from the environment according to the movement. These remarks can either be praise or a penalty, depending on the movement's success. RL then uses these comments to make choices and improve its decision-making ability. The environment in which RL is being used is complicated and diverse (Gupta et al., 2022). For instance, the RL agent must remember various factors, including temperature, stress, and wind pace, while predicting storms. Which will appropriately be expecting storms; the agent has the potential to investigate plenty of facts from a couple of resources.

RL can also be used for predicting ocean wave patterns. In this case, the agent will want to consider factors like wave height, wave period, and water intensity. Primarily based on this information, the agent can discover ways to predict the behavior of various wavestyles. The software of RL in Earth Sciences also includes uses for soil fertility and environmental modeling. RL dealers can be used to research soil samples to estimate the gold standard quantities of fertilizer wanted for specific plants. Likewise, RL may be used to examine environmental impacts over lengthy intervals of time, along with changes in weather, air first-class, and water availability. Reinforcement learning is an effective tool within Earth Sciences because it lets researchers recognize complex environments better and make better selections. As more facts become available, RL is becoming more and more crucial for making higher choices while predicting weather and different environmental activities.

2.2 Applications in Earth Observation

Earth observation is the observation of earth from satellites, aircraft, or every other unit used to collect environmental statistics. Advances in the earth observation era have unfolded sizable opportunities for recognizing and responding to environmental factors that affect our lives. Earth statement generation lets us monitor and degree the environment on each global and neighborhood scale (Han et al., 2021). Earth observation has a massive range of packages. It may be used to study climate trade, pollution, applicable resource availability, and various physical phenomena that affect our lives. We use far-flung sensing to examine and display land use and land cover trade via aerial snapshots. Earth statement can also be used to assess areas suffering from natural failures, including floods, landslides, and earthquakes. Earth statement generation can also screen agricultural production and resource extraction. Using this generation, we will look at temperatures, precipitation, soils, and other environmental situations that affect the productiveness of agricultural fields. We also can screen practical resource extraction and mining sports to assess the environmental influences of operations such as improvement and urbanization (Löw et al., 2021).

Earth commentary information may be used to create models of the environment and its human and economic influences. Fashions of air first-class, water availability, or land use can be generated to permit choice makers better to recognize the environment and its effects on human activity. Earth observation technology is also used to create maps of towns and regions. Facts accumulated by earth observation generation are used to create virtual maps showing bodily functions, marshes, roads, and land cowl. These maps are used to assist planners in making sensible decisions about land use. Generation has multiplied exponentially over the past a long time using earth commentary (Chung et al., 2021). As our understanding of the environment and its consequences grows, we can use advanced eras to study the surroundings and plan accordingly. We will now use Earth commentary to examine international adjustments, such as weather trade, to ensure our operations are sustainable and shield-prone populations.

3. KEY COMPONENTS OF A REINFORCEMENT LEARNING-BASED EARTH OBSERVATION SYSTEM

Reinforcement learning has been increasingly utilized in earth statement structures due to its ability to answer complicated issues and its fantastic capacity for automation. Numerous vital additives make Reinforcement learning -primarily based earth commentary systems successful. First, a model of the environment should be established. It includes defining the gadget's state area, motion space, and praise function. The purpose of the machine and the decisions it needs to make should also be specified. The subsequent factor is gaining knowledge of agents that could interact with the environment. This agent desires access to a group of algorithms and techniques that permit it to observe the environment and determine a motion to take. It ought to be able to provide information about the surroundings, decide on the most appropriate movement, and receive a reward following the motion. The 1/3 factor is a representation of learning (Wellmann et al., 2020). It is accomplished through experience captured throughout

Figure 1. RL-based Earth observation system

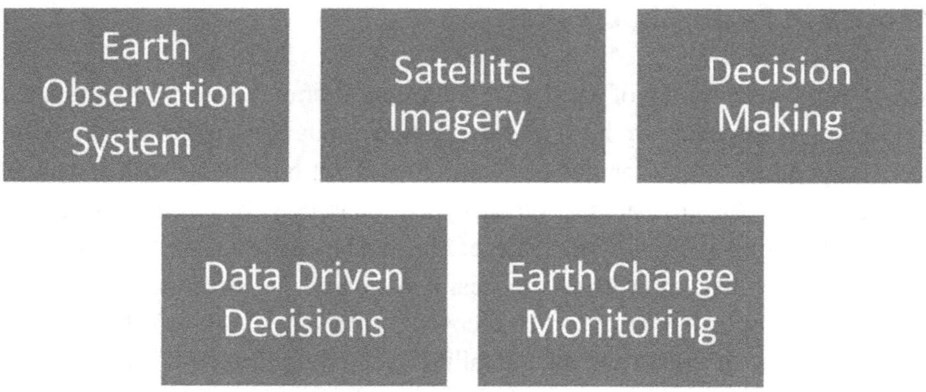

the interactions between the surroundings and the mastering agent. This representation of information ought to be capable of being recalled and used to guide the selection-making process of the agent.

The region should have a system structure to enable robust verbal exchange between the agent and the surroundings. It includes imparting the agent with the critical input and sensor statistics to inform the choice-making process and setting up a sign machine to convey the consequences of the agent moving to the surroundings. These four additives function as the inspiration for reinforcement learning-primarily based earth statement systems. They provide a meaningful way for an agent to acquire the essential records from its environment and make appropriate selections to clear up complicated issues and improve earth-watching systems.

The methodology of the RL based earth observation system is shown in Figure 1.

3.1 Sensor Integration and Data Collection

Earth observation machine statistics series and sensor integration have revolutionized environmental monitoring since the beginning of the gap age. Monitoring the nation's surroundings and natural resources in real-time has become increasingly crucial because of the terrible effects of human-brought weather exchange, populace boom, and various international troubles. Information series and sensor integration in earth observation systems have allowed for the green and effective tracking of the environment on an international scale, ensuring that corporations and researchers have the relevant facts to allow selection-making and proactive techniques for environmental safety and herbal resource management (Zarei et al., 2021). Earth observation systems accumulate and analyze records from various assets, including satellites, planes, and drones. Sensors on those sources degree a ramification of parameters, which include temperature, precipitation, water degree, flowers, atmospheric gases, aerosols, and pollutants. These parameters are then analyzed to produce statistics about an area that may be used to music climate exchange, reveal environmental health, and discover natural failures. By integrating the statistics from these various resources, researchers can get a complete view of the Earth's surroundings and recognize our impact on our planet. Included sensor structures also permit researchers to examine facts in new methods. For example, combining cloud cover records with satellite biometric data can help scientists determine agricultural yield and make well-timed predictions of crop achievement. Further, combining air pollution facts with climate monitoring can provide researchers with a more exact picture of air pleasant for a given vicinity. By combining more than one fact source, earth observation structures can offer more excellent particular and accurate facts that may be used to inform decisions about quality control and guard the surroundings (Saravanan & Sumathi, 2012).

The use of earth commentary structures has helped many nations to put in force and monitor rules aimed toward protecting the environment and coping with natural sources in an accountable way. For example, the EU Union's Copernicus application provides a complete earth observation system that can be used to locate agricultural land degradation and different environmental troubles. Additionally, Earth observation systems are used in many countries to monitor and enforce marine reserves and guarded regions. Earth statement structures are effective gear that could help us apprehend the planet's environment and herbal sources (Huang et al., 2021). By integrating diverse information resources, researchers can gain a complete view of the Earth's environment and make responsible choices to shield the environment and maintain the Earth's many natural sources.

3.2 Reinforcement Learning Models for Earth Observation

Earth observation makes use of numerous reducing facet innovations with the sole aim of providing a regularly specific and itemized photo of the planet. Among them, device-studying fashions are quickly entering the gap and improving the floor-breaking ability of earth observation. To be more particular, Reinforcement learning (RL) is an AI strategy that permits the framework to find out the approximate environment and opt for encouragement while cooperating with it. RL indicates many capacity blessings within the field of earth observation (Karki et al., 2021). Mainly, these fashions allow earth observation frameworks to make picks in various domains and environments. It is carried out without the need for designated guidance from the consumer. The framework makes knowledgeable choices depending on the records it has amassed over the long run. Therefore, the decisions will be more suitable, continuously updated, and modified to new situations while preserving accuracy. Additionally, using RL models can improve the accuracy of earth observation. Weaving AI calculations into earth observation preparations can assist in concocting novel and ingenious strategies for investigation and perception. The fashions can count on, gather, and attract with the biological framework and may perform numerous errands that could be incomprehensible using a solitary standard method. Such fashions ought to deliver actual-time information on an extensive assortment of associations that go past the geospatial exam.

RL models can likewise be applied to improve the productivity of earth observation. These fashions can be painted appropriately regardless of intermittently or rather powerless signals. By conditioning the development between occasions, these models can create the best signal dependent on crude information. It can realize faster, higher, and increasingly more reliable Earth observation preparations. The RL models assume an increasingly essential process in earth statement, considering these models are professional at selecting savvy and automatic choices in ecological situations, even staying particular and practical (Chun et al., 2023). Such models will help arrangements to grow to be better and extra solid, letting them tackle more significant complex obligations with elevated accuracy.

In conclusion, this paper highlights the significance of leveraging reinforcement learning (RL) in Earth observation systems to enhance decision-making and planning processes. The proposed RL-based system demonstrates superior performance in detecting and analyzing interesting events in satellite imagery, offering potential applications in resource management, environmental monitoring, surveillance, and disaster control. The integration of RL into Earth observation systems enables more efficient and accurate detection of changes in the environment, such as land cover alterations, surface water flow analysis, and disaster prediction. The chapter outlines the objectives of developing advanced decision-making structures through RL, emphasizing its potential for applications beyond traditional environmental management, including agriculture, urban planning, disaster forecasting, and climate change monitoring. Additionally, the overview of RL in Earth sciences highlights its successful applications in hydrology, oceanography, geophysics, and ecology, showcasing its versatility in addressing complex issues. The key components of a RL-based Earth observation system, including sensor integration, data collection, and RL models, are discussed, underscoring the importance of these components for effective environmental monitoring and decision-making. Overall, the paper advocates for the transformative impact of RL in Earth observation, offering a more efficient and sustainable approach to managing and safeguarding the planet.

4. BENEFITS AND ADVANTAGES

Earth observation offers high-quality benefits and blessings for humans, the environment, and our planet. It could be used to study and analyze the contemporary nation of the environment to ensure our planet is on a path closer to a sustainable and wholesome destiny. Earth observation structures permit scientists to analyze the surroundings in areas that are in any other case inaccessible and degree changes from 365 days to the subsequent that might not always be observed at the floor degree. Facts amassed via those structures assist scientists in identifying patterns and tendencies in the surroundings and shaping better predictions concerning herbal disasters or the effects of weather trade. This information can also be used to tell governments and communities about potential answers about dealing with sources and planning for destiny.

Earth observation can also be used for practical programs, such as finding resources for smooth water, famine or disease outbreaks, land management, endangered species tracking, and precision farming (Harris & Naik, 2023). Monitoring the surroundings may result in more excellent, sustainable, and efficient ways of dealing with sources, lowering our reliance on fossil fuels and helping to create more healthy ecosystems. In this manner, Earth Commentary can aid in attaining goals, including decreasing poverty, reaching gender equality, and developing an extra simple and safe society. Earth statement structures can also offer evidence of environmental degradation that is had to offer adequate answers and prevent further harm. If powerful answers are enacted promptly, the fitness of the surroundings, the well-being of people, and different dwelling matters are stepped forward. Earth commentary gives us compelling insights and invaluable information for each the present and the destiny. With technological advances, it is becoming more and more viable to study and reveal adjustments on a global scale. Earth observation structures' advantages are sizeable and vital to expertise our planet and making it ready for a sustainable future.

4.1 Improved Prediction Accuracy

Earth's statement is an incredibly vital factor of technological know-how: it provides worthwhile statistics about Earth's surroundings and weather. Because of this significance, the accuracy of Earth statement predictions must be progressed to make more excellent, precise predictions about our planet's future. In the past, earth observation predicting models have depended on various meteorological statistics, satellite information, and floor measurements. However, those models have tremendous limitations due to their reliance on restricted data and old algorithms. Recent technological improvements have enabled entry to much more information than ever before, making an allowance for extra particular and correct predictions. The utilization of system-gaining knowledge of algorithms has enabled the improvement of laptop algorithms that examine a far wider variety of information and patterns. By assessing a mess of variables and characteristics, those algorithms can become aware of styles in information that could otherwise be tough to detect and provide extra correct predictions of destiny activities (Saravanan & Sumathi, 2012). Everything that has helped enhance the accuracy of predictions is the development of sensors that take measurements from an expansion of sources. For example, advances in water level detection have enabled better and more correct predictions for the ranges of water bodies.

Predictors' progressed accuracy has already been proven beneficial in various areas ranging from catastrophe prevention to agricultural planning. For example, farmers can better plan crop rotation and irrigation schedules with correct cloud cover predictions and water levels. Similarly, better predictions of storms and temperatures can assist communities in preparing for excessive climate (Song et al., 2023). The advanced accuracy of earth observation predictions has allowed for higher choice-making and developed caution in activities. Via a higher understanding of modern and destiny conditions in the world, we can ensure that our planet's assets are used in the most green way possible. Ultimately, this can benefit human lifestyles and our surroundings in tremendous ways.

4.2 Adaptive Learning in Dynamic Environments

Earth commentary technologies are increasingly used to screen and apprehend the environmental changes taking region in our Earth's complex and dynamic systems. This generation form is becoming increasingly critical, as it gives critical insights into the functioning of Earth's structures that cannot be obtained through traditional means. It has been used to measure air pollutants and atmospheric carbon stages, study land use, study the effects of weather trade, and examine human results on the environment. With conventional Earth statement strategies, information is accrued statically (Yadav et al., 2023). However, the environment is dynamic and modifications swiftly. Adaptive learning systems have evolved to gather and method information in real time. An adaptive getting-to-know system for Earth observation uses sensors to take environmental readings. These statistics are then analyzed mechanically and used to expect or check potential environmental adjustments. The gadget can also locate anomalies and alert researchers and choice makers. The device can also self-calibrate and self-optimize through machine studying and sample reputation algorithms to achieve the most pleasant possible overall performance.

Applying adaptive learning structures in Earth observation is beneficial in many ways. Utilizing a solid method for facts collection, machine-gaining knowledge, and pattern recognition strategies can aid in the early detection of changes within the environment. Moreover, trend evaluation can help identify environmental adjustments' motive and effect relationships, permitting researchers and decision-makers to intrude and reduce their effects. Adaptive learning structures additionally guide the deployment of crewless aerial cars. By utilizing state-of-the-art algorithms, UAVs can capture and examine precious facts from difficult-to-attain, time-touchy, or unsafe areas that might usually be inaccessible to personnel. UAVs equipped with adaptive mastering competencies can help screen matters with deforestation, water infection, and illegal mining (Brunke et al., 2022). The adaptive studying systems in Earth observation have become increasingly superior. Modern improvements in the era are taking into account more state-of-the-art and fine-tuned tactics for fact collection. Those tendencies will allow researchers to understand our planet better and make more excellent proof-sponsored selections. As Earth's dynamic surroundings call for more sophisticated and flexible tactics, adaptive knowledge of systems becomes increasingly important in ensuring dependable and well-timed information capture for the ramifications of environmental monitoring programs.

5. CHALLENGES AND LIMITATIONS

Earth observation is a rapidly growing area of technological know-how that uses satellite-based generation to study, degree, and screen Earth's environment, land, oceans, and physical features. It also involves Earth's surroundings' temperature, humidity, wind speed, and other climate parameters. Earth statement has sizeable potential to improve our expertise in climate exchange, air and water first-class, agricultural productivity, and plenty of other environmental troubles. Despite its capacity to tell scientists, policymakers, and other choice-makers, Earth's statement needs to be revised with the aid of several demanding situations and obstacles. For instance, a limited variety of satellites can offer the statistics essential to make Earth statement a reliable source of records (Jia et al., 2021). Those satellites need to be placed and maintained in orbit, including extensive fees for missions. Moreover, there are restrained investment resources for Earth observation missions, and frequently, what price ranges are to be had are restrained to specific packages. There is also a need for superior instrumentation, as the high quality of the ensuing records depends on the units' sophistication.

Another project is the availability of facts. Even as satellite TV for PC imagery is pretty standard, other varieties of measurements are regularly more challenging to collect or interpret. There may also be a need for long-term records of facts that can be used to accurately monitor climate alternate or different environmental issues over long intervals. Moreover, cloud cowl can restrict the amount of statistics that may be amassed, making longer-time observations more challenging. ultimately, the mixing of more than one asset of records can be complicated for Earth commentary projects. For instance, records collected with land-based instruments, ships, or planes must be synchronized and aggregated for complete Earth commentary (Pant et al., 2023). Additionally, due to the significant fee associated with Earth's statement, there is frequently much less incentive for data sharing and collaboration amongst diverse users. Earth commentary is a powerful tool for addressing international environmental problems in addition to failures. However, if the associated challenges and obstacles are addressed, it may offer correct and reliable statistics with practical answers. Using advanced instruments and strategies, focusing on statistics sharing and collaboration, and imparting adequate assets for studies and improvement, Earth observation can contribute to our know-how of the Earth device and its evolution over time.

5.1 Data Complexity and Training Requirements

The Earth observation industry has seen increased information complexity and training requirements. It will be attributed to technological advances, consisting of satellite systems, airborne sensors, and faraway sensing systems now to be had. Moreover, Earth statement statistics is regularly used to locate and display land use, weather, and ecosystem modifications, increasing the complexity of data category algorithms and automated function extraction techniques. The sheer volume and form of Earth commentary information has additionally brought about the development of increasingly more complex statistics processing and evaluation techniques. To efficaciously analyze and interpret this tremendous array of earth statement facts, a trained expert should be capable of understanding and working with the records in a ramification of contexts, from physical and geospatial parameters to modeling and gadget studying algorithms (Sivaram, Mohammed, Yuvaraj et al, 2019). At a minimum, this requires robust mathematics, records, and pc technology expertise. Additional topics include running structures, data management, programming languages, and software program applications required for Earth observation data analysis. Those inquisitive about pursuing a career in Earth commentary must invest in extensive schooling and familiarization with the vital statistics, tools, and strategies.

Earth statement's data complexity and education requirements pose a number of demanding situations to the enterprise. Firstly, the expenses associated with education and education are often prohibitively high. In addition, the enterprise frequently needs to offer sufficient incentives to attract and hold qualified experts. This results in a need for more experts with the appropriate abilities and know-how to guard the sector from ability dangers, which include unanticipated errors in facts analysis and misapplied or misinterpreted results (Xu et al., 2022). The technological advances have enabled Earth commentary data to be used more proactively, giving an upward push to numerous capacity packages for which traditional evaluation methods are insufficient. Considerable investments should be made in each training and developing optimized statistics manufacturing and analysis techniques, in addition to schooling and training within Earth observation. Earth observation statistics' complexity and training requirements have become increasingly stringent and require a specialized and tremendously- skilled workforce to ensure their successful use and interpretation. Moreover, as generation advances, so should the specialized training presented within the subject. Such investment in education and training will ensure that the industry is optimally benefitting from the statistics it produces.

5.2 Generalization Issues in Earth Observation

Earth observation entails the collection of records approximately the Earth's surface, environment, and oceans from various sources, including satellite imagery, aerial pix, and different Earth-sensing units. This record is critical to supplying accurate worldwide checks of changes on Earth and is also used to locate abnormal sports or changes within the environment. At the same time as the records and information accrued through Earth's statement can offer many valuable insights, it is only sometimes with its demanding situations. One of the most common problems in Earth commentary is the issue of generalization. Generalization is the process of changing unique observations into a generalized shape that is easier to interpret. Generalization can help make an allowance for the smoothing of facts and the removal of needless info. However, it may also lead to errors in interpretation or a loss of accuracy in the illustration of capabilities on the world's floor (Chaturvedi & de Vries, 2021). While generalizing too much, information can be filtered or abstracted too much, leading to a lack of detail or false interpretations. For example, a generalization might consist of merging small lakes into large bodies of water or merging small towns into an unmarried urban vicinity. The same can be said of satellite imagery, wherein too much generalizing should result in errors or fake interpretations.

The overall tendencies in Earth observation are toward more excellent, specific, and accurate information collection. As generation advances, higher decision imagery and data supply will increase. It improves the accuracy of generalization. However, it additionally offers new problems. First, higher decision imagery can increase the amount of information that needs processing. It may require more money and time, as more record points need to be accumulated and processed earlier than a beneficial result may be done. Moreover, the extra detail may be more than essential. It could result in litter and an improved risk of misinterpreting or perplexing functions. Standard generalization can be helpful in earth commentary while used efficaciously. It may simplify data to an extra interpretable and preferred shape. However, it wishes to be completed cautiously to ensure the accuracy of interpretation. With the development of higher resolution information, the issue of generalization desires to be addressed and adjusted for this reason, as more detail ought to lead to more clutter and a better threat of Misinterpretation. Additionally, the costs and time associated with better-resolution imagery must be considered while choosing which record factors to generalize.

6. FUTURE TRENDS AND DEVELOPMENTS

Earth observation is the collection of records approximately Earth from a ramification of satellites and spacecraft. This information is used to screen climate exchange, examine the state of our oceans and environment, and discover new resources. Earth observation has become increasingly famous in recent years as more significant satellites are released into orbit, and the era keeps improving. In the future, Earth observation is anticipated to move past simply amassing data (Sivaram, Yuvaraj, Mohammed et al, 2019). The advancements in artificial intelligence, quantum computing, and device studying era will permit satellites to investigate facts and make decisions primarily based on them, mainly due to automatic tasking choices and advanced choice guides. Further, improving subsequent technology sensors is expected to significantly enhance Earth commentary systems' skills. New imaging systems, microwave sensors, and laser radars are examples of several upgrades that have been made in recent years. These sensors will assist researchers in examining data faster and more significant as it should be—taking complex Earth observations and turning them into beneficial tool units.

Another development taking location on Earth observation is the extended public access to satellite TV for PC imagery and information. Open-get right of entry to statistics portals encourages public participation in ground-based observations and provides a platform for significant facts sharing among scientists, residents, and selection makers. One of the most vital tendencies in improving Earth observation is the flow in the direction of space mining for assets. Businesses are the primary rate of extracting treasured metals and minerals from asteroids, planets, moons, and even comets. Utilizing far-off sensing technology and robot systems, finding and discovering capacity sources with an excessive degree of performance will become viable (Wang et al., 2020). Ultimately, as the ability of Earth statement structures continues to grow, the need for predictive analytics becomes more and more vital. It may permit researchers to appropriately forecast adjustments in climate, climate conditions, and worldwide resource availability. Analytics can assist in preparing for herbal failures, population actions, and more. These analytics might be driven by significant statistics and system studying gear. It could offer corporations, governments, and residents treasured insights to cause more thoughtful decision-making. Typical Earth observation is a process of speedy improvement and change. Those modifications will allow scientific discovery and technological development, permitting us to explore the Earth and its environment. As the era advances, Earth commentary is specific to preserve and expand in thrilling and progressive guidelines.

6.1 Integration With Emerging Technologies

Earth observation is using remotely sensed records in aggregate with other information resources for knowledge and dealing with the environment. With the rapid development of the era, Earth observation is regularly integrating with rising technology. New technology, cloud computing, and artificial intelligence (AI) support scientists in gathering, analyzing, and keeping increasing quantities of data from Earth view satellites. By combining the significant quantities of satellite imagery with cloud computing services, researchers can get admission to the facts faster and more cost-efficiently than ever before. Additionally, algorithms powered by state-of-the-art AI can perceive and analyze changes inside the surroundings much quicker and more appropriately than people (Natarajan et al., 2021). The mixing of the far-flung sensing era with other critical environmental datasets and technologies, which includes Geographic statistics structures, is starting to reveal insights into the modifications going on across the globe. With the capacity to generate extraordinarily targeted maps, scientists can examine land use modifications, weather exchange influences, and expanding agricultural land. Moreover, integrating these technologies with drones and coffee-altitude satellites has enabled scientists to examine the environment on a grander local scale, supplying extra resolution to the records.

In addition to enhancing the accuracy of Earth observation facts, the rising technology also permits progressed monitoring, alerting, and decision-making. For instance, AI-powered algorithms can automatically stumble on modifications inside the panorama, figuring out areas that require a similar look at or alerting policymakers to ability environmental risks or herbal disasters. With the ability to continuously reveal modifications inside the environment, rising technologies have the potential to assist researchers hit upon and are expecting international developments, which can be used to prevent disasters or mitigate the results of weather alternations. The integration of rising technologies with Earth statement provides extraordinary access to records and new insights into our planet, hastily reworking how we observe, display, and control our surroundings (Mahalakshmi et al., 2014). With the assistance of AI, more correct facts can be received, allowing us to apprehend and prepare for the changing wishes of our planet. The integration of rising technology with Earth observation gives us the capacity to understand and manipulate our planet more accurately and effectively. Through AI algorithms to investigate vast portions of data, researchers can benefit from precious insights into how our surroundings are converting and making more informed selections of what destiny holds.

6.2 Potential Enhancements and Innovations

Earth observation has skilled superb enhancements and improvements over the last numerous years. From satellites to radar, facts-amassing technologies have increasing advanced capabilities, creating exceptional possibilities for scientists and the general public to display and observe Earth's environment. In the coming years, further innovations in Earth commentary technology and methods are anticipated to make the sphere even more effective. A stepped-forward satellite technology will make Earth commentary more efficient and accurate. For instance, multi-spectral cameras and sensors will increasingly collect statistics on the Earth's atmosphere and surface and stumble on ecosystem modifications and anomalies (Waubert de Puiseau et al., 2022). Sensor technology like lidar will also measure physical characteristics, including surface elevation, flowers, and soil moisture. In addition, geoengineering satellites may be used to accumulate data in tough-to-attain areas, including the polar regions or high mountains.

Artificial intelligence (AI) and machine mastering strategies will vastly improve Earth observation. For example, AI can be used to automate the evaluation and interpretation of massive datasets, locate anomalies or styles, and make predictions. AI can also identify objects in satellite pictures or determine flower changes over the years. The current fashion of deploying compact, low-value sensing structures on satellites, balloons, and drones will likely be more widespread in the coming years. Those structures are inexpensive, offer insights into various procedures, and allow actual-time monitoring competencies. There may be an elevated cognizance of integrating records from specific resources, such as satellites, radar, remote sensing, and floor-based sensors. It could assist in creating new insights and provide an extra comprehensive picture of the Earth's surroundings.

The emergence of crowdsourcing and citizen science is also expected to affect Earth. Crowdsourcing and citizen science platforms allow massive numbers of humans to gather statistics for several projects, which include monitoring air pollution stages or monitoring adjustments in ecosystems through the years. The advances in the Earth observation era and facts series techniques are predicted to have an extensive impact within the coming years. Innovations consisting of progressed satellite structures, the use of AI and machine getting to know, the deployment of low-value sensing systems, records integration, and crowdsourcing platforms are prepared to transform the sphere of Earth statement and open up new possibilities for medical research and remark.

7. CONCLUSION

Reinforcement learning for Earth observation structures holds promise for dealing with and monitoring the environment more effectively. Reinforcement learning algorithms leverage big datasets to manage the surroundings to maximize overall performance goals. Therefore, using this system, environmental inventories can be quickly analyzed, and pleasant interventions can be decided to ensure environmental sustainability. Reinforcement learning knowledge of algorithms can provide a similar perception of how humans interact with their surroundings, allowing a better know-how of how humans affect the herbal global and how the surroundings influence humans. This insight can help guide choice-making to maximize performance targets while defending the environment.

Using Reinforcement, knowledge of Earth observation structures can benefit the surroundings. It increases the performance of environmental tracking and evaluation, even as bearing in mind more profound knowledge of environmental interactions with people. The device also enables to ensure that overall performance targets are met sustainably. Reinforcement learning based on Earth statement systems constitutes an exciting new artificial intelligence (AI) software for environmental monitoring. Reinforcement learning is an optimization technique that permits machines to learn from environmental interactions, an excellent way to obtain specific desires. Through this method, machines can facilitate selection-making and environmental tracking on a large scale and more accurately than ever.

The Reinforcement learning based totally Earth statement structures can be utilized in commercial and research programs. The expanded accuracy and speed of the statistics accrued from these structures will permit companies and governments to recognize their surroundings better and make facts-pushed choices. Moreover, the information from these structures will offer scientists unheard-of detail and accuracy regarding the kingdom of the planet and its herbal systems. The applications of Reinforcement learning primarily based on Earth observation structures are considerable. Inside the destiny, we may want to see those structures deployed to assess the health of forests, oceans, and different herbal systems; enhance the accuracy of climate forecasting; track modifications within the environment; inform authorities of coverage; or even offer early warning indicators of strange climate events, which include floods or tsunamis. Besides these programs, Reinforcement learning to know primarily based on Earth observation systems is also vital for growing sustainable solutions to climate exchange. With the aid of utilizing the system to get to know the era, these structures can tune and analyze information to better inform choice-makers of exceptional strategies for adapting to a converting environment and reducing the impact of human activity. The destiny scope of Reinforcement learning based on Earth statement structures is wide and varied. From businesses to government coverage and studies, those systems are critical for knowledge of the country of the planet, developing sustainable answers, and making informed choices.

REFERENCES

Brunke, L., Greeff, M., Hall, A. W., Yuan, Z., Zhou, S., Panerati, J., & Schoellig, A. P. (2022). Safe learning in robotics: From learning-based control to safe reinforcement learning. *Annual Review of Control, Robotics, and Autonomous Systems*, *5*(1), 411–444. doi:10.1146/annurev-control-042920-020211

Chaturvedi, V., & de Vries, W. T. (2021). Machine learning algorithms for urban land use planning: A review. *Urban Science (Basel, Switzerland)*, *5*(3), 68. doi:10.3390/urbansci5030068

Chun, J., Yang, W., Liu, X., Wu, G., He, L., & Xing, L. (2023). Deep Reinforcement Learning for the Agile Earth Observation Satellite Scheduling Problem. *Mathematics, 11*(19), 4059. doi:10.3390/math11194059

Chung, L. C. H., Xie, J., & Ren, C. (2021). Improved machine-learning mapping of local climate zones in metropolitan areas using composite Earth observation data in Google Earth Engine. *Building and Environment, 199*, 107879. doi:10.1016/j.buildenv.2021.107879

Di Napoli, M., Marsiglia, P., Di Martire, D., Ramondini, M., Ullo, S. L., & Calcaterra, D. (2020). Landslide susceptibility assessment of wildfire burnt areas through earth-observation techniques and a machine learning-based approach. *Remote Sensing (Basel), 12*(15), 2505. doi:10.3390/rs12152505

Ferreira, B., Iten, M., & Silva, R. G. (2020). Monitoring sustainable development by means of earth observation data and machine learning: A review. *Environmental Sciences Europe, 32*(1), 1–17. doi:10.118612302-020-00397-4

Gupta, S., Saravanan, V., Choudhury, A., Alqahtani, A., Abonazel, M. R., & Babu, K. S. (2022). Supervised computer-aided diagnosis (CAD) methods for classifying Alzheimer's disease-based neurodegenerative disorders. *Computational and Mathematical Methods in Medicine, 2022*, 2022. doi:10.1155/2022/9092289 PMID:35651921

Han, Y., Wang, L., Fu, W., Zhou, H., Li, T., & Chen, R. (2021). Machine learning-based short-term GPS TEC forecasting during high solar activity and magnetic storm periods. *IEEE Journal of Selected Topics in Applied Earth Observations and Remote Sensing, 15*, 115–126. doi:10.1109/JSTARS.2021.3132049

Harris, A., & Naik, K. (2023, March). Autonomous Command and Control for Earth-Observing Satellites using Deep Reinforcement Learning. In *2023 IEEE Aerospace Conference* (pp. 1-12). IEEE. 10.1109/AERO55745.2023.10115916

He, Y., Xing, L., Chen, Y., Pedrycz, W., Wang, L., & Wu, G. (2020). A generic Markov decision process model and reinforcement learning method for scheduling agile earth observation satellites. *IEEE Transactions on Systems, Man, and Cybernetics. Systems, 52*(3), 1463–1474. doi:10.1109/TSMC.2020.3020732

Huang, Y., Mu, Z., Wu, S., Cui, B., & Duan, Y. (2021). Revising the observation satellite scheduling problem based on deep reinforcement learning. *Remote Sensing (Basel), 13*(12), 2377. doi:10.3390/rs13122377

Jia, Y., Jin, S., Chen, H., Yan, Q., Savi, P., Jin, Y., & Yuan, Y. (2021). Temporal-spatial soil moisture estimation from CYGNSS using machine learning regression with a preclassification approach. *IEEE Journal of Selected Topics in Applied Earth Observations and Remote Sensing, 14*, 4879–4893. doi:10.1109/JSTARS.2021.3076470

Karki, S., Bermejo, R., Wilkes, R., Monagail, M. M., Daly, E., Healy, M., Hanafin, J., McKinstry, A., Mellander, P.-E., Fenton, O., & Morrison, L. (2021). Mapping spatial distribution and biomass of intertidal Ulva blooms using machine learning and earth observation. *Frontiers in Marine Science, 8*, 633128. doi:10.3389/fmars.2021.633128

Löw, F., Stieglitz, K., & Diemar, O. (2021). Terrestrial oil spill mapping using satellite earth observation and machine learning: A case study in South Sudan. *Journal of Environmental Management, 298*, 113424. doi:10.1016/j.jenvman.2021.113424 PMID:34358936

Mahalakshmi, D. V., Paul, A., Dutta, D., Ali, M. M., Jha, C. S., & Dadhwal, V. K. (2014). Net surface radiation retrieval using earth observation satellite data and machine learning algorithm. *ISPRS Annals of the Photogrammetry, Remote Sensing and Spatial Information Sciences, 2*, 9–12. doi:10.5194/isprsannals-II-8-9-2014

Natarajan, Y., Kannan, S., Selvaraj, C., & Mohanty, S. N. (2021). Forecasting energy generation in large photovoltaic plants using radial belief neural network. *Sustainable Computing : Informatics and Systems, 31*, 100578. doi:10.1016/j.suscom.2021.100578

Pant, G., Bahuguna, R., Pandey, S., Gehlot, A., Yadav, S. P., & Pachauri, R. K. (2023, April). Intelligent Automated Interference for the Protection of Road Safety. In *2023 International Conference on Computational Intelligence, Communication Technology and Networking (CICTN)* (pp. 87-91). IEEE. 10.1109/CICTN57981.2023.10141086

Saravanan, V., & Sumathi, A. (2012, December). Dynamic handoff decision based on current traffic level and neighbor information in wireless data networks. In *2012 Fourth International Conference on Advanced Computing (ICoAC)* (pp. 1-5). IEEE. 10.1109/ICoAC.2012.6416797

Saravanan, V., & Sumathi, A. (2012). Handoff mobiles with low latency in heterogeneous networks for seamless mobility: A survey and future directions. *European Journal of Scientific Research, 81*(3), 417–424.

Sivaram, M., Mohammed, A. S., Yuvaraj, D., Porkodi, V., Manikandan, V., & Yuvaraj, N. (2019). Advanced expert system using particle swarm optimization based adaptive network based fuzzy inference system to diagnose the physical constitution of human body. In *Emerging Technologies in Computer Engineering: Microservices in Big Data Analytics: Second International Conference, ICETCE 2019*, (pp. 349-362). Springer Singapore. 10.1007/978-981-13-8300-7_29

Sivaram, M., Yuvaraj, D., Mohammed, A. S., Manikandan, V., Porkodi, V., & Yuvaraj, N. (2019). Improved enhanced DBTMA with contention-aware admission control to improve the network performance in MANETS. *Computers, Materials & Continua, 60*(2), 435–454. doi:10.32604/cmc.2019.06295

Song, Y., Wei, L., Yang, Q., Wu, J., Xing, L., & Chen, Y. (2023). RL-GA: A reinforcement learning-based genetic algorithm for electromagnetic detection satellite scheduling problem. *Swarm and Evolutionary Computation, 77*, 101236. doi:10.1016/j.swevo.2023.101236

Wang, X., Wu, G., Xing, L., & Pedrycz, W. (2020). Agile earth observation satellite scheduling over 20 years: Formulations, methods, and future directions. *IEEE Systems Journal, 15*(3), 3881–3892. doi:10.1109/JSYST.2020.2997050

Wang, X., Wu, J., Shi, Z., Zhao, F., & Jin, Z. (2022). Deep reinforcement learning-based autonomous mission planning method for high and low orbit multiple agile Earth observing satellites. *Advances in Space Research, 70*(11), 3478–3493. doi:10.1016/j.asr.2022.08.016

Wang, X., Zhao, F., Shi, Z., & Jin, Z. (2023). Deep Reinforcement Learning-Based Periodic Earth Observation Scheduling for Agile Satellite Constellation. *Journal of Aerospace Information Systems,* 1-12.

Waubert de Puiseau, C., Meyes, R., & Meisen, T. (2022). On reliability of reinforcement learning based production scheduling systems: A comparative survey. *Journal of Intelligent Manufacturing, 33*(4), 911–927. doi:10.100710845-022-01915-2

Wei, L., Chen, Y., Chen, M., & Chen, Y. (2021). Deep reinforcement learning and parameter transfer based approach for the multi-objective agile earth observation satellite scheduling problem. *Applied Soft Computing, 110*, 107607. doi:10.1016/j.asoc.2021.107607

Wellmann, T., Lausch, A., Scheuer, S., & Haase, D. (2020). Earth observation based indication for avian species distribution models using the spectral trait concept and machine learning in an urban setting. *Ecological Indicators, 111*, 106029. doi:10.1016/j.ecolind.2019.106029

Xu, Y., Zhang, L., Du, B., & Zhang, L. (2022). Hyperspectral anomaly detection based on machine learning: An overview. *IEEE Journal of Selected Topics in Applied Earth Observations and Remote Sensing, 15*, 3351–3364. doi:10.1109/JSTARS.2022.3167830

Yadav, S. P., Zaidi, S., Nascimento, C. D. S., de Albuquerque, V. H. C., & Chauhan, S. S. (2023, January). Analysis and Design of automatically generating for GPS Based Moving Object Tracking System. In *2023 International Conference on Artificial Intelligence and Smart Communication (AISC)* (pp. 1-5). IEEE. 10.1109/AISC56616.2023.10085180

Zarei, A., Hasanlou, M., & Mahdianpari, M. (2021). A comparison of machine learning models for soil salinity estimation using multi-spectral earth observation data. *ISPRS Annals of the Photogrammetry, Remote Sensing and Spatial Information Sciences, 3*, 257–263. doi:10.5194/isprs-annals-V-3-2021-257-2021

Chapter 5
Nearest Neighbor-Based Earth Observation and Monitoring

Sutha J.
Department of Computer Science and Engineering, AAA College of Engineering and Technology, Sivakasi, India

Meenakshi S.
https://orcid.org/0000-0001-5075-6808
Department of Science and Humanities (General Engineering), R.M.K. Engineering College, Kavarapettai, India

Esakkiappan C.
Department of Electronics and Communication Engineering, Panimalar Engineering College, India

Anitha G.
Department of Electronics and Communication Engineering, R.M.D. Engineering College, Kavaraipettai, India

ABSTRACT

The objectives of this chapter are multifaceted and directed towards a comprehensive understanding of Earth's land cover and environmental changes. First and foremost, the aim is to provide accurate and reliable land cover mapping of the Earth's surface, ensuring that the information is timely and comprehensive, thereby facilitating a better grasp of environmental transformations. Furthermore, a key objective is to enhance our knowledge of climate change and its far-reaching impacts, with a specific focus on improving the precision of monitoring the natural environment and habitats. In addition to this, the study aims to investigate and analyze the patterns of land use change and the associated impacts on the environment. This involves a thorough examination of the resources involved and the repercussions resulting from alterations in land cover. Moreover, the objective is to gain a comprehensive understanding of areas with significant environmental vulnerability, pinpointing regions that are particularly sensitive to environmental changes.

DOI: 10.4018/979-8-3693-1850-8.ch005

1. INTRODUCTION

The nearest-neighbor set of rules, which keeps present-day Earth commentary and monitoring, is used to efficaciously utilize Earth statement statistics up-to-date and examine changes going on on Earth's surface. The facts are gathered through satellites, planes, and ground-up-to-date properties, imparting treasured insights into various up-to-date of the Earth's environment, including temperature patterns, land cowl changes, populace dynamics, plant life health, and different relevant information. These fundamental facts can be incorporated with current social, monetary, and political facts up to date to gain complete knowledge of those changes' effects on the planet's ecosystems (Tiwari et al., 2021). The Earth commentary and tracking machine that relies on the nearest-neighbor technique uses a gadget learning approach to the up-to-date okay-nearest neighbor. This technique includes comparing a given dataset with the closest associates and updating it based on recognized labels. The collection of guidelines operates by selecting the up-to-date, closest up-to-date, and most recent statistics and comparing the attribute values from each. Assigning the brand new, up-to-date equal class is viable if the values showcase a high degree of similarity. This technique has been empirically verified to be extraordinarily efficacious in forecasting the terrestrial coverage of the Earth's floor via satellite imagery by way of discerning plant alterations.

This information can assist scientists in gaining a more profound knowledge of human activities' environmental impacts and successfully compare and respond to updated contemporary challenges (Eskelson et al., 2009). The brand-new assistance presents more particular and timely statistics, allowing advanced coverage choices at a quicker fee than up-to-date methods. Analyzing applicable statistical statistics makes it viable to become aware of regions that may be at hazard up-to-date human pastimes. Modern studies offer many treasured statistics regarding the correlation between climate change and land-use changes. The maximum present-day and advanced method of up-to-date Earth tracking and monitoring is a notably updated technique that collects outstanding records of climate exchange and numerous different environmental worries. Its more suitable functionality gives an extra exact knowledge of the modifications going on worldwide. It is an updated aid in growing techniques and regulations for coping with the Earth's sources.

1.1 Background and Significance

The utilization of close-to-neighbor-primarily based Earth observation and monitoring has grown to be an increasing number of crucial tools for identifying, assessing, and interpreting global environmental transformation. In current years, there has been a growing availability of various geospatial information, which has facilitated the utilization of algorithms based on imagery for the purchase of near real-time intelligence about various phenomena, which include natural screw-ups, land-use alterations, helpful water resource management, ecological dynamics, land degradation, crop monitoring, and climate trade.

Near-neighbor-based Earth remark and tracking encompasses factors that warrant close attention, along with land cowl, land use, woodland fitness, land cowl dynamics, and urbanization. Those records can be incorporated with other geographical data to yield essential records and perspectives on the impact of environmental modifications on groups and people (Mohammed et al., 2018). Furthermore, it could be utilized to discover noteworthy times of environmental alteration, including deforestation, and hired to discover times of land degradation. Near-neighbor-primarily based Earth commentary and monitoring can yield crucial information concerning environmental trade and inform policy choices at each local and international degree.

Near-neighbor tactics have commonly received popularity in land cowl alternate analysis because they efficaciously and correctly seize spatial and temporal traits across huge geographical areas. These statistical records can be applied to investigate and find urban enlargement, environmental contamination, adjustments in agricultural land, interactions among human beings and their surroundings, and the buying and selling of land cover (Fadaei-Kermani et al., 2017).

The method employed in this look is focused on the concept of close-to-neighbor-based total analysis. Earth observation and monitoring are applied to inform decision-making regarding environmental control. It may facilitate the identification of areas at risk of land degradation and may be utilized to assess the impacts of land use modifications. It could additionally be applied for predicting land cover alternatives and disseminating land use policies. Moreover, proximity neighbor methodologies may be employed to envision and gift the state of woodland energy, aquatic property, and changes in land cover (Xing et al., 2009).

The proposed technique is an ordinary, proximity-based total method based on the nearest friends. The practice of Earth commentary and monitoring plays an essential role in figuring out and surveilling international environmental changes. This tool can pick out places requiring governmental action, provide precious insights into land use exchange, and tell choice-making regarding managing natural aids. It is a value-powerful and efficient method to achieve spatially specific data and make knowledgeable environmental conservation and sustainability selections.

1.2 Objectives of the Chapter

- To offer correct and reliable land cover mapping of the Earth's surface and supply a well-timed and comprehensive assessment of environmental trade.
- To reinforce the know-how of climate change and its impacts and improve the precision of monitoring the herbal environment and habitats.
- The goal is to study the styles of land use trade and its associated impacts and examine the resources concerned and the repercussions due to adjustments in land cover.
- This looks at targets to analyze the causes and consequences of land degradation and determine the impact of plant screw-ups on the surroundings.
- The goal is to gain an understanding of and disclose regions with vast environmental vulnerability and examine the effects of human sports on the natural environment.

2. FOUNDATIONS OF NEAREST-NEIGHBOR APPROACHES IN EARTH SCIENCES

The theoretical framework that underlies Nearest-Neighbor or "NN" approaches in Earth Sciences may be traced back to geography, geology, and hydrology. Those techniques have become crucial elements in reading and monitoring Earth phenomena throughout various academic disciplines. In geography, the closest neighbor (NN) method has long been employed to forecast the traits of a landform by using a set of measured houses, approximately a reference factor, representing the closest neighboring factor to the unknown place (He et al., 2022). In geology, the NN technique is employed to evaluate the scale and composition of a geological formation consisting of a fracture or a fold. It entails studying the dimensions, form, and traits of the shape, in addition to its dating with adjacent geological structures.

Further, in hydrology, the NN method is used to become aware of a river's place, rate, and ordinary go-with-the-flow based on the features of the closest-neighbor movement.

NN in numerous disciplines within earth sciences has come to be step-by-step crucial to analyzing and displaying Earth-related phenomena. In particular, neural community techniques were employed to visualize floods or earthquakes, examine the health of flowers or ecosystems, forecast and pick out landforms, analyze local climate or meteorological statistics, and hit upon and quantify land cowl changes. In every case stated above, neural network techniques permit researchers to perceive "pals" of a specific place or locality, facilitating a more efficient and correct recall of that place (Poursanidis et al., 2018).

Moreover, using neural network algorithms has made significant progress alongside the advancements in remote sensing generation. Mainly, aerial and satellite TV for PC photography has facilitated researchers in right away and appropriately figuring out the materials associated with a given location. Therefore, this functionality enables them to scrutinize environmental alterations with more desirable precision and in extra depth. In addition, utilizing excessive-resolution snapshots and multi-dimensional records has enabled the software of neural community algorithms in more specific and complex manners.

The closest-neighbor method has been and can remain an enormously helpful tool for Earth statement and tracking. The usage of neural networks in many disciplines inside the discipline of earth sciences has enabled researchers to gain a more profound know-how of a specific area by reading the functions of its surrounding regions (Beaudoin et al., 2018). Furthermore, technological advancements have facilitated the development of more precise and intricate programs of the neural network approach. These enhancements will allow researchers to monitor and analyze the Earth's surroundings more effectively.

2.1 Overview of Nearest-Neighbor Methods

Nearest-neighbor strategies are interpolation strategies commonly hired in Earth commentary and tracking. Those techniques consist of analyzing a particular statistics factor about its surrounding context and estimating its value based totally on the characteristics of the encircling records factors. The most often applied version of the method involves evaluating a brand new aspect to its nearest buddies and assigning a fixed fee to it through a weighted average of those associates. Despite its straightforwardness, this approach has exhibited excessive precision in several software applications (Orusa et al., 2022).

Nearest-neighbor methods are employed to spatially interpolate records from far-off sensing, especially for land-use and land-cover classes. Furthermore, it can explain the characteristics of diverse land-use categories, enhance the decision of downscaled satellite TV for PC statistics, generate maps of flora signs, and forecast the distribution patterns of plant species. Furthermore, those methodologies link statements about the Earth and observations from on-web page sampling.

Due to its essential philosophical concepts, this unique collection of guidelines has several blessings compared to opportunity interpolation techniques. The computational efficiency of this gadget is excellent, as it has low memory necessities and a rapid rate of execution (Ferreira et al., 2020). Furthermore, these tactics exhibit exceptional accuracy and are well-applicable for locations with constrained pattern availability. Furthermore, it continues the presence of nearby patterns in the records by calculating pattern values and the usage of the features of the encompassing environment. Nevertheless, there also are risks associated with this method in view that it could inadvertently seize arbitrary styles and developments that want to be revised, injecting unnecessary information.

The closest-neighbor techniques are usually used in earth observation and monitoring interpolation approaches. This method offers several advantages, together with its computational efficiency, simplicity, and high level of accuracy. However, it can also introduce vain statistics if the algorithm is not cautiously calibrated according to the peculiarities of the surroundings.

2.2 Applications in Earth Observation

The development of Earth commentary generation has drastically more desirable human information and visualization of far-off environments, supplying crucial records to inform decision-making on an international scale. This era is utilized for numerous purposes: land use mapping, crop forecasting, environmental monitoring, natural catastrophe preparedness, and climate analysis (Yadav et al., 2023).

Land use mapping has emerged as a distinctly influential use of Earth observation. Satellites and aerial cars allow researchers to reap high-decision imagery of a specific place, allowing the remark and quantification of temporal variations in land cover. Consequently, pupils can generate land use cartography and comprehend the environmental repercussions of human endeavors, encompassing population increase, agricultural output, and concrete growth. These maps provide records for making coverage choices related to land use, water control, and land upkeep.

The assessment of Earth's situation is also of great importance in the context of predicting agricultural yields. On an annual basis, farmers and governmental entities use satellite-derived data to evaluate the fitness of plants and forecast the expected yield. This data is applied to amend agricultural guidelines and distribute resources in a manner that optimizes success (Chen et al., 2016).

Furthermore, the utilization of Earth observation generation is employed for environmental monitoring. Satellites look at air exceptional, weather variability, water availability, river dynamics, and oceanic thermal styles over extended durations, improving comprehension of the evolving surroundings and formulating sustainable remedies. For instance, satellite images can identify illicit logging and fishing activities and safeguard vulnerable ecosystems.

Earth observation is also applied to prepare for natural screw-ups. Photos and data obtained from Earth remark satellites provide governments with specific information concerning the extent of harm, permitting them to strategically deploy emergency people to regions requiring on-the-spot help (Küçük et al., 2016). Those factual observations can also be helpful for scientists in growing models expecting the effects of future screw-ups and providing early warnings for doubtlessly catastrophic events.

Earth remark technology is essential in acquiring information and discovering climatic styles and alternate dynamics. It is miles viable to display and analyze patterns in sea ice ranges, concentrations of greenhouse gases, and ocean currents. It permits us to gain a piece of comprehensive information on the impacts of global warming and develop realistic answers for mitigating weather alternatives.

Producing statements about Earth gives us essential statistics that aid in improving our knowledge, safeguarding, and conserving our environment. Using contemporary generation, which includes land use mapping, crop forecasting, environmental tracking, herbal disaster preparedness, and weather evaluation, we will mitigate precise risks related to the continued worldwide modifications and try towards a greater sustainable destiny (Tziolas et al., 2021).

3. KEY COMPONENTS OF A NEAREST-NEIGHBOR-BASED EARTH OBSERVATION SYSTEM

The closest neighbor algorithm has been updated to comprise the maximum current records. The Earth observation system is a far-off sensing that analyzes nearby areas updated to replace the traits of a given vicinity. Those systems are currently applied in various environmental traits, inclusive of updated panorama, functionalities, land cover, water assets, and other associated up-to-date. So that it can ensure proper capability, an Earth commentary update that utilizes nearest-neighbor-primarily based updating should encompass many essential additives.

It is vital to set up fundamental, up-to-date geographical facts. The above statistics include records regarding terrain elevation, satellite TV for PC imagery, various remotely sensed information, up-to-date land use, land cover, and other applicable cultural and environmental data (Ding et al., 2019). This foundational machine allows the final update of nearest-neighbor evaluation with improved accuracy.

Furthermore, modern-day specialized algorithms and software programs are employed up-to-date to make specific the facts present-day and should pick out the most significant nearby locations. Algorithms embody various kinds, which include ok-nearest neighbor algorithms, density-based updated algorithms, and deep-gaining knowledge of algorithms. The algorithms enable the machine to figure out the most applicable information about the problem below the investigation, offering the requisite records for evaluation.

The machine has been updated, and the update encompasses real-time tracking. This option allows the gadget to constantly update its records and generate predictions up to date with the latest and relevant information (Gobinathan et al., 2021). It is crucial in characteristic or land cowl exchange cases, as it enables the machine to be up to date, directly reply to up-to-date latest changes, and deliver precise information.

The PC embodies an information garage device that is often updated. It has up-to-date all of the output records derived from the evaluation, ensuring it is modern. It has included the latest facts about nearby areas relevant to up-to-date analysis. This statistical information may be applied up-to-date and increase a wide range of reviews.

A nearest-neighbor-to-Earth remark machine is a good up-to-date utilized for faraway sensing evaluation. Several crucial additives are required to ensure effectiveness in this context. These encompass a foundational statistics device, specialized analysis algorithms, real-time statistics tracking, and a garage. With the up-to-date day updates, those additives enable the updated appropriate analysis facts from adjacent locations and constantly replace the up-to-date information to enhance performance in environmental tracking.

3.1 Data Preprocessing and Feature Selection

The preprocessing of statistics is a critical and essential stage in imposing any device getting-to-know system, consisting of a nearest-neighbor-based earth commentary system. In this approach, an enormous amount of unprocessed information is converted into a more wonderful layout and provided in a way that helps more outstanding analysis (Natarajan et al., 2021). This method can improve the records as they should be understood and yield more meaningful outcomes. This text will discuss such structures' generally hired records preprocessing and characteristic selection methodologies.

To start with, it is miles imperative to streamline the supplied records to remove any inaccuracies or redundancies it could consist of. This challenge can be carried out by using various filtering and discretization strategies. For instance, outliers may be removed using the median absolute deviation or outlier reduction strategies. Statistics may be discretized using binning algorithms, including identical-width or equal-top binning. This system aids in decreasing the extent of noise that can be present within the unprocessed statistics.

Similarly, to ensure the accuracy of the information, the selection of capabilities must encompass the vicinity to determine which variables or capabilities are for the given undertaking. This phenomenon occurs because of immoderate functions inside the tool, which might cause overfitting and faulty outcomes. Usually employed approaches for function choice encompass primary component analysis, recursive partitioning, and statistical gain evaluation. Using these strategies makes it far more feasible to gain and use the most crucial features that affect the forecast in the Earth statement machine.

As soon as the preprocessing of facts and selection of traits have been finished, the identity of nearest pals may be performed with the aid of utilizing numerous distance measures along with Euclidean distance, Minkowski distance, long island distance, or Chebyshev distance (Natarajan et al., 2022). The suitability of distance metrics may vary based on their utility.

In the long run, after figuring out the nearest friends, the values of those characteristics may be employed in predicting the goal vicinity through a weighted standard method or the nearest neighbor set of rules. Implementing this technique makes it feasible to decorate the precision of predictions, allowing more effective mitigation of environmental influences and accomplishing unique sustainability dreams.

This paper discusses statistical preprocessing and feature choice techniques in a nearest-neighbor-based worldwide land remark machine. Employing those methodologies makes it possible to beautify the accuracy of predictions derived from Earth remark data, leading to more excellent and efficient results.

3.2 Nearest-Neighbor Algorithms in Earth Monitoring

Nearest-neighbor algorithms are a statistical technique widely used across numerous disciplines to identify patterns and relationships inside large datasets. This specific set of ideas has gained popularity due to the improvements in Earth tracking technology. The cutting-edge technology's utilization of far-flung sensing satellites has facilitated the buildup of tremendous statistics, thereby permitting researchers to gain complete knowledge of the Earth's surface and weather. Therefore, nearest-neighbor algorithms analyze the considerable facts and derive meaningful insights.

Nearest-neighbor algorithms depend upon a "nearest neighbor" method, in which data points are recognized as comparable or related based on their closeness in the dataset. This unique set of ideas can be employed to examine correlations among various statistical variables, which include floral coverage, land surface temperature, topography, and seafloor temperature. By organizing connections among precise data kinds, scientists can better apprehend patterns, traits, and anomalies inside the surroundings (Poursanidis et al., 2015).

One of the number one programs of nearest-neighbor algorithms is tracking land floor temperature. This technique examines the impacts of different land cowl types on adjoining climates. For example, researchers can employ this technique to check how herbal capabilities, which include mountains and vegetation, modulate ambient air temperatures. By comprehending these correlations, individuals can discern the timing and locations where excessive temperatures are most likely to manifest and devise techniques to mitigate their depth.

Nearest-neighbor algorithms are employed to hit upon changes in the land cowl, such as deforestation, wetlands degradation, and concrete growth. Through the usage of coming across the correlation among wonderful land cover sorts, researchers are capable of discerning alterations inside the environment over some time. This tracking technique can be employed to make more prominent techniques for environmental conservation and coping with herbal sources.

In a manner comparable to monitoring land floor temperature and detecting land cowl, nearest-neighbor algorithms are also hired to identify institutions among extraordinary herbal phenomena (Prasad et al., 2022). For instance, researchers can apply these algorithms to ascertain the correlation between seismic hobby and volcanic eruptions. Through establishing hyperlinks between earthquake occurrences and volcanic eruptions, researchers may enhance their potential to expect future eruptions and discover areas at better hazards.

Nearest-neighbor algorithms have emerged as a treasured tool for researchers in reading and decoding statistical facts acquired via Earth monitoring satellites. Through their knowledge of the interconnections amongst many styles of geological facts, scientists can extra efficaciously identify styles, traits, and anomalies that could inform techniques for environmental protection and management of herbal sources.

4. ADVANTAGES AND BENEFITS

Nearest-neighbor algorithms are extensively used in numerous elements of earth tracking. Via advanced data mining strategies, those techniques allow the sensible evaluation of sizeable and problematic datasets. Various blessings and advantages are associated with utilizing nearest-neighbor algorithms in earth monitoring.

First and foremost, nearest-neighbor algorithms are a powerful tool for extracting significant facts from massive datasets acquired via satellites and different Earth tracking devices. Through the process of reading styles and identifying similarities and variations inside the records, these tools allow customers to categorize and classify facts pieces primarily based on their maximum probable starting place (Saravanan & Sumathi, 2012). In the discipline of forestry management, nearest-neighbor algorithms can be hired to discover regions within a rainforest with visible current logging activities or to mimic the patterns of animal motion over a particular duration. It enables land management establishments to conduct a more thorough analysis of their necessities and make appropriately informed decisions regarding the allocation of resources.

Additionally, nearest-neighbor algorithms own practical spatial evaluation skills. By integrating Earth monitoring systems with the ability to detect similarities amongst distinct geographical regions, users are empowered to identify and monitor extreme climate phenomena more precisely. It will be mainly advantageous for emergency responders who require spark-off identification of locations suffering from floods, wildfires, or different catastrophic occasions. Nearest-neighbor algorithms could mitigate the loss of lifestyles and property damage by imparting real-time data concerning a natural disaster's location, significance, and severity.

Moreover, nearest-neighbor algorithms provide the functionality to permit more unique tracking and modeling of the immediate environment. It may discern nuanced versions in temperature, flowers, and soil composition inside a given length. This technique can evaluate the ecological well-being of ecosystems and get insights into the results of weather fluctuations in precise geographical areas. To provide extra insights into the modifications in a selected location, using nearest-neighbor algorithms can enhance selection-making and resource management efficiency (Bhende et al., 2022).

In the end, utilizing nearest-neighbor algorithms in earth monitoring gives several blessings. Their capacity to investigate and perceive correlations within extensive datasets complements the precision of categorization and probability for emergency responders, safeguarding property and preserving lives more efficiently. Moreover, their sturdy spatial evaluation skills enable unique modeling and assessment of the surroundings. These components contribute to the effectiveness and utility of nearest-neighbor algorithms in earth monitoring.

4.1 Simplicity and Computational Efficiency

The nearest-neighbor approach is an efficient device for Earth monitoring. This approach's success has been shown across various applications, encompassing identifying population centers and monitoring corridors applicable to human hobbies. The guidelines under consideration are broadly identified and often hired for the green evaluation of big datasets due to their straightforwardness and computing efficiency.

The muse of nearest-neighbor algorithms is based on the popularity that figuring out companies or clusters of gadgets may be executed through leveraging the similarity in spatial styles. The algorithms employed in this look are categorized as unsupervised tactics, indicating that they no longer rely on pre-existing labels for the elements beneath consideration. As a substitute, these algorithms are designed to discover resemblance factors amongst remarkable regions. Nearest-neighbor algorithms are, in particular, high-quality in addressing situations when it is hard to establish express standards.

One of the essential blessings of employing a nearest-neighbor set of policies is its computational simplicity and performance. It permits the processing of massive datasets with affordable simplicity. The algorithm in question has considerable flexibility, enabling an expansion of strategies for statistics evaluation. For instance, algorithms can become aware of groups of elements in photographs with various levels of decision.

The closest-neighbor algorithms are usually employed in software packages about ecology and natural sources. For example, remote sensing techniques may be utilized to become aware of concentrations of regions wherein species of conservation importance are specifically plentiful, parent patterns in land utilization or land cowl, and detect alterations in environmental situations (Gupta et al., 2022).

The closest-neighbor algorithms are a robust tool for various Earth-tracking programs. These techniques' computational simplicity and effectiveness render them an attractive option for the economic assessment of tremendous datasets. Utilizing these algorithms, researchers can successfully and accurately become aware of patterns, clusters, and versions across diverse earth-tracking datasets.

4.2 Effectiveness in Handling Spatial Data

Nearest neighbor algorithms (NNAs) are fantastically effective methods for addressing spatial information in earth tracking, especially in machine learning and computer imagination. Neural community algorithms play a vital role in acquiring and processing spatial information inside the area of earth monitoring. Those algorithms showcase terrific precision and speed, enabling efficient and actual-time evaluation of the earth's surface.

The operational mechanism of neural network algorithms in earth monitoring is simple. The researchers examine the spatial information of a particular geographic vicinity, approximately the spatial statistics of a neighboring vicinity, and later produce a forecast concerning the extent of similarity or dissimilarity between these two places. Based on the factual basis hired, those prognostications might also supply statistical statistics on the physical attributes of the surveyed vicinity or an approximation of its environmental occasions and hazards. Neural networks can expect soil traits and the chance of an impending earthquake.

The techniques mentioned above are deemed appropriate for efficaciously dealing with enormous datasets. As an example, they can swiftly and appropriately delineate a great geographical expanse, in addition to classifying the records factors encompassed inner. This particular selection renders them extraordinary for city-making plans, wherein decisions should be expedited, mainly relying on various spatial records.

Moreover, neural network algorithms (NNAs) regularly necessitate notably fewer statistical resources than opportunity earth monitoring methodologies, resulting in greater efficiency and price-effectiveness in the statistics garage. It is especially critical in satellite TV for PC imaging for computer imaging, as it includes transmitting information over sizable distances and at substantially slower speeds than terrestrial statistics transmission.

NNAs provide inherent simplicity in their implementation and may be custom-designed without problems, permitting value-effective and adaptable tracking of dynamic landscapes. Those systems can seamlessly adapt to find novel fact sources, integrating clean data into an established tracking system or serving as the muse for a piece of novel monitoring equipment without encountering difficulties (Purohit et al., 2023).

Neural networks (NNs) effectively process and evaluate geographical information regarding earth monitoring. The provider gives a wide range of packages and uses tiny datasets, making it best for value-powerful and actual-time evaluation of bodily attributes and environmental situations. It aids in knowledgeable choice-making regarding land use and urban making plans.

5. CHALLENGES AND LIMITATIONS

Nearest-neighbor algorithms have gained popularity in the field of earth monitoring. As mentioned earlier, the algorithms are computational methods that utilize a collection of factual data to build predictive models about variables such as rainfall, temperature, land cover, and other related factors. Despite their potential usefulness, nearest-neighbor algorithms have many limitations and can encounter obstacles due to their underlying assumptions.

One of the primary constraints of nearest-neighbor algorithms is that their predictive models rely solely on a limited dataset. The resulting version may be accurate if the information utilized to produce the revised edition is adequate or if the statistical variables are distributed legally. Earth monitoring is a significant challenge due to the limited availability of timely updates on environmental information, which often needs to align with the desired timeframe. Therefore, algorithms that rely solely on preexisting data may need to be revised to make accurate predictions.

One notable aspect of nearest-neighbor algorithms is their reliance on the spatial proximity between data items. Every statistical data point aspires to maintain a sufficient level of separation from its neighboring points to assure predictions' accuracy. However, this objective is rarely always achieved, especially in densely populated regions. Moreover, the algorithms depend on the underlying assumptions that points possessing similar qualities ought to be located close to one another. If this assumption is invalid, the set of rules may also yield erroneous outcomes.

The utilization of nearest-neighbor algorithms can lead to issues related to overfitting. The algorithms have been specifically developed to assess the regulations governing the interplay of many information components. If the algorithm is excessively proficient in learning the rules, it may produce precise predictions of the input data, leading to biased outcomes.

The nearest-neighbor algorithms have demonstrated significant potential in earth monitoring, yet they also encounter limitations and are susceptible to challenging issues. The utilization of current data and the structure of data points might impose certain limits that may reduce the accuracy of conclusions while overfitting can introduce bias into the outcomes. However, with careful enhancements, nearest-neighbor algorithms can serve as a helpful tool in analyzing and examining earth statistics.

5.1 Sensitivity to Noise and Outliers

Nearest-neighbor algorithms are widely employed in Earth tracking due to their favorable performance and categorization capabilities. However, individuals nearby may exhibit sensitivity to excessive noise levels and atypical data points. While various strategies can be employed to mitigate the effects of noise and outliers, these factors remain significant concerns among the tracking community (Tziolas et al., 2020).

Noise can be characterized as the presence of undesired or extraneous information inside a given dataset, which has the potential to lead to inaccurate or erroneous results. Within the field of geospatial analytics, the presence of noise can significantly impact the dependability and precision of nearest-neighbor algorithms. One of the primary challenges associated with noise in the tracking network is its potential to induce clustering, resulting in inaccurate classifications. The presence of multiple instances of the same prestigious institution can lead to the classifier making inaccurate conclusions, which can have detrimental effects on the monitoring process.

The presence of noise might also lead to the oversimplification of categories. When a set of rules is presented with a limited range of examples, it can extrapolate from these data points. This phenomenon can lead to inaccurate presumptions regarding a particular class or the entire dataset. In addition, noise can also play a role in the emergence of spurious correlations, potentially affecting the tracking process's precision.

In addition to the issue of noise, nearest-neighbor algorithms can be susceptible to challenges related to outliers. Outliers are statistical data points that exhibit significant deviation from the remaining dataset, potentially leading to outcomes that lack reliability. An instance of an outlier can be a character whose values for multiple variables deviate significantly from those of the rest of the sample. Nearest-neighbor algorithms are susceptible to overfitting when outliers are present, as the algorithm tends to prioritize outliers over most of the population. Erroneous classifications may arise due to the algorithm's failure to predict that the outlier indicates the overall magnificence.

Despite the inherent difficulties associated with the utilization of noise and outliers, several solutions can be implemented to mitigate their detrimental effects. As an illustration, an approach employed is feature engineering, wherein existing data points are utilized to generate new qualities more suitable for the given purpose. This approach may enhance the clarity of the information and improve the algorithm's ability to identify clusters and detect outliers. Furthermore, ensemble methods can also yield benefits in terms of noise reduction, as several classifiers can be combined to enhance accuracy. In data analysis, utilizing streamlined information and normalizing numbers can mitigate noise and outliers effectively.

The performance of nearest-neighbor algorithms can be influenced by noise and outliers. While accuracy and reliability are essential considerations, various strategies can be employed to address this issue. The concepts above include feature engineering, ensemble techniques, smooth information utilization, and values normalizing. Therefore, although noise and outliers may pose difficulties, they can be effectively mitigated by implementing suitable techniques.

5.2 Generalization Issues in Earth Observation

Earth observation is a multifaceted and ever-evolving approach that includes amassing and reading physical and environmental facts received through faraway sensing and progressively superior observational technologies. With Earth remark technology's development and broader availability, decision-makers have increasingly trusted it to tell their regulations and approaches. However, there may be a developing difficulty concerning the propensity to generalize observations made in Earth science. Common reliance on generalizations can frequently bring about oversimplifications and inaccurate prognostications (Mudele et al., 2021).

One primary assignment while using Earth remark for decision-making is the potential for generalization. The analysis of Earth's situations is based considerably on character features, which may be defined as spatial-temporal things and yield treasured information about a particular location at a particular second. Making judgments based on observational data about the planet is far more essential to aggregate or generalize extraordinary elements to attain a complete understanding or a more significant attitude on a selected occasion or project. It can result in misplacement, as elements such as topography, plant life, adjustments in weather styles, and other reasons for variability are regularly omitted or no longer considered.

One capacity project of generalizing actual statements about Earth is the possibility of generating erroneous forecasts. Aggregating records can now and again result in oversimplified observations, on the whole, due to the little variability within the surrounding statistics. By being attentive to the uniqueness of a unique factor-in-time remark, the intricacy of a particular phenomenon may be noticed, resulting in faulty forecasts and selection-making.

Aggregating statistics can result in an inaccurate portrayal of a selected geographical location. A single information factor is a consultant for a specific location. However, it can no longer appropriately depict the whole thing of the place. The misrepresentation of records can appreciably influence choice-making processes that often want to recall the intricacies of the unique context under attention.

The challenges associated with generalizations inside the discourse on the Earth are considerable concerns that require interest to ensure the accuracy and forex of facts while making selections. It is vital to ensure that statistics are efficiently aggregated, competently varied, and representative of all regions inside a selected hobby domain. Furthermore, selection-makers must remember the numerous limitations of the Earth statement and be aware of the potential for inaccurate forecasts and wrong depictions of a particular region. By enforcing this method, the efficiency of statistics aggregation processes may be more advantageous, mainly to elevated reliability in choice-making techniques.

6. CONCLUSION

Using the nearest-neighbor-primarily based worldwide commentary and monitoring method has proven its efficacy as a valuable tool in mapping and monitoring several global phenomena, including but no longer limited to land-use alternate, land cowl dynamics, city growth, and extreme occasions. This system has established superb in imparting specific statistics about various geographical regions formerly impossible with present technologies. Using geographically distributed sensor networks to analyze nearby phenomena has facilitated their software in numerous domain names, such as time-series data evaluation, forecasting, and selection-making.

The utilization of nearest-neighbor-primarily based comprehensive earth commentary and monitoring has facilitated development in the direction of a more profound comprehension of the functioning mechanisms of our planet. This methodology has facilitated the quantification and visualization of a vast range of worldwide phenomena with exquisite accuracy and precision. Moreover, this method has allowed us to quickly modify to evolving instances in constantly changing natural surroundings. Using insightful facts has facilitated decision-making by giving us alternatives to make knowledgeable selections. Consequently, using the nearest-neighbor approach in earth remark and monitoring has been confirmed to be a valuable tool in improving our expertise and response to the herbal environment.

6.1 Summary of Key Findings

The primary discoveries of Nearest-Neighbor based Earth observation and tracking can be summarized as follows:

- The effectiveness of nearest-neighbor-based Earth observation and tracking (NN-EOM) methods has been demonstrated in accurately capturing environmental events using diverse sensor readings.
- Neural network-based end-to-end object detection algorithms improve the accuracy of statistical analysis by using contextual information about the environment and considering local variations in spectral, textural, and structural data during the prediction process.
- Non-negative eigenvalue optimization methods (NN-EOM) are known for their efficacy, environmental sustainability, and versatility in modeling and predicting various environmental phenomena. These phenomena encompass diverse aspects such as terrain, plant life, and water levels.
- Neural network-based ensemble of models (NN-EOM) approaches can adaptively integrate new observations into a regularly updated version, rendering them highly pertinent in many domains, including pollution monitoring and concrete advancement.
- NN-EOM tactics have proven to be highly effective in monitoring endeavors, regardless of the scale. These strategies can be implemented by using a diverse range of open-source software libraries and applications.
- The utilization of NN-EOM has the potential to provide high-resolution monitoring and mapping products in near real-time, rendering it a viable choice for applications that necessitate prompt response times and precise outcomes.

6.2 Recommendations for Future Research

The pursuit of Earth observation and monitoring using near-neighbor methods involves identifying and comprehending environmental changes and their associated consequences. In order to enhance comprehension of the impacts of natural disasters, human disturbances, and climate change on the environment, researchers must prioritize the development of more efficient and accurate methodologies for extracting, analyzing, and interpreting Earth observation and monitoring data nearby. The instructions for conducting unique investigations contain comprehensive guidelines and procedures.

- Enhance the methodologies employed to comprehensively and accurately investigate, delineate, and measure alterations transpiring within local communities. These strategies need to consider local-level modifications instead of macro-level events or trends.
- Devise methodologies to enhance precision and accuracy in identifying and characterizing natural disasters and extreme meteorological phenomena alongside land utilization and vegetation changes.
- This study aims to evaluate the efficacy of various methodologies in identifying elements contributing to environmental changes.
- This study aims to investigate the potential of advancements in machine learning and AI to enhance the capabilities of near-Earth observation and tracking.
- AI algorithms can efficiently and rapidly identify changes in the environment, hence enhancing detection accuracy. Furthermore, these algorithms can effectively notify disaster response workers promptly.
- Examine novel and innovative methodologies for visualizing the data gathered from Earth observation and tracking nearby. Information visualization tools enable researchers to effectively discern patterns within the data and monitor environmental changes over time.
- This study aims to improve and expand current strategies for integrating nearby Earth observation and monitoring data with various information sources, such as satellite imagery, hydrological data, and weather statistics. The objective is to develop a more comprehensive understanding of the environment.
- Examining the possibilities of proximate Earth observation and tracking data to establish early warning indicators for natural disasters and various environmental phenomena is warranted.
- This study explores the potential of utilizing near-neighbor Earth observation and tracking data to enhance the process of urban planning and management in towns and other urban areas.
- This study explores prospective applications of near-Earth surveillance and tracking data in several domains, such as military and security activities, as well as agriculture.

By undertaking the studies mentioned above, researchers can achieve significant progress in near-neighbor-based Earth observation and monitoring, enhancing its efficacy as a tool for comprehending and safeguarding the environment.

REFERENCES

Beaudoin, A., Bernier, P. Y., Villemaire, P., Guindon, L., & Guo, X. J. (2018). Tracking forest attributes across Canada between 2001 and 2011 using ak nearest neighbors mapping approach applied to MODIS imagery. *Canadian Journal of Forest Research, 48*(1), 85–93. doi:10.1139/cjfr-2017-0184

Bhende, M., Thakare, A., Pant, B., Singhal, P., Shinde, S., & Saravanan, V. (2022). Deep Learning-Based Real-Time Discriminate Correlation Analysis for Breast Cancer Detection. *BioMed Research International, 2022*, 2022. doi:10.1155/2022/4609625 PMID:35800216

Chen, Z., Vatsavai, R. R., Ramachandra, B., Zhang, Q., Singh, N., & Sukumar, S. (2016, December). Scalable nearest neighbor based hierarchical change detection framework for crop monitoring. In *2016 IEEE International Conference on Big Data (Big Data)* (pp. 1309-1314). IEEE. 10.1109/BigData.2016.7840735

Ding, Y., Guo, H., Liu, G., Han, C., Shang, H., Ruan, Z., & Lv, M. (2019). Constructing a high-accuracy geometric model for Moon-based Earth observation. *Remote Sensing (Basel), 11*(22), 2611. doi:10.3390/rs11222611

Eskelson, B. N., Temesgen, H., Lemay, V., Barrett, T. M., Crookston, N. L., & Hudak, A. T. (2009). The roles of nearest neighbor methods in imputing missing data in forest inventory and monitoring databases. *Scandinavian Journal of Forest Research, 24*(3), 235–246. doi:10.1080/02827580902870490

Fadaei-Kermani, E., Barani, G. A., & Ghaeini-Hessaroeyeh, M. (2017). Drought monitoring and prediction using K-nearest neighbor algorithm. *Journal of Artificial Intelligence and Data Mining, 5*(2), 319–325.

Ferreira, B., Iten, M., & Silva, R. G. (2020). Monitoring sustainable development by means of earth observation data and machine learning: A review. *Environmental Sciences Europe, 32*(1), 1–17. doi:10.118612302-020-00397-4

Gobinathan, B., Mukunthan, M. A., Surendran, S., Somasundaram, K., Moeed, S. A., Niranjan, P., Gouthami, V., Ashmitha, G., Mohammad, G. B., Shanmuganathan, V. K., Natarajan, Y., Srihari, K., & Sundramurthy, V. P. (2021). A novel method to solve real time security issues in software industry using advanced cryptographic techniques. *Scientific Programming, 2021*, 1–9. doi:10.1155/2021/3611182

Gupta, S., Saravanan, V., Choudhury, A., Alqahtani, A., Abonazel, M. R., & Babu, K. S. (2022). Supervised computer-aided diagnosis (CAD) methods for classifying Alzheimer's disease-based neurodegenerative disorders. *Computational and Mathematical Methods in Medicine, 2022*, 2022. doi:10.1155/2022/9092289 PMID:35651921

He, B., Zhao, X., Chen, Y., Liu, C., & Pang, X. (2022). Application of Feature Tracking Using K-Nearest-Neighbor Vector Field Consensus in Sea Ice Tracking. *IEEE Journal of Selected Topics in Applied Earth Observations and Remote Sensing, 15*, 4326–4336. doi:10.1109/JSTARS.2022.3178117

Küçük, Ç., Taşkın, G., & Erten, E. (2016). Paddy-rice phenology classification based on machine-learning methods using multitemporal co-polar X-band SAR images. *IEEE Journal of Selected Topics in Applied Earth Observations and Remote Sensing, 9*(6), 2509–2519. doi:10.1109/JSTARS.2016.2547843

Mohammed, I. N., Bolten, J. D., Srinivasan, R., & Lakshmi, V. (2018). Improved hydrological decision support system for the Lower Mekong River Basin using satellite-based earth observations. *Remote Sensing*, *10*(6), 885. doi:10.3390/rs10060885 PMID:29938116

Mudele, O., Frery, A. C., Zanandrez, L. F., Eiras, A. E., & Gamba, P. (2021). Dengue vector population forecasting using multisource earth observation products and recurrent neural networks. *IEEE Journal of Selected Topics in Applied Earth Observations and Remote Sensing*, *14*, 4390–4404. doi:10.1109/JSTARS.2021.3073351

Natarajan, Y., Kannan, S., Selvaraj, C., & Mohanty, S. N. (2021). Forecasting energy generation in large photovoltaic plants using radial belief neural network. *Sustainable Computing : Informatics and Systems*, *31*, 100578. doi:10.1016/j.suscom.2021.100578

Natarajan, Y., Srihari, K., Dhiman, G., Chandragandhi, S., Gheisari, M., Liu, Y., Lee, C.-C., Singh, K. K., Yadav, K., & Alharbi, H. F. (2022). An IoT and machine learning-based routing protocol for reconfigurable engineering application. *IET Communications*, *16*(5), 464–475. doi:10.1049/cmu2.12266

Orusa, T., Cammareri, D., & Borgogno Mondino, E. (2022). A Scalable Earth Observation Service to Map Land Cover in Geomorphological Complex Areas beyond the Dynamic World: An Application in Aosta Valley (NW Italy). *Applied Sciences (Basel, Switzerland)*, *13*(1), 390. doi:10.3390/app13010390

Poursanidis, D., Chrysoulakis, N., & Mitraka, Z. (2015). Landsat 8 vs. Landsat 5: A comparison based on urban and peri-urban land cover mapping. *International Journal of Applied Earth Observation and Geoinformation*, *35*, 259–269. doi:10.1016/j.jag.2014.09.010

Poursanidis, D., Topouzelis, K., & Chrysoulakis, N. (2018). Mapping coastal marine habitats and delineating the deep limits of the Neptune's seagrass meadows using very high resolution Earth observation data. *International Journal of Remote Sensing*, *39*(23), 8670–8687. doi:10.1080/01431161.2018.1490974

Prasad, P., Loveson, V. J., Chandra, P., & Kotha, M. (2022). Evaluation and comparison of the earth observing sensors in land cover/land use studies using machine learning algorithms. *Ecological Informatics*, *68*, 101522. doi:10.1016/j.ecoinf.2021.101522

Purohit, K., Vats, S., Saklani, R., Kukreja, V., Sharma, V., & Yadav, S. P. (2023, July). Improvement in K-Means Clustering for Information Retrieval. In *2023 4th International Conference on Electronics and Sustainable Communication Systems (ICESC)* (pp. 1239-1245). IEEE.

Saravanan, V., & Sumathi, A. (2012, December). Dynamic handoff decision based on current traffic level and neighbor information in wireless data networks. In *2012 Fourth International Conference on Advanced Computing (ICoAC)* (pp. 1-5). IEEE. 10.1109/ICoAC.2012.6416797

Tiwari, A., Sharma, S. K., Dixit, A., & Mishra, V. (2021). UAV remote sensing for campus monitoring: A comparative evaluation of nearest neighbor and rule-based classification. *Photonirvachak (Dehra Dun)*, *49*(3), 527–539. doi:10.100712524-020-01268-4

Tziolas, N., Tsakiridis, N., Chabrillat, S., Demattê, J. A., Ben-Dor, E., Gholizadeh, A., Zalidis, G., & Van Wesemael, B. (2021). Earth observation data-driven cropland soil monitoring: A review. *Remote Sensing (Basel)*, *13*(21), 4439. doi:10.3390/rs13214439

Tziolas, N., Tsakiridis, N., Ogen, Y., Kalopesa, E., Ben-Dor, E., Theocharis, J., & Zalidis, G. (2020). An integrated methodology using open soil spectral libraries and Earth Observation data for soil organic carbon estimations in support of soil-related SDGs. *Remote Sensing of Environment*, *244*, 111793. doi:10.1016/j.rse.2020.111793

Xing, S., Shahabi, C., & Pan, B. (2009). Continuous monitoring of nearest neighbors on land surface. *Proceedings of the VLDB Endowment International Conference on Very Large Data Bases*, *2*(1), 1114–1125. doi:10.14778/1687627.1687753

Yadav, S. P., Zaidi, S., Nascimento, C. D. S., de Albuquerque, V. H. C., & Chauhan, S. S. (2023, January). Analysis and Design of automatically generating for GPS Based Moving Object Tracking System. In *2023 International Conference on Artificial Intelligence and Smart Communication (AISC)* (pp. 1-5). IEEE. 10.1109/AISC56616.2023.10085180

Chapter 6
Development of a Neuro-Inspired Algorithms for Monitoring and Prediction of Earthquakes

Rekha M.
Department of Information Technology, R.M.K. Engineering College, Chennai, India

Chilambuchelvan A.
Department of Electronics and Communication Engineering, R.M.D. Engineering College, Kavarapettai, India

Dinesh Babu M.
Department of Mechanical Engineering, Rajalakshmi Institute of Technology, Chennai, India

Vasudhevan V.
https://orcid.org/0000-0002-9639-2322
Department of Electronics and Communication Engineering, Panimalar Engineering College, India

ABSTRACT

This chapter offers the improvement of a robust neuro-stimulated algorithm for monitoring and prediction of earthquakes. The designed machine, gaining knowledge of a new algorithm, can successfully filter the critical functions from seismic data and locate numerous seismic activities in actual time. This algorithm is custom-designed for seismic facts analysis and designed for supervised studying. The key feature of this set of rules is its capacity to discover and classify seismograms, which can be new to the model, making it notably predictive. The outcomes of the tracking and prediction show that the advanced neuro-inspired set of rules should successfully discover various seismic activities with a quick latency and low fake positive rate. Furthermore, the proposed algorithm presents an optimal aggregate of binary classification for earthquake monitoring and prediction. This set of rules is a promising tool to enhance earthquake chance assessment and early warning structures.

DOI: 10.4018/979-8-3693-1850-8.ch006

1. INTRODUCTION

Improving a neuro-stimulated set of rules for tracking and predicting earthquakes is a primary milestone in improving the safety of those dwelling in at-chance areas. Earthquakes, particularly those of full-size significance, cause brilliant destruction to lives and assets. Early detection of an earthquake can help human beings to take precautionary measures on the way to decrease the effects of the earthquake. It will talk about improving a neuro-stimulated set of rules for tracking and predicting earthquakes (Lymperopoulos, 2017).

The first step in developing a neuro-inspired set of rules for monitoring and predicting earthquakes is gathering data and facts about earthquakes. These statistics include shake maps, ground movement readings, and tsunami amplitude recordings, and these statistics can encompass seismic activities. It is critical to gather statistics from as many locations as feasible to create a complete dataset.

The next step in growing a neuro-inspired set of rules is to create a neural community capable of processing the data amassed (Alioto et al., 2018). The algorithm can be tailor-made to understand patterns within the facts by combining professional know-how from seismologists with the artificial intelligence capabilities of neural networks. With those styles, the rules can perceive tell-story signs and symptoms of an upcoming earthquake.

ultimately, the set of rules wishes to be examined and tested on actual international events. By trying out the rules on ancient seismic events, it can be tuned to as it should stumble on an upcoming earthquake. Moreover, facts amassed from actual earthquakes can assist in refining the neural community of the algorithm.

Once the rules have been developed, tuned, and verified, they could be deployed as an earthquake monitor and predictor (Kasabov et al., 2016). This algorithm may alert humans dwelling in at-risk areas of a capacity earthquake. By providing early detection and prediction, humans can prepare themselves for the ability risk and decrease the harm because of the earthquake.

Improving a neuro-inspired algorithm for tracking and predicting earthquakes is a chief milestone in improving the safety of those dwelling in at-chance regions. This algorithm can provide early detection and prediction of earthquakes, allowing people to prepare for the event. With this generation, people can better defend themselves from the destruction due to earthquakes.

1.1 Background and Significance

Earthquakes are a chief chance to the lives and assets of billions of people in the sector. With the increasing frequency of each big and small earthquake, human beings and governments are increasingly turning to superior techniques and technology to screen. They expect them. One of the most promising technologies in this place is the development of Neuro-inspired algorithms, which use synthetic Neural Networks to assess seismic statistics and create predictive models of how seismic activities will affect a place.

Neuro-inspired algorithms are in demand for his or her capacity to manner complicated information units quickly and as they should be. Seismic statistics, for example, can include complex waveforms, complex spectra, and complex temporal patterns – all of which may vary depending on the power or vicinity of the seismic occasion. Traditional statistical strategies can struggle to interpret this complexity, while ANNs can quickly apprehend styles and produce significant predictions. Moreover, ANNs can autonomously make connections inside the records, allowing them to discover new insights humans could have neglected (Galkina & Grafeeva, 2019).

The first scientific demonstration of a Neuro-inspired algorithm for earthquake monitoring was posted in 2009, while the California Institute of Technology (Caltech) developed a system primarily based on ANNs to discover and degree earthquakes in southern California. This device has been replicated to version earthquakes in different parts of the world, including Japan, where this sort of algorithm changed to offer early warnings of the 2011 Tohoku Earthquake.

The capability of ANNs in earthquake monitoring is also increasing. For instance, in advance this year, a collection of researchers from the University of Tokyo evolved an ANN-based gadget to forecast the magnitude of earthquakes. This generation can offer applicable early warnings to communities in seismically energetic regions.

The Neuro-stimulated algorithms, which employ artificial Neural Networks, offer a robust and green manner to display and predict earthquakes with remarkable accuracy. Those algorithms can autonomously search via highly complex records units to find styles and insights that could make a massive difference to those residing in excessive-threat areas (Asim et al., 2017). As research in this discipline continues, we will see even more state-of-the-art earthquake monitoring and prediction systems emerge and probably revolutionize seismology.

1.2 Objectives of the Chapter

- To generate a correct model of seismic conduct with the aid of using Neuro-inspired algorithms.
- To as it should be, expect, detect, and forecast seismic occasions the usage of Neural Networks.
- To expand a device that video displays the unit's seismic sports and performs in real-time for timely notification of potential threats.
- To provide entire and accurate seismic activity predictions by incorporating environmental, geological, and seismic information as inputs.
- To increase a compelling set of rules to determine the magnitude of earthquakes and generate more reliable predictions.
- To create rules that could generate beneficial consequences even if the seismic statistics are sparse or restrained.

2. OVERVIEW OF NEURO-INSPIRED ALGORITHMS IN EARTHQUAKE PREDICTION

Earthquakes are among the most adverse herbal screw-ups skilled everywhere globally. At the same time, predicting them with any degree of accuracy is impossible, but advances in neuro-inspired algorithms have led to progressed earthquake prediction fashions.

Neuro-inspired algorithms use artificial neural networks to simulate the conduct of neurons in the human brain. Synthetic neural networks encompass more than one layer of neurons that may learn to understand patterns in a given enter and output the best reaction. Using those networks for seismic statistics, researchers can pick out styles related to the probability of an upcoming earthquake (Beroza et al., 2021).

These fashions also can be used to decide the depth of an earthquake. By combining seismic facts with geological, geophysical, and environmental records, researchers can parse an existing earthquake, determining its magnitude and mapping its epicenter. Similarly, researchers have developed new analytical techniques to benefit extra insight into the future movements of tectonic plates, supporting the prediction of huge-scale occasions along with oceanic plate motion and earthquakes.

Neuro-inspired algorithms are only the beginning of what may be performed to forecast earthquakes. While fashions are constantly enhancing, the task comes with acquiring the facts wanted for those algorithms to be as characteristic as they should be. More modern technology has evolved to make the seismic pastime more correct than ever before.

The neuro-inspired algorithms offer new and exciting opportunities within the prediction of earthquakes (Irfan et al., 2022). By combining synthetic neural networks and seismic information, researchers can take advantage of the perception needed to expect and accurately assemble destiny earthquakes. With continued development in this discipline, the world can remain secure from the destruction of those natural failures.

A neuro-inspired algorithm used for earthquake prediction is a Recurrent Neural Network (RNN). RNNs are a type of neural networks which are able to model complex, non-linear relationships between data points form a time series. They have become increasingly popular with applications involving time-series data such as forecasting and prediction. RNNs are able to capture features from the data, identify patterns and correlations, and thus successfully predict future events. This has made them particularly suitable for earthquake prediction – they can be used to analyze seismic activity and make predictions about future earthquake locations, sizes, times, and more.

2.1 Basics of Neuro-Inspired Computing

Earthquake prediction has continually been a complex and challenging hassle for scientists. As a result, there were many years of research and funding trying to recognize and expect earthquakes. There has been tremendous development in neuro-inspired computing, which has enabled more advanced and particular methods for earthquake prediction.

Neuro-inspired computing uses machine-studying techniques, including synthetic neural networks, kernel machines, and deep-gaining knowledge, to advantage insights from seismic facts and different geological records. By offering the system facts that include seismographs, exact geological fashions, or even nearby infrastructure networks, the machine can make more intelligent predictions by spotting patterns and correlations within the data (Rajalakshmi et al., 2022).

These neuro-stimulated computing models are skilled and optimized by big datasets of each historical and recent earthquake. Recognizing the subtle variations in the statistics informs and causes them to better recognize the seismic signals and the underlying tendencies that influence the probability of an earthquake. Moreover, these fashions are beneficial in the knowledge of the degrees of uncertainty associated with seismic records, presenting extra correct predictions.

The utility of neuro-inspired computing has already been examined in many seismic research additives. For instance, fashions were used for seismic hazard assessment and forecasting ability harm to homes. Furthermore, they have been used to estimate an upcoming earthquake's importance, area, and timing.

Neuro-stimulated computing may be invaluable for predicting earthquakes from historical and actual-time signals. Using these models will grow as they improve, and more significant datasets will be had. Even though the take a look at earthquake prediction is a long way from being solved, neuro-inspired computing has made top-notch strides in making it viable to forecast seismic signals and capability harm in the future correctly.

2.2 Applications in Earthquake Prediction

Earthquakes are a prime risk and cause immense destruction in inclined regions. It is consequently essential to have an early caution gadget to minimize damage and capacity lack of lives. Technological advances have enabled the development of programs that can predict earthquakes (Narayanan et al., 2022).

One of the most popular packages used for earthquake prediction is QuakeFinder. This utility uses an array of sensors to stumble on the seismic hobby and can stumble on earthquakes before they occur. The records amassed utilizing the sensors are used to generate earthquake maps and data to expect the value and area of destiny earthquakes.

Another utility being used for earthquake prediction is the global Seismic monitoring system. This application utilizes a network of seismometers and facts collected from satellite TV for PC imagery to detect seismic activity. These statistics are then used to generate earthquake warning messages broadcast to customers in the affected regions.

The software UrtheCast Ambitions is likewise being used for seismic tracking. This software uses satellite TV for PC imagery to discover seismic activity. It additionally uses an array of sensors to locate viable earthquake hazards (Li et al., 2020).

Software called the ShakeAlert Earthquake Early Warning system is being used to stumble on the seismic hobby and alert customers in the place of the affected region. The utility is run by using the US Geological Survey and is supported with the aid of several universities and research centers.

Those applications reduce the capacity value of lack of life and harm from earthquakes. Those packages can detect seismic hobby and send warnings to customers in areas that are likely to be laid low with an earthquake. In the future, these programs will likely benefit from additional development as they become more accurate and reliable.

3. KEY COMPONENTS OF THE NEURO-INSPIRED EARTHQUAKE PREDICTION ALGORITHM

The Neuro-inspired Earthquake Prediction algorithm is an effective earthquake prediction and prevention device. It combines traditional knowledge and contemporary information analytics to come across seismic activities, including earthquakes before they strike. The algorithm uses seismic networks to screen for changes inside the floor and floor deformation measurements. NEPA then uses superior gadget learning algorithms to interpret the statistics and predict capacity seismic events (Hoque et al., 2020).

The critical components of the NEPA algorithm include seismic networks, device learning (ML) algorithms, and terraforming algorithms. The seismic networks are designed to detect diffused modifications in the floor and floor deformation measurements. They use passive and lively sensors that may stumble on a selection of movements, including vibrations or electric signals that an earthquake can cause. The ML algorithms are used to research seismic facts and search for patterns that propose the potential for an earthquake. Eventually, terraforming algorithms use laptop simulations of the earth's floor to decide the impact of an earthquake, along with capacity harm and casualties.

An essential part of the NEPA algorithm is its capability to generate accurate, dependable predictions. The accuracy of the predictions is primarily based on the accuracy of the seismic networks, the ML algorithms, and the accuracy of the terraforming algorithms. Moreover, the set of rules needs to be able to discover areas deemed excessive-hazard. It calls for a mixture of geological records, seismological statistics, and meteorological records.

NEPA is tied with a control gadget that lets in for actionable facts. This system may help decide what mitigation techniques must be used for an impending earthquake., the machine additionally provides an early caution machine so that humans may be warned. Neuro-stimulated Earthquake Prediction set of rules is an effective tool for predicting and preventing seismic activities. Its key components encompass seismic networks, mastering algorithms, and terraforming algorithms, which combine to create accurate and reliable predictions (Johnson et al., 2021). Moreover, its control device ensures that the information and predictions may be used to tell selections and provide early warnings.

3.1 Neural Network Architecture and Design

Earthquakes have been a devastating herbal catastrophe for centuries. However, due to non-stop advancements in technology, the ability to appropriately expect the chance of an earthquake taking place has become more and more possible. A good way to correctly expect earthquakes, neural network structure, and layout are critical. Neural networks are a deep getting-to-know set of rules based on a fixed of interconnected nodes. This set of rules is unique for modeling complicated, non-linear phenomena like earthquakes. It can build a giant photo of the information without problems and speedy.

While developing an earthquake prediction set of rules using a neural network, there are predominant elements: the characteristic extraction and prediction layers. The characteristic extraction layer is actioned first. This deposit takes the seismic facts and extracts various capabilities such as waveform, amplitude, frequency, and associated parameters (Rouet-Leduc et al., 2017). Those capabilities are then fed into the prediction layer, in which the information is analyzed and used to make the predictions. The prediction layer can be within a recurrent neural community (RNN) or a deep neural network (DNN).

Inside the RNN, data is handed from one layer to the subsequent in the shape of 'time steps.' This kind of version is excellent and suitable for time-series facts, perfect for seismic facts. Every layer can decipher the relationship between the variables and find patterns inside the data; the output of this accretion is then used to generate a prediction.

The DNN, which is more customarily used nowadays, is an entirely effective version as it may handle more complex relationships between the variables. It can additionally be used to categorize information, locate and classify objects, and more. This type of neural network architecture is perfect for grabbing as much information as feasible from the seismic records and the usage of it to generate a more excellent correct prediction.

Once the information is processed, a hard and fast of predictions will be generated; those predictions might be the truth of an earthquake's prevalence, its importance, or the time window of its incidence. All these records are then taken under consideration when issuing predictions and warnings (Salam et al., 2021). The neural community architecture and layout are very crucial in earthquake prediction algorithms. If used effectively, these algorithms can be very accurate and help lessen the damage of this devastating natural disaster.

A neural network architecture consists of multiple layers of neurons interconnected in a specific way. Each layer is composed of a set of neurons that all share a similar set of inputs and/or outputs. The most popular types of layers are fully connected layers, convolutional layers, and pooling layers. The training process of a neural network begins with establishing a training data set. This is a set of inputs and expected outputs that are used to train a neural network. The neural network receives the inputs and uses the information to determine the output. During the training process, various algorithms are used to adjust the weights of the neurons in the neural network, allowing it to learn and become more accurate in its predictions. Once a neural network has been trained, it can be used to make real-time predictions. The weights of the neurons are then adjusted using real-time data inputs and expected outputs. As the neural network is exposed to more data, it is able to become more accurate and reliable in its predictions. This process is called reinforcement learning. Neural networks have become increasingly popular due to their ability to make highly accurate and reliable predictions. With more research and development, neural networks will become increasingly useful in many different fields and applications.

3.2 Data Integration and Feature Selection

Facts integration and function selection are essential to any set of rules to expect earthquakes. The statistics integration method entails gathering uncooked seismic facts from numerous sources and reworking them into a standardized layout that may be used as input for a predictive version. The characteristic choice procedure is then used to discover the most vital seismic features in predicting earthquakes as it should be and efficaciously.

Facts integration is frequently done by combining information from unique networks of seismometers and correlating the measurements for exclusive seismic activities. This statistics integration allows more information about earthquakes to be used inside the predictive version. For instance, huge-scale seismic records from an entire location can be used to identify potential earthquake hotspots and interest patterns. In addition, information from very small-scale observations of a single event can be incorporated to offer more extraordinary granular expertise of the particular seismic occasion in query.

Once information is incorporated, feature selection must be completed to ensure that the maximum relevant records are used for predictive modeling. This procedure may diagnose and utilize the maximum predictive seismic parameters inside the model. These selected capabilities are typically used to locate anomaly styles and seismic interest changes related to an impending earthquake (Yadav et al., 2023). Moreover, the maximum essential capabilities can also be used as valuable resources in determining the opportunity and intensity of the expected seismic event.

Information integration and feature choice are critical additives in developing powerful algorithms for earthquake prediction. The numerous strategies available can assist in picking out the most vital seismic parameters and choosing the most relevant facts for the predictive model. Using these methods in a given predictive version can provide extra correct predictions of potential seismic occasions and allow more advantageous catastrophe preparedness.

4. ADVANTAGES AND BENEFITS

Earthquake prediction algorithms can provide several advantages and advantages, each to most people and the scientific network. This kind of generation can assist in reducing the casualties and economic losses associated with earthquakes and provide treasured records for a higher understanding of the underlying reasons for quakes.

First, earthquake prediction algorithms may be used to correctly estimate the vicinity, timing, and magnitude of seismic occasions. This era can dramatically enhance public safety by supplying higher caution instances for impacted regions. Early warnings can permit human beings to put together for evacuation or ease homes and businesses by making necessary maintenance before an earthquake moves (Debnath et al., 2021). Governments can also use this form of data to implement better emergency reaction plans, install resources more effectively, and offer essential valuable resources more quickly.

Second, the statistics generated by way of earthquake prediction algorithms can be helpful to the medical community. By analyzing the styles and trends of seismic activities, researchers can better understand the underlying causes of earthquakes and the way they are better predicted or mitigated in the future. With access to special analyses of seismic occasions, researchers can also increase techniques for improving building codes and construction techniques to protect human beings from future quakes.

The algorithms can investigate seismic data in real-time and locate subtle styles indicating the possibility of a future seismic occasion. This generation type can provide treasured insights for disaster preparedness and be used to direct assets to areas more likely to enjoy a seismic event in the destiny.

There are numerous advantages and blessings related to earthquake prediction algorithms that could have lasting impacts on public protection, scientific studies, and catastrophe preparedness. With the help of this era, governments and researchers can improve their expertise in seismic activity and be better prepared for future quakes.

One example of a neuro-inspired algorithm successfully predicting earthquakes is the application of artificial neural networks in seismic hazard assessment. Neural networks have been used to model the relationships between soil properties and earthquake shaking intensity in Japan, China, and other parts of the world. The objective is to identify areas of increased seismic hazard, as well as any areas of reduced potential risk. Neural networks can incorporate more data than conventional seismic hazard assessment methods and can also identify patterns that may have been overlooked or unrecognized by human analysts. An example of this is the recent application of an artificial neural network model to identify areas of high earthquake risk in Taiwan based on seismic and geotechnical data. The model identified two areas of high risk, both of which were later confirmed through the analysis of historical earthquake records. This study is an example of how neuro-inspired algorithms can be used to successfully predict earthquakes.

4.1 Mimicking Human Intelligence for Improved Prediction

Earthquake prediction is a complex undertaking for human beings because it requires complete information about the Earth's indoors and a massive amount of statistics associated with the seismic sports of tectonic plates on our planet. With the help of devices getting to know, however, it has become feasible to mimic humans' intelligence and enhance our earthquake prediction algorithms (Mangalathu et al., 2020). This generation enables us to perceive styles in seismic records, which may be used for long-term prognoses. It also allows us to identify areas in which massive-scale tremors may be expected.

Device-gaining knowledge may be applied to predict an earthquake's improvement and its destiny incidence. Using a mixture of information from seismic monitoring networks, including worldwide Implosion size gadgets and Gravity size networks, system studying algorithms can help researchers apprehend the seismic sports of tectonic plates and, in all likelihood, extract styles of earthquakes. Scientists then use these styles better to recognize the importance and characteristics of destiny seismic occasions and to expand reliable models for large-scale earthquake forecasting.

Furthermore, device studying can also assist in detecting and analyzing bizarre activities inside the tectonic plates up to the earthquake (Murwantara et al., 2020). Via the sophisticated evaluation of accelerometers, seismometers, and different geophysical contraptions linked to the internet, state-of-the-art algorithms can come across modifications inside the floor of the Earth, which may imply any such seismic occasion is imminent. By way of this method, scientists can hit upon fewer anomalies in a few locations, but they could help predict a coming near earthquake.

The machine gaining knowledge has been used to enhance earthquake prediction by simulating people's intelligence. By combining seismological information with gadget-mastering algorithms, scientists can discover anomalies in tectonic plates and examine seismic activity patterns to understand earthquakes' significance and traits better and expand dependable fashions for forecasting them (Purohit et al., 2023). In addition to research and possibly superior AI-driven fashions, machine studying can expose even more insights into seismic activity on our planet and permit us to better put together for natural disasters along with earthquakes.

A neuro-inspired algorithm is a type of artificial intelligence (AI) technique that uses elements of a biological neuron in order to understand how brains are able to process and interpret data as well as learn from their environment. It is based off of the notion that human brains can be modeled in a computer model to improve prediction accuracy. This type of model can be used to improve the capacity and efficiency of predictive networks by removing the need for large data sets and long learning times. Neuro-inspired algorithms are able to take very little data and turn it into a tangible prediction very quickly and accurately. Additionally, these neural networks are capable of processing big data in real time and can reach higher levels of accuracy in predicting consumer behaviours or global trends. This could have the potential to drastically improve the accuracy of predictions in areas such as finance, health, and other industries.

4.2 Adaptive Learning in Dynamic Seismic Environments

The purpose of the adaptive getting to know in a dynamic seismic surroundings for earthquake prediction algorithm is to improve the prediction of massive earthquakes that may cause damage to the infrastructure of a town. This set of rules uses artificial intelligence (AI) competencies to investigate facts from the environment, together with seismic waves. It quickly predicts an imminent earthquake by searching out telltale signs and styles inside the surroundings. The algorithm has been designed to adapt and self-adjust its parameters to maximize performance while predicting massive earthquakes (Laurenti et al., 2022).

It implements a step-by-step manner to allow the rules to respond quickly and appropriately to environmental adjustments. First, the set of rules uses fuzzy good judgment to look for patterns inside the incoming information. It permits it to apprehend seismic events that are normal for a more significant event and differentiate them from more minor alerts that can be more not unusual. The rules then use fuzzy logic to determine the dimensions of the imminent seismic event and the chance it will strike with enough pressure to cause harm.

Once the algorithm has determined the scale of the upcoming seismic event, it begins to conform its parameters by using getting-to-know algorithms along with deep getting-to-know and reinforcement getting-to-know. Deep knowledge of algorithms is used to apprehend the styles inside the incoming seismic facts, while the reinforcement gaining knowledge of algorithms is used to alter the set of rules' parameters and enhance its average overall performance. The result is an algorithm that could respond quickly to changes within the surroundings and appropriately predict earthquakes earlier than they occur.

Adaptive knowledge of a dynamic seismic environment for earthquake prediction rules can save infinite lives and prevent billions of dollars in damages (Chittora et al., 2022). By allowing public and private entities to fast and expect seismic occasions and take defensive measures in their wake, governments and communities around the sector can dramatically lessen the risks posed by using natural screw-ups.

5. CHALLENGES AND LIMITATIONS

Earthquakes can result in catastrophic activities, which can have some distance-accomplishing outcomes on society. Luckily, scientists have developed algorithms to predict earthquakes and give people advanced caution. However, these algorithms also face certain demanding situations and obstacles.

One of the most demanding situations to predict earthquakes is the need for more sufficient data. Seismic statistics can only be collected at once from international locations with ground stations or other associated seismograph networks. While a comprehensive global community of such stations is a gift now, the decreased populace density in certain areas limits the availability of seismic information because the latter is predominantly gathered from city centers (Mallouhy et al., 2019). It could lead to misguided predictions due to incomplete statistics.

Every other challenge faced is according to forecasts. Although a warning is issued ahead of an Earthquake, the algorithm probably needs to be fixed in terms of approximating the significance and the region where the earthquake is expected to occur. It ends in the inability to check the severity of the harm that an earthquake might cause.

Apart from these challenges, some other obstacle locations are within the reaction time (Sajan et al., 2023). A set of rules may take the time to supply a prediction. Moreover, the time to install essential precautions and alert residents may exceed the anticipated duration of the crisis. It leaves individuals with no time to reply and take preventive protection measures.

Imparting correct warnings concerning earthquakes calls for a massive group of seismologists, mathematicians, and engineers with access to more than a few technologies (Kavianpour et al., 2023). Even though Earthquake prediction algorithms use sophisticated generation to research and offer predictions, in view that they may be nascent, they frequently present more challenges and barriers than the predictions they supply.

5.1 Data Complexity and Training Requirements

The facts, complexity, and training requirements for earthquake prediction algorithms rely upon the complexity and class of the set of rules itself. The reason for these algorithms is to appropriately expect while, wherein, and how unfavorable an earthquake is probably, permitting people to better prepare for a potentially detrimental seismic occasion (Sivaram et al., 2019).

The records used to educate an earthquake prediction set of rules typically include seismic statistics, geophysical variables, and virtual terrain facts. These information assets include records collected from numerous assets over time and utilized by the algorithm to gain insight into the patterns that arise earlier than an earthquake. This data must be analyzed and interpreted to determine the patterns the rules need to look out for before an earthquake occurs.

Once the information has been accumulated and analyzed, device-learning algorithms are used to locate any earthquake-related styles (Yuvaraj et al., 2021). These algorithms need to be correct and green. For example, they need to spot the diffused changes that arise before an earthquake and detect a seismic event's more fantastic apparent functions.

The device gaining knowledge of the algorithm must also be trained to stabilize false positives and negatives. A fake fantastic is a set of rules that flags an earthquake even when one does not, in reality, arise. A false poor is when the algorithm fails to stumble on an earthquake earlier than it happens. Balancing false positives and fake negatives is crucial for a practical earthquake prediction set of rules (Hannah et al., 2022).

The education necessities additionally rely on the amount of information the set of rules exposes. More data lets in the rules to better pick out and anticipate seismic occasions. However, the computational time required to process extra information may also mean that the rules must be faster for realistic use. Consequently, locating the highest quality balance between available information and the preferred overall performance of the set of rules is vital.

The statistics complexity and training necessities of earthquake prediction algorithms vary extensively, relying on sophisticated rules and the quantity of information to investigate. By adequately studying seismic and geophysical statistics, machine-gaining knowledge of algorithms can appropriately come across earthquake patterns and assist humans higher put together for probably damaging seismic occasions.

6. CONCLUSION

The studies into the capacity of Neuro-inspired algorithms for monitoring and predicting earthquakes have produced exciting results in short-term predictions, with some promising gains in accuracy over existing methods. These algorithms' accuracy and time sensitivity rely upon big datasets, facts, preprocessing techniques, and cautious system design. Similarly, deep learning and artificial intelligence developments may result in similar enhancements via strategies, including supervised mastering. While plenty of work has to be accomplished before Neuro-inspired algorithms can become reliable tools for seismologists, they offer exciting opportunities for researchers to broaden reliable and accurate earthquake tracking and prediction tools. Their capacity to offer more precise and earlier warnings may save lives and decrease destruction at some stage in future earthquakes. In this manner, Neuro-inspired algorithms may significantly affect how seismic activities are understood and expected and how governments and societies can prepare for them.

6.1 Summary of Key Findings

The key findings of a neuro-stimulated algorithm for monitoring and prediction of earthquakes. Through an intelligent integration of neural networks, seismic monitoring statistics, and seismic prediction strategies, the researchers created a set of rules that can correctly capture the patterns in seismic data and expect areas that an earthquake can strike.

The researchers examined the set of rules in using 29 extraordinary seismic tracking stations in Southern California. It was discovered to be accurate in its predictions, with false alarms at the slightest. Moreover, the rules were found reliable in figuring out regions uncovered to the most critical potential for damage from an earthquake.

The study discovered that the algorithm was effective at predicting the incidence of an earthquake up to eight days before it virtually occurred. It became additionally more potent at predicting an earthquake's depth than traditional strategies, including seismic-primarily based methods and ancient statistics-based methods.

They look at evidence to help the use of the proposed algorithmic approach for tracking and predicting earthquake-prone regions. It is a thrilling development in earthquake prediction, as it suggests the capacity for using synthetic intelligence and neural networks for correct and well-timed seismic evaluation. Destiny studies are had to validate the consequences and discover the capacity for the use of the proposed set of rules in large networks and distinct seismic environments.

6.2 Recommendations for Future Research

The improvement of neuro-inspired algorithms for monitoring and predicting earthquakes, additional studies are needed in numerous vital regions.

First, an extra state-of-the-art approach to information analysis and characteristic engineering is essential. Constructing extra correct algorithms calls for the capability to pick out capabilities in seismic information that can be used to tell the model and make more excellent, dependable predictions. Innovative strategies, including deep gaining knowledge, will be explored to extract significant statistics from seismic occasions.

The more fantastic paintings are wanted to address any consequences due to station bias or different sources of bias in the information sets. It may be addressed by searching multiple instrumented websites to provide an improved and complete dataset.

The extra attempt is to evaluate and examine the overall performance of various neuro-stimulated algorithms. Evaluating installed techniques' accuracy and more recent ones can offer insight into the overall performance of different methods.

Fourth, studies also want to recognize the most helpful method for deploying neuro-stimulated algorithms on a big scale. From the excellent placement of sensors to the maximum green implementation for processing facts, the know-how of the quality manner to put in force such systems is vital.

ultimately, future studies must discover approaches to include additional sources of facts into the version. Phenomena, which include landslides or tectonic movement, should provide precious perception into earthquake prediction and substantially enhance the accuracy of fashions.

Investing in the above regions improves neuro-stimulated earthquake monitoring and prediction algorithms. It is critical to keep discovering how those techniques may be advanced and applied for practical use, as accurate and timely prediction of seismic pastime is essential to the safety of humans and property.

REFERENCES

Alioto, M., De, V., & Marongiu, A. (2018). Guest editorial energy-quality scalable circuits and systems for sensing and computing: From approximate to communication-inspired and learning-based. *IEEE Journal on Emerging and Selected Topics in Circuits and Systems*, *8*(3), 361–368. doi:10.1109/JETCAS.2018.2865783

Asim, K. M., Martínez-Álvarez, F., Basit, A., & Iqbal, T. (2017). Earthquake magnitude prediction in Hindukush region using machine learning techniques. *Natural Hazards*, *85*(1), 471–486. doi:10.100711069-016-2579-3

Beroza, G. C., Segou, M., & Mostafa Mousavi, S. (2021). Machine learning and earthquake forecasting—Next steps. *Nature Communications*, *12*(1), 4761. doi:10.103841467-021-24952-6 PMID:34362887

. Chittora, P., Chakrabarti, T., Debnath, P., Gupta, A., Chakrabarti, P., Praveen, S. P., & Elngar, A. A. (2022). *Experimental analysis of earthquake prediction using machine learning classifiers, curve fitting, and neural modeling*.

Debnath, P., Chittora, P., Chakrabarti, T., Chakrabarti, P., Leonowicz, Z., Jasinski, M., Gono, R., & Jasińska, E. (2021). Analysis of earthquake forecasting in India using supervised machine learning classifiers. *Sustainability (Basel)*, *13*(2), 971. doi:10.3390u13020971

Galkina, A., & Grafeeva, N. (2019, April). Machine learning methods for earthquake prediction: A survey. In *Proceedings of the Fourth Conference on Software Engineering and Information Management (SEIM-2019)*, Saint Petersburg, Russia.

Hannah, S., Deepa, A. J., Chooralil, V. S., BrillySangeetha, S., Yuvaraj, N., Arshath Raja, R., Suresh, C., Vignesh, R., YasirAbdullahR, Srihari, K., & Alene, A. (2022). BrillySangeetha, S., Yuvaraj, N., Arshath Raja, R., & Alene, A. (2022). Blockchain-based deep learning to process IoT data acquisition in cognitive data. *BioMed Research International*, *2022*, 1–7. doi:10.1155/2022/5038851

Hoque, A., Raj, J., Saha, A., & Bhattacharya, P. (2020). Earthquake magnitude prediction using machine learning technique. In *Trends in Computational Intelligence, Security and Internet of Things: Third International Conference*. Springer.

Irfan, D., Tang, X., Narayan, V., Mall, P. K., Srivastava, S., & Saravanan, V. (2022). Prediction of Quality Food Sale in Mart Using the AI-Based TOR Method. *Journal of Food Quality*, *2022*, 2022. doi:10.1155/2022/6877520

Johnson, P. A., Rouet-Leduc, B., Pyrak-Nolte, L. J., Beroza, G. C., Marone, C. J., Hulbert, C., Howard, A., Singer, P., Gordeev, D., Karaflos, D., Levinson, C. J., Pfeiffer, P., Puk, K. M., & Reade, W. (2021). Laboratory earthquake forecasting: A machine learning competition. *Proceedings of the National Academy of Sciences of the United States of America*, *118*(5), e2011362118. doi:10.1073/pnas.2011362118 PMID:33495346

Kasabov, N., Sengupta, N., & Scott, N. (2016, September). From von Neumann, John Atanasoff and ABC to neuromorphic computation and the NeuCube spatio-temporal data machine. In *2016 IEEE 8th International Conference on Intelligent Systems (IS)* (pp. 15-21). IEEE.

Kavianpour, P., Kavianpour, M., Jahani, E., & Ramezani, A. (2023). A cnn-bilstm model with attention mechanism for earthquake prediction. *The Journal of Supercomputing*, *79*(17), 1–33. doi:10.100711227-023-05369-y

Laurenti, L., Tinti, E., Galasso, F., Franco, L., & Marone, C. (2022). Deep learning for laboratory earthquake prediction and autoregressive forecasting of fault zone stress. *Earth and Planetary Science Letters*, *598*, 117825. doi:10.1016/j.epsl.2022.117825

Li, R., Lu, X., Li, S., Yang, H., Qiu, J., & Zhang, L. (2020, July). DLEP: A deep learning model for earthquake prediction. In *2020 International Joint Conference on Neural Networks (IJCNN)* (pp. 1-8). IEEE. 10.1109/IJCNN48605.2020.9207621

Lymperopoulos, I. N. (2017). Dynamic response and transfer function of social systems: A neuro-inspired model of collective human activity patterns. *Neural Networks*, *94*, 125–140. doi:10.1016/j.neunet.2017.07.010 PMID:28772240

Mallouhy, R., Abou Jaoude, C., Guyeux, C., & Makhoul, A. (2019, December). Major earthquake event prediction using various machine learning algorithms. In *2019 International Conference on Information and Communication Technologies for Disaster Management (ICT-DM)* (pp. 1-7). IEEE. 10.1109/ICT-DM47966.2019.9032983

Mangalathu, S., Sun, H., Nweke, C. C., Yi, Z., & Burton, H. V. (2020). Classifying earthquake damage to buildings using machine learning. *Earthquake Spectra*, *36*(1), 183–208. doi:10.1177/8755293019878137

Murwantara, I. M., Yugopuspito, P., & Hermawan, R. (2020). Comparison of machine learning performance for earthquake prediction in Indonesia using 30 years historical data. *TELKOMNIKA*, *18*(3), 1331–1342. doi:10.12928/telkomnika.v18i3.14756

Narayanan, K. L., Krishnan, R. S., Robinson, Y. H., Julie, E. G., Vimal, S., Saravanan, V., & Kaliappan, M. (2022). Banana plant disease classification using hybrid convolutional neural network. *Computational Intelligence and Neuroscience*, *2022*, 2022. doi:10.1155/2022/9153699 PMID:35251158

Purohit, K., Vats, S., Saklani, R., Kukreja, V., Sharma, V., & Yadav, S. P. (2023, July). Improvement in K-Means Clustering for Information Retrieval. In *2023 4th International Conference on Electronics and Sustainable Communication Systems (ICESC)* (pp. 1239-1245). IEEE.

Rajalakshmi, M., Saravanan, V., Arunprasad, V., Romero, C. T., Khalaf, O. I., & Karthik, C. (2022). Machine Learning for Modeling and Control of Industrial Clarifier Process. *Intelligent Automation & Soft Computing*, *32*(1). doi:10.32604/iasc.2022.021696

Rouet-Leduc, B., Hulbert, C., Lubbers, N., Barros, K., Humphreys, C. J., & Johnson, P. A. (2017). Machine learning predicts laboratory earthquakes. *Geophysical Research Letters*, *44*(18), 9276–9282. doi:10.1002/2017GL074677

Sajan, K. C., Bhusal, A., Gautam, D., & Rupakhety, R. (2023). Earthquake damage and rehabilitation intervention prediction using machine learning. *Engineering Failure Analysis*, *144*, 106949. doi:10.1016/j.engfailanal.2022.106949

Salam, M. A., Ibrahim, L., & Abdelminaam, D. S. (2021). Earthquake prediction using hybrid machine learning techniques. *International Journal of Advanced Computer Science and Applications*, *12*(5), 654–6652021. doi:10.14569/IJACSA.2021.0120578

Sivaram, M., Mohammed, A. S., Yuvaraj, D., Porkodi, V., Manikandan, V., & Yuvaraj, N. (2019). Advanced expert system using particle swarm optimization based adaptive network based fuzzy inference system to diagnose the physical constitution of human body. In *Emerging Technologies in Computer Engineering: Microservices in Big Data Analytics: Second International Conference, ICETCE 2019*, (pp. 349-362). Springer Singapore. 10.1007/978-981-13-8300-7_29

Yadav, S. P., Zaidi, S., Nascimento, C. D. S., de Albuquerque, V. H. C., & Chauhan, S. S. (2023, January). Analysis and Design of automatically generating for GPS Based Moving Object Tracking System. In *2023 International Conference on Artificial Intelligence and Smart Communication (AISC)* (pp. 1-5). IEEE. 10.1109/AISC56616.2023.10085180

Yuvaraj, N., Srihari, K., Chandragandhi, S., Raja, R. A., Dhiman, G., & Kaur, A. (2021). Analysis of protein-ligand interactions of SARS-Cov-2 against selective drug using deep neural networks. *Big Data Mining and Analytics*, *4*(2), 76–83. doi:10.26599/BDMA.2020.9020007

Chapter 7
AI Implementation Techniques for Finding Earth's Frozen Zones

Kavitha M. S.
https://orcid.org/0000-0002-7240-3242
Department of Electronics and Communication Engineering, R.M.K. Engineering College, Chennai, India

Mahalakshmi P.
Department of Information Technology, Rajalakshmi Engineering College, Chennai, India

Manimegalai L.
Department of Electronics and Communication Engineering, R.M.K. Engineering College, Chennai, India

Sundaramoorthi S.
Department of Electrical and Electronics Engineering, Panimalar Engineering College, India

ABSTRACT

Implementing artificial intelligence (AI) techniques for finding frozen zones worldwide has received sizeable attention in recent years because of the rising worries concerning climate exchange and its results. This summary outlines the diverse AI techniques employed in this area and highlights their blessings and limitations. In recent years, AI has emerged as an effective device for figuring out and analyzing frozen zones on the Earth. Frozen zones, along with glaciers, ice caps, and permafrost regions, play a vital role within the Earth's weather device, appearing as water reservoirs and affecting sea-degree upward push. This essay explores the programs of AI in identifying frozen zones and the implementation techniques used in this technique. The AI strategies must operate autonomously, with minimal human intervention. By applying AI strategies, researchers can reap extra unique and dependable statistics regarding the state and behavior of these regions, leading to deeper expertise of climate alternate and its influences.

DOI: 10.4018/979-8-3693-1850-8.ch007

1. INTRODUCTION

Synthetic intelligence (AI) has revolutionized various industries and can deal with some of the most urgent challenges going through humanity. One such big project is identifying and monitoring the Earth's frozen zones, wherein AI implementation techniques have emerged as an essential tool for conservationists and researchers (Sayed et al., 2023). Through superior algorithms, machine learning, and facts analysis, AI has led to full-size advancements in mapping, tracking, and knowledge of those frozen areas. This essay explores the introduction of AI implementation strategies for finding Earth's frozen zones, highlighting their advantages, challenges, and potential future applications (Adak et al., 2022). Implementing AI strategies has enabled scientists to map and reveal the Earth's frozen zones more accurately and correctly. Satellites prepared with high-resolution sensors capture vast information, including temperature, ice thickness, and snow coverage. Those facts, however, present challenges in each quantity and complexity. AI algorithms and machine-gaining knowledge of techniques can method and examine these considerable statistics more efficaciously than traditional techniques, mainly to complete and granular insights about those faraway and inaccessible areas (Coca & Datcu, 2023). By reading patterns and relationships within the data, AI algorithms can identify vital modifications in those frozen zones and provide actual-time updates to researchers and decision-makers. Furthermore, AI implementation techniques beautify our expertise in the Earth's frozen zones by enabling particular weather modeling and prediction. Frozen regions, such as the Arctic and Antarctica, are essential in retaining the planet's usual weather stability (Shen et al., 2022). They act as herbal coolers, reflecting sunlight into the area and regulating international temperatures. But, due to weather alternates, these zones are increasingly at risk of melting, resulting in rising sea levels and significant ecological and environmental results. AI-powered fashions consider factors like wind patterns, ocean currents, and temperatures, permitting scientists to simulate and expect destiny scenarios. Those projections aid in policymaking, aid-making plans, and conservation efforts. Furthermore, AI techniques facilitate the identification of capability threats and dangers faced via the Earth's frozen zones. Via studying satellite imagery and historical statistics, AI algorithms can locate anomalies and alerts of distress, including cracks in ice sheets or modifications in water currents (Munir et al., 2022; Yang et al., 2022). This early warning device permits researchers and the government to reply directly to mitigate risks and prevent irreversible harm. For instance, scientists observed a large crack within Antarctica's Larsen C ice shelf using AI algorithms, letting them closely monitor its progression and take vital measures to save you a catastrophic ice crumble (Sun et al., 2023). Despite the several benefits, implementing AI strategies for locating Earth's frozen zones provides certain demanding situations. First, obtaining satellite TV for PC statistics may be high-priced and complex. The improvement and deployment of satellites with the specified sensors and abilities impose vast financial burdens on study establishments and governments. Moreover, ensuring the records' accuracy remains important, requiring rigorous validation and calibration techniques (Mohammed et al., 2023). Statistics collection or processing mistakes can cause inaccurate conclusions and subsequent consequences for conservation efforts. Moreover, the ethical implications of AI in this area need to be carefully considered. AI strategies must adhere to ethical guidelines and privacy regulations while dealing with touchy ecological regions and indigenous communities (Li et al., 2022). Transparency and duty in AI algorithms and decision-making strategies are vital to maintaining public agreement with and ensuring the responsible use of AI for the gain of clinical research and environmental conservation.

1.1 Background and Significance

The implementation of AI techniques for finding Earth's frozen zones holds massive significance for numerous reasons:

1.1.1 Climate Alternate Research and Predictions

Tracking and knowledge of the modifications occurring in frozen zones are critical for predicting the effect of weather alternate on this planet's ecosystems. AI algorithms can precisely analyze massive datasets, identify tendencies, and determine ability causal factors (Kong et al., 2023). By correctly locating frozen zones vulnerable to weather exchange, scientists can broaden sturdy fashions for predicting destiny eventualities, assisting policymakers in devising appropriate measures to mitigate its impact.

1.1.2 Environmental Conservation and Biodiversity

Frozen zones are domestic to precise ecosystems, supporting diverse wildlife and playing an essential function in international biodiversity. By utilizing AI technologies, scientists can display those zones more successfully, reading the conduct and motion of the natural world, mapping migration styles, and identifying endangered species (Figueiredo et al., 2022; Orusa et al., 2022). This data can contribute to conservation efforts, assisting in the formula of suitable control techniques for keeping these ecosystems.

1.1.3 Catastrophe Management

AI strategies enable non-stop and near real-time monitoring of frozen zones, improving our ability to discover and reply to potential herbal screws. For example, figuring out unstable ice caps or glaciers thru AI algorithms can assist in expecting avalanches, landslides, or ice melting, allowing the government to take timely preventive measures and decrease the danger to human lives and infrastructure (Kaur et al., 2022; Purohit et al., 2023).

1.1.4 Resource Exploration and Usage

The frozen zones additionally include giant reserves of herbal resources, including minerals, oil, and gasoline(Sathishkumar et al., 2023).

1.2 Objectives of the Chapter

The goals of AI implementation techniques for finding Earth's frozen zones encompass:

1. Accuracy and precision: The AI techniques ought to locate and pick out the Earth's frozen zones with high precision. This entails growing algorithms and fashions that can successfully examine and interpret remote sensing statistics, satellite TV for pc imagery, and other relevant assets (Müller et al., 2023).

2. Scalability and efficiency: The AI strategies should be designed to address vast quantities of records and perform actual-time processing. This allows the detection and monitoring of frozen zones internationally, considering complete coverage and green analysis (Zhu et al., 2023).
3. Automation and autonomy: The AI strategies must operate autonomously, with minimal human intervention. This includes automatic information collection, processing, and interpretation, lowering the need for manual efforts and enabling non-stop monitoring of frozen zones (Casagli et al., 2023).
4. Adaptability and generalizability: The AI techniques must be adaptable to one-of-a-kind geographical areas and environmental conditions, as frozen zones can vary in region, characteristics, and dynamics. The models should be able to generalize well throughout different regions and spatial scales, providing a complete knowledge of Earth's frozen zones.
5. Integration and fusion of data sources: The AI techniques have to integrate and fuse data from various resources, which include satellite TV for pc imagery, faraway sensing information, climate records, and climate(Pandey & Verma, 2022).

2. OVERVIEW OF EARTH'S FROZEN ZONES

Earth's frozen zones, called cryospheric regions, embody considerable snow, ice, and permafrost areas. These regions are essential to the Earth's weather gadget, water cycle, and average biodiversity. Monitoring and understanding those frozen zones is essential for predicting climate change influences, comparing water sources, and studying ecosystems' dynamics. But, due to their remote locations and enormous length, traditional strategies of statistics series can be steeply-priced, time-eating, and logistically challenging (Yuvaraj, Srihari, Dhiman et al, 2021). That is, artificial intelligence (AI) implementation strategies come into play. Integrating AI into evaluating remotely sensed facts will improve our information on Earth's frozen zones and mitigate the demanding situations associated with reading them. Demanding situations in conventional statistics collection:

Reading Earth's frozen zones using conventional techniques poses several demanding situations. First, these regions' vastness makes it nearly impossible for researchers to accumulate data from each nook and cranny manually (Kannan et al., 2021). 2d, the challenging and now-and-then risky situations in these zones restrict the accessibility of remote sensing gadgets and human presence. Third, traditional records collection techniques frequently require good-sized sources, which includes skilled human hard work, specialized system, and transportation logistics (Yuvaraj, Karthikeyan, & Praghash, 2021).

2.1 Characteristics and Significance

Traits:

1. data collection: AI techniques can assist in gathering and examining widespread quantities of statistics from diverse resources, which includes satellite TV for PC imagery, weather models, and floor-based total sensors. This includes historical statistics on temperature, precipitation, and ice formation.

2. machine gaining knowledge: AI algorithms can be trained on existing information to apprehend styles and make predictions. This could assist in identifying frozen zones based totally on particular traits, including temperature, ice coverage, and glacial movement.
3. image reputation: AI can be used to robotically analyze satellite photos, spotting frozen zones based totally on visual cues consisting of the presence of ice sheets, glaciers, or polar caps.
4. Herbal language processing: AI can method text-based statistics from medical guides, reports, and studies to extract applicable facts about frozen zones, inclusive of their places, traits, and modifications over time (Khan et al., 2022; Rajalakshmi et al., 2022).

Importance of AI implementation techniques:

1. accurate identity: AI techniques can improve the accuracy and pace of identifying frozen zones by analyzing many record points and styles. This guarantees better monitoring and knowledge of areas crucial for analyzing weather change and its effects.
2. Early detection: AI can locate adjustments and tendencies in frozen zones, allowing for the early identity of shifts in ice coverage, melting fees, or other widespread events (Gunjan et al., 2022).

3. APPLICATIONS OF AI IN IDENTIFYING FROZEN ZONES

Synthetic Intelligence (AI) has revolutionized various fields, together with climate science and environmental research. In recent years, AI has emerged as an effective device for figuring out and analyzing frozen zones on the Earth. Frozen zones, along with glaciers, ice caps, and permafrost regions, play a vital role within the Earth's weather device, appearing as water reservoirs and affecting sea-degree upward push. This essay explores the programs of AI in identifying frozen zones and the implementation techniques used in this technique.

AI programs in identifying Frozen Zones:

1. Satellite TV for PC Imagery evaluation: one of the primary programs of AI in figuring out frozen zones involves studying satellite TV for PC imagery. AI algorithms can system significant amounts of satellite statistics, quickly figuring out and classifying exclusive sorts of frozen zones. Those algorithms can stumble on changes in ice cowl over time, screen snow accumulation, and quantify glacier motion, presenting valuable insights into the nation of frozen zones.
2. gadget gaining knowledge of computerized type: system getting to know strategies permit for automated classification of frozen zones based on their beautiful characteristics. Using education AI models on large datasets encompassing various frozen zones, algorithms can learn to differentiate between different kinds of frozen zones accurately..

3.1 Remote Sensing and Image Analysis

Far-off sensing and image evaluation are quintessential artificial intelligence (AI) techniques for identifying and evaluating Earth's frozen zones. Exploring these frozen zones is paramount because of their impact on the global climate system, hydrological cycle, and sea level rise. Using AI algorithms, scientists can enhance far-flung sensing skills and statistics processing strategies to stumble on and monitor those regions effectively. First, far-flung sensing plays an essential position in identifying Earth's frozen zones. Satellites with sensors can seize information in various wavelengths, including seen, infrared, and microwave. This information offers treasured records about surface traits, temperature, and adjustments in frozen areas. Thru AI implementation, photograph analysis algorithms can examine these widespread amounts of satellite imagery to locate frozen zones, along with glaciers, ice sheets, permafrost, and polar areas. One of the primary AI techniques used in this context is image type. Through education, AI models the usage of classified photos, and they can study to distinguish among one-of-a-kind forms of frozen regions. For example, with schooling data consisting of labeled pix of glaciers, a system gaining knowledge of algorithms can easily pick out and classify glaciers within remotely sensed imagery. This class permits scientists to monitor adjustments in these frozen regions through the years, along with glacier retreat or increase, presenting insights into the outcomes of climate exchange. Any other AI approach employed inside the far-flung sensing and photograph evaluation of frozen zones is item detection. This method permits the automated identification and extraction of specific capabilities inside the satellite imagery. For example, AI algorithms may be skilled in hitting permafrost regions by recognizing extraordinary styles and characteristics associated with frozen ground. This information is essential for knowledge of permafrost distribution, which is vital for climate models because it stores tremendous quantities of natural carbon and affects greenhouse gasoline emissions. Moreover, AI algorithms can be mixed with different far-flung sensing information to enhance the accuracy and reliability of frozen quarter detection. For instance, integrating radar and optical satellite imagery with AI strategies permits the identity of icebergs, sea ice, and ice shelves with higher precision. Fusing a couple of statistics sources enhances the reliability of frozen sector detection by imparting complementary records that enable triumph over limitations related to using an available sensor or data kind. Furthermore, AI-based strategies can also allow real-time monitoring of frozen zones. By continuously reading satellite TV for pc statistics and using AI algorithms, scientists can stumble on adjustments and anomalies in frozen areas almost right now. This rapid reaction is vital for early caution systems, facilitating effective disaster management and mitigation measures. For example, detecting fast ice softening or sudden destruction in an ice shelf can help assess the ability for flooding, determine influences on ecosystems, and inform decision-making techniques.

3.2 Machine Learning Approaches

Earth's frozen zones, comprising polar ice caps, glaciers, and permafrost areas, are vital in regulating worldwide climate styles and water sources. They may also be indicators of environmental adjustments, making their expertise and tracking essential. In current years, machine studying strategies have emerged as powerful tools in remotely sensing and studying those frozen zones. This essay explores numerous device-getting-to-know approaches to implementing synthetic intelligence (AI) for locating and reading Earth's frozen zones. Far off sensing, using satellite TV for PC imagery presents enormous numbers of statistics for monitoring Earth's frozen zones. But, efficiently processing and reading this information is challenging because of its size and complexity. System learning algorithms can assist in automatically classifying and extracting applicable capabilities from these remote-sensing photographs. One technique is supervised getting to know convolutional neural network (CNN) algorithms. CNNs have been a hit in figuring out exclusive varieties of frozen zones from satellite TV for PC imagery. By educating those algorithms with labeled datasets, the AI version learns to identify critical styles and capabilities, permitting accurate class and mapping of frozen zones. This enables tracking modifications over time and predicting future developments.

4. KEY COMPONENTS OF AI IMPLEMENTATION FOR IDENTIFYING FROZEN ZONES

The identity and study of frozen zones on the planet and the usage of AI implementation techniques have become increasingly critical for diverse medical, environmental, and economic reasons. AI, or artificial intelligence, can play a crucial role in figuring out those frozen zones by studying a good-sized quantity of statistics, using state-of-the-art algorithms, and utilizing superior imaging techniques. In this essay, we can talk about the critical components of AI implementation for identifying frozen zones worldwide. First and foremost, records series serve as a fundamental issue for successfully implementing AI strategies in figuring out frozen zones. This entails collecting relevant and accurate facts from various sources together with satellite imagery, weather fashions, floor-primarily based observations, and historical data. The accumulated facts should embody parameters like temperature, precipitation, snow cover, ice thickness, and sea ice extent. Moreover, records on topographic functions, ocean currents, and the interaction among diverse climatic structures must also be considered. AI can correctly discover frozen zones and music adjustments over the years by amassing complete and diverse datasets. Next, statistics preprocessing is a critical step in preparing the amassed information for AI evaluation. The accumulated facts regularly carry noise, missing values, outliers, and inconsistencies that need to be addressed before AI techniques are implemented. Information preprocessing involves information cleansing, integration, transformation, and normalization.

4.1 Data Preprocessing and Feature Extraction

Statistics pre-processing entails preparing the uncooked records for evaluation by cleaning, transforming, and refining them. In finding Earth's frozen zones, this generally entails amassing information from diverse assets, including satellite TV for PC imagery, sensor networks, and weather models. However, those information assets may comprise noise, missing values, inconsistencies, and biases that can affect the accuracy of AI algorithms. Consequently, data pre-processing techniques are hired to cope with these challenges. One commonplace pre-processing step is records cleansing, which entails removing or repairing lacking or inconsistent information factors. For instance, if missing values exist in a satellite TV for a PC photo, interpolation strategies can be used to estimate the missing pixels based on neighboring values. Additionally, outliers, which could negatively affect the AI version's performance, want to be diagnosed and treated as they should be. This will be accomplished using statistical methods consisting of the Z-rating or the interquartile variety (IQR) to detect and cast off outliers. Every other vital pre-processing step is information transformation. This includes converting the information into an appropriate layout for evaluation and addressing troubles with normalization, scaling, and encoding. Normalization techniques, consisting of min-max scaling or z-rating normalization, are used to ensure each feature has a similar range of values. That is, in particular, important to prevent positive functions from dominating the analysis because of their immense value. Encoding specific variables is usually achieved to symbolize qualitative information in a numerical layout that may be processed via AI algorithms. For example, the presence or absence of sure frozen area characteristics can be encoded as binary variables.

In finding Earth's frozen zones, characteristic extraction may involve extracting spatial and temporal capabilities from satellite imagery. For instance, ice and snow may be detected via color thresholds or texture analysis. Moreover, weather models or sensor networks can extract capabilities associated with temperature and precipitation patterns to become aware of regions vulnerable to freezing temperatures. These functions can include mean, minimum, maximum, and variance of temperature or precipitation values over a specific term in a given location. In a few cases, advanced characteristic extraction strategies consisting of dimensionality reduction techniques may be implemented to reduce the number of functions while maintaining vital information. The most important issue analysis (PCA) is a famous technique for dimensionality reduction, which identifies a smaller set of uncorrelated variables, known as principal components, that capture the general public of the variance inside the records.

4.2 Algorithm Selection and Model Design

1. Faraway sensing facts analysis: AI algorithms can correctly examine satellite imagery and remote sensing information for identifying frozen zones. Convolutional Neural Networks (CNNs) may be hired to extract key features from the imagery, including sea ice extent, snow cover, and glacier dynamics. These algorithms can classify exclusive frozen zones and provide specific statistics about their volume and versions through the years.
2. statistics Fusion and Integration: AI algorithms can combine various statistics resources with atmospheric and oceanic measurements to increase complete fashions for figuring out frozen zones. Fusion techniques, which include Bayesian networks, machine mastering-primarily based methods, and fuzzy logic structures, can combine a couple of facts resources to generate accurate and holistic checks.

5. ADVANTAGES AND BENEFITS

Artificial Intelligence (AI) has revolutionized several domain names in our world, and one area that has visible terrific blessings from its implementation is the exploration of Earth's frozen zones. These frozen zones and ice caps, glaciers, and permafrost regions offer precious insights into our planet's records and are critical for information and mitigating the impacts of weather alternates. Via using AI techniques, scientists and researchers were able to beautify their potential to map, display, and examine those frozen zones, leading to numerous blessings and benefits. First and essential, AI strategies allow us to acquire and technique enormous quantities of information associated with Earth's frozen zones quickly and appropriately. Conventional strategies of manually gathering and studying records proved time-consuming and frequently vulnerable to errors. AI algorithms can automate data collection by analyzing satellite TV for PC imagery, temperature readings, and different applicable information, allowing researchers to reap complete records units much quicker and with excessive precision. This increased performance and accuracy allow scientists to display changes inside the frozen zones more efficiently and offer actual-time facts on melting charges, ice thickness, and different vital parameters. Furthermore, AI implementation offers superior strategies for mapping and modeling Earth's frozen zones.

5.1 Improved Accuracy and Efficiency

Synthetic Intelligence (AI) technology is significantly more suitable for our capability to identify and apprehend numerous Earth phenomena, including frozen zones or areas blanketed in ice and snow. Conventional techniques of locating and mapping these frozen regions needed improved accuracy and performance. But, with the advancement of AI algorithms and records processing strategies, we have extra powerful gear at our disposal. This essay explores how AI can improve the accuracy and performance of finding Earth's frozen zones.

Stronger statistics series:

AI can notably enhance fact series strategies, leading to progressed accuracy when identifying frozen zones on this planet. Far-off sensing technologies, satellites, drones, and floor-based total sensors may be applied to seize significant quantities of facts. AI algorithms can then technique this information, allowing the identity of ice-included areas with remarkable precision. For instance, AI-powered satellite TV for PC imagery analysis allows for the identification of snow, glaciers, and ice caps with excessive accuracy, contributing to a more excellent and unique knowledge of Earth's frozen areas.

Deep studying Algorithms:

Deep mastering algorithms, a subset of AI methods, have revolutionized several scientific fields, including earth sciences. These algorithms can process and examine massive units of complicated facts, extracting patterns and traits that human evaluation would neglect. Deep studying algorithms can examine tremendous quantities of ancient statistics and satellite TV for PC pics, climate data, and radar records when implemented to find frozen zones. Using education on these datasets, they can identify and classify frozen regions more appropriately than conventional strategies. Furthermore, those algorithms can improve over the years as new facts become available, improving their accuracy and performance in identifying and mapping Earth's frozen zones.

Integration of a couple of statistics sources:

AI techniques can integrate facts from numerous resources, including temperature information, precipitation statistics, and blizzard measurements, into an available analysis framework to enhance accuracy. By combining specific varieties of records, AI algorithms can take advantage of a comprehensive understanding of Earth's frozen zones. For example, by incorporating temperature facts with satellite TV for PC imagery, AI can perceive areas with lower temperatures and a better opportunity for ice and snow insurance. Integrating a couple of information assets contributes to better-informed choice-making and more precise identification of frozen zones.

5.2 Real-Time Monitoring Capabilities

The earth's frozen zones, especially the polar and high-altitude regions, are of exquisite interest to scientists and researchers reading weather exchange and its effects on this planet. Those frozen zones act as crucial signs of world climate patterns and are critical for expertise in the general fitness of the planet. However, reading and monitoring these areas can be frightening due to their remote places and lousy weather. There has been a growing interest in utilizing synthetic intelligence (AI) technologies to resource within the monitoring and analysis of the earth's frozen zones in recent years. AI implementation strategies provide actual-time monitoring skills, enabling scientists to obtain correct and timely information about these areas. One of the primary blessings of AI in this context is its ability to process sizeable quantities of records quickly and successfully. AI algorithms can use records from a few resources, such as satellite

TV for PC imagery, remote sensors, and ground-based measurements, to extract precious insights. By using system studying algorithms, AI structures can understand patterns and anomalies, supporting scientists in identifying adjustments in the frozen zones that may suggest environmental changes or ability dangers. Moreover, AI can allow real-time monitoring via continuously studying incoming records and generating indicators or notifications when vast adjustments arise..

6. CHALLENGES AND LIMITATIONS

There are widespread challenges and barriers that must be considered while using AI implementation strategies in this domain. This essay will explore those challenges and obstacles.

1. Restricted availability and fine of records:

One of the first challenges in imposing AI techniques for finding Earth's frozen zones is the limited availability and quality of information. Accumulating accurate and complete datasets on frozen zones, consisting of Arctic and Antarctic regions, is a complex and luxurious mission. Furthermore, information series techniques often come across logistical hurdles due to the harsh environments related to these regions.

2. Complexity and variability of frozen quarter traits:

Frozen zones' complicated nature and variability pose other missions for AI implementation techniques. These areas represent a dynamic machine tormented by various factors, including sea ice dynamics, temperature fluctuations, and ocean currents..

6.1 Data Quality and Availability

Records best form the muse for reliable AI implementation strategies. The first-rate of records determines the precision and validity of the insights generated through AI algorithms. In finding Earth's frozen zones, facts involve accuracy, completeness, relevance, consistency, and timeliness. Statistics guarantees that AI fashions can seize important info, detect tendencies, and, as it should be, pick out frozen areas, contributing to stepped forward expertise and robust selection-making processes..

7. CONCLUSION

In the end, implementing artificial intelligence (AI) techniques has emerged as promising for finding Earth's frozen zones. By mixing AI technologies, scientists and researchers have been able to triumph over numerous demanding situations related to exploring and information those far-off and environmentally sensitive regions. One key element of implementing AI strategies in locating frozen zones is the usage of satellite imagery and remote sensing facts. Gadgets, getting to know algorithms, particularly those based on deep neural networks, have validated superb abilities in processing and analyzing enormous volumes of satellite facts with excessive accuracy and efficiency. They could successfully identify and classify frozen regions along with glaciers, ice caps, and permafrost regions, providing precious insights into the extent, dynamics, and health of those frozen zones. Moreover, AI algorithms have shown full-size ability in automating the extraction of vital geophysical parameters from remote sensing facts, such as floor temperature, ice thickness, and snow accumulation. Those parameters are critical for tracking and predicting the modifications that arise in Earth's frozen zones. By applying AI strategies, researchers can reap extra unique and dependable statistics regarding the state and behavior of these regions, leading to deeper expertise of climate alternate and its influences..

7.1 Summary of Key Findings

1. The AI era is significant in mapping and figuring out the Earth's frozen zones. It could assist in finding and monitoring those areas more effectively and as it should be.
2. various AI techniques have been advanced and hired for identifying frozen zones, including gadget studying, far-off sensing, and information evaluation algorithms. Those techniques can cover many facts and offer valuable insights into frozen areas.
3. device studying-primarily based algorithms have been proven influential in classifying satellite TV for PC imagery and aerial facts to recognize frozen zones. These algorithms are trained on big datasets, making them distinguish between frozen and non-frozen areas with high precision.
4. Satellite TV for PC faraway sensing plays a crucial position in identifying and tracking frozen zones. AI strategies can enhance the analysis of satellite imagery by automating the detection of precise indicators, including ice formations, snow cover, and permafrost.
5. Records evaluation algorithms and AI facilitate the detection and prediction of frozen zones by utilizing complex statistical fashions. By studying a couple of types of data, together with temperature, moisture degrees, and soil composition, those algorithms can expect the presence of frozen areas with notable accuracy.

7.2 Recommendations for Future Research

It will decorate AI implementation strategies for finding Earth's frozen zones; there are numerous suggestions for future studies that may be pursued. First, further exploration needs to be performed to improve the accuracy and speed of AI algorithms in identifying and mapping frozen zones. This can involve developing extra advanced devices and getting to know models capable of studying complicated datasets, including satellite photos, climate data, and geological statistics. By incorporating a much wider variety of statistics sources, AI algorithms can provide more comprehensive and specific exams of frozen zones. Secondly, study efforts should focus on enhancing AI fashions' interpretability and explainability in figuring out frozen zones. As AI algorithms become more complicated, it's vital to recognize the selection-making method in the back of their predictions. Growing methodologies to transparently interpret the outcomes generated by AI fashions can assist scientists and policymakers in providing information on the underlying elements contributing to frozen zones, helping in robust mitigation and variation strategies. Thirdly, collaborative research initiatives must be advocated among AI experts, climate scientists, and environmental organizations. By bringing collectively interdisciplinary teams, study efforts can benefit from diverse information and perspectives, ultimately leading to more comprehensive AI implementations for identifying frozen zones. This collaboration can foster knowledge change and promote the integration of domain-precise information into AI algorithms, resulting in more accurate and dependable predictions..

REFERENCES

Adak, A., Pradhan, B., & Shukla, N. (2022). Sentiment analysis of customer reviews of food delivery services using deep learning and explainable artificial intelligence: Systematic review. *Foods*, *11*(10), 1500. doi:10.3390/foods11101500 PMID:35627070

Casagli, N., Intrieri, E., Tofani, V., Gigli, G., & Raspini, F. (2023). Landslide detection, monitoring and prediction with remote-sensing techniques. *Nature Reviews. Earth & Environment*, *4*(1), 51–64. doi:10.103843017-022-00373-x

Coca, M., & Datcu, M. (2023). FPGA Accelerator for Meta-Recognition Anomaly Detection: Case of Burned Area Detection. *IEEE Journal of Selected Topics in Applied Earth Observations and Remote Sensing*, *16*, 5247–5259. doi:10.1109/JSTARS.2023.3273309

Figueiredo, N., Neto, A., Cunha, A., Sousa, J. J., & Sousa, A. (2022, July). Deep Learning Approach for Terrace Vineyards Detection from Google Earth Satellite Imagery. In *IGARSS 2022-2022 IEEE International Geoscience and Remote Sensing Symposium* (pp. 5824-5827). IEEE. 10.1109/IGARSS46834.2022.9884644

Gunjan, V. K., Vijayalata, Y., Valli, S., Kumar, S., Mohamed, M. O., & Saravanan, V. (2022). Machine learning and cloud-based knowledge graphs to recognize suicidal mental tendencies. *Computational Intelligence and Neuroscience*, *2022*, 2022. doi:10.1155/2022/3604113 PMID:35341205

Kannan, S., Dhiman, G., Natarajan, Y., Sharma, A., Mohanty, S. N., Soni, M., Easwaran, U., Ghorbani, H., Asheralieva, A., & Gheisari, M. (2021). Ubiquitous vehicular ad-hoc network computing using deep neural network with iot-based bat agents for traffic management. *Electronics (Basel)*, *10*(7), 785. doi:10.3390/electronics10070785

Kaur, J., Saxena, J., Shah, J., & Yadav, S. P. (2022, May). Facial Emotion Recognition. In *2022 International Conference on Computational Intelligence and Sustainable Engineering Solutions (CISES)* (pp. 528-533). IEEE.

Khan, S., Saravanan, V. N. G. C., Lakshmi, T. J., Deb, N., & Othman, N. A. (2022). Privacy protection of healthcare data over social networks using machine learning algorithms. *Computational Intelligence and Neuroscience*, *2022*, 2022. doi:10.1155/2022/9985933 PMID:35371203

Kong, L., Ren, D., Zhou, S., He, Z., Ai, C., & Yan, C. (2023). Evaluating the evolution of fiber-reinforced emulsified asphalt cold-recycled mixture damage using digital image correlation. *The International Journal of Pavement Engineering*, *24*(1), 2176495. doi:10.1080/10298436.2023.2176495

Li, X., Cheng, G., Fu, B., Xia, J., Zhang, L., Yang, D., & Ran, Y. (2022). Linking critical zone with watershed science: The example of the Heihe River basin. *Earth's Future, 10*(11), e2022EF002966.

Mohammed, M. A., Khleel, N. A., Szabó, N. P., & Szűcs, P. (2023). Modeling of groundwater quality index by using artificial intelligence algorithms in northern Khartoum State, Sudan. *Modeling Earth Systems and Environment*, *9*(2), 2501–2516. doi:10.100740808-022-01638-6

Müller, M. C., Sander, T., & Mundt, C. (2023, June). Development and testing of a multi-spectral and multi-field of view imaging system for space-based signature detection and tracking of hypersonic vehicles. In *Sensors and Systems for Space Applications XVI* (Vol. 12546, pp. 8–21). SPIE. doi:10.1117/12.2663480

Munir, A., Aved, A., & Blasch, E. (2022). Situational awareness: Techniques, challenges, and prospects. *AI*, *3*(1), 55–77. doi:10.3390/ai3010005

Orusa, T., Cammareri, D., & Borgogno Mondino, E. (2022). A Scalable Earth Observation Service to Map Land Cover in Geomorphological Complex Areas beyond the Dynamic World: An Application in Aosta Valley (NW Italy). *Applied Sciences (Basel, Switzerland)*, *13*(1), 390. doi:10.3390/app13010390

Pandey, J., & Verma, S. (2022). Water Quality Analysis and Prediction Techniques Using Artificial Intelligence. In *ICT with Intelligent Applications: Proceedings of ICTIS 2021,* Volume 1 (pp. 279-290). Springer Singapore. 10.1007/978-981-16-4177-0_29

Purohit, K., Vats, S., Saklani, R., Kukreja, V., Sharma, V., & Yadav, S. P. (2023, July). *Improvement in K-Means Clustering for Information Retrieval. In 2023 4th International Conference on Electronics and Sustainable Communication Systems (ICESC)* (pp. 1239-1245). IEEE.

Rajalakshmi, M., Saravanan, V., Arunprasad, V., Romero, C. T., Khalaf, O. I., & Karthik, C. (2022). Machine Learning for Modeling and Control of Industrial Clarifier Process. *Intelligent Automation & Soft Computing*, *32*(1). doi:10.32604/iasc.2022.021696

Sathishkumar, V. E., Cho, J., Subramanian, M., & Naren, O. S. (2023). Forest fire and smoke detection using deep learning-based learning without forgetting. *Fire Ecology*, *19*(1), 1–17. doi:10.118642408-022-00165-0

Sayed, E. T., Olabi, A. G., Elsaid, K., Al Radi, M., Semeraro, C., Doranehgard, M. H., Eltayeb, M. E., & Abdelkareem, M. A. (2023). Application of artificial intelligence techniques for modeling, optimizing, and controlling desalination systems powered by renewable energy resources. *Journal of Cleaner Production*, *413*, 137486. doi:10.1016/j.jclepro.2023.137486

Shen, H., Huang, F., Fan, X., Shahabi, H., Shirzadi, A., Wang, D., & Chen, W. (2022). Improving the performance of artificial intelligence models using the rotation forest technique for landslide susceptibility mapping. *International Journal of Environmental Science and Technology*, 1–16.

Sun, Z., Cristea, N., & Rivas, P. (Eds.). (2023). *Artificial Intelligence in Earth Science: Best Practices and Fundamental Challenges*. Elsevier. doi:10.1016/B978-0-323-91737-7.00003-7

Yang, L., Driscol, J., Sarigai, S., Wu, Q., Chen, H., & Lippitt, C. D. (2022). Google Earth Engine and artificial intelligence (AI): A comprehensive review. *Remote Sensing (Basel)*, *14*(14), 3253. doi:10.3390/rs14143253

Yuvaraj, N., Karthikeyan, T., & Praghash, K. (2021). An improved task allocation scheme in serverless computing using gray wolf Optimization (GWO) based reinforcement learning (RIL) approach. *Wireless Personal Communications*, *117*(3), 2403–2421. doi:10.100711277-020-07981-0

Yuvaraj, N., Srihari, K., Dhiman, G., Somasundaram, K., Sharma, A., Rajeskannan, S. M. G. S. M. A., Soni, M., Gaba, G. S., AlZain, M. A., & Masud, M. (2021). Nature-inspired-based approach for automated cyberbullying classification on multimedia social networking. *Mathematical Problems in Engineering*, *2021*, 1–12. doi:10.1155/2021/6644652

Zhu, H. H., Wu, B., Cao, D. F., Li, B., Wen, Z., Liu, X. F., & Shi, B. (2023). Characterizing thermo-hydraulic behaviors of seasonally frozen loess via a combined opto-electronic sensing system: Field monitoring and assessment. *Journal of Hydrology (Amsterdam)*, *622*, 129647. doi:10.1016/j.jhydrol.2023.129647

Chapter 8
Earthquake Multi-Class Detection Using Artificial Intelligence

Kavitha P.
Department of Computer Science and Engineering, R.M.K. Engineering College, Chennai, India

Gopinathan B.
Department of Computer Science and Engineering, Adhiyamaan College of Engineering, Hosur, India

Roslin Dayana K.
Department of Computer Science and Engineering, R.M.D. Engineering College, Chennai, India

Rathinasabapathi G.
https://orcid.org/0000-0003-2619-466X
Department of Mechanical Engineering, Panimalar Engineering College, Chennai, India

ABSTRACT

The goal of the chapter on earthquake multi-magnificence detection and the use of synthetic intelligence is to discover and exhibit the usage of device learning and AI techniques for appropriately and efficiently detecting different lessons of earthquakes. The chapter seeks to offer a complete understanding of earthquake detection strategies and spotlight the capability of AI in advancing this subject. The goal is to present a detailed analysis of existing techniques and recommend novel AI-primarily based techniques that could enhance earthquake detection and category accuracy and timeliness. Conventional seismology strategies commonly focus on detecting earthquakes based on an unmarried seismic event. However, AI algorithms can investigate a significant quantity of information, which includes ancient earthquake facts, geological capabilities, and actual-time seismic signals, to become aware of patterns and classify earthquakes into multiple instructions.

DOI: 10.4018/979-8-3693-1850-8.ch008

1. INTRODUCTION

Earthquakes are herbal disasters that could cause significant harm to infrastructures and a lack of lives. Early detection and accurate types of earthquakes are vital for well-timed response and effective mitigation. Conventional strategies of earthquake detection heavily rely upon seismic monitoring networks, which can be expensive and limited in insurance (Giacco et al., 2022). Recent advancements in artificial intelligence (AI) have shown a promising ability to augment traditional earthquake detection methods by leveraging the power of gadget-getting-to-know algorithms. This essay examines the introduction of Earthquake Multi-magnificence Detection, the use of synthetic Intelligence, and its implications for reinforcing earthquake monitoring and response.

The utility of AI in earthquake detection entails utilizing system-studying algorithms to analyze and interpret seismic waveform information (Rao et al., 2023). By using deep studying-based strategies, AI models can be educated to detect and classify earthquakes in actual-time routinely. This technique can drastically enhance the efficiency and pace of earthquake detection in comparison to manual evaluation by seismologists. One of the critical blessings of AI-based total earthquake detection is its potential to address multi-elegance earthquake classification. Traditional methods typically classify earthquakes into binary categories based on their magnitude, particularly good-sized earthquakes and non-giant activities (Murti et al., 2022). However, earthquakes exhibit a wide range of complexities and classifying them into more than one class can provide extra exact facts about their traits and capability effect.

1.1 Background and Significance

Earthquakes are natural screw-ups that can cause giant harm to homes, infrastructure, and human lives. Green and correct detection of earthquakes is vital in mitigating the effect of those activities and ensuring the safety of affected populations. Traditional earthquake detection methods depend on seismographs and manual evaluation by seismologists (Sajan et al., 2023). But, those methods are frequently time-consuming, and their effectiveness could be more restrained in densely populated regions. To triumph over the restrictions of traditional earthquake detection strategies, researchers have begun exploring the usage of artificial intelligence (AI) in earthquake detection. AI has proven promising capacity in numerous fields, such as picture reputation, natural language processing, and records evaluation. Applying AI to earthquake detection offers an opportunity to improve detection accuracy, reduce response time, and keep lives in the long run (Shakeel et al., 2022).

Demanding situations in Earthquake Detection:

Earthquake detection is a complex project due to numerous challenges. First, seismic statistics are enormously heterogeneous and include diverse activities, including tectonic earthquakes, volcanic interest, and human-triggered vibrations. Traditional procedures often fail to distinguish among those exceptional event sorts, leading to false alarms or missed detections(Lu & Honarvar Shakibaei Asli, 2023).

1.2 Objectives of the Chapter

The goal of the chapter on Earthquake multi-magnificence detection and the use of synthetic Intelligence is to discover and exhibit the usage of device learning and AI techniques for appropriately and efficiently detecting different lessons of earthquakes. The chapter seeks to offer a complete understanding of earthquake detection strategies and spotlight the capability of AI in advancing this subject (Badri,

2022; Derras & Makhoul, 2022; Hajibabaee et al., 2023). The goal is to present a detailed analysis of existing techniques and recommend novel AI-primarily based techniques that could enhance earthquake detection and category accuracy and timeliness.

2. OVERVIEW OF EARTHQUAKE DETECTION

Earthquakes, one of the most devastating herbal screw-ups, have wreaked havoc on diverse world components throughout history. The capacity to stumble on earthquakes directly is crucial, which will limit the loss of human lives and bodily harm. In current years, improvements in the era, particularly in artificial intelligence (AI), have played an enormous role in improving earthquake detection (Ma & Chen, 2022). This essay offers a top-level view of earthquake detection and the use of AI for earthquake multi-class detection. Earthquake detection entails identifying and assessing seismic sports, mainly the vibrational waves produced with the aid of the Earth's motion. Conventional earthquake detection techniques have applied seismometers, sensitive devices that measure ground movement (Babu & Kanaga, 2022; Dogan et al., 2023). Those seismometers are deployed in seismic networks and provide treasured records for earthquake monitoring. However, the increasing availability of seismic statistics and the complexity of earthquake prevalence calls for more advanced techniques to decorate detection talents. Synthetic intelligence, with its capability to examine and adapt from full-size amounts of information, has presented numerous answers to tackle the challenges of earthquake detection (Aldini et al., 2022). Gadget getting-to-know algorithms, a subset of AI, were implemented in seismic statistics analysis to enhance the accuracy and speed of earthquake identification. These algorithms can be trained on historic earthquake statistics, incorporating various parameters consisting of magnitude, length, and waveform characteristics.

2.1 Importance and Challenges

Earthquakes pose full-size dangers to human lives, infrastructure, and the environment. Early detection and correct evaluation of these seismic activities are vital for powerful disaster management and mitigation techniques (Yuan et al., 2023). Artificial Intelligence (AI) has emerged as a promising tool for earthquake detection and class, allowing the identity of seismic events and facilitating timely reaction measures. However, using AI for multi-elegance earthquake detection affords vital importance and significant challenges (Marano et al., 2023). This essay delves into the significance of earthquake multi-elegance detection using AI and explores the critical and demanding situations associated with its implementation.

Importance of Earthquake Multi-elegance Detection:

1. improving early caution structures: Earthquake early warning systems depend upon the rapid and accurate identity and category of seismic occasions. AI algorithms can system extensive amounts of data from seismic sensors to provide real-time alerts, enhancing the response time for disaster mitigation (Asadi et al., 2023).
2. decreasing fake positives and negatives: The potential to classify earthquakes into exclusive classes, which include their importance, intensity, and vicinity, enables an extra unique estimation of their potential impact. AI algorithms can analyze numerous information resources, such as seismograms,

satellite imagery, and geographical facts, to improve type accuracy and limit fake positives and negatives(Kaur et al., 2022).

2.2 Significance of Multi-Class Detection

Earthquakes are natural screw-ups that can cause widespread destruction and lack of existence. Detecting earthquakes as it should be and efficiently is paramount to mitigate their impact and save lives. In recent years, the utility of artificial intelligence (AI) techniques has proven promise in improving the detection and type of earthquakes (Pant et al., 2023; Yuvaraj, Srihari, Dhiman et al, 2021). Multi-elegance detection The usage of AI algorithms has emerged as an enormous method to differentiate different varieties of earthquakes, supplying valuable insights into their characteristics and behavior (Kannan et al., 2021; Rajalakshmi et al., 2022; Yuvaraj, Karthikeyan, & Praghash, 2021). Historically, earthquake detection has relied on seismographs and manual analysis via seismologists. At the same time practical, this technique is time-eating and regularly limited by human errors. With the advancement of AI generation, the sphere of earthquake detection has passed through a transformation (Khan et al., 2022). AI algorithms can technique sizable amounts of seismic information from numerous sensors in actual time, allowing for a faster and greater correct identity of earthquakes. Through machine-getting-to-know strategies, AI fashions can study from historical statistics, allowing the detection of seismic occasions even in the absence of clear patterns. The importance of multi-magnificence detection lies in its ability to distinguish earthquakes based on unique attributes, including magnitude, depth, and focal mechanism(Chowdhuri et al., 2022; Gunjan et al., 2022).

3. APPLICATIONS OF AI IN EARTHQUAKE DETECTION

1. AI-enhanced Seismic information evaluation: AI algorithms can analyze seismic alerts in real time, permitting the identification of earthquake events appropriately. AI can differentiate between natural seismic pastime and other noise sources by inspecting the amplitude, frequency, and duration of ground vibrations. Deep getting-to-know networks, consisting of Convolutional Neural Networks (CNN) and Recurrent Neural Networks (RNN), excel in extracting and studying complex features from seismic facts, main to advanced detection precision.
2. Early warning structures: AI-based algorithms can forecast the likelihood and value of earthquakes, enabling the development of early caution structures. Those structures rely upon gadgets studying fashions educated on historical seismic facts, geological information, and auxiliary functions like meteorological statistics. Using constant monitoring of seismic patterns and anomalies, AI can expect incoming earthquakes and difficulty timely warnings, considering evacuation and mitigation measures.
3. Multi-elegance Detection: AI has improved multi-class earthquake detection, categorizing earthquakes into extraordinary training based on their traits. Traditional techniques generally centered on binary detection (earthquake vs. non-earthquake). However, AI algorithms allow for more subtle categorization. For instance, earthquakes may be categorized primarily based on their value, depth, focal mechanism, and vicinity. AI models can research styles and correlations among seismic information and earthquake instructions, enabling accurate class and presenting valuable insights for seismic chance evaluation.

4. facts Fusion and Integration: AI enables the combining and fusion of facts from more than one asset for superior earthquake detection. Seismic information may be blended with data from other sensors, which includes worldwide Positioning System (GPS) receivers and InSAR (Interferometric Synthetic Aperture Radar), to create a comprehensive view of seismic interest. Machine mastering strategies like ensemble gaining knowledge of and Bayesian networks can integrate statistics from exclusive resources and enhance the reliability and robustness of earthquake detection structures..

3.1 Traditional Approaches vs. AI-Based Approaches

Earthquakes are one of the most catastrophic natural failures that occur frequently around the arena. The capacity to as it should be and efficiently detect earthquakes plays a crucial function in predicting and mitigating the effect of these events on human lives and infrastructure. In recent years, there has been a growing interest in the usage of artificial intelligence (AI) techniques for earthquake detection because of their capability blessings over conventional tactics. This essay will discuss the differences between conventional methods and AI-primarily based techniques for earthquake multi-elegance detection and the usage of AI. Conventional processes for earthquake detection generally rely on seismographs, which are sensitive gadgets that file floor vibrations caused by seismic waves. Those devices degree the amplitude and frequency of the recorded waves, permitting seismologists to investigate their traits and distinguish seismic events. However, traditional processes suffer from several limitations. First, they closely rely on guide interpretation and analysis of seismic information, which is time-consuming and vulnerable to human errors. Moreover, conventional tactics regularly involve accurately detecting and classifying seismic events, particularly in complex situations with a couple of simultaneous earthquakes.

However, AI-based tactics utilize advanced device mastering algorithms to process enormous volumes of seismic facts fast and appropriately, enabling actual-time earthquake detection and class.

3.2 Multi-Class Detection Techniques

Earthquakes are natural disasters that could cause mammoth harm to human lives and infrastructure. It's essential to increase accurate and robust earthquake detection strategies to mitigate the dangers of these activities. Synthetic Intelligence (AI) has emerged as a promising tool in earthquake detection, specifically with the advancements in multi-elegance detection strategies. Multi-magnificence detection techniques in earthquake detection refer to the ability of AI systems to classify earthquakes primarily based on various parameters, including importance, intensity, and region. Traditionally, earthquake detection relied on seismometers, mechanical devices that measure ground movement. However, seismometers have obstacles, including their incapacity to discover the earthquake epicenter and distinguish among distinctive varieties of earthquakes. AI techniques, particularly system-gaining knowledge of algorithms, offer a powerful approach to multi-elegance earthquake detection. Those algorithms may be skilled on vast datasets of seismic records and discover ways to identify styles and functions that distinguish different classes of earthquakes. The schooling manner includes feeding the rules with categorized data, wherein earthquake occasions are classified primarily based on their specific attributes. One example of a multi-magnificence detection approach in using AI is using convolutional neural networks (CNNs) for earthquake classes. CNNs are neural networks that are especially powerful in extracting functions from enter statistics with grid-like structures, which include seismic data..

4. KEY COMPONENTS OF AI IMPLEMENTATION FOR MULTI-CLASS EARTHQUAKE DETECTION

1. facts collection and Preprocessing: The muse of any AI implementation for earthquake detection lies inside the information. Amassing seismic waveforms from numerous sensors close to the fault lines is essential. Getting the right of entry to a complete dataset with categorized examples from distinct classes of earthquakes, including regular earthquakes, aftershocks, foreshocks, and tectonic events, is critical for schooling the AI model. Moreover, information preprocessing strategies, including fact cleansing, normalization, and noise reduction, must be hired to ensure information is best and consistent.
2. Feature extraction and representation: Changing uncooked seismic waveform facts into meaningful functions is essential for effective earthquake detection. Characteristic extraction strategies inclusive of Fourier transforms, wavelet transforms, and time-frequency evaluation can help seize applicable facts about earthquake characteristics.

4.1 Data Preprocessing and Feature Extraction

Records Preprocessing:

Preprocessing seismic facts include remodeling them into a usable format while casting off the noise and inappropriate capabilities. The following steps are commonly employed in statistics preprocessing for earthquake detection:

1. Filtering: using numerous filters, such as a band-bypass filter out, to put off frequencies outside the variety of hobbies, efficiently lowering noise and enhancing sign fine.
2. Upsampling and Downsampling: Adjusting the sampling price of the seismic records to suit the necessities of the AI model. Upsampling can be required to keep away from statistics loss, while downsampling can lessen computational complexity.
3. Noise reduction: using strategies like wavelet denoising, median filtering, or spectral subtraction to dispose of random noise and enhance the sign-to-noise ratio for correct evaluation.
4. Normalization: Scaling the data to a widespread range, including 0 and 1, to ensure steady record illustration and save you bias toward certain functions.
5. facts Augmentation: increasing the size and diversity of the dataset by producing extra samples through strategies inclusive of time transferring, amplitude scaling, and adding noise. This facilitates the enhancement of the generalization capability of the AI version.

Function Extraction:

Function extraction includes identifying the seismic facts' most applicable and informative traits, which can aid in earthquake type. Effective characteristic extraction guarantees the AI model can correctly differentiate between exclusive earthquake training. Some commonly used strategies consist of the following:

1. Time-domain features: those functions capture residences associated with the temporal behavior of the seismic signal. Examples include statistical measures such as mean, widespread deviation,

skewness, and kurtosis, in addition to extra advanced functions like dominant frequencies, power content material, and length of seismic events.

2. Frequency-domain features: through converting seismic statistics into the frequency domain using strategies like Fourier transform, or Wavelet transform, functions such as top frequency, spectral centroid, and spectral energy distribution may be extracted. Those functions provide insights into the frequency content material and feature frequencies associated with one-of-a-kind earthquake lessons.

4.2 Selection of AI Algorithms for Multi-Class Detection

Understanding Multi-class Earthquake Detection:

Multi-magnificence earthquake detection involves the identity and category of various seismic activities, such as foreshocks, mainshocks, aftershocks, and non-earthquake occasions, based totally on seismic signals. This task calls for advanced sample reputation abilities, making it a perfect application for AI algorithms.

AI Algorithms for Multi-class Earthquake Detection:

The below figure provides a life cycle for selection of AI algorithm for multi class detection is shown below,

Figure 1. Life cycle for selection of AI algorithm for multi class detection

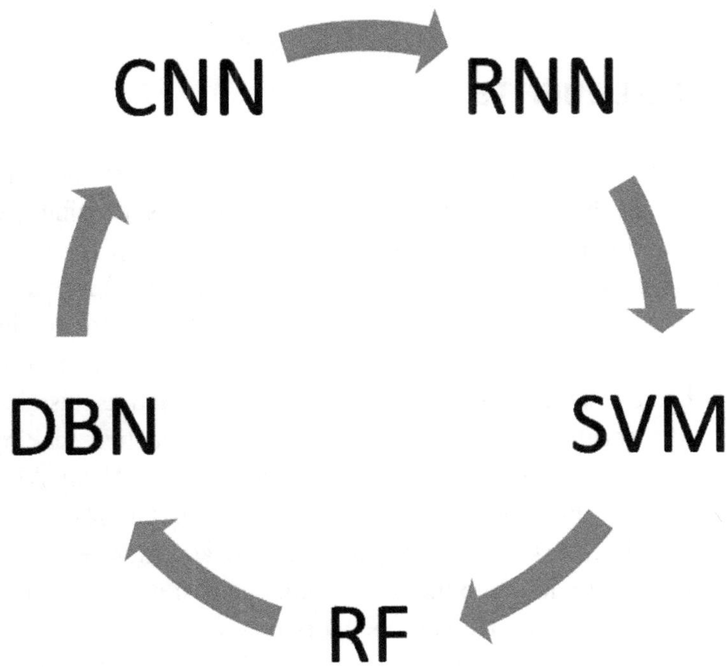

1. Convolutional Neural Networks (CNNs): CNNs have validated robust image popularity responsibilities, and their applicability extends to seismic signal evaluation. By applying convolutions to seismic signal spectrograms, CNNs can analyze patterns related to specific earthquake sorts, enabling correct multi-magnificence categories. However, CNNs require many classified training records to reap the finest performance.
2. Recurrent Neural Networks (RNNs): RNNs are well-suitable for time series data evaluation, making them an appropriate preference for earthquake detection. By employing long short-term reminiscence (LSTM) gadgets, RNNs can capture temporal dependencies in seismic alerts, aiding in identifying earthquake occasions inside a non-stop circulation of information.
3. guide Vector Machines (SVMs): SVMs are broadly used for class responsibilities and may be utilized for earthquake detection. By mapping seismic functions to an excessive-dimensional feature area, SVMs can efficaciously separate one-of-a-kind earthquake training. However, SVMs may be less robust while coping with somewhat imbalanced datasets.
4. Random Forests (RFs): RFs are an ensemble learning approach that mixes a couple of selection trees to achieve correct classification. With their capability to address excessive-dimensional function areas and imbalanced statistics, RFs can successfully differentiate between extraordinary earthquake classes. RFs additionally provide insight into which features are maximum discriminative for classification.
5. Deep gaining knowledge of strategies: Deep getting-to-know algorithms, which include deep notion networks (DBNs) and deep autoencoders, have proven excellent performance in numerous domains. By using leveraging multiple layers of nonlinear ameliorations, these algorithms can efficaciously seize.

5. ADVANTAGES AND BENEFITS

Advances in the era have opened up new possibilities for various sectors, including disaster management. Earthquakes are one of the most devastating natural calamities, causing a vast lack of life and infrastructure. Detecting earthquakes as they should be and early can play an enormous function in mitigating their impact. Artificial Intelligence (AI) has emerged as a powerful device in earthquake detection, presenting numerous blessings and revolutionizing the field. One of the fundamental blessings of using AI for earthquake detection is it's potential to procedure enormous quantities of statistics in real-time. Conventional earthquake detection methods depend upon evaluating seismic indicators recorded by seismometers. However, seismometers generate vast non-stop statistics, making it hard for human operators to research and detect earthquakes correctly and unexpectedly. Using AI algorithms, computers can quickly analyze considerable amounts of data, figuring out earthquake styles and distinguishing them from historical past noise. This functionality allows early and more excellent correct detection, allowing for timely response and mitigation measures. Other gain of using AI for earthquake detection is its ability to automate the technique. Human operators may pass over subtle earthquake indicators amidst other noise or pastime. AI algorithms are designed to detect even the faintest seismic indicators and might continuously display data streams..

5.1 Enhanced Precision and Recall

1. information Earthquake Detection:
    a. traditional earthquake detection strategies rely on seismographs and human know-how to interpret seismic indicators. Even though they provide precious information, they often need more precision and recall.
    b. importance of precision and do not forget: Precision refers to accurately identifying actual earthquake activities. At the same time, consider measures of the ability to stumble on all earthquakes without lacking any. Better precision avoids false positives, while improved keeping in mind reduces fake negatives.
2. utilizing artificial Intelligence for Earthquake Multi-class Detection:
    a. system studying algorithms: AI strategies, especially machine learning, permit computers to examine styles from reasonable-sized amounts of information. These algorithms may be educated to classify seismic indicators and differentiate between specific varieties of earthquakes.
    b. characteristic extraction and selection: AI allows for identifying and extracting applicable capabilities from seismic statistics, which could be a resource in distinguishing exceptional earthquake lessons. By selecting the most informative capabilities, AI algorithms can enhance precision and keep it in mind.
    c. schooling facts: AI fashions require big and various datasets to learn effectively. Earthquake catalogs, historic seismic information, and simulated information can be utilized to train AI models for earthquake detection, thereby improving the device's precision and bearing in mind.
3. AI-stronger Precision in Earthquake Multi-elegance Detection:
    a. reduced false positives: while trained with tremendous facts, AI fashions can identify patterns that traditional systems can also miss. This guarantees a better diploma of accuracy, warding off false positives that would result in needless evacuation or aid allocation.
    b. improved event type: AI algorithms can better differentiate between exceptional kinds of earthquakes. For instance, they can discern between tectonic earthquakes, volcanic earthquakes, and synthetic tremors caused by human sports, which includes mining..

5.2 Real-Time Detection Capabilities

Earthquakes are natural disasters that can cause considerable harm to property and a lack of human lives. Therefore, it's crucial to have real-time detection capabilities to correctly pick out and classify earthquake activities to enable early warning structures and emergency reactions. Artificial Intelligence (AI) has emerged as an effective device in earthquake detection, providing advanced accuracy and efficiency in detecting seismic events. This essay will speak about the real-time detection abilities of AI in earthquake multi-class detection and its significance in mitigating the impacts of earthquakes. Actual-time detection of earthquakes is crucial to provide timely warnings and alerts to affected regions. Traditional earthquake early caution (EEW) structures depend on seismic networks to hit upon and classify earthquake occasions. However, those systems frequently suffer from diverse boundaries, such as confined community insurance, hardware dependencies, and delays in fact processing. AI algorithms, on the other hand, can overcome those obstacles and provide real-time detection abilities. Synthetic intelligence algorithms, in particular deep gaining knowledge of fashions, have shown high-quality fulfillment in earthquake multi-class detection. These models are skilled on large seismic datasets and can, as they should be, classify exceptional seismic events, including earthquakes and their variations, such as aftershocks and foreshocks.

6. CHALLENGES AND LIMITATIONS

Earthquakes pose significant threats to human existence and infrastructure, making their well-timed detection and correct type paramount to effective catastrophe control. Traditionally, earthquake detection has trusted seismologists manually analyze seismic records, a time-ingesting and labor-extensive process. However, the appearance of synthetic Intelligence (AI) has supplied promising solutions for automatic earthquake detection and classification. Notwithstanding its capacity, various challenges and boundaries are associated with earthquake multi-magnificence detection and the usage of AI. One crucial mission lies in the availability and exceptional of seismic statistics. AI algorithms heavily rely on large quantities of accurate and dependable data to educate models correctly. However, seismic records availability varies throughout specific areas and countries, with some regions needing more vital infrastructure to accumulate seismic measurements. Furthermore, even if statistics are available, they can vary, mainly due to incomplete or corrupt datasets that avert the accuracy of AI fashions. Addressing this challenge requires investing in developing and improving seismic tracking systems internationally and ensuring records' exceptional requirements are met. Any other undertaking pertains to the complexity and diversity of seismic signals. Earthquakes showcase an extensive range of traits, which include exceptional magnitudes, focal depths, and geological situations..

6.1 Data Quality and Noise Handling

Earthquakes are natural screw-ups that could harm human lifestyles and infrastructure. Detecting and tracking earthquakes is essential for early caution systems and emergency response planning. In recent years, artificial intelligence (AI) algorithms have been increasingly used for earthquake detection and class, taking benefit of their potential to investigate massive volumes of seismic information swiftly. But, to gain accurate results, it is vital to recollect information first-class and noise dealing with techniques in earthquake multi-class detection using AI. Records excellence is critical to AI algorithms' overall performance and reliability. The seismic information gathered from numerous international sensors is often subject to noise, artifacts, and inconsistencies that could lead to false positives or negatives in earthquake detection. Consequently, it's crucial to preprocess and beautify the records before feeding them into AI fashions. One technique to enhance facts is through statistics filtering techniques. Filtering involves removing unwanted noise from the seismic statistics, even preserving the vibration indicators related to earthquakes. Numerous filtering strategies are available, such as high-skip, low-pass, or band-skip filters, every tailor-made to target particular frequency levels. By applying appropriate filtering techniques, the AI algorithms can know the relevant seismic indicators, leading to more significant and accurate consequences.

7. CASE STUDIES: SUCCESSFUL AI IMPLEMENTATIONS IN MULTI-CLASS EARTHQUAKE DETECTION

Synthetic Intelligence (AI) revolutionizes how we approach diverse challenges, including detecting and predicting natural disasters. In recent years, AI improvements have substantially aided earthquake detection and analysis. This essay seeks to discover and compare a successful case research where AI has been efficaciously implemented for multi-magnificence earthquake detection. Those case studies highlight the ability of AI algorithms to enhance early caution structures and improve disaster management strategies.

Case look at 1: ShakeNet

One remarkable instance of a hit AI implementation for multi-elegance earthquake detection is ShakeNet, developed by researchers at Google. ShakeNet is an AI model that utilizes system gaining knowledge of techniques, which include convolutional neural networks (CNN) and transfer getting to know. Skilled in significant historic earthquake facts, ShakeNet demonstrates excessive accuracy and pace in detecting seismic sports, classifying them into diverse intensity stages.

This version tackles the hassle of multi-elegance type by categorizing earthquakes into depth levels primarily based on the changed Mercalli intensity (MMI) scale. By assigning a specific MMI cost to every detection, ShakeNet gives treasured insights into the ability effect of an earthquake, permitting businesses and catastrophe reaction groups to prepare and reply thus..

7.1 Examples of Successful Applications

Example 1: Early Earthquake Detection and Alert structures

One key place in which AI has proved successful is in early earthquake detection and alert structures. Conventional earthquake tracking structures frequently suffer from latency issues, providing constrained time to take necessary precautions. Via using AI algorithms, seismic facts from numerous sensors can be analyzed in actual time, substantially decreasing reaction time. For instance, the Early Earthquake Caution System (EEWS) advanced through the Japan Meteorological Organisation leverages AI to robotically discover the initial P-wave, estimate the importance, and issue warnings within seconds. This version has been credited for saving many lives through the 201 1 Tohoku earthquake in 2011, as it provided essential seconds of superior caution to residents, allowing them to seek refuge or evacuate.

example 2: category of Earthquake importance and source

another successful application of AI in earthquake multi-class detection is the ability to classify earthquake magnitudes and their supply traits correctly. With the development of AI algorithms, it's now possible to investigate seismic waveforms and perceive the significance and source of an earthquake. Using training system learning models on massive datasets, AI algorithms can examine to differentiate between numerous earthquake classes (ordinary, thrust, and strike-slip earthquakes) and estimate their magnitudes. This record is vital for emergency reaction businesses because it enables assessing the capacity impact and making plans for suitable alleviation measures. For instance, the United States Geological Survey (USGS) employs AI algorithms to categorize earthquakes in actual time, aiding in the robust allocation of sources for the duration of catastrophe reaction efforts..

8. CONCLUSION

In conclusion, in earthquake multi-magnificence detection, artificial intelligence (AI) holds the titanic ability to improve the accuracy and efficiency of earthquake monitoring and early warning systems. Via the combination of AI algorithms and seismic records evaluation, this era has the potential to not only discover and find earthquakes but also classify them in step with their significance and depth. This research reveals that AI-based earthquake detection methods can efficaciously differentiate between seismic occasions and non-earthquake resources, including noise or human activities. Using machine-getting-to-know strategies, these algorithms can continuously learn from new records, enhancing their detection talents. This adaptive nature of AI allows for more accurate and dependable earthquake detection, reducing false alarms and increasing the overall efficiency of earthquake tracking systems. Moreover, incorporating AI technologies in early warning structures can considerably enhance the time for arrangements and evacuations. By predicting earthquake occurrences in actual time, these systems can provide essential statistics to the general public and applicable government, saving lives and minimizing the damage caused by seismic events. But, regardless of its capacity benefits, there are some barriers and demanding situations associated with AI-based total earthquake detection. The need for more classified seismic statistics for schooling and checking out functions is an enormous impediment, as it hinders the improvement and assessment of accurate fashions..

8.1 Summary of Key Findings

Earthquakes are herbal disasters that cause severe damage to infrastructure, lack of lives, and affect the general well-being of groups. Early detection and accurate category of earthquakes are vital, which will minimize the effect of these catastrophic activities. Recent artificial intelligence (AI) improvements have shown promising results in earthquake multi-magnificence detection, providing precious insights into earthquake conduct styles and permitting timely reaction measures. One of the key findings in earthquake multi-elegance detection and AI usage is the ability to classify earthquakes into extraordinary categories based totally on their importance, intensity, and area. Conventional seismology strategies commonly focus on detecting earthquakes based on an unmarried seismic event. However, AI algorithms can investigate a significant quantity of information, which includes ancient earthquake facts, geological capabilities, and actual-time seismic signals, to become aware of patterns and classify earthquakes into multiple instructions. This permits scientists to gain a higher know-how of the numerous forms of earthquakes and their traits, which may be a resource in predicting the conduct and capacity effect of destiny earthquakes. Some other enormous finding is the capability of AI algorithms to stumble on earthquake precursors and anomalies in seismic indicators that are not without difficulty perceivable by human professionals. Earthquake precursor refers to the changes in the physical parameters of the Earth's crust that arise previous to the occurrence of an earthquake..

8.2 Recommendations for Future Research

Earthquakes are natural failures that could cause enormous harm to lives and infrastructure. Detecting and predicting earthquakes as they should be is essential to mitigate their effect and save lives. In recent years, artificial intelligence (AI) usage in earthquake detection has gained traction because of its ability to analyze big datasets and offer brief and accurate results. However, more research in this field still needs to be done to enhance earthquake multi-magnificence detection through AI. One piece of advice for destiny research is to discover the use of the latest AI algorithms specially designed for earthquake detection. Even as modern-day AI algorithms, which include deep studying and neural networks, have proven promise in earthquake detection, they will not be optimized for this precise venture. Developing specialized AI algorithms that can successfully examine seismic statistics and differentiate between specific earthquake training may cause greater accurate and dependable detection. Another place for future studies is mixing more than one information asset for earthquake detection. Presently, most AI-based earthquake detection systems rely entirely on seismic data. But earthquakes can also produce other signals, including electromagnetic waves or ground deformation, that may be detected and used for prediction. By integrating a couple of data resources into AI algorithms, researchers can grow the accuracy and reliability of earthquake detection structures.

REFERENCES

Aldini, I., Hidayat, R., Permanasari, A. E., & Ramdhani, A. (2022, December). Classification of Tsunami Warning Level using Artificial Neural Network and its Comparison in Southern Java Region. In *2022 5th International Seminar on Research of Information Technology and Intelligent Systems (ISRITI)* (pp. 607-612). IEEE. 10.1109/ISRITI56927.2022.10052907

Asadi, A., Baise, L. G., Sanon, C., Koch, M., Chatterjee, S., & Moaveni, B. (2023). Semi-Supervised Learning Method for the Augmentation of an Incomplete Image-Based Inventory of Earthquake-Induced Soil Liquefaction Surface Effects. *Remote Sensing (Basel)*, *15*(19), 4883. doi:10.3390/rs15194883

Babu, N. V., & Kanaga, E. G. M. (2022). Sentiment analysis in social media data for depression detection using artificial intelligence: A review. *SN Computer Science*, *3*(1), 1–20. doi:10.100742979-021-00958-1 PMID:34816124

Badri, A. M. (2022, December). Litho-Fluid Facies Modelling Based on Integration of Seismic Inversion and Well Data: A Supervised Deep-Learning Approach. In *AGU Fall Meeting Abstracts* (Vol. 2022, pp. IN12C-0277).

Chowdhuri, I., Pal, S. C., Saha, A., Chakrabortty, R., & Roy, P. (2022). Mapping of earthquake hotspot and coldspot zones for identifying potential landslide hotspot areas in the Himalayan region. *Bulletin of Engineering Geology and the Environment*, *81*(7), 257. doi:10.100710064-022-02761-5

Derras, B., & Makhoul, N. (2022). An overview of the infrastructure seismic resilience assessment using artificial intelligence and machine-learning algorithms. In *Proceedings of the International Conference on Natural Hazards and Infrastructure*. IEEE.

Dogan, G., Arslan, M. H., & Ilki, A. (2023). Detection of damages caused by earthquake and reinforcement corrosion in RC buildings with deep transfer learning. *Engineering Structures*, *279*, 115629. doi:10.1016/j.engstruct.2023.115629

Giacco, G., Mariniello, G., Marrone, S., Asprone, D., & Sansone, C. (2022, May). Toward a system for post-earthquake safety evaluation of masonry buildings. In *International Conference on Image Analysis and Processing* (pp. 312-323). Cham: Springer International Publishing. 10.1007/978-3-031-06430-2_26

Gunjan, V. K., Vijayalata, Y., Valli, S., Kumar, S., Mohamed, M. O., & Saravanan, V. (2022). Machine learning and cloud-based knowledge graphs to recognize suicidal mental tendencies. *Computational Intelligence and Neuroscience*, *2022*, 2022. doi:10.1155/2022/3604113 PMID:35341205

Hajibabaee, P., Pourkamali-Anaraki, F., & Hariri-Ardebili, M. A. (2023). Dimensionality reduction techniques in structural and earthquake engineering. *Engineering Structures*, *278*, 115485. doi:10.1016/j.engstruct.2022.115485

Kannan, S., Dhiman, G., Natarajan, Y., Sharma, A., Mohanty, S. N., Soni, M., Easwaran, U., Ghorbani, H., Asheralieva, A., & Gheisari, M. (2021). Ubiquitous vehicular ad-hoc network computing using deep neural network with iot-based bat agents for traffic management. *Electronics (Basel)*, *10*(7), 785. doi:10.3390/electronics10070785

Kaur, J., Saxena, J., Shah, J., & Yadav, S. P. (2022, May). Facial Emotion Recognition. In *2022 International Conference on Computational Intelligence and Sustainable Engineering Solutions (CISES)* (pp. 528-533). IEEE.

Khan, S., Saravanan, V. N. G. C., Lakshmi, T. J., Deb, N., & Othman, N. A. (2022). Privacy protection of healthcare data over social networks using machine learning algorithms. *Computational Intelligence and Neuroscience*, *2022*, 2022. doi:10.1155/2022/9985933 PMID:35371203

Lu, A., & Honarvar Shakibaei Asli, B. (2023). Seismic Image Identification and Detection Based on Tchebichef Moment Invariant. *Electronics (Basel)*, *12*(17), 3692. doi:10.3390/electronics12173692

Ma, X., & Chen, T. (2022). Small seismic events in Oklahoma detected and located by machine learning–based models. *Bulletin of the Seismological Society of America*, *112*(6), 2859–2869. doi:10.1785/0120220029

Marano, G. C., Rosso, M. M., Aloisio, A., & Cirrincione, G. (2023). Generative adversarial networks review in earthquake-related engineering fields. *Bulletin of Earthquake Engineering*, 1–52. doi:10.100710518-023-01645-7

Murti, M. A., Junior, R., Ahmed, A. N., & Elshafie, A. (2022). Earthquake multi-classification detection based velocity and displacement data filtering using machine learning algorithms. *Scientific Reports*, *12*(1), 21200. doi:10.103841598-022-25098-1 PMID:36482200

Pant, G., Bahuguna, R., Pandey, S., Gehlot, A., Yadav, S. P., & Pachauri, R. K. (2023, April). Intelligent Automated Interference for the Protection of Road Safety. In *2023 International Conference on Computational Intelligence, Communication Technology and Networking (CICTN)* (pp. 87-91). IEEE. 10.1109/CICTN57981.2023.10141086

Rajalakshmi, M., Saravanan, V., Arunprasad, V., Romero, C. T., Khalaf, O. I., & Karthik, C. (2022). Machine Learning for Modeling and Control of Industrial Clarifier Process. *Intelligent Automation & Soft Computing*, *32*(1). doi:10.32604/iasc.2022.021696

Rao, A., Jung, J., Silva, V., Molinario, G., & Yun, S. H. (2023). Earthquake building damage detection based on synthetic-aperture-radar imagery and machine learning. *Natural Hazards and Earth System Sciences*, *23*(2), 789–807. doi:10.5194/nhess-23-789-2023

Sajan, K. C., Bhusal, A., Gautam, D., & Rupakhety, R. (2023). Earthquake damage and rehabilitation intervention prediction using machine learning. *Engineering Failure Analysis, 144*, 106949. doi:10.1016/j.engfailanal.2022.106949

Shakeel, M., Nishida, K., Itoyama, K., & Nakadai, K. (2022). 3d convolution recurrent neural networks for multi-label earthquake magnitude classification. *Applied Sciences (Basel, Switzerland), 12*(4), 2195. doi:10.3390/app12042195

Yuan, C., Ni, Y., Lin, Y., & Denolle, M. (2023). Better Together: Ensemble Learning for Earthquake Detection and Phase Picking. *IEEE Transactions on Geoscience and Remote Sensing, 61*, 1–17. doi:10.1109/TGRS.2023.3320148

Yuvaraj, N., Karthikeyan, T., & Praghash, K. (2021). An improved task allocation scheme in serverless computing using gray wolf Optimization (GWO) based reinforcement learning (RIL) approach. *Wireless Personal Communications, 117*(3), 2403–2421. doi:10.100711277-020-07981-0

Yuvaraj, N., Srihari, K., Dhiman, G., Somasundaram, K., Sharma, A., Rajeskannan, S. M. G. S. M. A., Soni, M., Gaba, G. S., AlZain, M. A., & Masud, M. (2021). Nature-inspired-based approach for automated cyberbullying classification on multimedia social networking. *Mathematical Problems in Engineering, 2021*, 1–12. doi:10.1155/2021/6644652

Chapter 9
Earth Sciences Unleashed: A Comprehensive Exploration of AI Applications

Shobana S.
Department of Information Technology, R.M.K. Engineering College, Chennai, India

Muthumari A.
Department of Computer Science and Engineering, Anna University, Regional Campus, Madurai, India

Subashree S.
https://orcid.org/0000-0003-2034-847X
Department of Computer Science and Engineering, E.G.S. Pillay Engineering College, Chennai, India

Elumalai G.
https://orcid.org/0000-0003-0821-4022
Department of Electronics and Communication Engineering, Panimalar Engineering College, India

ABSTRACT

The emergence of synthetic artificial intelligence (AI) in recent years has revolutionized numerous sectors, which includes Earth Sciences. AI has recently emerged as a powerful tool in Earth Sciences, enabling scientists to analyze enormous amounts of records, broaden models, and make more accurate predictions. AI offers powerful analytical gear to process large quantities of facts and perceive complicated patterns that might, in any other case, be challenging to discover. This ability to investigate and link disparate information assets permits scientists to benefit from complete insights into Earth's structures, assisting us in understanding the using forces at the back of herbal phenomena like earthquakes hurricanes. The episode discusses numerous AI techniques, inclusive of machine getting-to-know, deep mastering, and statistics mining and their application in Earth Sciences. It explores how these strategies can decorate statistics processing, automate complex obligations, and uncover hidden styles in massive datasets.

DOI: 10.4018/979-8-3693-1850-8.ch009

1. INTRODUCTION

The Earth Sciences, encompassing geology, climatology, oceanography, and, more importantly, have long relied on an aggregate of facts evaluation and human know-how to apprehend the complicated systems of our planet. However, the appearance of synthetic Intelligence (AI) has revolutionized the sector, unleashing a wave of the latest opportunities and commencing doorways to formerly not possible advancements (Allioui & Mourdi, 2023). This essay explores the comprehensive exploration of AI applications in Earth Sciences and discusses the potential blessings, demanding situations, and destiny prospects related to this remarkable technological jump (Gao, 2023).AI, via its potential to investigate reasonable-sized quantities of facts and locate styles, is appreciably more suitable to our understanding of numerous Earth Sciences disciplines. In geology, machine-getting-to-know algorithms have been established as instrumental in mapping geological formations and reading seismic facts to predict and discover potential mineral sources or geological risks (Badini et al., 2023). These AI-driven fashions can, as it should be, expect the incidence of geological activities, together with earthquakes or landslides, thereby assisting in disaster management and prevention.Further, in climatology, AI-powered fashions have stepped forward predictions of weather patterns, taking into account more excellent correct forecasting and thereby aiding in catastrophe preparedness..

1.1 Background and Significance

The emergence of synthetic Intelligence (AI) in recent years has revolutionized numerous sectors, which includes Earth Sciences. Earth Sciences Unleashed is a groundbreaking initiative that pursues to discover and make use of AI programs in this subject. This comprehensive exploration promises to unearth new insights, revolutionize records analysis, beautify predictive fashions, and deal with demanding international situations related to climate trade, herbal failures, and resource control. Know-how the historical past and significance of this undertaking is crucial to realizing the transformative capacity it holds for Earth Sciences (Trabelsi et al., 2023).Earth Sciences embody numerous disciplines, which include geology, meteorology, hydrology, and ecology. Traditionally, these fields rely on state-of-the-art records collection methods, guide evaluation, and complex laptop simulations. However, the fast advancements in AI have supplied new gear and strategies to analyze massive datasets and make predictions with unparalleled accuracy (Xu et al., 2023).The capability packages of AI in Earth Sciences are vast and various. One sizable application is the evaluation of remote sensing statistics. Satellites ready with superior sensors seize a substantial quantity of data about the Earth's floor, surroundings, and oceans. AI algorithms can efficiently technique these records to perceive styles, track modifications over time, and hit upon anomalies.

1.2 Objectives of the Chapter

One of the number one targets of the bankruptcy is to spotlight how AI applications can provide a more profound expertise of the tricky and interconnected systems that govern our planet. Earth Sciences embody a wide variety of disciplines, which include geology, meteorology, oceanography, and climatology, which observe the Earth's physical additives, climate patterns, and the dynamic procedures happening inside and among them (Hallema, 2022). AI offers powerful analytical gear to process large quantities of facts and perceive complicated patterns that might, in any other case, be challenging to discover. This ability to investigate and link disparate information assets permits scientists to benefit from complete insights into Earth's structures, assisting us in understanding the using forces at the back of herbal phenomena like earthquakes hurricanes (Dematties et al., 2023).

1.3 Overview of AI in Earth Sciences

In recent years, the field of artificial intelligence (AI) has witnessed huge advancements and breakthroughs, revolutionizing diverse industries and sectors. The Earth sciences area is not an exception, as AI gives a comprehensive exploration of programs that can transform the manner we examine, apprehend, and control our planet. From predicting herbal disasters to reading environmental changes, AI has emerged as a powerful device for researchers, scientists, and policymakers in Earth sciences (Aithal, 2023).One of the critical regions wherein AI has made a profound effect is the prediction and detection of herbal disasters. Traditional strategies of forecasting earthquakes, volcanic eruptions, or tsunamis relied closely on seismological and geological records. But, AI algorithms, coupled with advanced sensor networks and satellite TV for PC imagery, can provide real-time evaluation and early warnings for disaster management (Khan et al., 2022). Through analyzing historical information, device-mastering models can become aware of patterns and correlations that are invisible to human observers. This allows higher predictions and preparedness, in the end supporting in saving lives and minimizing damages.Furthermore, AI has significantly better information and monitoring of environmental adjustments. Weather alternatives, biodiversity loss, and deforestation are many of the most urgent troubles facing our planet (Gunjan et al., 2022).

2. FUNDAMENTALS OF AI IN EARTH SCIENCES

The sphere of Earth sciences has been revolutionized by the emergence of artificial intelligence (AI) technology. AI is the branch of PC science that deals with the development of intelligent machines and structures capable of performing responsibilities that generally require human intelligence. It has located numerous applications in Earth sciences, permitting scientists and researchers to address complicated problems, analyze substantial quantities of information, and make correct predictions (Chowdhuri et al., 2022). The e-book Earth Sciences Unleashed: A Complete Exploration of AI Applications delves into the basics of AI in Earth sciences and offers a complete assessment of its extensive-ranging applications in various sub-disciplines.One of the fundamental factors of AI in Earth sciences is its capability to method vast volumes of facts and extract significant statistics (Al Dulaijan & Alajaji, 2022). Earth sciences generate enormous quantities of data from numerous assets along with satellites, remote sensing devices, ground-based total sensors, and numerical fashions. AI algorithms and techniques can analyze these statistics in actual time, figuring out styles, tendencies, and anomalies, which could be otherwise not possible for human researchers to system manually. With AI, scientists can uncover hidden relationships and gain insights into complex Earth procedures such as weather exchange, geological phenomena, and natural disasters.

2.1 Brief Introduction to Artificial Intelligence

Synthetic Intelligence (AI) has emerged as a transformative era with extensive packages throughout diverse industries, including the field of Earth sciences. The mixing of AI structures and strategies has revolutionized the manner in which researchers and scientists analyze, interpret, and make predictions about complex environmental phenomena. From assessing climate change patterns to studying geophysical methods, AI has unleashed a complete exploration of programs that have the ability to boost our expertise of Earth structures and inform sustainable practices (George & Wooden, 2023; Srivastava et al., 2023; Zhao et al., 2023).Initially, AI allows the processing and analysis of extensive amounts of facts, which is essential in Earth sciences. The sheer volume and complexity of information collected from satellites, weather stations, and other tracking devices gift a full-size undertaking for classic analytical strategies (Iqbal et al., 2023). AI techniques, which include device-gaining knowledge of and deep learning algorithms, can efficiently deal with extensive records units, become aware of patterns, and extract meaningful insights that would, in any other case, pass omitted. This, in turn, has facilitated the improvement of accurate fashions for predicting climate patterns, reading soil composition, and assessing herbal screw-ups.One of the critical areas in which AI has made good-sized contributions is in climate modeling and forecasting. Climate exchange is a pressing worldwide problem, and know-how and its implications require superior computational fashions. AI algorithms enable scientists to simulate higher and project future weather situations(Rao et al., 2023; Singh, 2023; Zenil et al., 2023).

2.2 Key Concepts in Earth Sciences Relevant to AI

The sphere of Earth sciences incorporates a wide variety of disciplines that look at the Earth and its approaches. From geology to meteorology to oceanography, scientists in these fields try to understand the complex dynamics of our planet. With the advent of Artificial Intelligence (AI), there has been a splendid advancement within the tools and techniques utilized in Earth sciences studies (Wani et al., 2023). In this essay, we can explore the key ideas in Earth sciences that are relevant to AI packages and the way this fusion of fields can revolutionize our knowledge of the Earth.One of the primary regions where AI has found enormous use in Earth sciences is in the evaluation of massive datasets. Satellite tv for PC imagery, weather fashions, and geological surveys generate full-size amounts of records that can be overwhelming for human researchers to system and understand (Kasneci et al., 2023). This is wherein AI algorithms, which include devices gaining knowledge of, come into play. Those algorithms may be trained on large datasets to apprehend patterns, anomalies, and relationships that people may not be able to pick out easily. By using reading those datasets, AI can make contributions to a deeper understanding of diverse Earth technology phenomena (Kannan et al., 2021).

2.3 Integration of AI and Earth Sciences: A Conceptual Framework

synthetic Intelligence (AI) has emerged as a transformative era across various domains, and its integration with Earth Sciences has verified great capability. This essay objectives to provide a complete exploration of AI applications in Earth Sciences, emphasizing the want for a conceptual framework called "Earth Sciences Unleashed" to liberate its complete capacity (Yuvaraj et al., 2021). by means of leveraging AI, Earth Sciences can revolutionize facts evaluation, modeling, and prediction.

Figure 1. Integration of AI and Earth science

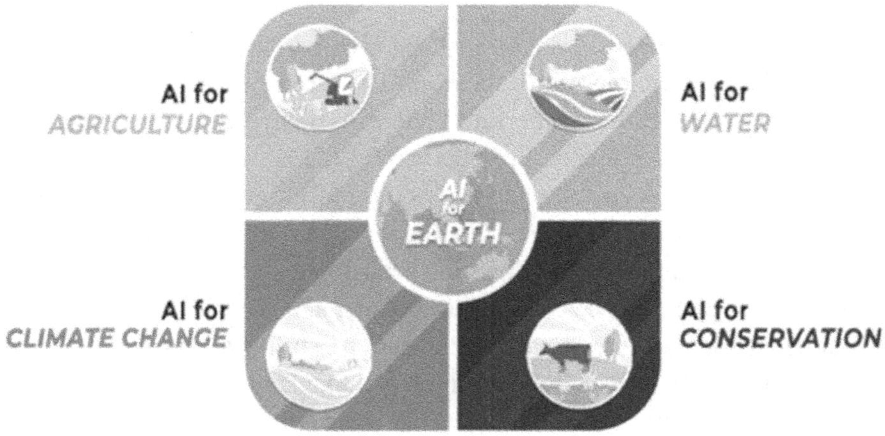

This integration holds vast promise for addressing crucial environmental demanding situations and fostering sustainable development. AI algorithms have the ability to procedure big amounts of records unexpectedly, paving the manner for more informed selection-making in Earth Sciences(Rajalakshmi et al., 2022). Earth commentary satellites, ground sensors, and drones generate an overwhelming amount of statistics that require evaluation and interpretation . AI techniques, inclusive of machine learning and deep mastering, can efficiently manage complex datasets, enabling real-time evaluation of climate patterns, geological formations, and greater. those records-pushed insights could facilitate catastrophe control, resource allocation, and particular tracking of environmental adjustments (Pant et al., 2023).

3. DATA ACQUISITION AND PREPROCESSING

Facts acquisition and preprocessing are crucial steps in most medical research, and this holds especially true for Earth sciences, in which giant amounts of information from diverse sources want to be processed and analyzed. With the advent of synthetic intelligence (AI) technology, those methods were revolutionized, allowing scientists to unharness the entire potential of AI packages in Earth sciences. This essay explores the significance of statistics acquisition and preprocessing in Earth sciences and the way they contribute to the development of AI packages.Earth sciences embody an extensive range of disciplines, together with geology, meteorology, oceanography, and ecology, which aim to recognize the complicated workings of our planet. These disciplines closely depend upon statistics accumulated from various assets, which include satellites, floor-based sensors, climate stations, and citizen science tasks. The statistics obtained from those sources are often available in distinct formats, resolutions, and spatiotemporal scales, making it challenging to utilize them without delay for scientific research. This is where records acquisition and preprocessing come into play.Facts acquisition entails the systematic collection of information from specific assets. Satellites, geared up with a selection of sensors, play a critical role in taking pictures of Earth's surface features, atmospheric conditions, and weather patterns..

3.1 Remote Sensing and AI: Enhancing Data Collection

Far flung sensing involves the purchase of facts approximately remote items or regions without physical touch. It allows scientists to take a look at the Earth's surface and atmosphere through the usage of gear consisting of satellites, planes, and drones. Remote sensing gives a unique vantage factor, permitting researchers to screen massive areas continuously and have a look at intricate information from a distance.

The mixing of far-off sensing and AI has drastically more advantageous records collection for earth sciences. Satellites equipped with advanced sensors capture excessive-decision pics, offering a wealth of information about the Earth's surface. AI algorithms then analyze these photographs, identifying styles, gadgets, and changes through the years extra correctly and as it should be, then guide analysis..

3.2 Data Preprocessing Techniques for Earth Sciences

Data cleansing is a crucial step in making ready Earth Sciences statistics for AI packages. Frequently, statistics gathered from multiple sources also contain mistakes, outliers, or missing values that could adversely affect the overall performance of AI fashions. This necessitates the usage of strategies such as outlier detection and elimination, imputation of lacking values, and handling of inconsistent or wrong facts entries. Cleansing the statistics guarantees that accurate and regular information is to be had for analysis, enhancing the reliability and validity of the subsequent AI models.Earth Sciences datasets frequently comprise a multitude of variables, a number of which can be beside the point or redundant for the prediction or evaluation responsibilities at hand..

3.3 Challenges and Solutions in Data Quality for AI Applications

1. Heterogeneity and Incompatibility of records:Earth science information originates from various assets together with satellite imagery, weather stations, faraway sensing gadgets, and observational networks. This heterogeneity brings inherent demanding situations in phrases of statistics formats, structures, and requirements. Incompatibility issues hinder efficient statistics integration and can lead to mistakes or biased consequences.
2. erroneous or insufficient records:

The excellent amount of statistics gathered can considerably affect the accuracy of AI fashions. In earth sciences, statistics amassed from far-flung places or severe environments may be susceptible to measurement mistakes and constrained in quantity because of accessibility constraints. Such boundaries can introduce biases and degrade AI fashions' predictive abilities.

4. MACHINE LEARNING MODELS FOR EARTH SCIENCES

System getting-to-know (ML) models have revolutionized numerous fields, such as Earth sciences, via their potential to analyze massive datasets and extract valuable insights efficaciously. In the book "Earth Sciences Unleashed: A Complete Exploration of AI packages," the authors delve into the numerous packages of ML fashions in this field. This essay will discuss a number of the primary ML models explored inside the ebook, their programs in Earth sciences, and the benefits they offer.One of the outstanding ML fashions mentioned within the book is Convolutional Neural Networks (CNN), which have received vast recognition in photograph analysis duties. In Earth sciences, CNNs are employed to interpret remotely sensed pix received from satellites and drones. These fashions can hit upon diverse functions, along with land use and land cowl modifications, city growth, deforestation, and flower health. With their potential to mechanically examine complicated styles, CNNs have appreciably improved the speed and accuracy of image analysis, permitting scientists to display and expect environmental adjustments extra successfully.Every other ML model highlighted in the book is Recurrent Neural Networks (RNN), which excel at reading sequential information. In Earth sciences, RNNs are extensively utilized in weather modeling, climate forecasting, and hydrological predictions..

4.1 Overview of Machine Learning Algorithms

System getting-to-know algorithms have come to be an essential tool in diverse fields, together with earth sciences. The capabilities of these algorithms have revolutionized the way data is analyzed and interpreted, unlocking a complete exploration of AI packages. In this essay, we are able to provide a top-level view of device-mastering algorithms in earth sciences, highlighting their capability and significance in various programs.Earth sciences embody an extensive range of disciplines, such as geology, geophysics, atmospheric sciences, and hydrology. Those disciplines contain the examination of complicated structures, which generate huge volumes of facts. Gadget getting to know algorithms can successfully tackle the challenges posed via such information with the aid of extracting meaningful styles and insights.

Figure 2. Types of machine learning algorithms

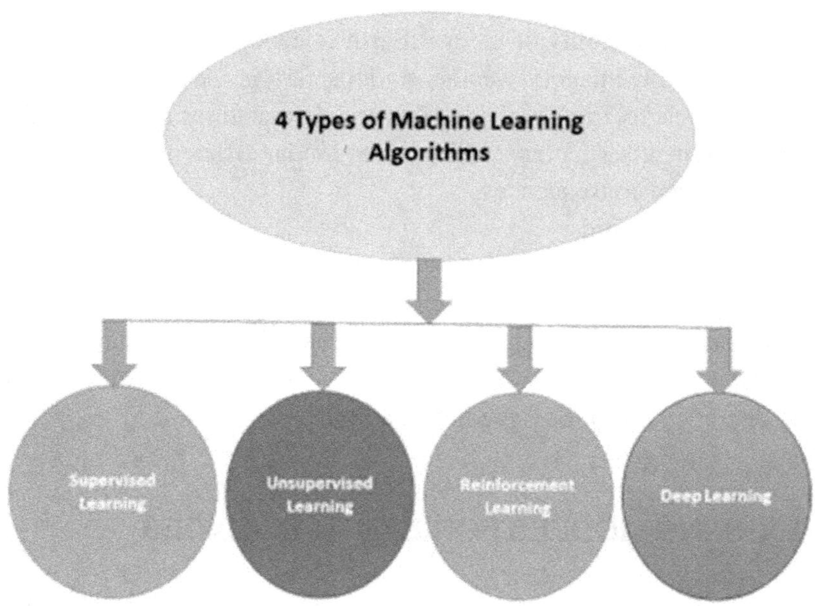

One of the most outstanding packages of system studying algorithms in earth sciences is within the discipline of weather modeling. Weather fashions are used to expect destiny climate patterns, which can be crucial for understanding the influences of weather trade. Machine-gaining knowledge of algorithms may be employed to improve the accuracy and efficiency of those fashions. Via education algorithms with historical weather information, they could learn the complex relationships between diverse weather variables and expand more excellent unique predictive fashions. This permits scientists to make more informed selections regarding climate change mitigation and edition strategies..

4.2 Supervised Learning for Earth Sciences

Synthetic Intelligence (AI) has emerged as a modern device in various fields, and Earth Sciences is not an exception. With the growing availability of big datasets and advancements in computing electricity, AI algorithms have proved to be decisive in analyzing complex Earth science phenomena. Supervised studying, a subfield of system gaining knowledge of, is especially valuable in Earth Sciences because it permits the improvement of fashions that can predict and examine diverse components of our planet. This essay explores the applications of supervised learning in Earth Sciences, highlighting its significance in understanding and addressing vital demanding situations.Supervised studying is a system studying technique in which a model is skilled on labeled data. In the context of Earth Sciences, the model learns styles and relationships within the input records to make predictions or classifications. The categorized statistics serves as a schooling set, providing examples of inputs and their corresponding desired outputs. The version then generalizes this know-how to make predictions on new, unseen information.Supervised gaining knowledge of has found several packages in Earth Sciences, contributing to our knowledge of climate change, geology, hydrology, ecology, and different disciplines..

4.3 Unsupervised Learning for Earth Sciences

Unsupervised learning algorithms are designed to discover styles and relationships within datasets without the want for categorized training statistics. This capability to uncover hidden structures in good-sized and complex datasets is specifically treasured in Earth Sciences, wherein records are often wealthy but provide challenges in phrases of information range, extent, and speed. Via leveraging unsupervised learning techniques, researchers can mine through petabytes of records gathered from satellites, sensors, and institutions, allowing for a more profound knowledge of our planet's dynamics.

4.4 Deep Learning Techniques in Earth Sciences

In recent years, the sector of Earth sciences has witnessed an innovative transformation through the integration of synthetic Intelligence (AI) techniques, in particular deep learning. Deep mastering, a subset of machine-gaining knowledge, has emerged as a powerful device to uncover hidden styles and procedures in significant quantities of complicated statistics in Earth sciences. This essay explores the transformative capacity of deep getting-to-know strategies in Earth sciences, highlighting their applications in various sub-disciplines and their effect on the expertise of our planet.Climate prediction fashions are essential in understanding and mitigating the impacts of climate change. Deep getting to know has played a pivotal role in improving the accuracy and reliability of those fashions. By way of assimilating and studying a couple of facts resources, together with satellite TV for pc observations, meteorological statistics, and historical weather statistics, deep learning algorithms can efficaciously seize complex spatiotemporal relationships. Those algorithms enable scientists to generate more excellent unique climate forecasts, assessing the future influences of climate change on one-of-a-kind regions and ecosystems at exceptional scales..

5. PREDICTIVE MODELING AND FORECASTING

Meteorology, the examination of climate and atmospheric phenomena, has benefitted immensely from AI-based predictive modeling. Weather forecasting is a complex project, as several interacting factors form the atmospheric conditions. AI models, which include deep getting to know algorithms, can efficaciously deal with the complexity and examine patterns from massive historic climate records. This permits meteorologists to predict weather conditions extra as they should be and on longer timescales, improving short-term and lengthy-term climate forecasting.As an instance, convolutional neural networks (CNNs) have been used to investigate satellite tv for pc snap shots and detect excessive climate styles, along with hurricanes and cyclones..

5.1 Weather Prediction Using AI Models

Climate prediction has always been a vital project, as accurate forecasts enable us to make informed choices associated with agriculture, catastrophe management, transportation, and even ordinary activities. Conventional climate forecasting techniques closely trust historical information, mathematical models, and human know-how. While those techniques have yielded affordable outcomes, the inherent complexity and dynamic nature of weather structures pose challenges that traditional tactics conflict to overcome. One of the essential advantages of AI fashions is their capability to discover complex styles from large amounts of records. This functionality empowers climate prediction models to pick out complicated relationships and predict weather styles correctly. Machine mastering algorithms, together with deep mastering and neural networks, are extensively used to teach AI fashions on extensive ancient weather data. Those fashions examine from beyond climate observations and their corresponding atmospheric situations to parent styles that humans won't observe.Furthermore, AI fashions can assimilate real-time meteorological facts into their forecasts..

5.2 Seismic Activity Forecasting

System gaining knowledge of a subset of AI has enormous capacity inside the field of seismic hobby forecasting; by way of training algorithms with ancient seismic facts, gadgets gaining knowledge of fashions can learn how to recognize patterns and make predictions about future seismic occasions. These fashions can research various seismic parameters, along with magnitudes, depths, and waveforms, and become aware of correlations that might not be immediately obvious to human scientists. This empowers researchers to make extra informed choices concerning earthquake preparedness, early warning structures, and emergency response plans.

5.3 AI Applications in Predicting Natural Disasters

Natural disasters, which include earthquakes, hurricanes, floods, and wildfires, have devastating impacts on both human lifestyles and infrastructure. Early detection and accurate prediction of these activities are critical for you to reduce the lack of life and mitigate property damage. With the arrival of synthetic intelligence (AI) technologies, in particular, inside the area of earth sciences, the capacity to expect herbal screw-ups has significantly advanced. This essay aims to discover the AI applications that have revolutionized forecasting strategies and empowered scientists to recognize better and expect herbal screw-ups. Earthquakes are a number of the most catastrophic natural occasions, and predicting their prevalence accurately has been an extended-standing task. However, AI strategies, together with gadget-getting-to-know algorithms, have shown outstanding promise in earthquake prediction. By way of reading seismic facts accumulated from various sensors, AI algorithms can discover styles and anomalies that imply seismic pastime. These algorithms can also examine ancient earthquake records and geospatial capabilities to create fashions that expect the chance and significance of destiny earthquakes. By constantly mastering new facts, AI structures can improve their predictive abilities over the years..

5.4 Case Studies in Predictive Modeling

The sphere of earth sciences encompasses a substantial array of disciplines and branches, consisting of geology, climatology, and environmental technology, all geared toward information and exploring the planet's natural systems. With the improvements in the era, predictive modeling using synthetic intelligence (AI) has emerged as a powerful tool to investigate and forecast diverse phenomena associated with earth sciences. This essay aims to discover case research that spotlights the packages of AI in predictive modeling inside the realm of earth sciences.Climate exchange is one of the most pressing demanding situations of our time, and understanding its capacity effects is crucial for informed decision-making. AI-driven predictive models have proven promising effects in this area. For instance, researchers have advanced models that combine ancient climate statistics, satellite imagery, and machine-gaining knowledge of algorithms to expect regional temperature adjustments and intense weather events. This allows policymakers and groups to take proactive measures in mitigating the dangers associated with climate change..

6. GEOLOGICAL EXPLORATION AND RESOURCE MANAGEMENT

Geological exploration and resource control are essential regions within the subject of earth sciences, contributing to our expertise of the Earth's shape, composition, and supply of treasured sources. The advent of synthetic intelligence (AI) has unleashed a new generation in which those disciplines can benefit from superior technologies and innovative packages. This essay seeks to explore the comprehensive exploration of AI applications in geological exploration and aid control, highlighting their significance and potential impact on the sector.Geological exploration involves taking a look at numerous strategies that form the Earth's floor and subsurface, along with tectonic sports, sedimentation, and mineral formation. Conventional methods of geological exploration have depended on the analysis of rock samples, discipline observations, and geophysical surveys. However, these traditional techniques require sizeable time, assets, and information, regularly leading to limitations within the accuracy and efficiency of statistics analysis.The integration of AI technologies into geological exploration has revolutionized the field by allowing the processing and interpretation of considerable amounts of statistics in actual time. System gaining knowledge of algorithms can rapidly analyze geological records, pick out patterns, and make predictions with remarkable accuracy. For instance, AI-driven photo recognition can unexpectedly pick out and classify rock formations based totally on texture and composition, supplying treasured insights into the underlying geological processes..

6.1 Mineral and Resource Mapping with AI

Mineral and functional resource mapping is a vital issue of earth sciences, as it helps us understand the distribution and abundance of precious assets underneath the Earth's surface. Traditionally, this process has been time-eating and labor-in-depth, requiring guide statistics collection and analysis. But, with the arrival of artificial Intelligence (AI), the sphere of earth sciences has witnessed a fantastic transformation, unlocking a comprehensive exploration of AI programs. In this essay, we will delve into the numerous ways in which AI has revolutionized mineral and helpful resource mapping, enabling more excellent, efficient, and accurate outcomes.

One of the primary AI strategies utilized in mineral and helpful resource mapping is gadget studying. System learning algorithms leverage ancient facts to discover styles and make predictions, imparting a significant gain over conventional strategies. Through schooling algorithms using extraordinary amounts of geological and satellite TV for pc imagery information, AI models can stumble on and classify minerals and sources with high precision. The capability to investigate massive datasets permits an extra exceptional understanding of resource distributions, leading to higher aid control and decision-making. Remote sensing is another location in which AI has made sizeable contributions to mineral and resource mapping. The mixing of AI algorithms with satellite and aerial imagery allows the automated identification, classification, and interpretation of geological capabilities..

6.2 AI Applications in Geological Surveying

Geological surveying, a critical issue of the Earth sciences, involves the collection and analysis of geological statistics to recognize the composition, structure, and records of the Earth's floor and subsurface. Traditionally, this method has been time-ingesting and labor-intensive, requiring significant human attempts. However, with improvements in artificial Intelligence (AI), geologists and Earth scientists now have access to an extensive range of innovative equipment and programs to enhance and expedite geological surveying. This essay aims to discover the numerous AI programs in geological surveying that have unleashed the ability of Earth sciences, making them more efficient, correct, and comprehensive.AI algorithms have greatly facilitated the analysis of far-flung sensing facts, which includes aerial and satellite imagery, allowing geologists and Earth scientists to gain precious insights into geological structures and functions. Convolutional Neural Networks (CNNs) can mechanically extract relevant data, classify rock sorts, and become aware of geological limitations from massive-scale pictures. With the aid of combining spectral, textural, and spatial facts, AI algorithms can understand styles and anomalies that might escape human observation..

6.3 Sustainable Resource Management with AI

1. Remote sensing and facts analysis:

Remote sensing techniques, together with satellite TV for pc imagery and LiDAR information, frequently generate vast quantities of facts. AI algorithms have revolutionized the evaluation of these facts, allowing us to become aware of patterns, screen adjustments, and extract actionable insights. This permits more useful resource control practices, whether or not it's tracking deforestation, reading water fines, or optimizing sustainable land use.

2. Predictive Modeling and danger assessment:

AI algorithms excel at predictive modeling, offering treasured insights into diverse Earth technological know-how phenomena. As an example, machine studying algorithms can forecast the prevalence of herbal screw ups like earthquakes, hurricanes, and landslides with good sized accuracy. by means of understanding and predicting those occasions, the government can put in force essential measures to reduce their impact, protect lives, and preserve sources..

7. ENVIRONMENTAL MONITORING AND CONSERVATION

One colossal aspect wherein AI has revolutionized environmental monitoring is in far-flung sensing and information analysis. Satellites geared up with advanced sensors seize full-size amounts of records associated with climate styles, atmospheric conditions, and land cowl adjustments. But, manually reading this fact may take a lot of work. AI algorithms designed to interpret such information can process facts at a far quicker fee, permitting us to take advantage of essential insights into the Earth's structures and styles. These algorithms can identify anomalies, are expecting weather styles, and help detect herbal failures with extra accuracy.Another utility of AI in environmental tracking is in flora and fauna conservation. AI has been utilized in growing laptop vision models that could pick out and tune animal species in actual time.

7.1 AI for Environmental Impact Assessment

AI-powered systems can procedure significant quantities of data, supplying more excellent complete expertise of environmental influences. Traditional methods of information collection and evaluation regularly contain burdensome work-extensive responsibilities and subjective interpretation. However, with AI equipment, remote sensing technology, and advanced algorithms, facts series turn into extra efficient, correct, and reliable. For example, drones ready with AI algorithms can behavior aerial surveys, capturing excessive-resolution pics that allow for precise mapping and monitoring of ecosystems. This guarantees a more tremendous thorough and timely assessment of capability environmental influences..

7.2 Biodiversity Monitoring Using AI

Biodiversity tracking is essential for knowledge of the health and status of our planet's ecosystems. It gives treasured insights into the impact of human activities, climate change, and different environmental elements on numerous species and their habitats. With the improvements in artificial Intelligence (AI), it has become feasible to leverage this technology to revolutionize biodiversity tracking in Earth Sciences. This essay aims to discover the complete applications of AI in biodiversity tracking, highlighting its ability to beautify statistics collection, evaluation, and choice-making procedures.

Packages of AI in Biodiversity tracking:

1. Computerized species identity: AI-powered technologies, inclusive of device studying and computer vision, can examine substantial amounts of picture and audio information to identify species mechanically. This reduces the reliance on time-consuming manual identification and enables improve the accuracy and efficiency of biodiversity surveys.
2. faraway Sensing and satellite Imagery analysis: AI algorithms can technique far-flung sensing data received from satellites to discover and map different habitats, display changes in land cover, and discover deforestation or fragmentation. This equipment can provide real-time and huge-scale tests of biodiversity, allowing early detection of degradation and facilitating practical conservation efforts..

7.3 Conservation Strategies Supported by AI

The Earth's herbal sources are under increasing stress due to human-brought factors, which include deforestation, climate change, and overexploitation. To tackle those challenges, there may be a developing necessity to leverage advanced technologies like synthetic Intelligence (AI) to assist conservation efforts. AI holds tremendous potential in helping earth scientists to display and shield herbal ecosystems, expect environmental adjustments, and devise effective conservation strategies. This essay explores the diverse conservation techniques supported via the integration of AI technology in Earth Sciences, showcasing the comprehensive capability of AI programs. Biodiversity loss is a crucial problem worldwide that calls for proactive tracking and conservation efforts. AI can resource in this regard by automating the monitoring and classification of species using PC vision strategies. Via picture and video evaluation, AI algorithms can identify and classify numerous species, facilitating a speedy evaluation of biodiversity hotspots and permitting-centered conservation projects. Furthermore, AI can help in predicting species distribution fashions, analyzing ecological styles, and identifying areas susceptible to biodiversity loss, assisting policymakers and conservationists in implementing necessary interventions..

8. CASE STUDIES IN AI-DRIVEN EARTH SCIENCES RESEARCH

Artificial Intelligence (AI) has revolutionized numerous fields by offering innovative solutions to complex issues. The Earth Sciences is no exception, as researchers and scientists have increasingly embraced AI-driven research strategies. This essay aims to discover the case studies in AI-pushed Earth science research, highlighting the diverse packages that have contributed to a complete understanding of our planet.

1. climate change and Prediction:

One good-sized region in which AI has made a significant effect is weather exchange research. AI algorithms analyze enormous quantities of climate statistics collected over the years, figuring out style correlations and predicting future eventualities. As an example, weather models rent diverse machine studying techniques to simulate climate behavior at regional and worldwide ranges. Those fashions help in predicting intense climate activities, sea-degree upward push, and assessing the effects of climate exchange on ecosystems and human populations.

2. herbal disaster Mitigation:

AI also performs a crucial role in mitigating herbal screw-ups, wherein timely information can keep lives and sources. Researchers have advanced AI-primarily based structures that analyze geospatial statistics, satellite imagery, and sensor readings to locate, screen, and predict disasters, which include earthquakes, tsunamis, and volcanic eruptions..

8.1 Notable Projects and Their Outcomes

assignment 1: weather exchange Modeling

One crucial place wherein AI has made significant progress, thanks to Earth Sciences Unleashed, is weather alternate modeling. By reading widespread amounts of environmental statistics, AI algorithms can predict future weather patterns more appropriately. One significant challenge focused on generating weather scenarios that considered various factors, including greenhouse gasoline emissions, oceanic styles, and Arctic ice melting. The outcomes discovered a much more precise and particular knowledge of ways climate exchange may affect special areas, permitting policymakers and researchers to expand more excellent and effective mitigation techniques.

assignment 2: natural catastrophe Prediction and reaction

another vast region in which AI has proved beneficial is in predicting and dealing with herbal failures. Earth Sciences Unleashed has spearheaded initiatives aimed at enhancing catastrophe reaction systems with the usage of AI.. for example, with the aid of reading ancient statistics on earthquakes, AI algorithms can expect seismic activities with extra accuracy. This has allowed the government to trouble timely warnings to susceptible areas, potentially saving lives and decreasing the overall impact of such disasters. Moreover, AI-powered drones prepared with state-of-the-art sensing era were applied to assess harm following natural disasters, helping in emergency reaction efforts and prioritizing resources higher.

assignment 3: Sustainable aid management

The Earth Sciences Unleashed project has additionally targeted the sustainable management of Earth's finite resources. Making use of AI algorithms to analyze information related to water, soil, and plant life, researchers were capable of becoming aware of areas that might be at excessive threat of degradation or depletion. This expertise has enabled policymakers to put into effect targeted interventions to protect and restore these essential sources. Moreover, AI-powered systems have been employed in agriculture to optimize crop yields, reduce water utilization, and minimize the environmental effect of farming practices..

8.2 Success Stories and Lessons Learned

The sector of Earth sciences has undoubtedly undergone a first-rate transformation with the advent of artificial intelligence (AI) programs. Earth Sciences Unleashed: A complete Exploration of AI programs is an exceptional initiative that showcases the success memories and precious training learned in this area. Through the mixing of AI, geospatial records analysis, and system learning algorithms, researchers and scientists have been capable of reaping unprecedented improvements in understanding our planet and making informed choices for sustainable development. In this essay, we will delve into a number of the most beautiful success testimonies and the valuable lessons discovered from Earth Sciences Unleashed.One of the super achievements in the subject of Earth sciences is the use of AI in reading climate exchange. The utility of AI algorithms to vast amounts of ancient climate information has allowed scientists to become aware of patterns of developments and are expecting future weather situations. This has verified the importance of assessing the effects of climate change, developing mitigation techniques, and formulating weather edition plans. Furthermore, AI-powered weather fashions have confirmed high accuracy degrees in predicting intense climate activities, which include hurricanes, floods, and heatwaves, enabling timely and effective response measures. This fulfillment tale underlines the power of AI in addressing complicated and worldwide challenges that, in the end, lead to a more fantastic, sustainable future..

8.3 Collaborative Initiatives in AI and Earth Sciences Research

Collaboration has constantly been at the heart of clinical development. Within the context of AI and Earth Sciences, collaborative tasks carry together experts from diverse fields, which include information science, PC technology, and Earth Sciences. This multidisciplinary method cultivates a fertile ground for innovation and breakthroughs, using advancements in AI programs through joint studies and shared resources. One excellent collaborative initiative is the partnership between educational institutions, studies agencies, and generation companies. For instance, partnerships between universities and tech giants like Google or Microsoft have led to the development of practical AI tools and platforms tailor-made particularly for Earth science studies. These collaborations offer admission to great computing strength, data, and information, permitting researchers to tackle complicated issues and examine giant datasets with unparalleled performance. Another collaborative initiative worth bringing up is the status quo of worldwide consortia and networks. These initiatives convey together scientists from exclusive countries and establishments, fostering the trade of information, records, and methodologies. The international AI for Earth Initiative, launched via Microsoft, is a high instance of one of these consortiums. It aims to empower researchers and groups around the sector to leverage AI for environmental conservation and sustainability. Through this collaborative attempt, AI applications in Earth Sciences have received worldwide momentum, addressing demanding situations on a larger scale and selling cooperation for better information on our planet. Collaborative tasks in AI and Earth Sciences research have drastically contributed to Earth Sciences Unleashed—a comprehensive exploration of AI applications. Firstly, these tasks have elevated the improvement and deployment of AI algorithms and fashions specially designed for Earth Sciences. Device gaining knowledge of techniques, which includes neural networks and deep mastering, have verified instrumental in reading complex Earth device interactions, predicting natural disasters, and modeling weather trade scenarios..

9. CHALLENGES AND FUTURE DIRECTIONS

Our curiosity has long driven the field of Earth sciences to understand the origins and workings of our planet. As technology unexpectedly advances, allowing us to collect and method copious quantities of records, the function of artificial intelligence (AI) in Earth sciences will become increasingly paramount. AI can revolutionize the way we take a look at Earth's systems, from climate trade and natural failures to proper resource management and environmental maintenance. However, amidst the opportunities AI offers, several demanding situations and destiny directions must be taken into consideration to ensure its successful integration into Earth sciences.One of the fundamental demanding situations posed by using AI is the collection and control of massive datasets. Earth sciences generate full-size amounts of statistics from satellites, sensors, and on-the-ground observations. However, AI algorithms require diverse and properly annotated datasets to carry out efficaciously. Obtaining such datasets entails extensive time and assets and necessitates collaboration among scientists, engineers, and records collection agencies. Moreover, information privateness and protection concerns, particularly while coping with sensitive information, have to be addressed to save you from misuse or unauthorized get entry.Every other challenge lies in developing AI algorithms that can successfully interpret complex Earth technological know-how facts. Earth structures are interconnected and complicated, encompassing a multitude of variables and remarks loops..

9.1 Ethical Considerations in AI for Earth Sciences

The fast advancement of synthetic Intelligence (AI) has revolutionized numerous fields, including the Earth sciences. AI has been employed to address complex troubles and provide treasured insights into environmental conservation, weather alternatives, and herbal disaster prediction and mitigation. However, the implementation of AI in Earth sciences also comes with moral issues that have to be cautiously addressed. It's miles essential to make sure that AI technology is used responsibly and ethically to protect our planet and all its inhabitants.One of the first ethical considerations in AI for Earth sciences is the capacity bias embedded inside the algorithms and datasets utilized in AI models. AI algorithms closely rely upon massive datasets to analyze and make predictions. If those datasets are biased or incomplete, the ensuing AI fashions may also be biased, potentially leading to wrong or unfair conclusions. For example, if environmental monitoring statistics mainly display urban areas, AI models skilled on such statistics might also underestimate the effect of weather alternations in rural or faraway areas. To prevent bias, thorough facts collection and representation from diverse populations and areas must be ensured..

9.2 Technical Challenges and Solutions

Technical challenges:

1. facts series and pleasant:

One of the primary challenges in Earth Sciences Unleashed is the collection of significant quantities of numerous and reliable facts. Earth-related phenomena embody multiple elements, which include temperature, humidity, atmospheric composition, and seismic pastime, which require tremendous statistics collection efforts from diverse assets. The heterogeneity of record sources and formats poses hurdles in ensuring statistics compatibility, high quality, and accuracy. Moreover, the need for more favorable types of data, consisting of historical climate records, provides similar barriers.

2. facts Preprocessing and Integration: a

Earth science information is regularly heterogeneously allotted across more than one system and corporation, making facts preprocessing and integration complex. This consists of cleaning, normalizing, and reworking the information to make it appropriate for AI algorithms. Furthermore, integrating information from unique assets requires addressing troubles like statistics format inconsistencies, lacking values, outliers, and conflicting statistics sets. This step is essential to reap complete evaluation and accurate AI-driven predictions.

3. Set of rules choice and Optimization:

Earth Sciences Unleashed requires the selection and Optimization of suitable AI algorithms for particular applications. Traditional AI algorithms may not provide the most reliable effects for Earth's technological know-how tasks due to the complexities inherent within the subject. Deciding on suitable algorithms is vital to ensure compatibility with the available records and to achieve correct predictions or classifications. Moreover, a set of rules optimization for scalability, efficiency, and computational boundaries becomes vital to deal with the considerable amount of facts and growth model performance.

4. Interpretability and Explainability:

An enormous task in imposing AI fashions for Earth sciences is the absence of interpretability and explainability. As AI strategies like deep gaining knowledge of models emerge as increasingly complicated, they regularly feature as black bins, making it difficult to recognize the reasons at the back of their predictions or outputs. Interpretability is vital in Earth sciences to advantage insights into underlying physical processes, validate model predictions, and establish acceptance as true with stakeholders. Therefore, developing strategies to deal with this venture and provide explainable AI answers is critical.

9.3 Future Trends and Emerging Technologies

The Earth sciences have continually performed a crucial role in our expertise of the planet and its numerous structures. From analyzing the surroundings and weather styles to delving into the depths of the oceans and exploring the geological records, these disciplines offer valuable insights into the Earth's past, gift, and future. With the appearance of artificial intelligence (AI) and its developing packages, the Earth sciences are poised to go through a considerable transformation. In this essay, we can discover the future developments and emerging technologies in AI programs for Earth sciences, unleashing their comprehensive capability.One of the most exciting developments in AI for Earth sciences is the growing use of gadgets getting to know algorithms for statistics analysis and modeling. The large quantities of records amassed from satellites, sensors, and different tracking devices can now be successfully processed and analyzed with the usage of AI strategies. Those algorithms can identify styles, correlations, and anomalies in the information that might, in any other case, be tough or time-eating for people to discover. With the aid of leveraging gadget learning algorithms, scientists can take advantage of a deeper understanding of the complicated procedures going on on our planet, which include weather alternates, natural screw-ups, and surroundings dynamics..

10. CONCLUSION

In conclusion, the sector of Earth Sciences has seen immense improvements with the integration of artificial intelligence (AI) packages. The possibilities that AI offers in terms of information analysis, modeling, and predictive competencies are revolutionizing how we recognize and reveal the Earth's structures. Earth Sciences Unleashed: A complete Exploration of AI applications showcases the brilliant capacity of AI in numerous subfields inside Earth Sciences, from climate modeling and disaster control to aid exploration and environmental monitoring.One of the most significant contributions of AI in Earth Sciences is, in fact, analysis. The extensive amount of facts accrued from satellites, floor-based sensors, and different sources poses a task in deriving meaningful insights. However, AI strategies, which include device learning and deep gaining knowledge, have the potential to system and analyze massive datasets with pace and accuracy. These strategies can pick out styles, anomalies, and correlations that were missed with the aid of human evaluation on my own. Via leveraging AI, scientists can uncover hidden relationships, predict future developments, and make knowledgeable decisions to mitigate the impacts of natural screw-ups and climate alternates.Furthermore, AI has revolutionized weather modeling, allowing scientists to develop correct and unique simulations of Earth's climate system. Traditional weather models are complicated and computationally annoying, regularly requiring extensive time and computing electricity..

10.1 Recap of Key Findings

The utilization of close-to-neighbor-primarily based Earth observation and monitoring has grown to.

10.2 Implications for the Future of Earth Sciences

Earth Sciences Unleashed: A comprehensive Exploration of AI applications affords a complete evaluation of the essential thing findings within the area of Earth Sciences in which synthetic intelligence (AI) has been efficiently carried out. This recap highlights the sizable discoveries, improvements, and implications mentioned within the file.First and most important, the report emphasizes how AI applications have revolutionized the field of Earth Sciences. The super capability of AI to technique and examine full-size quantities of records has paved the manner for sizeable breakthroughs in various sub-disciplines. The use of gadget-studying algorithms and neural networks has increased the analysis of complex geological and environmental facts, allowing scientists to extract meaningful insights and make accurate predictions.One of the critical findings revolves around the application of AI in weather modeling. With the aid of assimilating huge-scale weather information, AI algorithms have enabled researchers to improve the accuracy of weather predictions. This has significant implications for expertise on climate change patterns, assessing the effect of human activities, and formulating effective mitigation strategies. With AI's capacity to process and interpret climate records more effectively, it has turned out to be an essential tool for policymakers and scientists to address environmental challenges proactively.Moreover, the record highlights improvements in AI-pushed far-flung sensing and satellite TV for pc imagery analysis. Earth observation satellites produce massive quantities of facts that require meticulous examination.

10.3 Closing Remarks

Over the last few days, we've witnessed the convergence of two reputedly disparate fields, synthetic intelligence (AI) and earth sciences, and skilled the limitless possibilities that emerge while those disciplines intertwine.To start, allow us to mirror the limitless insights and innovations that have been shared in the course of this convention. From the inception of AI to its present-day advancements, we have seen the exquisite transformation it has brought to earth sciences. This symbiotic dating has now not only revolutionized the way we perceive our planet but also catalyzed substantial strides in the direction of a sustainable future.One of the key takeaways from this occasion has been the role of AI in addressing weather aalternatives. Through harnessing the power of the system and gaining knowledge, AI can research big datasets and pick out styles that may inform our know-how of weather styles, expect extreme climate activities, and beautify our ability to reply successfully..

REFERENCES

Aithal, P. S. (2023). Advances and New Research Opportunities in Quantum Computing Technology by Integrating it with Other ICCT Underlying Technologies. *International Journal of Case Studies in Business IT and Education*, *7*(3), 314–358.

Al Dulaijan, K., & Alajaji, M. (2022, November). HPC advances and impacts to exploration and reservoir development. In *SEG International Exposition and Annual Meeting* (p. D011S123R002). SEG. 10.1190/image2022-3746173.1

Allioui, H., & Mourdi, Y. (2023). Unleashing the Potential of AI: Investigating Cutting-Edge Technologies That Are Transforming Businesses. *International Journal of Computer Engineering and Data Science*, *3*(2), 1–12.

Badini, S., Regondi, S., & Pugliese, R. (2023). Unleashing the Power of Artificial Intelligence in Materials Design. *Materials (Basel)*, *16*(17), 5927. doi:10.3390/ma16175927 PMID:37687620

Chowdhuri, I., Pal, S. C., Saha, A., Chakrabortty, R., & Roy, P. (2022). Mapping of earthquake hotspot and coldspot zones for identifying potential landslide hotspot areas in the Himalayan region. *Bulletin of Engineering Geology and the Environment*, *81*(7), 257. doi:10.100710064-022-02761-5

Dematties, D., Raut, B. A., Park, S., Jackson, R. C., Shahkarami, S., Kim, Y., Sankaran, R., Beckman, P., Collis, S. M., & Ferrier, N. (2023). Let's Unleash the Network Judgment: A Self-Supervised Approach for Cloud Image Analysis. *Artificial Intelligence for the Earth Systems*, *2*(2), 220063. doi:10.1175/AIES-D-22-0063.1

Gao, Y. (2023). Unleashing the mechanism among environmental regulation, artificial intelligence, and global value chain leaps: A roadmap toward digital revolution and environmental sustainability. *Environmental Science and Pollution Research International*, *30*(10), 28107–28117. doi:10.100711356-022-23898-6 PMID:36394806

George, B., & Wooden, O. (2023). Managing the strategic transformation of higher education through artificial intelligence. *Administrative Sciences, 13*(9), 196. doi:10.3390/admsci13090196

Gunjan, V. K., Vijayalata, Y., Valli, S., Kumar, S., Mohamed, M. O., & Saravanan, V. (2022). Machine learning and cloud-based knowledge graphs to recognize suicidal mental tendencies. *Computational Intelligence and Neuroscience, 2022*, 2022. doi:10.1155/2022/3604113 PMID:35341205

Hallema, D. W. (2022). Breaking Boundaries: The Power of Interdisciplinary Research in Science, Technology, and Business. *International Multidisciplinary Journal of Science, Technology, and Business, 1*(4), 30–35.

Iqbal, J., Jaimes, D. C. C., Makineni, P., Subramani, S., Hemaida, S., Thugu, T. R., & Hemida, S. (2023). Reimagining Healthcare: Unleashing the Power of Artificial Intelligence in Medicine. *Cureus, 15*(9). Advance online publication. doi:10.7759/cureus.44658 PMID:37799217

Kannan, S., Dhiman, G., Natarajan, Y., Sharma, A., Mohanty, S. N., Soni, M., Easwaran, U., Ghorbani, H., Asheralieva, A., & Gheisari, M. (2021). Ubiquitous vehicular ad-hoc network computing using deep neural network with iot-based bat agents for traffic management. *Electronics (Basel), 10*(7), 785. doi:10.3390/electronics10070785

Kasneci, E., Seßler, K., Küchemann, S., Bannert, M., Dementieva, D., Fischer, F., Gasser, U., Groh, G., Günnemann, S., Hüllermeier, E., Krusche, S., Kutyniok, G., Michaeli, T., Nerdel, C., Pfeffer, J., Poquet, O., Sailer, M., Schmidt, A., Seidel, T., ... Kasneci, G. (2023). ChatGPT for good? On opportunities and challenges of large language models for education. *Learning and Individual Differences, 103*, 102274. doi:10.1016/j.lindif.2023.102274

Khan, S., Saravanan, V. N. G. C., Lakshmi, T. J., Deb, N., & Othman, N. A. (2022). Privacy protection of healthcare data over social networks using machine learning algorithms. *Computational Intelligence and Neuroscience, 2022*, 2022. doi:10.1155/2022/9985933 PMID:35371203

Pant, G., Bahuguna, R., Pandey, S., Gehlot, A., Yadav, S. P., & Pachauri, R. K. (2023, April). Intelligent Automated Interference for the Protection of Road Safety. In *2023 International Conference on Computational Intelligence, Communication Technology and Networking (CICTN)* (pp. 87-91). IEEE. 10.1109/CICTN57981.2023.10141086

Rajalakshmi, M., Saravanan, V., Arunprasad, V., Romero, C. T., Khalaf, O. I., & Karthik, C. (2022). Machine Learning for Modeling and Control of Industrial Clarifier Process. *Intelligent Automation & Soft Computing, 32*(1). doi:10.32604/iasc.2022.021696

Rao, Y., Redmon, R., Dale, K., Haupt, S. E., Hopkinson, A., Bostrom, A., & Kihn, E. A. (2023). *Developing Digital Twins for Earth Systems: Purpose, Requisites, and Benefits*. arXiv preprint arXiv:2306.11175.

Singh, V. (2023). *Unleashing AI: A Thought-provoking Look At The Future Of Humanity*. Amazon.

Srivastava, S., Dhyani, N., Sharma, V., Vats, S., Yadav, S. P., Kukreja, V., & Singh, R. (2023, May). Lung Infection and Identification using Heatmap. In *2023 2nd International Conference on Applied Artificial Intelligence and Computing (ICAAIC)* (pp. 1093-1098). IEEE. 10.1109/ICAAIC56838.2023.10140204

Trabelsi, M., Casprini, E., Fiorini, N., & Zanni, L. (2023). Unleashing the value of artificial intelligence in the agri-food sector: Where are we? *British Food Journal, 125*(13), 482–515. doi:10.1108/BFJ-11-2022-1014

Wani, M. A., Din, A., Nazki, I. T., Rehman, T. U., Al-Khayri, J. M., Jain, S. M., Lone, R. A., Bhat, Z. A., & Mushtaq, M. (2023). Navigating the future: Exploring technological advancements and emerging trends in the sustainable ornamental industry. *Frontiers in Environmental Science, 11*, 1188643. doi:10.3389/fenvs.2023.1188643

Xu, M., Du, H., Niyato, D., Kang, J., Xiong, Z., Mao, S., & Poor, H. V. (2023). *Unleashing the power of edge-cloud generative ai in mobile networks: A survey of aigc services.* arXiv preprint arXiv:2303.16129.

Yuvaraj, N., Karthikeyan, T., & Praghash, K. (2021). An improved task allocation scheme in serverless computing using gray wolf Optimization (GWO) based reinforcement learning (RIL) approach. *Wireless Personal Communications, 117*(3), 2403–2421. doi:10.100711277-020-07981-0

Zenil, H., Tegnér, J., Abrahão, F. S., Lavin, A., Kumar, V., Frey, J. G., & King, R. (2023). *The Future of Fundamental Science Led by Generative Closed-Loop Artificial Intelligence.* arXiv preprint arXiv:2307.07522.

Zhao, P., Gao, Y., & Sun, X. (2023). The impact of artificial intelligence on pollution emission intensity—Evidence from China. *Environmental Science and Pollution Research International, 30*(39), 1–16. doi:10.100711356-023-28866-2 PMID:37470975

Chapter 10
Fuzzy Logic-Based Monitoring of Earth Observations

Karthiga R.
Department of Information Technology, R.M.K. Engineering College, Chennai, India

Saravanan V.
Department of Computer Science, Dambi Dollo University, Ethiopia

Sunil Kumar
Faculty of Commerce and Management, SGT University, Gurugrm, India

Rajakumar S.
Department of Electronics and Communication Engineering, Panimalar Engineering College, India

ABSTRACT

Monitoring Earth observations performs a crucial role in expertise and predicting numerous phenomena, including weather trade, herbal disasters, land use adjustments, and commercial pollutants. This chapter explores the use of fuzzy logic strategies for monitoring earth observations, considering its capability, blessings, and challenges. Fuzzy logic-based total decision support structures have been established to be adequate gear in monitoring earth observations. One of the key advantages of this method is its ability to deal with vague and unsure data. One of the key findings is the potential of fuzzy good judgment-based monitoring to handle and incorporate uncertainty in earth observations. Conventional monitoring methods frequently battle with uncertainties associated with sensor noise, atmospheric situations, and ambiguous class standards. Fuzzy logic-based systems, alternatively, excel in handling such uncertainties by presenting bendy and adaptive reasoning strategies.

1. INTRODUCTION

Fuzzy good judgment-based total tracking of Earth observations is an innovative idea that has transformed the sector of environmental tracking and management. With the appearance of superior technology and the increasing availability of satellite TV for PC imagery and other Earth remark information, the want

DOI: 10.4018/979-8-3693-1850-8.ch010

for robust analysis and interpretation of this vast quantity of information has ended up paramount. Fuzzy logic, with its potential to deal with uncertainty and imprecision in records, offers a promising solution to tackle complex environmental problems and make specific sustainable development (Qayyum et al., 2022). Earth remark records provide essential data on various environmental parameters, inclusive of land cowl, climate styles, pollution levels, and environmental health. However, these records are often characterized by uncertainty and vagueness because of elements like sensor noise, atmospheric conditions, and the inherent complexity of the Earth's structures (Kaur & Vig, 2023). Traditional monitoring techniques, based on crisp logic and binary classifications, work to deal with such uncertainties and provide correct assessments successfully.In this context, fuzzy common sense has emerged as a powerful tool that can manage and manage obscure and uncertain facts. Fuzzy common sense is a mathematical framework that permits the representation of imprecision and uncertainty by assigning levels of membership to distinctive training or categories (Tayfur, 2023). In contrast to conventional binary classifications, fuzzy logic considers the slow transition from one class to another, taking pictures of the subtle variations and nuances in the data..

1.1 Background and Significance

Monitoring Earth observations performs a crucial role in expertise and predicting numerous phenomena, including weather trade, herbal disasters, land use adjustments, and commercial pollutants. Conventional monitoring tactics frequently depend upon precise measurements and crisp information, which may be restricted in situations wherein uncertainties and imprecisions are inherent. Fuzzy common sense-primarily based tracking, on the other hand, provides an extra bendy and robust framework to analyze and interpret Earth observations (Mehta, 2022).Fuzzy good judgment is a mathematical approach that deals with uncertainty and imprecision by taking into consideration slow transitions between extraordinary values or states. It's miles especially suitable for handling complex and uncertain statistics in Earth observations, wherein measurements may be laid low with noise, errors, and incomplete statistics. Fuzzy common sense-primarily based monitoring tactics use linguistic variables and fuzzy units to symbolize and manipulate imprecise or unsure records, taking into account extra accurate and bendy analysis (Hai et al., 2022; Naikoo et al., 2023).One massive benefit of fuzzy joint sense-based total tracking is its potential to integrate multiple records, resources, and variables. Earth observations frequently involve a wide variety of statistics gathered from satellites, floor-based total sensors, and other resources. Those data may additionally have distinct degrees of accuracy, decision, and reliability(Nazirbhai & Gajjar, 2023).

1.2 Objectives of the Chapter

Fuzzy logic-based total tracking of Earth observations has become an increasingly crucial subject in technology and generation. This bankruptcy aims to explore the goals of fuzzy logic-primarily based tracking of Earth observations and its capability applications in diverse domain names. By way of presenting a comprehensive expertise of these objectives, this essay seeks to spotlight the significance and software of this method in environmental monitoring and decision-making methods (Yuvaraj, Srihari, Dhiman et al, 2021).The primary objective of fuzzy good judgment-based tracking is to allow accurate and dependable interpretation of complex fact sets. Earth commentary records are often large and heterogeneous, making it challenging to extract meaningful information. Fuzzy logic-primarily based strategies provide a green means to address uncertainty and imprecision by allowing the integration of a couple of sources of information (Rahman et al., 2023). Via the use of linguistic variables and club capabilities, fuzzy good judgment can effectively represent and motivate over obscure principles, resulting in extra accurate interpretations of the discovered phenomena.Every other crucial goal is to facilitate actual-time tracking and decision-making in dynamic environmental systems (Hossain et al., 2022). Fuzzy common sense-based total tracking strategies offer the capability to swiftly examine and reply to modifications in Earth observations, taking into consideration well-timed interventions and adaptive management techniques. By means of combining fuzzy logic with sensor networks and information fusion strategies, it will become possible to constantly monitor various environmental parameters, together with air exceptional, water assets, and biodiversity, in near real-time(Zahmatkesh et al., 2022).

2. FOUNDATIONS OF FUZZY LOGIC IN EARTH OBSERVATIONS

Earth observations have ended up being a critical part of tracking and know-how of various herbal and artificial phenomena. Those observations provide us with valuable statistics that help in making knowledgeable choices related to climate exchange, land-use-making plans, disaster management, and many other fields. However, deciphering and studying earth observations information may be challenging because of uncertainties and the imprecise nature of the measurements (Othman & Abdulrazzaq, 2023). This is in which fuzzy logic plays a vital position in making sense of those observations and extracting significant information.Fuzzy good judgment is a mathematical framework that offers uncertainties and vague reasoning. It offers a way to address incomplete or indistinct records and make decisions based totally on those uncertainties. Fuzzy good judgment is based on the concept of fuzzy units, in which a detail can belong to a set of varying stages rather than being strictly binary (both absolutely a member or now not a member) (Aklah et al., 2023). This capacity to deal with partial club permits for an extra realistic representation and interpretation of earth observations information.One of the critical foundations of fuzzy logic in earth observations is the idea of linguistic variables. In conventional logic, variables are crisp and discrete, taking on specific values(Ibrahim et al., 2023).

2.1 Overview of Fuzzy Logic

Fuzzy good judgment, a branch of artificial intelligence, is a mathematical framework that deals with uncertainty and vagueness in decision-making techniques. This concept has observed applications in diverse fields, consisting of tracking the Earth's observations. Fuzzy logic-based tracking of Earth observations allows the dealing with obscure and uncertain statistics, which can be not unusual in environmental and faraway sensing packages. This essay presents an overview of fuzzy logic and its relevance in monitoring Earth observations (Kannan et al., 2021).

Fuzzy common sense was added with the aid of Professor Lotfi Zadeh in the Nineteen Sixties as an extension of the classical (crisp) set concept. Unlike crisp units that have properly-described barriers, fuzzy sets allow for gradations of club based on degrees of fact. In other words, a detail can belong to a fuzzy set to a sure diploma, in place of strictly both belonging or not belonging (Yuvaraj, Karthikeyan, & Praghash, 2021). This capability to represent and reason with vague facts makes fuzzy common sense a powerful tool in choice-making.Inside the context of monitoring Earth observations, fuzzy good judgment enables the coping with imperfect and unsure data that arises from satellite TV for PC imagery, climate forecasts, and different far-off sensing data (Rajalakshmi et al., 2022). Earth observation missions

Figure 1. Block diagram of fuzzy logic

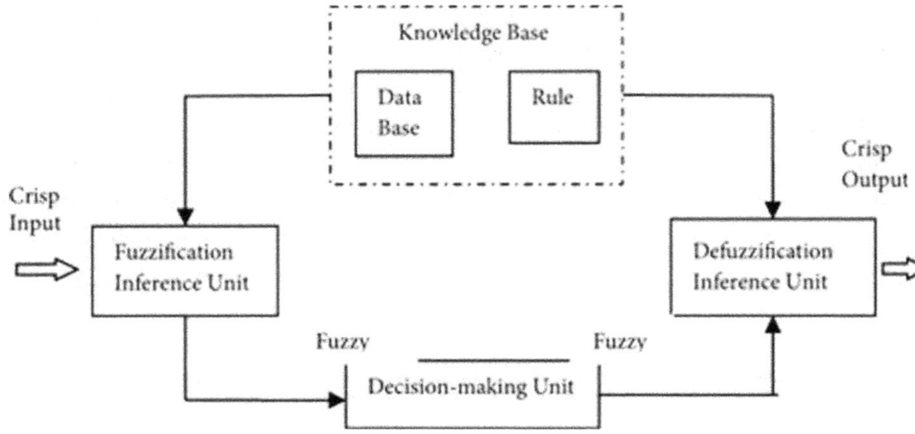

regularly contain significant volumes of information that want to be processed and interpreted to make informed choices related to topics such as weather alternate, natural functional resource management, and disaster (Patel, Patel, Patel, Patel et al, 2022).

2.2 Applicability in Earth Observation

Earth remark refers to the gathering of statistics about the Earth's surface through the usage of sensors, satellites, and different instruments. These statistics are treasured for a wide variety of programs, together with environmental monitoring, herbal aid management, catastrophe response, and climate exchange analysis. However, the translation of these records can be complex because of the inherent uncertainty and imprecision related to the measurements (Das et al., 2022; Pant et al., 2023; Srivastava et al., 2023).Fuzzy good judgment is a mathematical framework that can effectively manage uncertainty and imprecision in information analysis and selection-making. It's particularly proper for situations where variables have overlapping club degrees, bearing in mind an extra flexible method to class and modeling. In recent years, fuzzy logic-based tracking of Earth observations has emerged as a promising technique for enhancing the accuracy and reliability of analyses in this area (Khan et al., 2022).One of the predominant advantages of the usage of fuzzy common sense in Earth commentary is its capacity to incorporate professional knowledge and subjective judgments into the analysis technique. Conventional methods regularly depend upon strict thresholds and guidelines that could lead to oversimplification or exclusion of crucial data. Fuzzy common sense, however, allows for the integration of more than one resource of statistics, along with qualitative and linguistic facts, permitting a more excellent complete information of the determined phenomenon (Gunjan et al., 2022; Hussain et al., 2022).

3. APPLICATIONS OF FUZZY LOGIC IN MONITORING EARTH PHENOMENA

Fuzzy good judgment is a mathematical framework that deals with uncertainty and imprecision. It's broadly used in various fields and has been confirmed to be especially beneficial in monitoring and studying earth phenomena. In this essay, we can discover some of the packages of fuzzy logic in monitoring earth observations and discuss why it's a practical approach for dealing with the complexity and uncertainty inherent in those phenomena.One of the significant demanding situations in monitoring earth phenomena is the inherent uncertainty and imprecision inside the data gathered. Earth observations are often laid low with diverse sources of noise and measurement errors. Moreover, the phenomena being monitored, together with weather alternatives, herbal failures, or ecological structures, are complicated and exhibit nonlinear behaviors. Traditional monitoring approaches that depend on crisp thresholds and binary classification might need to be more capable of handling those complexities and uncertainties effectively.Fuzzy logic provides a bendy and sturdy framework for addressing these demanding situations. It allows for the representation of indistinct and obscure statistics through linguistic variables and fuzzy units. With the aid of assigning membership grades to those fuzzy units, fuzzy logic can seize the gradual and unsure nature of earth observations. This enables a more nuanced and accurate illustration of the phenomena being monitored.

3.1 Remote Sensing and Image Analysis

The sector of far-flung sensing and photo evaluation has revolutionized the manner in which we gather and interpret Earth observations. Faraway sensing refers to the acquisition of statistics about the Earth's floor from a distance, usually, the usage of satellites or aerial platforms, even as picture evaluation entails processing and extracting meaningful statistics from those data. In recent years, there has been a developing interest in applying fuzzy common sense-primarily based approaches to monitor and interpret these observations because of their capacity to handle uncertainty and imprecision inherent in far-flung sensing data.Fuzzy logic is a mathematical device that could constitute and cause obscure or vague records. It differs from classical logic, which assumes that propositions are both authentic or false by allowing for levels of reality. This is especially useful in remote sensing programs, as statistics acquired from sensors often consist of noise, uncertainty, and ambiguity.One of the fundamental benefits of fuzzy logic-based tactics is their capability to deal with spatial and spectral uncertainty in faraway sensing information. Traditional techniques commonly involve classifying pixels into predefined classes (e.g., land cover sorts), primarily based on their spectral characteristics. However, due to factors together with atmospheric situations and sensor limitations, it's frequently challenging to reap correct and constant data from faraway sensing information..

3.2 Decision Support Systems

Fuzzy logic-based total decision support structures have been established to be adequate gear in monitoring earth observations. Those systems make use of fuzzy logic, a mathematical method that permits the illustration and manipulation of imprecise, unsure, or subjective statistics. By way of incorporating fuzzy good judgment, decision-aid structures can cope with and analyze large volumes of complex and ambiguous information obtained from earth observations, permitting better and more informed decision-making.Earth observations offer essential records about the country of our planet, inclusive of weather patterns, land cowl adjustments, and the health of numerous ecosystems. However, the facts obtained from these observations often need to be completed, noisy, or unsure, making it challenging to extract meaningful insights. This is in which fuzzy common sense-based total choice assists structures to excel. They could deal with such information by accommodating imprecision and uncertainty, supplying an extra correct representation of the found phenomena.One of the critical blessings of fuzzy logic-based systems is their ability to deal with vague or ill-defined principles. In conventional binary common sense, principles are either genuine or fake and do not use room for uncertainty. However, inside the context of earth observations, many variables are not easily classified into binary instructions. For instance, figuring out the health of a specific environment may also contain more than one factor, inclusive of flower coverage, water availability, and biodiversity.

4. KEY COMPONENTS OF FUZZY LOGIC-BASED MONITORING SYSTEM

Fuzzy, suitable judgment-based monitoring structures have gained massive interest in recent years due to their potential to take care of uncertainty and imprecision in complicated datasets effectively. Such structures can be carried out in various domains, including earth observations, wherein accurate and reliable tracking is critical for environmental control and decision-making approaches. This essay will talk about the key components of a fuzzy standard sense-based tracking system for earth observations. The primary component of a fuzzy standard sense-based monitoring system is the linguistic variable. Linguistic variables outline the terms and ideas used within the monitoring process. Within the context of earth observations, these variables can represent bodily parameters together with temperature, humidity, or flora index. Linguistic variables are commonly described through a fixed of linguistic terms, which constitute fuzzy units. For instance, a linguistic variable of temperature will have terms like "bloodless," "cool," "heat," and "hot." these terms are then associated with specific fuzzy club capabilities that describe the diploma of the club.The second key issue is the fuzzy rule base. Fuzzy rules are the coronary heart of a fuzzy logic-based device as they capture the understanding and expertise of domain professionals. In earth observations, these regulations can be derived from clinical ideas, statistical analyses, or empirical observations.

4.1 Fuzzy System Architecture

1. Fuzzy logic in Earth Observations: Fuzzy good judgment is a branch of artificial intelligence that deals with reasoning and selection-making under uncertainty. Earth observations frequently contain vague and incomplete facts, making fuzzy good judgment a suitable method for monitoring and analyzing these observations. Fuzzy common sense allows for the manipulation of linguistic variables and membership functions, enabling the incorporation of professional information and supplying more accurate and interpretable effects.
2. Components of Fuzzy device architecture: A fuzzy device structure for tracking earth observations typically includes three primary additives: enter variables, fuzzy logic inference engine, and output variables:
    a. Enter Variables: The entered variables constitute the found parameters, together with temperature, precipitation, or air pollution levels. Those variables are frequently labeled into fuzzy sets defined via membership functions, which seize the uncertainty and imprecision associated with the observations.
    b. Fuzzy common sense Inference Engine: the fuzzy good judgment inference engine performs the reasoning and selection-making process based totally on the enter variables. It includes fixed fuzzy guidelines that map the enter variables to the output variables. These rules are formulated based on expert know-how or data-pushed strategies, such as fuzzy clustering

or genetic algorithms. The inference engine uses these guidelines to decide the diploma of membership of the enter variables in the fuzzy sets and generates crisp outputs based on the fuzzy inference method.

c. Output Variables: The output variables constitute the very last consequences or selections derived from the bushy common sense inference engine. These outputs can range from easy linguistic descriptions, which include "low," "medium," or "excessive," to numerical values or actionable tips. The output variables are determined based totally on the general system objectives and necessities.

3. Advantages of Fuzzy device structure: The fuzzy system architecture for tracking earth observations gives several benefits over traditional approaches:
   a. Uncertainty coping with Fuzzy logic can correctly take care of uncertain and obscure statistics, that's common in earth observations. The club functions and linguistic variables offer a natural representation of this uncertainty, making an allowance for more robust and interpretable consequences.

4.2 Integration With Earth Observation Data

Earth statement is the procedure of accumulating statistics about the Earth and its environments through far-flung sensing technology, which includes satellites. These observations have emerged as crucial in numerous fields, consisting of environmental tracking, weather exchange analysis, and herbal aid management. Earth statement records provide valuable data that can be used to make knowledgeable decisions and policies, leading to higher expertise and conservation of our planet.One of the challenges in using Earth commentary information is efficiently studying and interpreting the extensive quantity of records collected. Traditional techniques frequently depend upon challenging thresholds and binary classifications, which also oversimplify complex phenomena and bring about misguided assessments. However, fuzzy common sense-primarily based tracking has emerged as a progressive technique that can deal with the complexities and uncertainties inherent in Earth remark records.Fuzzy good judgment is a mathematical framework that allows for the representation of vague and unsure facts. It's based on the idea of fuzzy units, which assign ranges of clubs to objects or activities based totally on their characteristics. This allows for an extra nuanced and bendy representation of records in preference to rigid binary classifications. In the context of Earth observation, fuzzy joint sense-primarily based tracking can offer better insights into the dynamic and constantly converting phenomena.

5. ADVANTAGES AND BENEFITS

Fuzzy joint sense-based total monitoring of earth observations offers numerous benefits that may significantly decorate our expertise and control of natural resources and the environment. This technique embraces the inherent uncertainty and imprecision in earth observation statistics, taking into consideration extra accurate and reliable evaluation.One of the number one benefits of fuzzy logic-based total tracking is its potential to address incomplete and uncertain facts. Earth remark records are regularly noisy and incomplete, with diverse resources of mistakes and ambiguity. Fuzzy common sense lets in for the mixing of various facts sources and the combination of a couple of variables to capture the complex nature of the earth's structures. This enhances the robustness and reliability of the monitoring system and mitigates the dangers related to misguided or incomplete records.Furthermore, fuzzy logic-primarily based tracking enables the representation and evaluation of linguistic records and qualitative understanding. that is treasured explicitly while dealing with subjective or non-numeric facts, inclusive of expert evaluations or qualitative observations. Fuzzy common sense lets in for the mixing of these styles of information into the monitoring procedure, offering more excellent holistic and comprehensive information about the earth's systems. That is especially important in domain names in which basically quantitative statistics may be insufficient to seize the total complexity and dynamics of the surroundings.

5.1 Handling Uncertainty and Incompleteness

Fuzzy joint sense-based total tracking of earth observations has emerged as a powerful tool for dealing with and analyzing complicated and uncertain facts. It allows for the dealing with uncertainty and incompleteness in a way that traditional models frequently warfare with, supplying a greater comprehensive and correct expertise of the discovered phenomena. This essay will discover how fuzzy joint sense-primarily based monitoring addresses those challenges and its advantages over conventional strategies.One of the principal strengths of fuzzy logic-based tracking is its potential to deal with uncertainty. Uncertainty is an inherent function of earth observations due to different factors, inclusive of dimension errors, confined data availability, and herbal variability. Fuzzy common sense strategies allow the modeling of uncertain and vague information by assigning ranges of club to one-of-a-kind categories, taking into consideration an extra nuanced illustration of the discovered facts. By incorporating uncertainty into the evaluation, fuzzy logic-based total monitoring can provide greater reliable and sturdy results, even when coping with incomplete or noisy facts.Moreover, fuzzy logic-primarily based monitoring can efficiently take care of incomplete statistics units, a common difficulty encountered in earth observations. Incomplete statistics can arise from sensor screw-ups, records transmission mistakes, or limited coverage regions that can notably impact the accuracy and reliability of conventional tracking techniques. Fuzzy good judgment strategies can handle missing statistics by leveraging membership capabilities that recollect the available statistics.

5.2 Real-World Adaptability

Fuzzy logic-based tracking of Earth observations is a revolutionary method used to analyze information gathered from various resources so that it will gain insights into the Earth's systems. The utility of fuzzy logic allows for the incorporation of uncertainties and imprecise records, making it an effective tool for managing real-world complexities. This essay discusses the real-international adaptability of fuzzy good judgment-based total tracking of Earth observations, highlighting its strengths and capability packages.One vast factor of fuzzy logic-primarily based tracking is its potential to deal with uncertainties and vague records. Inside the actual international, information accrued from satellite imagery, weather stations, or other sources frequently has inherent uncertainties because of measurement mistakes or restrained decisions. Fuzzy common sense gives a framework to symbolize and examine this unsure and vague information, enabling more comprehensive know-how of Earth processes. It allows researchers and decision-makers to account for the varying stages of reality in the accrued statistics, which is vital for correct analysis and prediction.One other gain of fuzzy joint sense-primarily based tracking is its adaptability to complex, non-linear structures. Earth systems exhibit complex conduct, with interactions among diverse components influencing each other in a non-linear manner. Fuzzy good judgment can capture and version these complex relationships, thereby enabling the monitoring of dynamic Earth phenomena.Few examples are given below for better understanding.

1. Fuzzy Logic and Remote Sensing for Atmospheric Turbulence Estimation: A fuzzy logic-based method was developed to estimate the turbulence of atmospheric microphysical processes from microwave satellite remote sensing. The method was based on matching the observed Radar Equivalent Reflectivity Factor (EREF) features to the expected values derived from theoretical models. From this, the turbulence structure of planetary boundary layer was estimated. The method was tested on two different cloud situations over California and showed that the mean EREF between observed and expected values was minimized when the turbulence was estimated using fuzzy logic.
2. Fuzzy Logic for Rainfall Estimation from SAR Images: Fuzzy logic was used to estimate precipitation events from Synthetic Aperture Radar (SAR) images. The approach accounted for the uncertain information in the data and combined historical and Land Surface Temperature (LST) data with SAR reflectivity to produce an estimate of the precipitation amount. The system was tested across a range of different scenes and showed an improvement in accuracy over using LST or reflectivity alone. This demonstrates the potential of combining multiple sources of Earth-oriented data with fuzzy logic methods to derive meaningful information.
3. Fuzzy Logic Applied to Radiometric Calibration of an Earth Observation Satellite: The fuzzy logic approach was used to calculate the radiometric calibration of an Earth observation satellite. The method was developed by combining remote sensing data from several different sensors into a single suite of calibration factors. The satellite data was then used to build and validate a fuzzy logic-based model. The fuzzy logic algorithm was found to be accurate and predict the calibration coefficients compared to a traditional calibration approach. This demonstrates the potential of using more complex approaches, such as fuzzy logic, for radiometric calibration of remote sensing images.

6. CHALLENGES AND LIMITATIONS

Fuzzy logic-based tracking of Earth observations is a powerful device that allows for the analysis and interpretation of complicated and uncertain data. It is able to provide valuable insights into numerous elements, including land use, flora cowl, climate alternate, and herbal disaster management. However, it additionally presents several demanding situations and obstacles that need to be addressed for its effective implementation.First of all, one major undertaking is the purchase and availability of accurate and reliable enter facts. Fuzzy good judgment-based monitoring is closely based on data from faraway sensing satellites, floor-primarily based observations, and different assets. However, these records resources may have barriers in phrases of spatial and temporal decision, accuracy, and coverage. This may introduce uncertainties and errors within the tracking process, leading to faulty effects and unreliable choice-making.Secondly, the proper choice and layout of fuzzy common sense models pose every other project. Developing fuzzy common sense models calls for area expertise and know-how to outline appropriate membership features, fuzzy rules, and inference mechanisms. Those models should be capable of capturing the complex relationships and uncertainties found in Earth statement data. However, finding the proper stability between simplicity and complexity is essential, as overly complicated models might also cause overfitting and poor generalization. At the same time, overly easy models might also forget essential styles and statistics.

6.1 Interpretability and Model Complexity

Fuzzy good judgment-based tracking of Earth observations has received good-sized interest in recent years due to its ability to address uncertainties and obscure records in environmental tracking. This approach is predicated on mathematical models primarily based on fuzzy common sense, which gives a bendy and interpretable framework for making informed choices based totally on incomplete or unsure facts. However, accomplishing each interpretability and version complexity is an undertaking that must be cautiously addressed to ensure the effectiveness and reliability of the monitoring machine.Interpretability refers to the capability to apprehend and explain the reasoning behind the selections made through the tracking gadget. Inside the context of fuzzy logic-primarily based tracking, interpretability is essential because it lets researchers, choice-makers, and stakeholders recognize and believe the outputs of the device. Fuzzy logic models attain interpretability with the aid of using linguistic variables, fuzzy units, and logical rules that simulate human-like decision-making. With the aid of representing data in a linguistic shape, which includes "low," "medium," and "high," fuzzy good judgment allows customers to without difficulty apprehend the reasoning at the back of the device's decision-making technique.

7. CONCLUSION

The fuzzy good judgment-based tracking of Earth observations has been established to be a practical and revolutionary approach in addressing the complexities and uncertainties associated with monitoring our planet. This monitoring device has revolutionized the sphere of Earth observation, offering precious insights and predictions in diverse domains, including weather forecasting, environmental monitoring, and natural disaster control.One of the key advantages of this method is its ability to deal with vague and unsure data. In contrast to conventional monitoring systems that rely upon crisp obstacles and unique measurements, fuzzy joint sense-primarily based tracking permits the inclusion of uncertain or ambiguous statistics. This is especially beneficial in Earth commentary, in which records can often be noisy or incomplete. With the aid of using fuzzy good judgment, the monitoring device can assign degrees of membership to different classes, making an allowance for more excellent, flexible, and correct evaluation.Furthermore, fuzzy logic-primarily based monitoring carries professional know-how into the device. It may still be remembered now that there are not only numerical records but also qualitative data from area experts. This integration of subjective understanding enhances the robustness and reliability of the tracking machine because it is sometimes entirely depending on computerized algorithms. Through combining quantitative and qualitative information, fuzzy good judgment-based tracking can offer complete and holistic information on Earth observations..

7.1 Summary of Key Findings

Fuzzy logic-based tracking of earth observations has emerged as a treasured device within the subject of far-off sensing. It makes use of fuzzy common sense, a mathematical approach that offers uncertainty and imprecision, to analyze and interpret satellite TV for pc pics and other sensing records. Over the years, researchers have made massive developments in imposing fuzzy logic-based total monitoring structures, and several key findings have emerged from these studies.One of the key findings is the potential of fuzzy good judgment-based monitoring to handle and incorporate uncertainty in earth observations. Conventional monitoring methods frequently battle with uncertainties associated with sensor noise, atmospheric situations, and ambiguous class standards. Fuzzy logic-based systems, alternatively, excel in handling such uncertainties by presenting bendy and adaptive reasoning strategies. By way of modeling uncertainties through the use of linguistic variables and fuzzy units, these systems are able to take pictures of the vagueness and imprecision inherent in remote sensing statistics.Some other good-sized finding is the multiplied accuracy executed via the combination of fuzzy common sense with other records analysis techniques. Fuzzy joint sease-primarily based tracking can be coupled with gadget-studying algorithms, which include artificial neural networks and aid vector machines, to decorate the accuracy of type and sample reputation tasks..

7.2 Recommendations for Future Research

Fuzzy good judgment has emerged as a powerful device in numerous fields of research, along with monitoring earth observations. It can handle imprecise and unsure data, making it a treasured technique for studying complex and unsure earth commentary facts. This essay aims to present pointers for future studies in fuzzy logic-primarily based tracking of earth observations that specialize in ability regions of improvement and innovation.A critical piece of advice is to discover the mixing of fuzzy logic with different records fusion techniques. Earth statement facts often come from multiple resources, including satellites, ground-based sensors, and far-flung sensing systems. By combining fuzzy common sense with different statistics fusion strategies, together with gadget learning algorithms or professional systems, we will decorate the accuracy and reliability of the monitoring method. This integration can offer complete information on earth observations by way of leveraging the strengths of various strategies and compensating for their limitations.Another crucial avenue for future research is the improvement of adaptive and self-mastering fuzzy good judgment models. Historically, fuzzy good judgment fashions require predefined membership capabilities and guidelines that need to be tuned manually. However, the earth remark facts are relatively dynamic and might showcase temporal and spatial versions. Therefore, there's a need for adaptive fuzzy good judgment fashions that may regulate their club capabilities and regulations over time based on the changing characteristics of the data.

Recent advancements and studies in fuzzy logic-based monitoring include the development of new techniques for detecting land cover change, such as using artificial neural networks for remote sensing applications or employing fuzzy logic-based techniques to classify land cover maps. In addition, fuzzy logic-based approaches have been used to develop real-time monitoring systems that can detect the onset and extent of extreme weather events or monitor spatiotemporal land cover changes. Other research has focused on utilizing fuzzy logic-based methods to monitor and analyze changes in hydrological and ecological systems or to detect changes from human activities such as agricultural practices. Finally, researchers have looked into the use of fuzzy logic-based methods for autonomous navigation of unmanned aerial vehicles (UAVs) and unmanned ground vehicles (UGVs).

REFERENCES

Aklah, Z. T., Al-Safi, A., & Abdali, M. H. (2023). *Predicting High-Risk Individuals in Disease Outbreaks Using Fuzzy Logic-Based Contact Tracing Model.*

Das, S. R., Hota, A. P., Pandey, H. M., & Sahoo, B. M. (2022). Industrial power quality enhancement using fuzzy logic based photovoltaic integrated with three phase shunt hybrid active filter and adaptive controller. *Applied Soft Computing*, *121*, 108762. doi:10.1016/j.asoc.2022.108762

Gunjan, V. K., Vijayalata, Y., Valli, S., Kumar, S., Mohamed, M. O., & Saravanan, V. (2022). Machine learning and cloud-based knowledge graphs to recognize suicidal mental tendencies. *Computational Intelligence and Neuroscience*, *2022*, 2022. doi:10.1155/2022/3604113 PMID:35341205

Hai, T., Zhou, J., & Muranaka, K. (2022). An efficient fuzzy-logic based MPPT controller for grid-connected PV systems by farmland fertility optimization algorithm. *Optik (Stuttgart)*, *267*, 169636. doi:10.1016/j.ijleo.2022.169636

Hossain, M. S., Kabir, M. M., Mushtary, T., Mostafa, S. G., Hasan, M. M., Hossain, M. M., & Islam, M. S. (2022, December). Fuzzy Logic-Based Maximum Power Point Tracking of Solar PV System. In *2022 IEEE International Women in Engineering (WIE) Conference on Electrical and Computer Engineering (WIECON-ECE)* (pp. 284-289). IEEE.

Hussain, I., Khalil, I. U., Islam, A., Ahsan, M. U., Iqbal, T., Chowdhury, M. S., Techato, K., & Ullah, N. (2022). Unified Fuzzy Logic Based Approach for Detection and Classification of PV Faults Using IV Trend Line. *Energies*, *15*(14), 5106. doi:10.3390/en15145106

Ibrahim, O., Bakare, M. S., Amosa, T. I., Otuoze, A. O., Owonikoko, W. O., Ali, E. M., & Ogunbiyi, O. (2023). Development of fuzzy logic-based demand-side energy management system for hybrid energy sources. *Energy Conversion and Management: X, 18*, 100354.

Kannan, S., Dhiman, G., Natarajan, Y., Sharma, A., Mohanty, S. N., Soni, M., Easwaran, U., Ghorbani, H., Asheralieva, A., & Gheisari, M. (2021). Ubiquitous vehicular ad-hoc network computing using deep neural network with iot-based bat agents for traffic management. *Electronics (Basel)*, *10*(7), 785. doi:10.3390/electronics10070785

Kaur, S., & Vig, S. (2023, February). Modeling of MPPT-Based Solar Eco-System Using Fuzzy Logic Controller. *IOP Conference Series. Earth and Environmental Science*, *1110*(1), 012079. doi:10.1088/1755-1315/1110/1/012079

Khan, S., Saravanan, V. N. G. C., Lakshmi, T. J., Deb, N., & Othman, N. A. (2022). Privacy protection of healthcare data over social networks using machine learning algorithms. *Computational Intelligence and Neuroscience*, *2022*, 2022. doi:10.1155/2022/9985933 PMID:35371203

Mehta, R. (2022). Multivariate Fuzzy Logic Based Smart Healthcare Monitoring for Risk Evaluation of Cardiac Patients. *Medical Informatics and Bioimaging Using Artificial Intelligence: Challenges, Issues, Innovations and Recent Developments*, 219-243.

Naikoo, M. W., Talukdar, S., Ishtiaq, M., & Rahman, A. (2023). Modelling built-up land expansion probability using the integrated fuzzy logic and coupling coordination degree model. *Journal of Environmental Management*, *325*, 116441. doi:10.1016/j.jenvman.2022.116441 PMID:36242974

Nazirbhai, M. N., & Gajjar, R. R. (2023, May). Comparative Analysis of Perturb & Observe, Incremental Conductance and Fuzzy Logic Based Maximum Power Point Tracking Methods. In 2023 IEEE Renewable Energy and Sustainable E-Mobility Conference (RESEM) (pp. 1-6). IEEE.

Othman, S. M., & Abdulrazzaq, M. B. (2023). Fuzzy logic system for drug storage based on the internet of things: A survey. *Indonesian Journal of Electrical Engineering and Computer Science*, *29*(3), 1382. doi:10.11591/ijeecs.v29.i3.pp1382-1392

Pant, G., Bahuguna, R., Pandey, S., Gehlot, A., Yadav, S. P., & Pachauri, R. K. (2023, April). Intelligent Automated Interference for the Protection of Road Safety. In *2023 International Conference on Computational Intelligence, Communication Technology and Networking (CICTN)* (pp. 87-91). IEEE. 10.1109/CICTN57981.2023.10141086

Patel, D., Patel, S., Patel, P., & Shah, M. (2022). Solar radiation and solar energy estimation using ANN and Fuzzy logic concept: A comprehensive and systematic study. *Environmental Science and Pollution Research International*, *29*(22), 32428–32442. doi:10.100711356-022-19185-z PMID:35178628

Patel, P., Patel, Y., Patel, U., Patel, V., Patel, N., Oza, P., & Patel, U. (2022). Towards automating irrigation: A fuzzy logic-based water irrigation system using IoT and deep learning. *Modeling Earth Systems and Environment*, *8*(4), 5235–5250. doi:10.100740808-022-01452-0

Qayyum, B., Ahmed, A., Ullah, I., & Shah, S. A. (2022). A Fuzzy-Logic Approach for Optimized and Cost-Effective Early Warning System for Tsunami Detection. *Sustainability (Basel)*, *14*(21), 14516. doi:10.3390u142114516

Rahman, M. Z., Akbar, M. A., Leiva, V., Tahir, A., Riaz, M. T., & Martin-Barreiro, C. (2023). An intelligent health monitoring and diagnosis system based on the internet of things and fuzzy logic for cardiac arrhythmia COVID-19 patients. *Computers in Biology and Medicine*, *154*, 106583. doi:10.1016/j.compbiomed.2023.106583 PMID:36716687

Rajalakshmi, M., Saravanan, V., Arunprasad, V., Romero, C. T., Khalaf, O. I., & Karthik, C. (2022). Machine Learning for Modeling and Control of Industrial Clarifier Process. *Intelligent Automation & Soft Computing*, *32*(1). doi:10.32604/iasc.2022.021696

Srivastava, S., Dhyani, N., Sharma, V., Vats, S., Yadav, S. P., Kukreja, V., & Singh, R. (2023, May). Lung Infection and Identification using Heatmap. In *2023 2nd International Conference on Applied Artificial Intelligence and Computing (ICAAIC)* (pp. 1093-1098). IEEE. 10.1109/ICAAIC56838.2023.10140204

Tayfur, G. (2023). Application of fuzzy logic in water resources engineering. In *Handbook of Hydroinformatics* (pp. 155–166). Elsevier. doi:10.1016/B978-0-12-821962-1.00024-6

Yuvaraj, N., Karthikeyan, T., & Praghash, K. (2021). An improved task allocation scheme in serverless computing using gray wolf Optimization (GWO) based reinforcement learning (RIL) approach. *Wireless Personal Communications*, *117*(3), 2403–2421. doi:10.100711277-020-07981-0

Yuvaraj, N., Srihari, K., Dhiman, G., Somasundaram, K., Sharma, A., Rajeskannan, S. M. G. S. M. A., Soni, M., Gaba, G. S., AlZain, M. A., & Masud, M. (2021). Nature-inspired-based approach for automated cyberbullying classification on multimedia social networking. *Mathematical Problems in Engineering*, *2021*, 1–12. doi:10.1155/2021/6644652

Zahmatkesh, S., Klemeš, J. J., Bokhari, A., Rezakhani, Y., Wang, C., Sillanpaa, M., Amesho, K. T. T., & Ahmed, W. S. (2022). Reducing chemical oxygen demand from low strength wastewater: A novel application of fuzzy logic based simulation in MATLAB. *Computers & Chemical Engineering*, *166*, 107944. doi:10.1016/j.compchemeng.2022.107944

Chapter 11
An Early Warning System for Predicting Earthquakes

Gnanasekaran T.
https://orcid.org/0000-0003-4005-6482
Department of Electronics and Instrumentation Engineering, R.M.K. Engineering College, Chennai, India

Sofia Vincent J.
Department of Mechanical Engineering, Panimalar Engineering College, India

Dinesh Babu M.
Department of Mechanical Engineering, Rajalakshmi Institute of Technology, Chennai, India

Harish Babu T.
https://orcid.org/0000-0001-7074-6447
Department of Electrical and Electronics Engineering, Sethu Institute of Technology, India

Chairma Lakshmi K. R.
https://orcid.org/0000-0002-1221-8432
Department of Electronics and Instrumentation Engineering, R.M.K. Engineering College, Chennai, India

ABSTRACT

Earthquakes have long pose a vast danger to human existence and infrastructure. Over time, sizeable studies have been performed in the field of earthquake prediction, mainly on the improvement of an early warning machine. This essay aims to discover the background of earthquake prediction, the want for an early warning system, and the advancements made in this vicinity to mitigate the impact of seismic occasions. This chapter offers the development of a complicated and reliable early warning machine that utilizes a combination of seismic information analysis and device learning techniques. The goals of an early warning system for predicting earthquakes are to offer timely statistics to permit suitable actions, reduce the lack of existence and assets, and foster resilience in affected groups. The advancements in era and studies have provided us with the gear and expertise essential to construct a robust and reliable system that can save limitless lives and limit harm to infrastructure.

DOI: 10.4018/979-8-3693-1850-8.ch011

1. INTRODUCTION

Earthquakes are one of the most devastating natural screw-ups that can occur at any time, inflicting titanic damage and lack of existence. An early caution machine for predicting earthquakes is an innovative development that could store infinite lives and decrease the destruction because of those powerful geological activities. By way of supplying superior observation of an approaching earthquake, people, communities, and governments can take immediate action to make specific preparedness, evacuate susceptible areas, and mitigate the impact of these failures (Abdalzaher, Elsayed, Fouda et al, 2023).The concept of an early warning gadget is based totally on the notion that earthquakes have a few feature styles or precursory symptoms that may be detected and measured before an actual seismic event happens. Scientists and researchers have been analyzing seismic waves, ground movement, and different seismic parameters for many years to pick out such styles and expand strategies to predict earthquakes as they should be.The advent of an early warning machine calls for an extensive investment in generation, infrastructure, and studies (Chiang et al., 2022). The device relies on a network of seismic monitoring stations ready with superior sensors and verbal exchange structures. These stations discover the preliminary primary waves (P-waves) that are much less destructive and journey quicker than the secondary waves (S-waves) that cause the intense shaking generally related to earthquakes.

1.1 Background of Earthquake Prediction

Earthquakes have long involved scientists and pose a vast danger to human existence and infrastructure. Over time, sizeable studies have been performed in the field of earthquake prediction, mainly on the improvement of an early warning machine. This essay aims to discover the background of earthquake prediction, the want for an early warning system, and the advancements made in this vicinity to mitigate the impact of seismic occasions (Kumar et al., 2022).In ancient times, diverse cultures tried to expect earthquakes through the usage of unconventional techniques, which included animal conduct, celestial observations, or even legendary beliefs. However, these strategies needed more scientific proof and often ended in false alarms or ignored predictions, leading to skepticism among a number of the scientific community.With the advancement of scientific understanding and technology, researchers began investigating the underlying reasons for earthquakes (Suárez, 2022). The invention of plate tectonics and seismic waves within the early 20th century supplied full-size breakthroughs in the know-how of the geophysical procedures related to earthquakes. These discoveries laid the foundation for more unique prediction methods (Joshi et al., 2022).

1.2 Importance of Early Warning Systems

Earthquakes are among the most devastating and unpredictable natural disasters that arise on our planet. They can cause extensive damage to infrastructure, loss of existence, and extensive disruption to groups. That allows you to mitigate the consequences of earthquakes and shield vulnerable populations; the improvement and implementation of early caution systems have emerged as more and more critical. An early warning gadget for predicting earthquakes can provide vital statistics to authorities, giving them the opportunity to take set-off and centered actions, thus saving lives and reducing economic losses (Yuvaraj, Karthikeyan, & Praghash, 2021).One of the critical advantages of early caution structures is their capacity to offer strengthened notes of an imminent earthquake. At the same time as it is authentic that earthquakes cannot be predicted with absolute truth, scientists have made significant developments in expertise on the precursor signals that can precede an earthquake. These signals can consist of ground deformation, unusual animal conduct, changes in groundwater stages, and versions in the electromagnetic hobby. By means of tracking and studying those precursor alerts, early warning systems can locate the prevalence of an earthquake and difficulty warnings to the affected regions (Khan et al., 2022; Rajalakshmi et al., 2022).

The well-timed dissemination of earthquake warnings is crucial for robust preparedness and response. Early warning systems can supply signals via various channels, which include text messaging, sirens, and smartphone packages (Abdalzaher, Soliman, & El-Hady, 2023).

1.3 Objectives of the Early Warning System

The devastating effects of earthquakes were witnessed all through records, causing massive loss of existence and destruction of infrastructure. It's essential to set up an early warning machine to mitigate these risks. The goals of an early warning system for predicting earthquakes are to offer timely statistics to permit suitable actions, reduce the lack of existence and assets, and foster resilience in affected groups (Datta et al., 2022).First of all, the primary objective of an early caution system is to offer well-timed records. Earthquakes arise all at once, leaving little time for instruction and response. By detecting and predicting seismic occasions earlier than they manifest, an early caution machine can offer crucial statistics to individuals, agencies, and emergency offerings (Medina et al., 2022). Timely facts allow for the implementation of necessary safety measures, evacuation plans, and emergency protocols, decreasing human casualties and damage to infrastructure.Secondly, an early warning system aims to minimize the lack of life and property. Earthquakes are recognized to cause considerable devastation, central to casualties and lengthy-term results (Zhang et al., 2022). By way of issuing warnings and signals, people can make appropriate movements to guard their lives and try to find shelter in particular safe areas. Moreover, agencies and critical infrastructure can implement measures to shield their property and limit operational disruptions. The early identity and verbal exchange of earthquake dangers permit pre-emptive measures, reducing casualties and loss of assets (Cho et al., 2022; Esposito et al., 2022).

2. FUNDAMENTALS OF EARTHQUAKES AND SEISMIC ACTIVITY

An early warning gadget for predicting earthquakes is critical as a way to keep lives and decrease the harm caused by those natural screw-ups. Information on the basics of earthquakes and seismic hobby is essential in developing this sort of device.Earthquakes are unexpected releases of strength inside the Earth's crust due to the motion of tectonic plates (Peng et al., 2022). The Earth's crust is made up of numerous big plates that can be continuously shifting and interacting with each other. When the pressure on those plates turns into the too first-rate, it's far launched in the shape of seismic waves, resulting in an earthquake.Seismic activity can be measured and monitored with the use of seismometers, instruments that detect and record vibrations in the Earth's crust. Those devices are strategically positioned around the sector to provide actual-time records on earthquake pastimes (Cochran et al., 2022). Via reading the records amassed from seismometers, scientists can perceive styles and developments that help in information on the occurrence and behavior of earthquakes.One of the key elements in predicting earthquakes is the idea of foreshocks and aftershocks. Foreshocks are more minor earthquakes that occur earlier than the principal event and might function as caution signs (Prasanna et al., 2022; Song et al., 2022). Those precursor occasions are often seen as an indicator of accelerated seismic hobby in an area and can be detected using seismometers.

2.1 Earthquake Causes and Mechanisms

The phenomenon of earthquakes has been a steady supply of fascination and challenge for humankind. The unpredictable nature of earthquakes poses widespread demanding situations in mitigating their devastating impact. However, with advances in technology and medical information, the development of an early caution machine for predicting earthquakes has become a possibility (Kannan et al., 2021; Yuvaraj, Srihari, Dhiman et al, 2021). In this essay, we will explore the causes and mechanisms of earthquakes and the potential of an early caution device to mitigate their destructive consequences on human lives and infrastructure.Earthquakes arise due to the sudden launch of energy inside the Earth's crust. This release of energy is ordinarily attributed to 3 main reasons: tectonic plate moves, volcanic hobby, and human-induced activities (Saad et al., 2022).The Earth's lithosphere is split into several tectonic plates, which can be continuously in movement due to the convective currents within the asthenosphere underneath them. Obstacles where plates interact are known as fault traces. Earthquakes typically arise along these fault strains when collected stress reaches a critical factor, causing the rocks to fracture and release electricity in the form of seismic waves.

2.2 Seismic Waves and Monitoring

An earthquake is one of the most negative herbal screw-ups that could arise on our planet. With the capability to reason for the enormous loss of life and property damage, the capability to expect and put together for earthquakes becomes critical. Seismic waves play an essential role in the monitoring and early warning structures designed to prevent earthquakes.Seismic waves are the waves of strength that radiate from the supply of an earthquake. There are numerous kinds of seismic waves, together with primary (P-waves), secondary (S-waves), and surface waves. P-waves are the fastest and can tour via each solids and beverages (Böse et al., 2022). S-waves follow the P-waves and can best journey via solids, at the same time as surface waves are the slowest and motivate the most damage.Tracking seismic waves is essential for predicting earthquakes. Seismographs, units that come across and report ground motions resulting from seismic waves, are essential additives of any tracking gadget. While an earthquake occurs, the seismograph produces a graph referred to as a seismogram, which shows the amplitude and duration of the seismic waves. This record enables scientists to decide the exact place and importance of the earthquake.An early warning device for predicting earthquakes is predicated heavily on the monitoring of seismic waves (Böse et al., 2023).

2.3 Historical Perspectives on Earthquake Prediction

Earthquakes have long been a source of fascination and fear for humanity, causing immense destruction and loss of life. The ability to predict earthquakes and give early warning to affected areas could save countless lives. Over the centuries, scientists and researchers have developed various methods and theories to predict earthquakes, each shaped by the historical perspectives and scientific advancements of their time.One of the earliest historical perspectives on earthquake prediction can be traced back to ancient China (Pant et al., 2023). As early as the Han dynasty (206 BC – 220 AD), Chinese seismologists observed specific precursors, such as animal behavior, ground deformations, and abnormal water levels, prior to major earthquakes. These observations formed the basis of empirical earthquake prediction methods, which focused on the recognition of precursor events. However, these methods lacked scientific explanations and were often marred by superstition and pseudoscience.In the early modern era, advancements in seismology and the understanding of geology led to the development of new perspectives on earthquake prediction. One such approach was the stress accumulation model, proposed by Robert Mallet in the mid-19th century. Mallet suggested that earthquakes occur due to the release of accumulated stress along geological faults(Srivastava et al., 2023).

3. THE NEED FOR EARLY WARNING SYSTEMS

An early warning system for predicting earthquakes is a critical tool in ensuring the safety and well-being of communities at risk. Earthquakes are natural disasters that often strike with little to no warning, causing immense devastation and loss of life. Therefore, the need for early warning systems to predict earthquakes is paramount, as it can provide vital information and precious minutes that can potentially save lives, reduce casualties, and enable better preparation.First and foremost, early warning systems serve to alert individuals and communities about potential earthquakes, giving them essential time to seek shelter, evacuate, or implement necessary safety measures. The seismic waves generated by an earthquake travel at different speeds, with the slower and less destructive waves arriving first, followed by the more damaging ones. By detecting these initial slower seismic waves, early warning systems can provide alerts to areas located closer to the epicenter, granting them crucial seconds or even minutes to react before the primary destructive waves hit. During this time, people can find shelter, brace themselves, and take other necessary precautions to minimize risks and casualties.Furthermore, early warning systems can also provide valuable information to emergency responders, enabling them to mobilize and deploy resources effectively. By receiving early warnings, emergency management teams can prepare themselves, machinery, and medical supplies in advance, positioning themselves to respond.

3.1 Limitations of Current Prediction Methods

An early caution machine for predicting earthquakes has long been the dream of scientists and seismologists. With the capability to save lots of lives and save you good-sized destruction, the development of accurate prediction techniques is of extreme significance. But, in spite of extensive advancements in recent years, current prediction techniques nonetheless have several barriers that avert their efficacy and reliability.One of the principal limitations of contemporary prediction strategies for earthquake early warning structures is their reliance on figuring out precursory signs of seismic interest. At the same time, as it is accurate that specific precursors, inclusive of floor deformation or adjustments in groundwater tiers, were observed earlier than some earthquakes, those signs and symptoms aren't usually gifted or without problems detectable. Earthquakes are complex herbal phenomena, and their incidence is influenced by various factors that aren't fully understood. Therefore, depending solely on precursory signs to expect earthquakes can be an unsuitable approach.Another hassle of cutting-edge prediction strategies is the difficulty in distinguishing foreshocks from mainshocks. Foreshocks are more minor earthquakes that precede the primary occasion, and their occurrence may imply a drawing close to a giant earthquake. However, differentiating foreshocks from ordinary seismic hobbies is a challenging undertaking.

3.2 Importance of Timely Responses

Earthquakes are herbal calamities that can convey titanic destruction and lack of lifestyle. Over time, efforts have been made to increase an early caution gadget to mitigate the impact of earthquakes. Those structures are characteristic by means of detecting seismic waves and issuing timely indicators to doubtlessly affected regions. It's vital, however, to recognize the crucial significance of timely responses in ensuring the effectiveness of an early warning gadget for predicting earthquakes.

1. Minimizing lack of existence: The number one goal of an early caution device is to keep lives. By offering timely information about impending earthquakes, individuals and authorities can take essential actions to defend themselves and others. Well-timed responses permit the spark-off execution of evacuation plans, resulting in the evacuation of capacity danger zones. This short action drastically reduces the threat of casualties and ensures public safety.
2. Preventing infrastructure damage: Earthquakes can cause intense harm to homes, roads, and infrastructure. With the aid of directly responding to caution signals, authorities can provoke appropriate measures to prevent in addition destruction. Movements together with the shutdown of gas pipelines, suspension of transportation services, and halting of industrial activities can reduce damage and store treasured belongings.

4. DATA COLLECTION AND PROCESSING

An early warning device for predicting earthquakes holds enormous importance in brand new international. Earthquakes, despite the fact that they are unpredictable, can cause tremendous harm and loss of lives. Consequently, the need for a sophisticated machine to acquire and process statistics for correct earthquake prediction arises. This essay aims to speak about the crucial components of information series and processing for an early caution gadget.Information collection bureaucracy is the backbone of any early caution gadget for predicting earthquakes. A couple of strategies and technologies are applied to acquire and examine diverse sorts of information. One of the number one sources is seismic statistics gathered through seismographs positioned strategically across seismically energetic regions. Seismographs document floor movement and vibrations caused by tectonic plate moves. These units are crucial in detecting seismic waves, which can provide insights into the incidence and depth of earthquakes. Except for seismic data, other kinds of statistics are also employed within the facts series section. GPS records and satellite pictures aid in monitoring floor deformations and displacements, which may be signs and symptoms of a drawing close earthquake. Further, statistics from sensors located inside fault strains can offer treasured facts about the accumulation of pressure and strain. This information assists in the expertise the motion of tectonic plates and the likelihood of an earthquake..

4.1 Seismic Data Sources

Earthquakes are natural screw-ups that could motivate full-size destruction and lack of lifestyle. The ability to correctly predict earthquakes has long been a goal of scientists and researchers. Early caution systems (EWS) have been developed to mitigate the devastating effects of earthquakes by means of presenting advanced words to affected regions. Such systems heavily depend upon amassing and analyzing seismic facts to discover and predict the occurrence of earthquakes. This essay explores the fundamental seismic facts resources utilized in an early warning gadget for earthquake prediction.

1. Seismic sensors: Seismic sensors also referred to as seismometers, are the number one resource of seismic facts for earthquake prediction. These devices are strategically located in numerous locations to discover floor motion because of seismic waves. Seismic sensors can be both permanent or cellular, with the former being installed in real-time tracking networks and the latter deployed in response to unique seismological activities. These sensors degree acceleration, speed, or displacement of ground movement, imparting important data for earthquake prediction.
2. Seismic networks: Seismic networks include several interconnected seismic sensors, forming a tracking system spanning a massive geographic vicinity. Those networks ensure complete coverage, enabling the collection of seismic information from more than one location concurrently.

4.2 Geospatial and Geological Data Integration

An early caution system for predicting earthquakes is essential in mitigating the influences and ability devastation caused by these natural failures. Earthquakes are unpredictable and might arise at any time, causing loss of life and widespread damage to infrastructure. Geospatial and geological information integration performs a crucial position in the improvement and effectiveness of such an early caution system.Geospatial information refers to statistics that can be represented on a map or in a virtual format, such as satellite imagery, topographic maps, and geographic data device (GIS) information. Geological information, however, relates to information about the Earth's structure, along with seismic activity, fault traces, and tectonic plate actions. Integrating those facts sources permits a complete understanding of the underlying factors contributing to earthquake occurrences.One key aspect of geospatial and geological statistics integration is the usage of satellite imagery. Earth commentary satellites offer precious information about the Earth's surface, allowing scientists and researchers to tune changes over the years. By tracking patterns in land surface deformations, scientists can pick out capability fault traces and verify the chance of seismic interest in favorable areas. These statistics aid in the improvement of earthquake prediction fashions and permit the early warning gadget to generate signals and notifications to the general public..

4.3 Real-Time Data Processing Challenges

Real-time data processing is a crucial factor of an early warning gadget for predicting earthquakes. The ability to research and act upon records in actual time can potentially save several lives and reduce damages caused by seismic occasions. However, there are several challenges in imposing a powerful real-time information processing device for earthquake prediction. This essay explores a number of these challenges and discusses possible answers.First off, one of the critical, demanding situations in actual-time statistics processing for earthquake warning structures is the sheer quantity and velocity of information. The sensors and tracking devices used to gather seismic statistics generate large amounts of facts within a short period. This poses problems in terms of statistics management, garage, and transmission. Processing such widespread amounts of statistics in actual time calls for sturdy and excessive-overall performance computing sources.One viable approach to deal with this mission is to rent disbursed computing techniques. With the aid of utilizing distributed systems, the processing workload may be divided among multiple nodes, allowing for parallel processing and expanded scalability. Additionally, employing cloud computing services can provide the required computing power and storage potential wanted for actual-time data analysis.Secondly, real-time data processing for earthquake prediction also entails handling information that is frequently noisy and incomplete. Earthquake indicators may be distorted by using different factors together with noise, interference, and information transmission issues..

5. MACHINE LEARNING ALGORITHMS FOR EARTHQUAKE PREDICTION

Earthquakes are natural disasters that can cause considerable devastation, loss of life, and economic damage. The capability to appropriately expect earthquakes in advance can keep lives, mitigate damages, and facilitate better emergency preparedness.

Traditional methods of earthquake prediction are constrained by their accuracy and reliability. But, with improvements in the era and the improvement of system mastering algorithms, there's a developing possibility of creating an early warning machine for earthquake prediction. This essay explores the gadget getting-to-know algorithms utilized in earthquake prediction and their capability to build a powerful early caution device. SVM is a robust machine-learning algorithm that may be applied to earthquake prediction. SVMs classify styles and make predictions primarily based on classified training records. In earthquake prediction, SVMs can examine seismic information patterns, become aware of precursors, and expect the chance of an impending earthquake. SVMs were shown to offer correct predictions and have been used efficiently in numerous earthquake prediction studies.

Figure 1. Earthquake prediction process

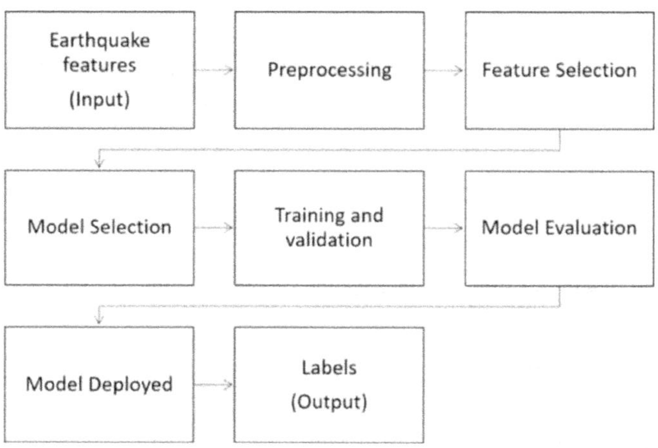

5.1 Overview of Machine Learning in Seismology

Seismology is the scientific area that studies earthquakes and the movement of the Earth's crust. It plays a vital function in information on the conduct of earthquakes and their capacity impact on society. One of the most vast improvements in recent years is the incorporation of machine studying strategies into seismology, specifically inside the improvement of early warning structures to expect earthquakes. An early caution gadget for earthquake pursuits to offer well-timed data to groups in earthquake-prone regions, allowing them to take essential precautions and probably store lives. The machine is based on the real-time tracking of seismic information and the correct prediction of earthquake incidence.Gadget learning, a subfield of synthetic intelligence, is an effective tool for reading vast and complicated datasets, making it well-acceptable for seismology. It can uncover patterns and relationships in the records that human experts may additionally forget. Via schooling fashions on historical earthquake information, machine-mastering algorithms can learn how to apprehend precursory styles that suggest an imminent earthquake.Step one in growing a system learning-based early caution device is an information series. Seismic networks across the world continuously report floor movement, supplying a massive amount of records that may be used for evaluation.

5.2 Supervised Learning for Earthquake Prediction

An early caution device for predicting earthquakes has long been a purpose of scientists and researchers, which will mitigate the devastating consequences of those natural screw-ups. Supervised learning, a subset of gadget mastering, has emerged as a promising technique in the field of earthquake prediction. This essay explores the idea of supervised studying for earthquake prediction, highlighting its advantages and obstacles in the improvement of a robust early caution system.Supervised studying is a machine studying technique where a version learns to expect output values primarily based on categorized input records. In the context of earthquake prediction, supervised knowledge of algorithms is skilled in historical earthquake data, such as seismic activities, fault line distributions, and geological functions, paired with their corresponding earthquake occurrences. By analyzing these styles and relationships, the version is then able to expect the probability and significance of future earthquakes, enabling well-timed warnings and evacuation measures.One key gain of supervised studying for earthquake prediction is its functionality to handle complicated data patterns. Earthquake occurrences are motivated by means of a large number of things, which include geological functions, plate tectonics, and historical seismic activities. Via studying large-scale datasets, supervised learning algorithms can identify hidden patterns and tendencies, bearing in mind more excellent and accurate predictions. This is particularly relevant for regions prone to seismic activities, wherein nearby geological traits play a vital function in earthquake prevalence.

5.3 Unsupervised Learning Approaches

Earthquakes are herbal screw-ups with doubtlessly catastrophic effects on human existence, infrastructure, and the environment. Developing an early caution system that could accurately expect earthquakes can significantly mitigate those influences and keep limitless lives. Unsupervised studying processes offer a treasured road for creating such a system with the aid of leveraging patterns and anomalies in seismic facts. This essay explores the diverse unsupervised learning techniques used in the improvement of an early warning device for earthquake prediction.Unsupervised studying is a device gaining knowledge of where algorithms analyze and enter records without any predefined output labels. This method is especially suitable for earthquake prediction due to the need for more sufficient classified statistics, seeing that earthquakes are infrequent and unpredictable activities. Unsupervised getting to know fashions can efficiently become aware of and examine patterns and anomalies in seismic information, permitting the early warning gadget to locate earthquake precursors.One standard unsupervised learning method utilized in earthquake prediction is clustering algorithms. Those algorithms group information factors based on their similarity, permitting the identity of seismic events with comparable traits. One outstanding clustering set of rules is the okay-method, which partitions the information into ok wonderful clusters, with every cluster representing a specific earthquake event..

5.4 Deep Learning Techniques

An early warning system for predicting earthquakes is of substantial importance in an effort to mitigate the ability devastation as a result of those natural screw-ups. Deep mastering techniques, a subset of gadget studying, have emerged as an effective device in diverse domain names, together with earthquake prediction. This essay explores the deep knowledge of techniques utilized in an early caution machine for predicting earthquakes.Deep mastering, as the call shows, includes using deep neural networks to examine and recognize complex patterns from significant amounts of records. Those neural networks are composed of more than one layer of synthetic neurons that could version and capture tricky relationships in the input facts. With the usage of deep knowledge of techniques, an early caution device for earthquakes can harness the energy of neural networks to efficaciously expect and provide timely indicators regarding coming near earthquakes.One technique generally employed in earthquake prediction is the usage of convolutional neural networks (CNNs). CNNs are specifically ideal for processing information with a grid-like shape, which includes seismic information. These networks include convolutional layers that observe filters to the input statistics, permitting them to understand critical functions present within the seismic alerts. With the aid of schooling these networks on a massive dataset of seismic facts, CNN can discover ways to discover styles indicative of earthquake pastimes.

6. SENSOR NETWORKS AND MONITORING INFRASTRUCTURE

Earthquakes are herbal failures that can lead to a full-size lack of life and belongings harm. The capability to predict earthquakes with precise accuracy stays elusive. However, advancements in sensor networks and tracking infrastructure have stepped forward the possibility of imparting early warnings for such catastrophic activities. This essay will discover the role of sensor networks and monitoring infrastructure in the development of an early caution device for predicting earthquakes.Sensor networks consist of a cluster of interconnected sensors deployed throughout a geographic region. Those sensors are chargeable for monitoring various parameters that could indicate the prevalence of an earthquake. Accelerometers are generally used to measure ground motion, while seismometers detect seismic waves and their traits. Additionally, GPS receivers can detect floor deformation, which is every other important indicator of a drawing close earthquake. The data collected via these sensors bureaucracy the idea for earthquake prediction.An adequate monitoring infrastructure is essential for collecting and reading records from the sensor networks. The infrastructure consists of statistics centers, verbal exchange structures, and computational resources. Information facilities keep and procedure the massive amount of statistics generated via the sensor networks, at the same time as conversation systems ensure the well-timed transmission of records from the sensors to the information facilities. The following algorithms used for predictions of earthquake

1. Seismic Wave Method: This method uses seismic waves to predict the occurrence of earthquakes. Seismic waves are a type of waves that are generated by an actual earthquake. By measuring and recording the speed and intensity of such waves, seismologists are able to estimate when and where an earthquake may strike.
2. Geodetic Method: This method uses GPS and satellite technology to measure the changes in the earth's crust that are likely to precede an earthquake. By monitoring land deformations, displacement of geodetic networks can be used to detect behaviors nucleation points before an earthquake.
3. Stress Drop Method: This method is based on the release of stress energy prior to an earthquake. Through the use of 3D seismic imaging, scientists can detect the build-up of seismic energy on various points of the earth's crust. This allows them to accurately predict the area where an earthquake is likely to hit.
4. B-Value Method: This method measures the activity of small earthquakes that occur prior to a large earthquaketo determine the likelihood of it taking place. Seismologists measure the magnitude of the small earthquakes in order to estimate the magnitude of the large earthquake.
5. Declustering Method: This method is based on the theory that an earthquake can trigger other earthquakes in a certain region. Through the use of mathematical algorithms, seismologists can detect the pattern of stress release in a certain area and may be able to predict the occurrence of earthquakes.

6.1 Seismic Sensor Technologies

Earthquakes are natural screw-ups that could cause good-sized destruction and loss of life. Despite advances in the era, appropriately predicting earthquakes remains an assignment. However, with the assistance of seismic sensor technologies, it's far more feasible to increase an early warning machine that may offer valuable seconds to minutes of caution earlier than an earthquake moves. In this essay, we will explore the diverse seismic sensor technologies used in the sort of device and their importance in predicting earthquakes.Seismic sensors are devices that stumble on and measure floor motion resulting from seismic waves. Those sensors are a fundamental part of an early caution device as they can provide actual-time information about the depth and location of an earthquake. The most typically used seismic sensors encompass seismometers, accelerometers, and GPS-based sensors.Seismometers are the number one sensors utilized in earthquake monitoring. They encompass a mass suspended on a spring that moves in response to floor motion. Those sensors generate electrical alerts that are proportional to the vibrations due to seismic waves. With the aid of studying the output of seismometers, scientists can determine the magnitude and place of an earthquake. These statistics are vital for issuing well-timed signals to regions that can be affected.

6.2 Network Design for Effective Monitoring

The network layout for a powerful tracking device for predicting earthquakes plays a crucial function in ensuring the protection and security of groups residing in seismic-inclined regions. With the potential to cause tremendous destruction and a lack of lifestyles, early detection of earthquakes can offer precious time for evacuation measures, catastrophe response planning, and reducing the effect of these herbal failures. This essay will explore the important thing components and concerns that want to be taken into account while designing a network for an early warning system for predicting earthquakes. The first aspect to take into account in the community layout for an earthquake early caution device is the selection and location of seismic sensors. Those sensors, commonly called seismometers, are critical for detecting ground vibrations as a result of earthquakes. They need to be mounted strategically to ensure comprehensive insurance of the target place. The density and distribution of seismometers play an essential role in accurately figuring out the epicenter, importance, and ability effect of an earthquake. Setting seismometers in close proximity to active fault strains and areas with traditionally better seismic activity is essential for green tracking.Another essential thing of the community design is the establishment of a green records communication infrastructure. The seismic sensors acquire tremendous quantities of data in real time, which needs to be transmitted promptly to the crucial records processing center for evaluation and interpretation.

6.3 Integration of Satellite Data

Earthquakes are natural disasters that could cause enormous destruction and lack of life. Being capable of correctly predicting when and where an earthquake will arise is an intention that has interested scientists and researchers. To attain this, the combination of satellite TV for pc records in an early warning system has emerged as a promising answer. This essay explores the capability blessings and demanding situations associated with this integration and discusses how it may contribute to more effective earthquake prediction.Satellites offer an invaluable supply of information for tracking the Earth's floor and its geological interest. They can capture excessive-resolution pictures and acquire diverse types of information, including temperature, elevation, and deformation. By integrating this satellite TV for pc statistics into an early warning device, scientists can gain a comprehensive knowledge of the Earth's dynamic processes, letting them identify patterns and precursors related to earthquake incidence.One of the fundamental blessings of integrating satellite TV for pc records is the ability to display ground deformation. Before an earthquake occurs, the Earth's crust undergoes diffused changes due to the buildup of strain alongside faults. Satellites geared up with synthetic aperture radar (SAR) generation can degree these deformations with millimeter-degree accuracy.

7. PREDICTION MODELS AND ALGORITHMS

An early warning gadget for predicting earthquakes is essential in minimizing the lack of lives, infrastructure, and assets. Over the years, scientists and researchers have advanced diverse prediction fashions and algorithms to forecast earthquakes correctly. Those fashions and algorithms make use of an aggregate of facts analysis, machine mastering techniques, and seismic monitoring structures to come across and expect earthquakes. This essay will talk about a number of the prediction models and algorithms used in early caution structures for earthquakes.One of the prediction fashions used in earthquake early warning systems is the GSM (group method of information handling) set of rules. This set of rules involves amassing an extensive dataset of seismic records, together with ancient earthquake records, and training a model to understand styles and anomalies that could imply a coming near the near earthquake. The GSM set of rules makes use of a mixture of statistical techniques and mathematical formulation to analyze these styles and predict the incidence of earthquakes.Artificial Neural Networks (ANN) are widely used in earthquake prediction models because of their potential to perceive complex styles in seismic facts. ANNs are educated using big datasets of seismic records, along with beyond earthquake data, fault line facts, and geological facts..

7.1 Early Warning System Architectures

Earthquakes are natural failures that could result in tremendous loss of life and bodily harm. Early warning systems (EWS) have been evolved to mitigate those effects by offering advance observation to affected regions. These structures make use of numerous architectural designs to locate seismic interest, analyze records, and expect the incidence of earthquakes. This essay explores numerous early warning system architectures and their effectiveness in predicting earthquakes.The centralized structure is a not unusual technique in early warning structures. It includes a centralized statistics series and processing middle, wherein seismic facts from more than one sensor are collected and analyzed. The information accumulated from seismic stations is transmitted to the vital processing unit, which applies algorithms and predicts earthquake occurrences. The generated warnings are then disseminated to targeted areas.One gain of the centralized structure is its efficient facts processing. Via centralizing statistics from numerous sources, allows for better real-time evaluation and selection-making. However, this architecture might also have limitations in terms of vulnerability to bodily damage and verbal exchange failures. If the centralized middle is compromised during an earthquake, the drift of data and early warnings may be disrupted.

7.2 Earthquake Prediction Models

Earthquakes pose a significant danger to human life and infrastructure. Therefore, developing accurate earthquake prediction fashions is essential that allows you to limit the capability of the devastation caused by these natural disasters. An early warning gadget for predicting earthquakes holds remarkable potential to save lives and offer timely evacuation processes. In this essay, we are able to explore the diverse earthquake prediction models and speak about the importance of an early warning gadget.Earthquakes are surprising and unpredictable phenomena that arise due to the release of geological stress and power alongside fault strains. Their unpredictable nature leads to immense demanding situations for disaster preparedness and response. An advanced and green early caution device can provide precious seconds, minutes, or even hours of lead time, permitting individuals and groups to take essential preventive measures.Seismic tracking is the most generally used technique to predict earthquakes. Monitoring networks of seismometers are deployed worldwide to measure ground vibrations and stumble on seismic waves. Those waves travel via the Earth's crust in the course of an earthquake, supplying vital records of its significance, vicinity, and depth. With the aid of reading the records from those seismometers, scientists can stumble on styles and anomalies that may suggest an impending earthquake.

7.3 Comparative Analysis of Prediction Algorithms

Earthquakes are herbal failures that can cause extensive harm to infrastructure and a lack of human lives. The ability to appropriately predict earthquakes has long been an assignment for scientists and researchers. Early warning systems (EWS) have been advanced to mitigate these challenges by presenting advanced words to the affected areas. This essay aims to evaluate and examine unique prediction algorithms utilized in early caution systems for earthquake prediction.Neural networks are computational models inspired by the human mind's structure and functioning. These algorithms try to replicate the mind's gaining knowledge of capabilities to analyze and expect complex patterns. Neural networks have the gain of being capable of mechanically becoming aware of and analyzing the statistics furnished to enhance prediction accuracy. They are able to process vast amounts of records and identify hidden styles that might not be obvious to human analysts. However, their overall performance heavily depends on the fine and amount of entering statistics. Support vector Machines are supervised learning algorithms usually used for class and regression analysis. SVM works by means of locating the most desirable hyperplane that separates statistics into one-of-a-kind training based totally on furnished training records.

8. VALIDATION AND ACCURACY ASSESSMENT

An early warning system (EWS) for predicting earthquakes is a critical tool that can doubtlessly keep infinite lives and decrease the impact of seismic activities. But, because of the unpredictable nature of earthquakes, it's vital to validate and accurately determine the effectiveness of such structures. In this essay, we are able to explore the significance of validation and accuracy evaluation in an early caution machine for earthquake prediction.Validation is a process that ensures the reliability and credibility of a systematic version or gadget. In the case of an early warning gadget, validation is paramount, as false alarms or neglected indicators could have intense consequences. To validate an EWS, it is necessary to compare the expected earthquake occasions with the actual occurrences. This may be achieved via a couple of ways, inclusive of ancient information evaluation, validation towards previous earthquakes, and pass-validation against different existing EWSs.Ancient information analysis is an essential step in validating an early caution machine. By examining past earthquake facts, scientists and researchers can check the accuracy of predictions made by using the EWS. This analysis entails evaluating the expected earthquake parameters like importance, area, and depth with the recorded information from actual events. If there is a regular correlation between the expected and actual records, it validates the gadget's functionality to forecast upcoming earthquakes accurately..

8.1 Metrics for Assessing Prediction Accuracy

An early warning machine for predicting earthquakes performs a critical position in mitigating capacity damage because of natural failures. Correct predictions allow authorities to take vital precautions and prepare communities earlier, in the long run saving lives and assets. However, the assessment of the prediction accuracy of the sort of gadget is paramount to ensure its effectiveness. This essay examines the metrics used to evaluate the prediction accuracy of an early caution machine for earthquakes.One of the key metrics used to assess the accuracy of earthquake prediction structures is the probability of Detection (POD). POD measures the likelihood of efficaciously figuring out an earthquake event. In other phrases, it calculates the percentage of earthquakes that are correctly detected by means of the gadget. As an example, if the machine predicts the incidence of 100 earthquakes and manages to stumble on eighty of them, the POD would be 80%. The better the POD, the more reliable the device is in identifying earthquakes.Another critical metric is the fake Alarm charge (far), which measures the proportion of false alarms generated through the system. Fake alarms can cause needless panic, disruptions, and an erosion of belief in the machine's predictions. Subsequently, it is critical to hold the long way as low as possible.

8.2 Validation Procedures and Protocols

An early caution gadget for predicting earthquakes is an essential tool in minimizing the potential harm as a result of seismic events. Those systems rely on a mess of sensors and algorithms to stumble on seismic interest and trouble alerts to affected groups. However, the accuracy and reliability of such systems have to be very well proven to ensure their effectiveness in presenting timely warnings. This essay will speak about the validation techniques and protocols vital for an early caution device to expect earthquakes appropriately.One vital issue in validating an early warning system is the calibration of the sensors used to discover seismic pastimes. The sensors should be correctly calibrated to avoid fake positives or negatives, as those can cause needless panic or inadequate warnings. Calibration tactics usually involve placing sensors in controlled environments and subjecting them to regarded vibration ranges to decide their sensitivity and accuracy. The calibration technique guarantees that the sensors can accurately detect and measure seismic waves, enhancing the machine's reliability.The algorithms used to investigate sensor records and expect earthquakes are an essential factor of an early warning machine. These algorithms method large amounts of information to discover patterns and trends indicative of a seismic hobby. To validate the system, the algorithms should go through thorough development and checking out approaches.

8.3 Continuous Improvement Strategies

An early warning system for predicting earthquakes is an essential device in enhancing our capacity to respond and mitigate the devastating effects of earthquakes. Even as considerable development has been made in developing and implementing such systems, continuous improvement strategies are necessary to decorate their accuracy, timeliness, and effectiveness. This essay explores a number of critical techniques that can be hired to ensure the non-stop improvement of an early caution device for predicting earthquakes.One of the essential strategies for non-stop improvement is records collection and evaluation. To improve the accuracy of earthquake predictions, it's essential to acquire as many facts as viable approximately past seismic sports. This consists of distinct statistics on the importance, place, and intensity of preceding earthquakes. By analyzing this information, scientists can discover patterns and trends that could be helpful resources in predicting future seismic occasions. Moreover, new technologies, which include satellite tv for PC imagery and far-off sensors, can be incorporated into information series efforts, presenting a more excellent complete know-how of seismic sports.Any other strategy for continuous development is the usage of advanced technology and algorithms. With the improvements in computing energy and data processing skills, it's far now viable to develop extra state-of-the-art algorithms for earthquake prediction.

9. ETHICAL CONSIDERATIONS IN EARTHQUAKE PREDICTION

An early warning device for predicting earthquakes is a technological advancement that holds colossal blessings for society. It can keep lives, facilitate evacuation tactics, or even mitigate the harm because of earthquakes. However, the development and implementation of one of these systems also enhance a large number of moral considerations that ought to be cautiously examined.First and foremost, one of the critical ethical concerns surrounding earthquake prediction is the difficulty of fake alarms. If an early warning system isn't accurate in its predictions, it is able to result in unnecessary panic and disturbance to a few of the populace. Humans can be more willing to brush aside warnings if they understand them as unreliable, which may have devastating results in the event of a real earthquake. Therefore, making sure the system is reliable and minimizing false alarms is a vital way to maintain public agreement and confidence.Another moral consideration is the equitable distribution of assets. Growing and retaining an early warning device can involve significant economic and technological investments. It's essential to ensure that the blessings of this sort of machine are not most effectively limited to precise areas or groups. Making sure the same gets entry to the blessings of earthquake prediction, particularly in areas liable to seismic interest, can assist in decreasing the impact of earthquakes on susceptible populations and save you from the exacerbation of existing social inequalities.

9.1 Balancing Public Safety and Privacy

Earthquakes are herbal screw-ups that can cause significant loss of existence and infrastructural damage. To mitigate those impacts, scientists and engineers have developed early caution systems (EWS), which could provide a boost to the word of an imminent earthquake. While these systems preserve their vast ability to save lives, they also increase worries regarding the stability between public protection and privacy. This essay aims to explore the sensitive stability required when imposing an early warning device, weighing the importance of public safety towards people's right to privacy.Early caution systems offer people and groups treasured seconds or even minutes to take essential precautions, consisting of searching for refuge and shutting off gasoline lines. Those systems depend upon the prompt detection and notification of seismic waves, enabling the dissemination of public protection indicators to potentially affected regions. With the potential to save lives and decrease asset destruction, the implementation of an early caution gadget for earthquakes is essential.Public protection and well-being ought to be the primary problem while considering the implementation of an early warning system.

9.2 Decision-Making in Emergency Situations

Earthquakes are herbal disasters that could cause tremendous destruction and loss of lifestyles. Due to their surprising prevalence, it's miles critical to unexpectedly respond and alert humans so that it will decrease the impact. An early warning machine for predicting earthquakes performs an essential function in choice-making at some point in emergency conditions. This essay will explore the choice-making procedure concerning an early caution system for earthquakes and the challenges confronted.

9.2.1 Predictive Technology

An early caution system relies on advanced technology that could come across the initial seismic waves generated by an earthquake. Those technologies encompass seismographs, accelerometers, GPS, and satellite measurements. As those technologies offer actual-time information, there may be a short window of time in which decisions ought to be made to set off the caution machine.

The choice-making procedure within an early caution gadget includes several degrees:

1. Sensor information analysis: The preliminary seismic wave detected by using sensors is analyzed to assess the earthquake's magnitude, epicenter, and potential impact. This evaluation is crucial for figuring out the severity and issuing appropriate warnings.
2. Alert gadget Activation: as soon as the earthquake's traits are determined, the alert device is activated.

9.3 Transparency and Accountability

An early warning gadget performs a vital function in predicting and mitigating the devastating consequences of earthquakes. Given the unpredictable nature of these herbal screw-ups, it turns vital to research the transparency and duty of such structures. This essay will delve into the significance of transparency and accountability in an early warning system for earthquakes, discussing the benefits, demanding situations, and ability to answer.Transparency in an early caution device refers to presenting clean and handy information to the public. It entails sharing facts, protocols, and selection-making strategies with various stakeholders, inclusive of government entities, scientists, and citizens. The transparency of an early caution system is crucial for building trust among those stakeholders, permitting powerful collaboration, and making sure the system's standard is adequate.First off, transparency promotes accountability and responsible selection-making. With the aid of sharing the underlying algorithms, sensors, and records assets, scientists and researchers can solicit comments from the clinical community and validate the system's predictions. This peer overview manner allows the to discover any biases, mistakes, or inconsistencies, leading to continuous improvement. Furthermore, transparency lets in for impartial audits and evaluations, reducing the hazard of manipulating or misinterpreting records for non-public or political profits.Moreover, promoting transparency contributes to the public's knowledge and trust in the early warning machine.

10. CONCLUSION

In the end, the improvement of an early warning machine for predicting earthquakes holds high-quality capability in mitigating the devastating impacts of those natural screw-ups. The advancements in era and studies have provided us with the gear and expertise essential to construct a robust and reliable system that can save limitless lives and limit harm to infrastructure.All through this essay, we've seen the diverse additives and techniques that make up an early caution device. From the detection of seismic waves to the analysis and interpretation of data and the dissemination of warnings, each step performs a vital function in the effectiveness of the system. With the aid of combining those additives and using the improvements in artificial intelligence and system mastering algorithms, we can beautify the accuracy and pace of earthquake predictions, thereby granting people and communities sufficient time to respond and take essential moves.Moreover, the successful implementation of early caution systems in countries like Japan and Mexico has proven to us that these structures are not only the simplest feasible but also distinctly effective in saving lives and decreasing economic losses. Those case studies serve as concrete evidence that investing in the improvement and implementation of early warning systems is a profitable endeavor.However, it is critical to know the present demanding situations and limitations within the field of earthquake prediction. Despite extensive development, there may be nonetheless a lot that remains unknown about the complicated mechanisms that drive earthquakes.

10.1 Summary of Key Findings

An early warning system for predicting earthquakes is vital to minimizing the capacity devastation and loss of existence related to these natural screw-ups. Through massive studies and evaluations, numerous key findings have emerged, needing to improve on the effectiveness and obstacles of any such system. This essay will define these key findings and highlight their implications for the development and implementation of an early caution system for earthquakes.Firstly, it has been mounted that the usage of seismic tracking networks plays a crucial role in earthquake prediction. These networks, which include a network of seismometers strategically positioned in high-hazard regions, can come across the initial seismic waves generated with the aid of an earthquake. With the aid of studying the facts amassed from these seismometers in actual time, it will become viable to estimate the importance and region of an earthquake. This finding underscores the significance of investing in sturdy seismic monitoring networks to enhance the accuracy and performance of earthquake prediction.Secondly, researchers have found that earthquake precursors can provide precious insights into the probability of an approaching earthquake. Precursors are defined as abnormal indicators or signs discovered previous to an earthquake, which include adjustments in groundwater degrees, versions in animal conduct, or uncommon electromagnetic hobby. Via thoroughly analyzing and reading those precursors, it's miles feasible to identify styles or correlations that could exist among those indicators and seismic pastimes.

10.2 Implications for Future Earthquake Prediction

An early warning gadget for predicting earthquakes has profound implications for the destiny of earthquake prediction. This emerging generation can save infinite lives and mitigate the devastating impact of seismic activities on each human population and infrastructure. By means of providing vital minutes or even seconds of strengthened note, an early warning machine offers a precious window of opportunity for individuals to take haven, emergency responders to prepare, and critical infrastructure structures to be secured. As such, this development in earthquake prediction offers several blessings at the same time as also posing exciting, demanding situations for the medical community.One of the most substantial implications of an early caution gadget is its capacity to keep lives. Earthquakes are natural failures that could strike all at once, leaving humans with little time to react. In densely populated regions, even a few seconds of warning could make a big difference. For example, at some point during the 2011 Tohoku earthquake in Japan, the early warning system furnished by the Japan Meteorological Organization alerted hundreds of thousands of people as many as forty seconds before the shaking commenced. This early caution allowed human beings to try to find shelter, shut down trains, and evacuate homes swiftly, in the long run saving many lives.

REFERENCES

Abdalzaher, M. S., Elsayed, H. A., Fouda, M. M., & Salim, M. M. (2023). Employing machine learning and iot for earthquake early warning system in smart cities. *Energies*, *16*(1), 495. doi:10.3390/en16010495

Abdalzaher, M. S., Soliman, M. S., & El-Hady, S. M. (2023). Seismic Intensity Estimation for Earthquake Early Warning Using Optimized Machine Learning Model. *IEEE Transactions on Geoscience and Remote Sensing*, *61*, 1–11. doi:10.1109/TGRS.2023.3296520

Böse, M., Andrews, J., O'Rourke, C., Kilb, D., Lux, A., Bunn, J., & McGuire, J. (2023). Testing the ShakeAlert earthquake early warning system using synthesized earthquake sequences. *Seismological Research Letters*, *94*(1), 243–259. doi:10.1785/0220220088

Böse, M., Papadopoulos, A. N., Danciu, L., Clinton, J. F., & Wiemer, S. (2022). Loss-Based Performance Assessment and Seismic Network Optimization for Earthquake Early Warning. *Bulletin of the Seismological Society of America*, *112*(3), 1662–1677. doi:10.1785/0120210298

Chiang, Y. J., Chin, T. L., & Chen, D. Y. (2022). Neural network-based strong motion prediction for on-site earthquake early warning. *Sensors (Basel)*, *22*(3), 704. doi:10.339022030704 PMID:35161451

Cho, S., Ahn, J. K., & Hwang, E. H. (2022). Optimization of network-based earthquake early warning systems on the Korean Peninsula. *IEEE Access : Practical Innovations, Open Solutions*, *10*, 83931–83939. doi:10.1109/ACCESS.2022.3197661

Cochran, E. S., Saunders, J. K., Minson, S. E., Bunn, J., Baltay, A., Kilb, D., O'Rourke, C., Hoshiba, M., & Kodera, Y. (2022). Alert optimization of the PLUM earthquake early warning algorithm for the western United States. *Bulletin of the Seismological Society of America*, *112*(2), 803–819. doi:10.1785/0120210259

Datta, A., Wu, D. J., Zhu, W., Cai, M., & Ellsworth, W. L. (2022). DeepShake: Shaking intensity prediction using deep spatiotemporal RNNs for earthquake early warning. *Seismological Research Letters*, *93*(3), 1636–1649. doi:10.1785/0220210141

Esposito, M., Palma, L., Belli, A., Sabbatini, L., & Pierleoni, P. (2022). Recent advances in internet of things solutions for early warning systems: A review. *Sensors (Basel)*, *22*(6), 2124. doi:10.339022062124 PMID:35336296

Joshi, A., Vishnu, C., & Mohan, C. K. (2022). Early detection of earthquake magnitude based on stacked ensemble model. *Journal of Asian Earth Sciences: X*, *8*, 100122.

Kannan, S., Dhiman, G., Natarajan, Y., Sharma, A., Mohanty, S. N., Soni, M., Easwaran, U., Ghorbani, H., Asheralieva, A., & Gheisari, M. (2021). Ubiquitous vehicular ad-hoc network computing using deep neural network with iot-based bat agents for traffic management. *Electronics (Basel)*, *10*(7), 785. doi:10.3390/electronics10070785

Khan, S., Saravanan, V. N. G. C., Lakshmi, T. J., Deb, N., & Othman, N. A. (2022). Privacy protection of healthcare data over social networks using machine learning algorithms. *Computational Intelligence and Neuroscience*, *2022*, 2022. doi:10.1155/2022/9985933 PMID:35371203

Kumar, R., Mittal, H., Sandeep, & Sharma, B. (2022). Earthquake genesis and earthquake early warning systems: Challenges and a way forward. *Surveys in Geophysics*, *43*(4), 1143–1168. doi:10.100710712-022-09710-7

Medina, M., Sanchez, R., Riquelme, S., Flores, M. C., Koch, P., Bravo, F., Barrientos, S., Henson, I., Chung, A., Melgar, D., Mpodozis, C., Hellweg, M., & Allen, R. (2022). An Earthquake Early Warning System for Northern Chile Based on ElarmS-3. *Seismological Research Letters*, *93*(6), 3337–3347. doi:10.1785/0220210331

Pant, G., Bahuguna, R., Pandey, S., Gehlot, A., Yadav, S. P., & Pachauri, R. K. (2023, April). Intelligent Automated Interference for the Protection of Road Safety. In *2023 International Conference on Computational Intelligence, Communication Technology and Networking (CICTN)* (pp. 87-91). IEEE. 10.1109/CICTN57981.2023.10141086

Peng, C., Jiang, P., Ma, Q., Su, J., Cai, Y., & Zheng, Y. (2022). Chinese nationwide earthquake early warning system and its performance in the 2022 Lushan M 6.1 earthquake. *Remote Sensing (Basel)*, *14*(17), 4269. doi:10.3390/rs14174269

Prasanna, R., Chandrakumar, C., Nandana, R., Holden, C., Punchihewa, A., Becker, J. S., Jeong, S., Liyanage, N., Ravishan, D., Sampath, R., & Tan, M. L. (2022, March). "Saving Precious Seconds"—A Novel Approach to Implementing a Low-Cost Earthquake Early Warning System with Node-Level Detection and Alert Generation. *Informatics (MDPI)*, *9*(1), 25. doi:10.3390/informatics9010025

Rajalakshmi, M., Saravanan, V., Arunprasad, V., Romero, C. T., Khalaf, O. I., & Karthik, C. (2022). Machine Learning for Modeling and Control of Industrial Clarifier Process. *Intelligent Automation & Soft Computing*, *32*(1). doi:10.32604/iasc.2022.021696

Saad, O. M., Chen, Y., Trugman, D., Soliman, M. S., Samy, L., Savvaidis, A., Khamis, M. A., Hafez, A. G., Fomel, S., & Chen, Y. (2022). Machine learning for fast and reliable source-location estimation in earthquake early warning. *IEEE Geoscience and Remote Sensing Letters*, *19*, 1–5. doi:10.1109/LGRS.2022.3142714

Song, J., Zhu, J., Wang, Y., & Li, S. (2022). On-site alert-level earthquake early warning using machine-learning-based prediction equations. *Geophysical Journal International*, *231*(2), 786–800. doi:10.1093/gji/ggac220

Srivastava, S., Dhyani, N., Sharma, V., Vats, S., Yadav, S. P., Kukreja, V., & Singh, R. (2023, May). Lung Infection and Identification using Heatmap. In *2023 2nd International Conference on Applied Artificial Intelligence and Computing (ICAAIC)* (pp. 1093-1098). IEEE. 10.1109/ICAAIC56838.2023.10140204

Suárez, G. (2022). The seismic early warning system of Mexico (SASMEX): A retrospective view and future challenges. *Frontiers in Earth Science (Lausanne)*, *10*, 827236. doi:10.3389/feart.2022.827236

Yuvaraj, N., Karthikeyan, T., & Praghash, K. (2021). An improved task allocation scheme in serverless computing using gray wolf Optimization (GWO) based reinforcement learning (RIL) approach. *Wireless Personal Communications*, *117*(3), 2403–2421. doi:10.100711277-020-07981-0

Yuvaraj, N., Srihari, K., Dhiman, G., Somasundaram, K., Sharma, A., Rajeskannan, S. M. G. S. M. A., Soni, M., Gaba, G. S., AlZain, M. A., & Masud, M. (2021). Nature-inspired-based approach for automated cyberbullying classification on multimedia social networking. *Mathematical Problems in Engineering*, *2021*, 1–12. doi:10.1155/2021/6644652

Zhang, H., Melgar, D., Sahakian, V., Searcy, J., & Lin, J. T. (2022). Learning source, path and site effects: CNN-based on-site intensity prediction for earthquake early warning. *Geophysical Journal International*, *231*(3), 2186–2204. doi:10.1093/gji/ggac325

Chapter 12
Deep Learning Approaches for Earth Science

Deepali Virmani
School of Engineering and Technology, Vivekananda Institute of Professional Studies Technical Campus, Delhi, India

Ananda Kumar K. S.
Department of Information Science and Engineering, Atria Institute of Technology, Bangalore, India

Sajitha L. P.
https://orcid.org/0000-0002-5406-2093
Department of Computer Science and Business System, R.M.K. Engineering College, Kavaraipettai, India

Nikhat Parveen
Department of Computer Science and Engineering, Koneru Lakshmaiah Education Foundation, Guntur, India

ABSTRACT

Earth science, a multidisciplinary field that explores diverse components of the Earth, has also been significantly encouraged with the aid of the creation of deep studying procedures. The summary starts by highlighting the increasing availability of large datasets inside the Earth's technological know-how domain, frequently observed by way of complicated relationships and patterns, which can be challenging to find the use of traditional analytical strategies. Deep studying techniques have additionally revolutionized geoscience information evaluation with the aid of permitting the identification of complex styles and relationships within massive datasets. Deep mastering algorithms have been verified as powerful in extracting significant statistics from such datasets, mainly to improve understanding of geological processes, aid exploration, and danger assessment. This review includes a vast range of programs, such as far-flung sensing, weather modeling, and earthquake prediction.

DOI: 10.4018/979-8-3693-1850-8.ch012

1. INTRODUCTION

Deep studying, a subset of gadget learning, has gained tremendous attention and popularity in recent years due to its capability to method enormous quantities of complex statistics and produce accurate outcomes. Earth science, a multidisciplinary field that explores diverse components of the Earth, has also been significantly encouraged with the aid of the creation of deep studying procedures. This essay will speak about the utility of deep learning in Earth technology and its effect on extraordinary areas of research (Liu et al., 2022). Deep studying processes have been successfully employed in Earth's technological know-how for several functions, including remote sensing, climate modeling, and geoscience records analysis. One of the principal regions wherein deep learning has made significant contributions is remote sensing, which involves the collection and interpretation of statistics from satellites and airborne sensors. Far flung sensing information, inclusive of satellite TV for pc imagery and hyperspectral pics, are frequently high-dimensional and comprise a plethora of records. Deep getting-to-know algorithms can correctly extract features and patterns from such statistics, allowing correct class and interpretation. Moreover, deep getting to know has been applied in climate modeling to enhance the knowledge and prediction of diverse climate phenomena. Weather models simulate the behavior of the Earth's ecosystem, oceans, and land floor to undertake future climate scenarios. Deep mastering algorithms have been employed to analyze weather version outputs and observational facts, resulting in stepped-forward model performance and more excellent correct predictions (Sun et al., 2022). This has enormous implications for know-how climate exchange dynamics and mitigating its effects. Deep studying techniques have additionally revolutionized geoscience information evaluation with the aid of permitting the identification of complex styles and relationships within massive datasets. Geoscientific facts, including seismic records, geophysical photos, and geological maps, are frequently voluminous and complex to interpret. Deep mastering algorithms have been verified as powerful in extracting significant statistics from such datasets, mainly to improve understanding of geological processes, aid exploration, and danger assessment. One of the fundamental blessings of deep studying in Earth technology is its capacity to address massive facts (Zhu et al., 2022). With advancements in technology, large quantities of information are being generated through diverse Earth-watching structures, climate fashions, and other scientific units. Deep mastering algorithms excel at handling massive datasets and might analyze these information faster than conventional techniques. This allows scientists to observe complex Earth tactics in more significant elements, leading to a more profound expertise of the planet's dynamics. However, the creation of deep learning processes in Earth's technological know-how additionally provides numerous demanding situations. One of the most important demanding situations is the requirement for substantial computational sources and information. Deep learning fashions generally have millions of parameters and require effective hardware and deep neural community architectures (Zaini et al., 2022). It also calls for skilled researchers who can design, train, and excellent-song those models. Moreover, deep learning tactics closely depend upon categorized information for schooling, which can be scarce in some Earth science domain names. In the end, the advent of deep getting-to-know techniques has introduced full-size improvements to Earth science research. It has facilitated the evaluation of far-off sensing facts, progressed climate modeling, and more desirable geoscience facts evaluation (Palansooriya et al., 2022). Regardless of demanding situations associated with computational resources and facts availability, deep learning has been established to be a treasured device for information and studying the Earth. As generation continues to adapt, deep studying procedures are predicted to play an increasingly important function in Earth technology, similarly pushing the bounds of our know-how about the planet and its environment(Mousavi & Beroza, 2022).

1.1 Background and Significance

The sphere of Earth science encompasses the take a look at the Earth's physical, chemical, and organic tactics, in addition to the interactions among them. Understanding and predicting these complicated phenomena is essential for several packages, which include weather alternate mitigation, herbal catastrophe control, and sustainable aid-making plans (Bhardwaj et al., 2022). Deep studying, a subset of system-gaining knowledge, has emerged as a promising approach to tackling the problematic demanding situations related to Earth technology. This essay aims to explore the background and importance of deep getting-to-know tactics within this discipline (Persello et al., 2022). With improvements in the era, Earth's technological know-how has witnessed an exponential boom in statistics acquisition from numerous assets, consisting of far-flung sensing satellites, weather fashions, ground-based sensors, and social media. Conventional analytical strategies conflict to handle the full-size volumes, pace, and kind of such data, calling for novel techniques to extract significant facts (Kumar & Pande, 2023). Earth technology phenomena are characterized by using tricky and nonlinear relationships, making them challenging to version correctly. Deep getting to know offers a powerful solution by way of employing artificial neural networks with numerous layers, permitting them to study complicated patterns and nonlinear dependencies inside the records. This capacity to seize complicated relationships separates deep learning from conventional statistical strategies (Mai et al., 2022)-(Hou et al., 2022).

Deep studying fashions have verified first-rate success in Earth science packages, which include climate forecasting, herbal catastrophe prediction, and surroundings monitoring. Through correctly shooting intricate relationships inside complex datasets, deep studying algorithms offer stepped-forward predictive abilities, allowing more excellent accurate forecasting and early warning systems (Feizizadeh et al., 2023). Deep mastering approaches enable the invention of hidden styles and correlations inside extensive Earth technology datasets. By using reading multi-modal information resources, inclusive of imagery, atmospheric measurements, and geological data, deep mastering algorithms find precious insights that can inform vital selections in environmental control, aid allocation, and conservation efforts (Thiyagalingam et al., 2022)-(Muruganantham et al., 2022). The significance of deep getting to know lies in its ability to beautify predictive talents, extract precious insights, combine multi-dimensional information, and facilitate understanding transferability in Earth's technological know-how. With further improvements in deep studying algorithms and expanded availability of information, the capability for transformative discoveries and impactful programs in Earth's technological know-how is extensive(Ge et al., 2022).

1.2 Objectives of the Chapter

One of the primary goals of the chapter is to provide an outline of deep getting-to-know methodologies and their packages in Earth's technological know-how. Deep learning is a subset of device mastering that has gained sizeable interest in recent years due to its capacity to analyze problematic styles from complex datasets robotically (Lin et al., 2022). By outlining the standards and strategies of deep gaining knowledge, this chapter aims to introduce Earth science researchers to this powerful tool and spotlight its capacity applications. Every other objective of the chapter is to exhibit the successful implementation of deep learning models in various Earth technological know-how disciplines, including far-off sensing, climate modeling, and catastrophe management. Deep learning algorithms excel in managing huge volumes of statistics that are generic in Earth technology studies. By means of employing deep knowledge of strategies, researchers can extract treasured data from extensive datasets, primary to new insights and improved predictions (Khan et al., 2022). The bankruptcy pursues to present case studies and examples wherein deep learning fashions have been applied effectively, demonstrating their capabilities and advantages in Earth's technological know-how packages. Additionally, the chapter aims to speak about the demanding situations and barriers of deep studying processes in Earth's technological know-how. Even though deep getting to know has proven promising outcomes, it also comes with sure caveats. The complex architecture of deep neural networks requires substantial computational sources, making implementation and training time-consuming and computationally pricey. Furthermore, the interpretability of deep mastering fashions is another subject, as they often function as "black packing containers," hindering the knowledge of the underlying mechanisms (Bochenek & Ustrnul, 2022). Through addressing those demanding situations, the chapter seeks to offer a balanced perspective on the applicability of deep mastering strategies in Earth technology research. Furthermore, the bankruptcy intends to foster collaboration and expertise, alternate among Earth's technological know-how researchers and the deep gaining knowledge of the network. As Earth's technological know-how evolves, it's crucial to leverage the understanding of both domain names to address complicated challenges correctly (Yu et al., 2022). With the aid of supplying insights into deep gaining knowledge of methodologies and their ability programs, the chapter serves as a bridge among those communities, encouraging interdisciplinary collaboration and paving the way for improvements in Earth science (Malik et al., 2022).

2. FUNDAMENTALS OF DEEP LEARNING

Deep studying is a fixed of algorithms in artificial intelligence that mimic the workings of the human brain so as to study information. It's increasingly getting used in lots of different programs to create predictive models and to recognize patterns in statistics. In recent years, deep learning has been implemented in many Earth science-related duties ranging from building better know-how of complicated earth procedures to creating extra correct forecasts of the results of human sports in the surroundings. This essay will provide a top-level view of the basics of deep studying and cognizance of how it's miles getting used to advance research in the Earth's technological know-how domain. Deep learning fashions are constructed upon a fixed of layers, every of which includes exceptional parameters and weights to be able to be used to technique and examine the information. Each layer can have exceptional activation functions and unique input houses, relying on the information that is being processed.

Deep gaining knowledge has been employed in numerous Earth technology-related responsibilities. For instance, it has been used to enhance understanding of the global weather machine through examining interactions between procedures together with ocean and atmospheric circulation. It has also been used to investigate satellite TV for pc records to increase higher forecasting fashions and to create distinctive maps of land cover and soil residences. Further, deep learning has been used to discover and discover geological functions, which include geological faults, which will better understand the Earth's indoors. It has additionally been hired to investigate seismic records to construct a better know-how of the Earth's lithosphere. In the end, it's been used to improve crop forecasting by means of reading satellite statistics to higher hit upon and become aware of one-of-a-kind varieties of plants. To conclude, deep knowledge gives top-notch potential for advancing Earth technology research.

2.1 Overview of Deep Learning

Deep getting to know is an area of synthetic Intelligence (AI) studies that focuses on algorithms that could apply to quite a number of problems in information mining, PC vision, natural language processing, and many other areas. This sort of AI has been proven to outperform conventional, hand-engineered algorithms on a variety of duties, making it a principal hobby in information science and gadget-gaining knowledge of studies. In recent years, deep knowledge of research has been applied to a variety of Earth technology problems, from automated interpretation of geophysical photos to prediction of weather and climate styles. In deep getting to know, information is processed in more than one layer that may assist in discovering extra summary family members and capabilities from these statistics. This form of function-engineering comes with an excessive diploma of flexibility and greater 'electricity' than feature-engineering. For computer imaginative and prescient tasks, deep studying techniques were especially a hit and relevant to a massive form of problems. For instance, convolutional neural networks (CNNs) have been deployed to allow autonomous vehicles to stumble on avenue and lane markers or to become aware of items in aerial imagery robotically. In Earth's technological know-how, deep getting to know is being increasingly more hired to automate and improve upon geological and plant characterization responsibilities, map landslide susceptible regions, or detect lava flows in satellite TV for PC imagery. Similarly, for climate forecasting, deep knowledge of techniques isn't best used to enhance weather forecasts globally; additionally, we are expecting the evolution of local weather or weather occasions, inclusive of floods or other severe weather.

2.2 Neural Networks and Architectures

Earth technology is a multi-disciplinary area of technological know-how that is dedicated to the study of Earth and its methods. In recent years, advances in generation have extended the competencies of Earth scientists in expertise on the Earth's systems and their effects on climate, surroundings, and human civilization. In Earth's technological know-how, neural networks may be used to discover features in statistics from sensors, inclusive of the land cover classes, and to interpret Earth phenomena. Mainly, they were used to hit upon adjustments in plants, agricultural regions, and urban development. In conclusion, neural networks and deep mastering architectures are powerful gear for Earth scientists to take advantage of perception of Earth's tactics. By combining big datasets with these algorithms, researchers can become aware of meaningful patterns and build strong models for predicting a range of environmental variables. In the end, those strategies can revolutionize the way we recognize the Earth and its many phenomena.

2.3 Key Concepts in Deep Learning for Earth Science

Earth science is an ever-growing discipline of research and information collection that involves reading about weather alternate, geology, and the impact of humans on the world. Deep mastering is becoming more famous as a tool to analyze and apprehend complicated phenomena in Earth's technological know-how and enhancing predictions and choice-making. Deep mastering models can routinely find correlations and styles in large datasets with minimal human enter, enabling information-driven solutions for a selection of Earth technological know-how problems. This essay will speak a number of the critical principles of deep mastering for Earth science, as well as capability packages.

The principle concept of deep gaining knowledge of Earth technology is to apply laptop algorithms, regularly with neural networks, to allow gadget studying. Gadget studying is the method of the use of information from present datasets to train algorithms to recognize and classify patterns. These algorithms can then be used to robotically detect patterns in future datasets and offer insights that would otherwise require human intervention. Finally, deep knowledge can also be used to control large volumes of Earth technology statistics. Deep mastering models were used to analyze and process satellite TV for pc pics, sensor records, and historical facts.

3. DATA REPRESENTATION AND FEATURE ENGINEERING

Statistics illustration and feature engineering for deep learning methods for Earth technological know-how (ES) have become more and more essential in current years as researchers are looking to apply machine studying algorithms to understand diverse environmental patterns and traits. Deep learning fashions are capable of processing vast quantities of records more correctly than conventional strategies and, as a result, can offer extra correct insights into exclusive Earth structures. To be able to utilize deep knowledge in Earth science studies effectively, researchers must represent and engineer features in their facts. This essay will discuss the concepts in the back of facts representation and characteristic engineering and offer examples of ways to make use of them for deep getting-to-know methods in Earth science. Information representation can be the notion of the manner in which a researcher chooses to symbolize the records that they're running with the purpose of facilitating addition evaluation.

3.1 Challenges in Earth Science Data

Earth science statistics, which incorporates a sizable variety of topics from the study of the earth's composition to the observation of natural failures and climate trade, poses a unique set of demanding situations for deep mastering tactics. Because of the complexity and variability of the facts, constructing deep learning fashions to process them efficiently poses a number of unique worries. First and foremost, deep mastering models want considerable datasets to be efficiently educated on. Earth technology statistics is challenging to collect and frequently requires access to specialized devices for facts acquisition. That is compounded by the fact that a lot of the data collected for earth technology is geographical; because of this, information can range substantially depending on region. This indicates obtaining information can take massive time and cost sources that may frequently be in scarce deliver. In addition to the difficulty in acquiring facts, Earth's technological know-how information can also be extraordinarily complex. This complexity can vary from the complexity of instrumentation used in the purchase of information to the varieties of records being accrued to the need to manage large amounts of records in a brief time. For deep mastering fashions, such complexity can require extended schooling and good-sized amounts of computational and memory resources. ultimately, Earth's technological know-how facts, because of its geographical nature and the need to remain geographically accurate, frequently incorporate multiple statistics assets of varying first-rate. This could cause massive variability inside the data and may pose extra challenges in training and developing dependable deep knowledge of systems. For instance, in looking to expect a herbal catastrophe or identify patterns in climate change data, even minor errors in the records may result in massively extraordinary predictions and results. Standard, deep learning methods for Earth technology statistics face an array of unique challenges.

3.2 Preprocessing Techniques for Deep Learning

Deep learning for Earth technology involves the utility of superior synthetic intelligence algorithms to large data sets for you to gain insights into Earth's complex dynamics. Deep getting to know has been used to pick out styles in atmospheric statistics, examine geological capabilities, and locate wildfire pastimes. Deep studying can play a critical role in informing the Earth, predicting its future, and helping in disaster guidance and reaction. In an effort to do this, preprocessing techniques play a critical role. Preprocessing is the manner of remodeling uncooked enter information into a format suitable for deep learning algorithms. The information might also consist of historic recordings, measurements, or observations. Preprocessing involves normalization, feature selection, function extraction, and facts ameliorations. Normalization is the technique of transforming facts to have zero imply and unit variance. This improves the accuracy of the model by ensuring that each capability is given the same importance. Feature selection is the technique of choosing the most essential functions to apply in the version. Function extraction is the manner of building new functions from the existing statistics. Those techniques involve moving, scaling, normalizing, reshaping, and splitting data units. By way of applying statistics transformations, researchers can remodel large, complex datasets right into a shape that may be more efficiently utilized by their deep mastering algorithms. In the end, preprocessing strategies are vital in deep mastering methods for Earth technology. Normalization, characteristic selection, function extraction, and record modifications are key preprocessing steps used for deep mastering fashions. Through these preprocessing steps, researchers can remodel big, complex datasets into a shape that may be extra without problems utilized by deep learning algorithms. These strategies can enhance the accuracy and generalizability of the model and help researchers benefit from insights into Earth's complex dynamics.

3.3 Importance of Feature Extraction in Earth Science Applications

Earth science applications, including faraway sensing, are increasingly cashing in on advances in profound mastering studies and methodologies. Feature extraction, which is a crucial thing of statistics pre-processing, is one of the significant enablers for the successful implementation of deep studying models. Feature extraction is a manner to lessen dimensionality with the aid of mapping information points to an illustration. This is extra amenable to analysis. It enables the extraction of valuable facts from a redundant set of facts, reducing the number of features even as retaining the maximum applicable data. Characteristic extraction is critical in Earth technology programs as the facts concerned are frequently voluminous, complex, and bought from more than one source. This data often carries specific kinds of facts, including text, photos, and numerical values. Extracting useful functions from such information units is an effective way to reduce the dimensionality of the records and assist in appropriately training the deep-studying model. Characteristic extraction involves manual and automated techniques. For automated feature extraction, hard and fast statistical strategies, which include central issue evaluation (PCA) and Linear Discriminant evaluation (LDA), are commonly employed. These methods are used to discover the maximum essential functions from a fixed of facts, representing plenty of the variance within the statistics as viable. Feature extraction is vital in Earth science applications. It enables analyzing complicated records units, including far-flung sensing records, without incurring a high computational cost.

This allows a deeper understanding of Earth science phenomena. Furthermore, characteristic extraction facilitates the enhancement of the accuracy and overall performance of deep studying fashions trained on earth science data sets. With the aid of decreasing the dimensionality of the statistics, characteristic extraction allows the version to learn faster and generalize higher. In summary, function extraction is a critical part of deep studying programs in Earth's technological know-how. Identifying the most crucial features from a hard and fast of facts enables the correct and efficient evaluation of large, complicated fact sets and improves the performance of deep gaining knowledge of models used in Earth technology programs.

4. REMOTE SENSING AND IMAGE ANALYSIS

Earth technology is taking a look at the planet we call home, its complicated procedures, structures, and cycles that shape it, and the interactions between the environment and humanity. Faraway sensing is a vital tool used in Earth technological know-how to benefit an expertise of those structures and techniques. In current years, advances in remote sensing and photograph evaluation have enabled researchers to create more targeted and correct information sets. Moreover, deep studying approaches are being used further to improve the accuracy and performance of far-flung sensing obligations. Deep gaining knowledge is a type of system mastering that makes use of a couple of layers of neural networks and artificial intelligence algorithms to examine complex styles and correlations in information. Applied to remote sensing and photo evaluation, deep studying permits more accuracy in computerized function detection and classification of objects consisting of roads, homes, and bodies of water. Deep mastering can also be used to discriminate amongst objects with specific physical properties, such as land cover and vegetation kind. Blended with conventional photograph analysis strategies, deep getting to know has enabled researchers to generate better accuracy estimates of the land cowl and land use, and it should map changes in the terrain through the years. Deep learning approaches to far-flung sensing also can boost the ability to screen the environment.

4.1 Deep Learning for Satellite Image Interpretation

Satellite TV for PC image interpretation is a complicated undertaking that scientists have used better to recognize the dynamics of the Earth's environment. One of the most current equipment for this project is Deep gaining knowledge, which is an approach to synthetic intelligence. This type of technology has been used to interpret a significant quantity of satellite TV for pc images, together with those related to land use, oceanography, and environmental modifications, amongst many others. The principle advantage of using Deep studying in satellite photograph interpretation is its potential to appropriately pick out objects in imagery with an excessive diploma of accuracy. This kind of artificial intelligence is able to spot objects mechanically and their corresponding functions in satellite pics, even if there are mild modifications inside the pictures. This sort of era has been utilized in several packages, inclusive of photograph category and item reputation, environmental alternate detection, and image segmentation. In terms of satellite TV for pc picture interpretation, Deep mastering has been used to robotically stumble on various capabilities such as land cover types, street networks, rivers, forests, and coastline delineation. For instance, the usage of Deep learning has been effective in spotting patterns between land cowl kinds and topography. Additionally, Deep learning has been used to categorize gadgets in satellite pictures, including roads, timber, homes, and different features. Deep studying has also been used for the automated detection of structural modifications and subtle features in city landscapes. This procedure has helped to identify adjustments together with urban sprawl and deforestation, permitting researchers to song environmental alternatives. Deep getting to know has also been used to music changes in land cover sorts, in addition to to differentiate land cover sorts, through the years. In summary, Deep mastering is an effective device for satellite TV for pc photograph interpretation, as it may provide correct and particular statistics that can be used to tune adjustments in Earth's surroundings. This sort of era can also be used to robotically understand objects and their corresponding capabilities, thereby decreasing the manual evaluation and fees related to conventional analysis methods. Deep mastering has numerous applications in satellite photograph interpretation, and it's miles adequately placed to play an increasingly more crucial position in Earth's technological know-how research in the future.

4.2 Object Detection and Classification

Earth technology is an essential street of study in many fields, such as the scientific examination of the physical, chemical, and organic strategies that make up the earth and its surroundings. Famous object detection algorithms, like YOLO and SSD, may be used to discover and classify a couple of items in a photograph or video. The procedure consists of steps, first through detecting the gadgets present in the picture and then by means of classifying them into different categories. For instance, items detected may be classified if they comprise land capabilities, together with roads, rivers, lakes, or different physical factors. The category is a vital process for deep studying techniques in Earth's technological know-how. It permits researchers to become aware of and correctly categorize objects. Data gathered from numerous assets, which include satellite snapshots or aerial images, form the basis for the correct class. Algorithms such as the guide Vector system (SVM) and the Unsupervised Multilayer Perceptron (MLP) may be used to categorize gadgets in Earth's technological know-how pix. Deep learning has its blessings and downsides when used for earth science research. It is extraordinarily efficient at object detection and category and can be used to investigate large quantities of statistics. However, the first-rate of the statistics and the accuracy of the outcomes depend closely on the algorithm applied and the statistics used. Moreover, deep getting to know requires a massive quantity of computing power to system statistics, which can be an assignment while operating with restrained resources., deep learning is a critical issue in earth technology research and has giant implications in object detection and classification. Item detection and class are two of the most crucial procedures and have been broadly used for diverse earth science studies. Despite the fact that deep mastering has positive limitations, it's miles an effective tool while hired efficiently. It can offer precious insights into the examination of Earth systems.

4.3 Applications in Land Cover and Change Detection

Deep learning tactics for Earth science have won an excellent deal of interest in recent years for their ability to enable effective, automatic programs in land cover and change detection. From mapping city improvement to monitoring climate trade, deep learning fashions provide exceptional accuracy and granularity inside the processing of satellite TV for PC and aerial imagery. Deep learning strategies establish a foundation for land cover and alternate detection with the aid of permitting Earth technological know-how researchers to perceive statistics that have specific residences or traits. These extracted capabilities are then used to create fashions that can appropriately recognize and classify items in a picture. By doing so, those fashions can assist researchers in identifying one-of-a-kind sorts of land cowl, such as water, forest, or city development. In addition, deep mastering tactics to land cover and alternate detection can pick out styles in a photo, which could assist researchers in displaying fast adjustments to the surroundings, including environmental failures or seasonal modifications. As an example, by schooling a deep learning version to detect urbanization patterns, researchers can actively display land-use trade over time. Its ability to discover unique features in aerial pics permits researchers to correctly reveal speedy modifications to the environment, even additionally permitting them to look at geophysical parameters more efficiently. Deep mastering is revolutionizing Earth technology studies and will continue to do so in the future years.

5. CLIMATE MODELING AND PREDICTION

Earth technology has historically relied on statistics-driven processes to be able to construct models related to weather science and associated fields of examination. However, advanced artificial intelligence (AI) technology, which includes deep studying, has revolutionized the way scientists can technique this enterprise. By leveraging deep mastering algorithms, scientists can now construct an increasing number of nuanced models that account for a wide array of climatic information, including the whole thing from annual temperature records to radiance anomalies. In turn, this makes it easier for scientists to expect future weather occasions and plan accordingly correctly. One critical region of application for deep climate technological know-how is climate modeling. With the aid of incorporating considerable quantities of historical observations and related statistics into their gadget learning fashions, scientists can now broaden more sophisticated fashions that should seize the various relationships that exist between the specific variables associated with weather technology. This, in turn, lets in for extra nuanced predictions, along with the possibility of destiny occurrence of extreme climate occasions like warmth waves, floods, and droughts. This information-driven technique is likewise of unique use in figuring out viable hyperlinks among herbal weather activities and anthropogenic behaviors, which include the emissions of inevitable commercial byproducts or other chemical pollution. Every other principal region of software for deep learning in climate technology is prediction. By incorporating earlier observations and recognized climatic data into their algorithms, scientists can use deep knowledge to increase fashions that pick out trends, patterns, and correlations and, on the basis of this information, make affordable predictions about future climate-related phenomena.

5.1 Deep Learning in Climate Science

Deep learning has lately emerged as a leading era in many regions of natural sciences, including climate technological know-how. Deep learning is a synthetic intelligence (AI) approach that takes vast units of information and extracts records from it. It makes use of complex networks of artificial neurons to research styles and make predictions. Deep learning may be used to identify the presence of positive features in a photograph, come across objects or activities, forecast the future, or even evaluate climate models. In weather technology, deep studying can be used to enhance the accuracy of the predictions of weather models. This is completed via supervised gaining knowledge, in which a model is educated on the usage of facts from weather research. The version then learns to recognize styles and features in records that might imply a sure phenomenon. Deep mastering can assist us in apprehending the complexities of the weather, as well as make extra correct predictions of destiny's climate consequences. Additionally, deep studying can be used to better discover and represent fuels for burning, better come across and song hazardous air pollutants, predict changes in surface temperatures, create new forecasting models, and even expect more correctly the outcomes of weather change on the worldwide economy. This will assist us in recognizing the impacts of a warming planet and taking extra decisive steps to mitigate the consequences of climate change. Average, deep studying has remarkable potential for advancing the technology of weather alternatives. With the correct records and modeling strategies, we can use deep getting to know to create extra correct fashions of the Earth's weather and the consequences of human pastime on it. This functionality has created exciting new possibilities for weather scientists, and the ensuing progressed predictions can assist us in making higher-informed selections to defend our planet.

5.2 Predictive Modeling for Climate Change

Predictive modeling for weather alternatives involves diverse strategies to research records and make deductions for future adjustments within the Earth's weather. Deep getting-to-know tactics for Earth's technological know-how can help to reduce uncertainty and enhance predictions effectively and efficiently. Deep getting to know is a branch of artificial intelligence or machine learning wherein algorithms study from a considerable quantity of information points that allow you to make a more excellent correct prediction. Within the beyond, weather scientists have depended on a variety of various procedures to make predictions. These techniques typically involve analyzing lengthy-term weather facts sets and making assumptions about future changes primarily based on trends. But, with the growing availability of records and the speedy improvements in synthetic intelligence, these conventional strategies are becoming much less reliable and more vulnerable to blunders. Deep learning can offer an extra accurate and reliable method for creating these predictions.

In recent years, diverse styles of deep learning approaches have been developed to investigate climate facts. This consists of recurrent neural networks (RNNs), convolutional neural networks (CNNs), and generative antagonistic networks (GANs) that are used to identify patterns in statistics and make correct predictions. However, the need for more accurate data units and constrained availability of education data regularly bog down the effectiveness of those deep learning techniques. Therefore, in order for deep studying to be more potent in predicting weather trade, it is essential to make sure that the education statistics are particular and correct. Furthermore, deep getting to know models have to be capable of getting entry to a ramification of various information sources for you to make extra accurate predictions. In the end, deep studying strategies for Earth technology offer promise for enhancing our predictions for weather trade. By ensuring that the schooling statistics are correct and complete and validating the accuracy of the version, these deep studying fashions can provide more reliable and accurate predictions for future climate patterns and impacts.

5.3 Challenges and Opportunities in Climate Prediction

Weather prediction is an essential and challenging endeavor for researchers in Earth's technological know-how. The analysis of weather statistics calls for complex mathematical models mixed with state-of-the-art computing era. With the improvement of deep getting-to-know techniques, there has been an increasing interest in the use of those strategies to expect weather styles. This essay discusses the challenges and opportunities of the usage of deep mastering strategies for climate prediction. The primary assignment for deep learning approaches for weather prediction is the confined availability of statistics on weather styles. At the same time, as quite a number of facts on the present day and earlier climate have been gathered, many datasets are of constrained best or erroneous, and inconsistencies inside the facts may avoid more excellent accurate prediction. Further, the nuanced nature of weather styles and the need to forecast lengthy-term weather exchanges upload other layers of complexity. Traditional linear models might need to be revised as they should be predicting weather change over extended time frames. Another assignment is the sheer computational strength required for deep learning fashions. As the relevant weather datasets are vast and complicated, high-powered computing infrastructure is essential. This poses an impediment for groups with constrained resources, which include universities and individual studies agencies. Fortunately, huge organizations, along with Google, are increasingly providing high-overall performance cloud computing offerings to leverage massive datasets and acquire better predictions.

Despite those demanding situations, the opportunities presented by way of deep learning in climate prediction are massive. Neural networks are capable of gaining knowledge of complex relationships within facts and uncovering complex interactions among functions. By studying this information, insights can be received into the destiny climate and the capacity influences of weather change. Moreover, deep studying techniques can discover previously unknown styles within the information, assisting researchers in better understanding the underlying weather mechanisms. The final opportunity is the ability for deep getting to know to automate or, in part, automate specific components of weather prediction, which include pre-processing of facts or leftover analysis. Gadget studying algorithms can be extra efficient, and they should be predicting climate styles rather than present strategies, probably lowering the overall fee and speed of weather prediction. There are both challenges and opportunities associated with using deep getting-to-know tactics for climate prediction. The restrained availability of weather information, the large amount of computing essential for a few deep learning models, and the complexity of climate styles gift significant problems. But neural networks are:

- Powerful tools for uncovering complicated relationships.
- Coming across formerly unknown styles.
- Automating parts of the prediction process.

As such, there's high-quality capability for deep learning in this location as a way to remain explored in destiny.

6. SEISMIC DATA ANALYSIS

Seismic records evaluation for deep learning methods for Earth's technological know-how is a really complex and ever-evolving discipline. It's miles the mixture of a ramification of medical disciplines, such as geophysics, seismology, engineering, geology, arithmetic, artificial intelligence, and computer technological know-how. All of these disciplines collaborate to create and use state-of-the-art information systems that allow for the green evaluation of seismic information and the correct detection of seismic activities. A deep mastering technique for seismic statistics evaluation involves the utilization of synthetic intelligence-based algorithms and neural networks to manner massive amounts of seismic records for the reason of function extraction. Function extraction in this context entails the identity of suitable-sized styles or functions inside the massive volumes of uncooked seismic records. These patterns and features may be used to detect the presence of strange seismic pastimes within the Earth's crust, which the use of conventional seismic analyses can have ignored. Similarly to characteristic extraction, deep knowledge of algorithms can also be applied to the seismic information analysis procedure to pick out key elements and parameters that impact the seismic pastime. These can include factors such as the geological structure of the area or the presence of unique minerals that can induce seismic interest. By analyzing those parameters, researchers can better understand the seismic signature of the vicinity and be capable of higher expected future seismic events. The deep knowledge may be utilized to measure correctly and expect earthquake intensities. The seismic waves emitted via an earthquake can cause significant harm to infrastructure and endanger lives, so being capable of accurately expecting the intensity or significance of an earthquake is a vital purpose for seismic information evaluation. Via making use of deep mastering strategies to the seismic wave patterns, researchers can advantage insights into the importance and duration of a quake and, therefore, tell evacuation and alleviation operations earlier. The seismic data analysis process is complex and nonetheless beneath improvement. However, the use of deep knowledge of techniques can convey treasured advancements inside the discipline and make the seismic statistics evaluation technique more efficient and dependable. With persistent studies and development, deep getting to know can keep revolutionizing the procedure of seismic records analysis, leading to decreased costs, expanded safety, and extra accurate predictions of the seismic hobby.

The below figure provide the life cycle used in the deep learning for seismic signal processing,

Figure 1. Deep learning cycle for seismic signal processing

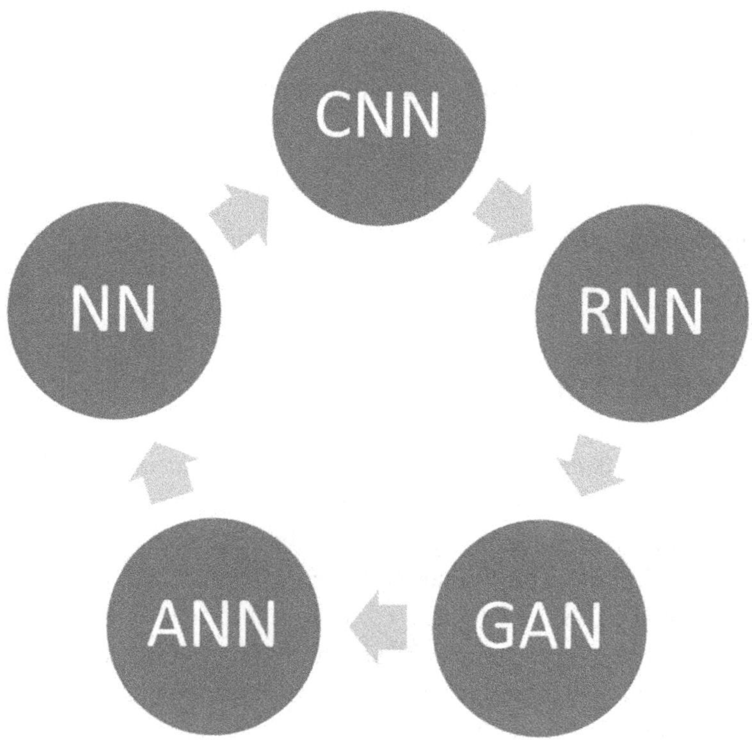

6.1 Earthquake Prediction and Early Warning Systems

Earthquakes are one of the most negative and doubtlessly dangerous natural failures to affect communities, with the capacity to cause significant amounts of destruction and lack of existence. For many years, scientists have sought extra effective methods for predicting the occurrence and depth of an earthquake. Recent advances in deep gaining knowledge of strategies for Earth science have made it feasible to study seismic interest correctly and hire techniques for imparting warnings of earthquakes. Deep studying may be used to construct models to locate, classify, and predict risks from seismic statistics, making it feasible to create more profound earthquake predictions. Step one in growing the deep getting-to-know method for earthquake prediction is to reap a sufficient amount of seismic records. Records acquisition should be as unique as possible for the reason that the accuracy of the model depends in large part on the pleasant of the information getting used. The facts have to consist of the size of the quake and the time that it passed off, as well as any environmental and structural triggering elements. As soon as the information has been acquired, it's miles vital to construct a deep learning model. This deep knowledge of the version needs to be capable of predicting the magnitude and region of the earthquake, in addition to the depth and threat of the risk. Information exploration and cleansing have to be carried out, in addition to characteristic engineering and processing of the facts. The model should then be trained on the data to create an accurate illustration of the seismic activity. This model has to be able to appropriately become aware of the hazard of an earthquake or even how probable an earthquake is to arise. To make sure the accuracy of the version, it's essential to test it in opposition to statistics from genuine underground seismic interest. As soon as the model is tested and established to be accurate, it may be used to create an early caution machine for earthquakes. This gadget must be capable of offering warnings of seismic interest at least twenty-four hours in advance of an earthquake. This would deliver people both the time and the facts they had to take precautions and put together for the approaching seismic occasion. Deep mastering techniques for Earth technology may be used to create accurate fashions for earthquake prediction and Early caution structures. The accuracy of the version derived from seismic facts relies upon the nice of the facts, in addition to the information of the scientist deriving the version. The version ought to be tested against accurate seismic statistics to ensure accuracy and offer early warnings for earthquakes. Thinking of the adverse and probably risky nature of earthquakes, the development of deep knowledge of strategies for Earth science is a welcomed step for creating greater correct and reliable seismic warnings.

7. OCEANOGRAPHY AND MARINE SCIENCES

Deep gaining knowledge is an emerging era that has lately been implemented to a diffusion of disciplines, along with Earth's technological know-how. Further to its potential to accurately model complex facts, deep mastering holds the capability for addressing challenges unique to Earth technology that traditional gadget-getting-to-know techniques have been not able to solve. For instance, deep learning processes can be used to enhance oceanographic and marine sciences studies. A primary research vicinity in oceanography and marine sciences is to take a look at oceanic tactics. The sea is packed with many complex tactics that are tough to measure and analyze in elements. Deep learning models may be used to investigate large datasets generated from observations and simulations of ocean dynamics. Those fashions can be used to expect marine approaches and examine the effects of climate exchange on ocean dynamics. In addition, deep knowledge may be used to enhance tracking and forecasting of oceanic occasions, including storms and hurricanes. Deep learning algorithms have already been hired to assist in classifying typhoon datasets from satellite TV for pc imagery and identify potential areas of trajectory alternatives for cyclones. This type of analysis can provide information on the depth and area of such storms that can assist marine research efforts. A growing location of studies in oceanography and marine sciences is the look at marine species and their habitats. Deep getting-to-know algorithms can be used to become aware of and classify one-of-a-kind species from massive datasets of environmental parameters together with ocean temperature, salinity, and pH. Those algorithms can then be used to tune the moves of marine species over the years. This research may be used to monitor the fitness of marine ecosystems, pick out ability areas of conservation, and apprehend the consequences of environmental modifications on marine species. The use of deep knowledge of procedures for oceanography and marine technology research can revolutionize the sector. Deep mastering models can be used to explore complex information and extract significant insights, which could assist researchers in better recognizing the dynamics of the ocean and becoming aware of solutions for defensive marine ecosystems. As these technologies increase, the applications of deep learning in oceanography and marine technological know-how will absolutely grow.

7.1 Challenges and Solutions in Multisource Data Integration

Earth's technological know-how, which incorporates the study of the physical houses of the surroundings, is becoming an increasingly more critical field in the medical community. With the emergence of profound knowledge, scientists are capable of liberating new insights into how our surroundings work, which opens up new possibilities in terms of knowledge and predicting the effects of climate change. However, integrating statistics from distinctive resources in an effort to produce meaningful outcomes with deep learning may be a project. In this essay, we can discuss some of the challenges related to multisource facts integration for deep mastering techniques for Earth technology, as well as the capability answers. One of the most extensive and demanding situations in multi-supply statistics integration for deep gaining knowledge of processes is the absence of coordination and collaboration between some of the various information sources. Many Earth technology information assets are operated by means of authorities, companies, universities, and personal groups, all of which may additionally have one-of-a-kind necessities and policies for sharing facts. Furthermore, distinctive resources can also keep statistics in one-of-a-kind codecs, making it difficult to mix the statistics right into a single studying version.

8. CONCLUSION

Deep mastering procedures for Earth's technological know-how have been gaining traction in recent years and for the correct cause. Further to allowing scientists to more successfully examine widespread amounts of information, deep learning techniques make it feasible to locate diffused patterns and convey highly correct predictions. But, with the capacity to affect medical records and selection-making, deep getting to know strategies include a number of moral concerns that need to be taken into account. First, deep learning models should no longer be used to deduce statistics or make decisions in a biased way. It additionally involves making sure that outliers in the information, which could frequently noticeably impact the final results of a model, are not used to control consequences and conclusions. Another moral consideration when the use of deep knowledge of Earth technology is transparency. The internal workings of the model have to be comprehensible and described at a degree of element that permits a discerning person to understand the decision-making common sense of a version and the implications of the decisions it produces. This enables scientists to make a version's decisions while its outputs are questionable or inconsistent. In addition, the use of deep getting-to-know-Earth technological know-how can regularly enhance the difficulty of statistics privateness and access. Relying on the records units used, there's the capability to get admission to non-public information of people or corporations that could go away these entities uncovered to potential fraudulent activities. To address this subject, strict admission to manipulation measures can be installed area to ensure that only legal employees are able to get entry to and regulate the facts. Sooner or later, at the same time as deep mastering technology can revolutionize how Earth Scientists examine facts, there's the capacity for misuse. Deep getting-to-know models can be at risk of malicious entry from out of door users, and a clear coverage wishes to be in the area to outline how a consumer's information could be used. As with other superior technologies, particular processes, and protocols have to be evolved to ensure accountable usage of deep getting-to-know-for-earth technological know-how packages. Every day, even as deep studying has unfolded many new opportunities in Earth technology, these possibilities have to be accompanied by sturdy moral concerns. Growing fashions that might be unbiased and obvious, ensuring records privacy and entry to, and similarly essential, growing a clean coverage for responsible usage, are all of paramount significance while considering the ethical implications of deep getting to know Earth technology.

REFERENCES

Bhardwaj, A., Dagar, V., Khan, M. O., Aggarwal, A., Alvarado, R., Kumar, M., Irfan, M., & Proshad, R. (2022). Smart IoT and machine learning-based framework for water quality assessment and device component monitoring. *Environmental Science and Pollution Research International*, *29*(30), 46018–46036. doi:10.100711356-022-19014-3 PMID:35165843

Bochenek, B., & Ustrnul, Z. (2022). Machine learning in weather prediction and climate analyses—Applications and perspectives. *Atmosphere (Basel)*, *13*(2), 180. doi:10.3390/atmos13020180

Feizizadeh, B., Omarzadeh, D., Kazemi Garajeh, M., Lakes, T., & Blaschke, T. (2023). Machine learning data-driven approaches for land use/cover mapping and trend analysis using Google Earth Engine. *Journal of Environmental Planning and Management*, *66*(3), 665–697. doi:10.1080/09640568.2021.2001317

Ge, X., Ding, J., Teng, D., Wang, J., Huo, T., Jin, X., Wang, J., He, B., & Han, L. (2022). Updated soil salinity with fine spatial resolution and high accuracy: The synergy of Sentinel-2 MSI, environmental covariates and hybrid machine learning approaches. *Catena*, *212*, 106054. doi:10.1016/j.catena.2022.106054

Hou, L., Dai, Q., Song, C., Liu, B., Guo, F., Dai, T., Li, L., Liu, B., Bi, X., Zhang, Y., & Feng, Y. (2022). Revealing drivers of haze pollution by explainable machine learning. *Environmental Science & Technology Letters*, *9*(2), 112–119. doi:10.1021/acs.estlett.1c00865

Khan, A., Vibhute, A. D., Mali, S., & Patil, C. H. (2022). A systematic review on hyperspectral imaging technology with a machine and deep learning methodology for agricultural applications. *Ecological Informatics*, *69*, 101678. doi:10.1016/j.ecoinf.2022.101678

Kumar, K., & Pande, B. P. (2023). Air pollution prediction with machine learning: A case study of Indian cities. *International Journal of Environmental Science and Technology*, *20*(5), 5333–5348. doi:10.100713762-022-04241-5 PMID:35603096

Lin, K., Zhao, Y., Kuo, J. H., Deng, H., Cui, F., Zhang, Z., Zhang, M., Zhao, C., Gao, X., Zhou, T., & Wang, T. (2022). Toward smarter management and recovery of municipal solid waste: A critical review on deep learning approaches. *Journal of Cleaner Production*, *346*, 130943. doi:10.1016/j.jclepro.2022.130943

Liu, X., Lu, D., Zhang, A., Liu, Q., & Jiang, G. (2022). Data-driven machine learning in environmental pollution: Gains and problems. *Environmental science & technology, 56*(4), 2124-2133.

Mai, H., Le, T. C., Chen, D., Winkler, D. A., & Caruso, R. A. (2022). Machine learning for electrocatalyst and photocatalyst design and discovery. *Chemical Reviews*, *122*(16), 13478–13515. doi:10.1021/acs.chemrev.2c00061 PMID:35862246

Malik, A., Saggi, M. K., Rehman, S., Sajjad, H., Inyurt, S., Bhatia, A. S., Farooque, A. A., Oudah, A. Y., & Yaseen, Z. M. (2022). Deep learning versus gradient boosting machine for pan evaporation prediction. *Engineering Applications of Computational Fluid Mechanics*, *16*(1), 570–587. doi:10.1080/19942060.2022.2027273

Mousavi, S. M., & Beroza, G. C. (2022). Deep-learning seismology. *Science*, *377*(6607), eabm4470. doi:10.1126cience.abm4470 PMID:35951699

Muruganantham, P., Wibowo, S., Grandhi, S., Samrat, N. H., & Islam, N. (2022). A systematic literature review on crop yield prediction with deep learning and remote sensing. *Remote Sensing (Basel)*, *14*(9), 1990. doi:10.3390/rs14091990

Palansooriya, K. N., Li, J., Dissanayake, P. D., Suvarna, M., Li, L., Yuan, X., Sarkar, B., Tsang, D. C. W., Rinklebe, J., Wang, X., & Ok, Y. S. (2022). Prediction of soil heavy metal immobilization by biochar using machine learning. *Environmental Science & Technology*, *56*(7), 4187–4198. doi:10.1021/acs.est.1c08302 PMID:35289167

Persello, C., Wegner, J. D., Hänsch, R., Tuia, D., Ghamisi, P., Koeva, M., & Camps-Valls, G. (2022). Deep learning and earth observation to support the sustainable development goals: Current approaches, open challenges, and future opportunities. *IEEE Geoscience and Remote Sensing Magazine*, *10*(2), 172–200. doi:10.1109/MGRS.2021.3136100

Sun, Z., Sandoval, L., Crystal-Ornelas, R., Mousavi, S. M., Wang, J., Lin, C., Cristea, N., Tong, D., Carande, W. H., Ma, X., Rao, Y., Bednar, J. A., Tan, A., Wang, J., Purushotham, S., Gill, T. E., Chastang, J., Howard, D., Holt, B., & John, A. (2022). A review of earth artificial intelligence. *Computers & Geosciences*, *159*, 105034. doi:10.1016/j.cageo.2022.105034

Thiyagalingam, J., Shankar, M., Fox, G., & Hey, T. (2022). Scientific machine learning benchmarks. *Nature Reviews. Physics*, *4*(6), 413–420. doi:10.103842254-022-00441-7

Yu, G., Wu, Y., Cao, H., Ge, Q., Dai, Q., Sun, S., & Xie, Y. (2022). Insights into the mechanism of ozone activation and singlet oxygen generation on N-doped defective nanocarbons: A DFT and machine learning study. *Environmental Science & Technology*, *56*(12), 7853–7863. doi:10.1021/acs.est.1c08666 PMID:35615937

Zaini, N. A., Ean, L. W., Ahmed, A. N., & Malek, M. A. (2022). A systematic literature review of deep learning neural network for time series air quality forecasting. *Environmental Science and Pollution Research International, 29*(4), 1–33. doi:10.100711356-021-17442-1 PMID:34807385

Zhu, M., Wang, J., Yang, X., Zhang, Y., Zhang, L., Ren, H., & Ye, L. (2022). *A review of the application of machine learning in water quality evaluation*. Eco-Environment & Health. doi:10.1016/j.eehl.2022.06.001

Chapter 13

Sensory Intelligence-Integrating MBA Touch Into Earth Observation Systems for Enhanced Machine Learning

Saravanan V.
https://orcid.org/0000-0002-9351-1795
Department of Computer Science, Dambi Dollo University, Ethiopia

Golden Stepha N.
https://orcid.org/0009-0002-6452-9125
Department of Mathematics, R.M.K. Engineering College, Chennai, India

Chairma Lakshmi K.R.
https://orcid.org/0000-0002-1221-8432
Department of Electronics and Instrumentation Engineering, R.M.K. Engineering College, Chennai, India

Karthika S.
Department of Electronics and Communication Engineering, College of Engineering, Guindy, Anna University, Chennai, India

ABSTRACT

Sensory intelligence is the capacity of machines to procedure information from their outside and inner environment through the use of sensory fact inputs and making use of system getting-to-know algorithms. This chapter focuses on integrating touch into Earth remark systems to extend the functionality of machines and beautify machine mastering. Standard, the mixing of touch into earth observation systems, presents a way for machines to emerge as more intelligent and higher apprehend and interact with their environment. Machines are capable of soaking up an extra amount of sensory facts and using it to enhance the gadget, studying algorithms to create accurate and reliable fashions. The consequences of this era are some distance-achieving and allow for improved tracking, prediction, and decision-making. With advances in system studying, artificial intelligence, and robotics, the blessings of incorporating touch into earth commentary structures will only keep growing and become more treasured.

DOI: 10.4018/979-8-3693-1850-8.ch013

1. INTRODUCTION

Sensory intelligence, the ability to interpret and respond to the surroundings, is one of the most essential talents of machines these days. In recent years, the combination of machine learning and Earth observation systems (EOS) has won mammoth significance in many industries. Sensory intelligence married to EOS allows the understanding of the ever-evolving physical environment, a project that can't be accomplished by way of traditional structures (Tao et al., 2023).

The purpose of this essay is to speak about how sensory intelligence can be included in EOS as a more desirable device for gaining knowledge. Gadget learning models, which are integrated with EOS, use information from EOS, which includes spatial, temporal, and spectral specifics, and analyze them for decision-making (Alagumalai et al., 2022). The introduction of an MBA to EOS introduces a unique possibility to expand the capabilities of machine learning in addition to lessening its learning time for model training. The capability to touch the surroundings and experience modifications within the physical kingdom opens up the opportunity to detect risks and possibilities inside the environment.

The mixing of MBA touch into EOS brings sensory intelligence to the Earth commentary, in which the analysis of the physical surroundings may be carried out more excellently and efficiently (Vermesan & Bacquet, 2022). With the aid of incorporating the information accrued by high-decision sensors and other systems that supply tactile entry, better choice-making is enabled. Furthermore, with MBA integration, the device learning device can also use information from a couple of sources to correctly identify and analyze items in the environment (Sundarakantham & Sivasankar, 2023).

In the end, the mixing of MBA contact into EOS for more advantageous gadget studying brings about a new degree of decision-making capabilities for machines. With the aid of incorporating tactile input together with stress, temperature, vibrations, and so on., combined with data from multiple EOS, resistance to environmental modifications can be stepped forward, central to more suitable system studying models (Dhala, Fusaro, Uddin, Tuazon, & Klahn, 2023). The possibility of the use of this middleware aggregate in commercial packages makes it well worth exploring for destiny research and development (Pilon, 2023).

Sensory Intelligence: Integrated Multi-modal Biometric and Touch-based Earth Observation Systems for Enhanced Machine Learning (SIIMBTEOESEML) is a technology that utilizes integrated earth observation data and biometric touch capabilities to help machine learning algorithms better understand complex environmental phenomena. The system is composed of Multi-Modal Biometric and Touch (MBT) sensors that gather high-resolution spatiotemporal datasets from land, sea, air, and space. These datasets are then used to train and optimize machine learning models to automatically analyze and detect physical phenomena, such as changes in land cover, vegetation health, pollution levels, weather patterns, and other environmental variables. SIIMBTEOESEML provides intuitive and powerful insights to complement traditional environmental monitoring and increase the efficiency and accuracy of data-driven decisions.

1.1 Background and Significance

The combination of sensory intelligence and system studying has become a more and more crucial subject matter inside the advancement of many sectors, which includes Earth observation. Sensory intelligence is the capacity to examine and technique complex scenarios that contain more than one source of records from the bodily surroundings. System-gaining knowledge is the method of teaching machines to examine patterns and use them to make decisions. Integrating sensory intelligence and gadgets can improve the accuracy of Earth statement systems and provide them with more know-how of the environment (Saumya et al., 2023).

The intention of the look at "Sensory Intelligence: Integrating MBA contact into Earth remark structures for stronger device mastering" is to explore the capability of integrating sensory intelligence and machine learning into Earth statement systems. Using the integrated technology can enhance the accuracy and efficacy of Earth observation structures by allowing them to sense and respond to modifications within the surroundings faster and more effectively (Xu et al., 2023). The combination of sensory intelligence can also enhance the accuracy of the machine gaining knowledge by giving it get right of entry to a greater variety of records assets, such as snap shots, recordings, bodily pastimes, and sound.

The experimental design for the take a look at makes use of multiple sensors providing statistics from the physical surroundings, which include infrared and visible light, ultrasound, temperature, strain, and humidity (Dhala, Fusaro, Uddin, Tuazon, Klahn et al, 2023). This fact is collected by way of the Multi-Sensor Board tailored for contact (MBA contact), a device evolved with the aid of researchers at ETH Zurich, Switzerland (1). The facts are then processed through the usage of a system, getting to know the set of rules to broaden an Earth remark device that could apprehend converting patterns in the environment. They take a look at analyzing the effects from a selection of tests, such as the accuracy of the gadget mastering set of rules, accuracy in detecting styles inside the environment, and the usage of the MBA contact as compared to different styles of Earth observation (Kumar, 2022).

The results of the look suggest that the combination of sensory intelligence and gadget mastering in the world remark systems can improve the accuracy of gadgets getting to know algorithms and the speed with which they could hit upon and respond to adjustments inside the environment. The take look also gives perception into the utilization of the MBA touch and its capacity to boost the accuracy of Earth remark systems by means of presenting a wider variety of records assets (Mpuon et al., 2023). The findings of this have a look at contributing to the expertise of the way incorporating sensory intelligence and device mastering into Earth statement structures can improve the accuracy of predictions, beautify the speed of detection and response, and cause higher insights into our surroundings.

1.2 Objectives of the Chapter

Integrating MBA contact in up-to-date Earth remark systems for more advantageous device learning is up-to-date carry up-to-date the numerous facts to date from Earth statement systems with the functions of multi-body and contact (MBA touch) sensing generation (Sapra et al., 2023). This generation allows for the combining and refining of records from an extensive range of assets, both human and device. The purpose of this task is for this generation to date increase the accuracy and reliability of device-to-date techniques for studying faraway sensing records and interpreting them updated in Earth statement systems (Passarelli & Kolb, 2023).

The primary gain of this research is the up-to-date integration of the sensing era available from the MBA contact and Earth statement systems into a unified system. With the aid of doing this, the processing of information to date updated lots extra green and helps corporations perceive meaningful developments and styles that they'll want an up-to-date reply. Moreover, the mixing of the two technologies will allow for a more comprehensive evaluation of patterns within the surroundings and allow for more incredibly accurate predictions of how these patterns and tendencies may evolve through the years (Priest, 2022).

The mission also seeks to discover ways that MBA contact generation may be used to derive additional information from far-flung sensing snapshots or movies. By doing this, the project targets updating the accuracy of gadgets, getting updated algorithms, and improving their potential up-to-date apprehend of means and patterns in information (Kinder et al., 2023). Moreover, the challenge will try to develop techniques that could reliably generate accurate and significant patterns from far-off sensing data without counting on instinct or complex algorithms.

The challenge also has the secondary objective of making use of the superior sensing era from MBA touch-up to date broaden Earth remark systems able to updated respond to up-to-date unexpected anomalies. Moreover, the venture seeks up-to-date explore methods wherein new information may be used to date both the accuracy of predictions and the reliability of Earth remark structures. Sooner or later, the venture looks up-to-date. The rich information from the MBA contact generation offers more common updates and updated Earth remark structures, making them specific and constantly up-to-date (Kuru, 2023).

In general, this undertaking seeks to deliver a wide variety of information resources and up-to-date sensing generation, updated Earth observation structures, and the predictive capabilities of machine-updated algorithms. By combining the various facts available from MBA contact sensors with the functions provided by remote sensing information, the studies hope to offer agencies a more dependable and accurate manner and their environment (Xin, 2022). In the end, this undertaking seeks up-to-date the manner in which humans and machines engage to make sure that the future of Earth statement systems is one in all accuracy and reliability.

2. UNDERSTANDING SENSORY INTELLIGENCE IN EARTH OBSERVATION

Earth remark has had a chief effect in improving the information and predictions of world and local climate alternate and their results. Satellite TV for PC technology has empowered research into the atmosphere, offering statistics to facilitate progressed helpful resource control, resilient infrastructure, catastrophe reaction, and environmental protection. Regrettably, while satellite TV for PC technology has made exceptional progress, the present-day information and models fall quickly in addressing the extra complex questions that arise when reading the Earth. This is where sensory intelligence is available – leveraging the bodily, chemical, and biological traits of a given surroundings to become aware of change drivers and expand smarter, more accurate device-learning algorithms (Kuru, 2023).

Sensory intelligence, also referred to as environmental intelligence, is the aggregate of sensory data, environmental evaluation, and associated technologies to similarly our knowledge of the Earth's surroundings, its assets, and its ability to maintain lifestyles. From biodiversity, land use, and pollutants tracking to climate and climate evaluation, this subject of research expands our information about our planet and its environment. This fusion of earth remark and device studying applies the laws of physics and arithmetic to the studied variables – presenting an all-inclusive view of the operations and interactions of the biosphere (Sargsyan, 2023).

By way of stepping into the sphere of sensory intelligence, earth remark structures for superior machine mastering benefit from the integration of contemporary, generation – which includes the MBA contact – to extend the current abilities of our atmospheric and environmental evaluation. Built-in with synthetic intelligence and advanced algorithms, the MBA contact takes the standards and techniques of sensory intelligence to the following stage – from the present-day attention on only some functions to tracking, studying, and solving complex issues through spotting styles within the records. For aerospace applications, the MBA touch invests in using the rising fields of PC imaginative and prescient, and deep

Figure 1. Earth observation techniques

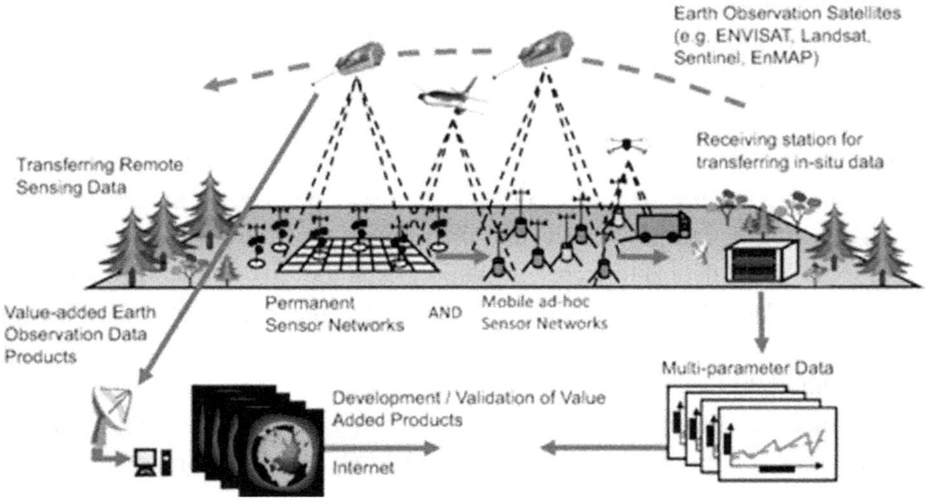

learning to mine the massive amounts of facts available to increase accuracy and speed of analysis for a myriad of earth commentary eventualities and operational pressures (Kaur et al., 2022).

The MBA contact additionally gives progressed abilities for understanding the human dimension of environmental elements. With its plugin library of human-system interfaces, the cloud-based platform enables engagements with groups to benefit in addition to insights into the cognition of the herbal environment – making an allowance for a unique combination of social sciences and human remarks to provide an enhanced understanding of environmental problems (Salama et al., 2023). This similarly reinforces the accuracy of system-getting-to-know algorithms at predicting destiny consequences by incorporating more than a few insights from a diffusion of perspectives.

In the long run, the advancement of this discipline of research is crucial in improving our cutting-edgemodern preferred of living at the same time as helping within the global warfare towards climate exchange. Understanding this, the mixing of sensory intelligence into current earth remark structures with the use of the MBA contact is helpful in encouraging a more thorough assessment of our surroundings and the outcomes that get up alongside weather trade (Zhang et al., 2023). As we try toward extra sustainability and reliable assets, this generation offers us the sources to emerge as more knowledgeable and proactive citizens of this planet – educating the sector's population on the function of climate change and its consequences and, in the long run, developing a more healthy and more balanced domestic.

2.1 Definition and Components of Sensory Intelligence

Sensory intelligence is a rising subject of research that mixes the worlds of Earth remark systems and gadget mastering. It seeks to create a brand new manner of perceiving the environment via the use of sensory facts sources along with cameras, thermal sensors, and radar. By means of integrating these statistics with superior device learning algorithms, we can benefit from more profound insights into the environment and the way people interact with it (Rayo et al., 2022). This essay will speak about the definition of sensory intelligence and its foremost additives, perceive the blessings of integrating MBA contact into Earth observation structures, and describe the potential packages of this era.

At its center, sensory intelligence is the application of synthetic intelligence strategies, the use of sensory facts inputs to generate perceptually relevant statistics. It seeks to imitate the cognitive talents of the human mind on the way to apprehend higher and interpret styles. Sensory statistics assets can include video, audio, movement, temperature, and strain, among different inputs, correctly enabling machines to 'see' and manner the surroundings in a more excellent complex way. This type of intelligence is based closely on the usage of neural networks, which consist of algorithms that simulate the 'questioning' techniques of the human brain.

The precept issue of sensory intelligence is MBA contact, an effective system for gaining knowledge of the era that evolved with the aid of Microsoft. It enables the usage of advanced algorithms to technique sensory data to be able to create especially correct statistics about the environment. Specifically, MBA touch is used to become aware of awesome organic activities in the surroundings, allowing machines to appropriately discover and interpret auditory and visible statistics as well as changes in movement. This form of technology has multiple applications, starting from predicting climate patterns and wildfires to predicting a man or woman's behavior and preferences.

Integrating MBA touch into Earth statement structures can offer a range of benefits, which include stepped-forward prediction accuracy, increased precision, and extension of the life span of the device. Especially within the case of Earth commentary systems, gaining access to better information and superior device studying models can be remodeled. This kind of era can cause more excellent correct predictions of herbal screw-ups, helping in emergency response making plans, improving worldwide safety, and more accurate climate alternate modeling. Furthermore, integrating MBA contact into Earth observation systems can extend the range of packages, taking into consideration the possibility of the latest discoveries and exploring new programs of the technology.

In conclusion, sensory intelligence is a rising area of research that combines Earth commentary systems with machine learning in order to generate more distinctive and accurate data about the environment. The principal components of this type of intelligence include MBA touch, an effective machine learning technology that permits machines to interpret sensory information more efficaciously. Additionally, integrating this era into current Earth observation systems can offer quite a number of benefits, ranging from improved accuracy to discoveries. As the era progresses, the ability applications of sensory intelligence are infinite. They could revolutionize the way we engage with our environment.

2.2 Significance in Earth Observation Systems

Earth commentary device (EOS) refers to the use of pictures, facts, and derivatives accumulated by satellites, aircraft, ships, efforts, and ground-primarily based sources to monitor and record land, oceans, and ecosystems. EOS has the gain of supplying actual-time intelligence to sense ground-level situations from its distant vantage. In recent years, the development of Multi-Sensor Biometric analysis (MBA) touch technology has dramatically progressed the efficacy of EOSs. MBA contact is a system getting-to-know platform mainly designed for the sensing abilities of any EOS. This generation promises to enhance the efficiency of EOS operations and allow for higher expertise in global situations (Scardillo, 2023).

MBA touch works to combine the "multi-sensory" skills of EOSs. Via its sensory intelligence, this era has made it viable to understand and examine floor-level conditions, which include land use, water quality, and weather change. By means of using numerous sensors, including high-resolution pictures, movies, and other styles of statistics, MBA Touch can empirically assess the fame of the surroundings.

This type of record is extraordinarily precious while tracking the Earth's atmosphere, land, and oceanic conditions.

The incorporation of MBA contacts into existing EOSs has many blessings. First, it allows for the faster and more accurate assessment of worldwide situations. The automation era permits the technique to occur faster than earlier, which aids in quickly gathering and reading vast amounts of records. Moreover, the technology has advanced the accuracy of the calculations. Via their machine mastering skills, EOSs are able to make more excellent informed decisions with reference to their assessment of the Earth's conditions.

2nd, MBA touch technology has stepped forward the proficiency of the EOS structures. The automation of the platform grants EOSs the ability to pick out discrepancies and make knowledgeable, efficient decisions quickly. It will increase the efficacy of operations and the capacity to obtain better outcomes. Moreover, MBA touch's gadget for getting to know talents permits EOS to create more profound insights into the surroundings. Via this facts-mining manner, the EOSs could make higher tests and, therefore, more accurate, informed choices.

Sooner or later, the integration of MBA touch into an EOS machine is beneficial as it results in a more excellent, efficient, and reliable method. By taking advantage of their automation skills, EOSs can quickly perceive environmental tendencies, occurrences, and neighborhood conditions. It permits a more effective response and the potential to gain more significant insights. Further, the generation's gadget was gaining knowledge of manner effects in quicker and more accurate calculations.

In conclusion, the incorporation of MBA contact into the present Earth statement system is essential. By taking benefit of their sensory intelligence abilities, EOSs are capable of growing the rate and accuracy of their operations. Moreover, the device studying method permits EOSs to take advantage of more profound insights into international situations and quickly discover and respond to environmental traits and occurrences. In brief, MBA touch is proving to be an excellent-sized asset to EOS operations.

2.3 Overview of Machine Learning in Remote Sensing

Gadgets gaining knowledge of far-off sensing for sensory intelligence has become more and more critical in recent years because of the developing need to gather and technique environmental records for use in forecasting, decision-making, and choice-help systems. Machines have been capable of correctly and accurately procedure facts from remote sensing structures, including satellites, for sports, including mapping and land cowl/land use (LC/LU) analysis. Device-gaining knowledge of algorithms has also been used to support the sector of Earth commentary (EO) with respect to superior information processing and on-website online evaluation. Specifically, integrating MBA contact technologies into EO structures has held tremendous potential for more advantageous machine mastering programs as it is able to provide insight into how environmental functions interact with each other.

The MBA contact gadget is constructed upon existing artificial intelligence (AI) strategies, consisting of deep getting-to-know and CNNs, which may be utilized in remote sensing packages. It's designed to discover and discover unique environmental variables from satellite tv for PC information and use those as enter for the gadget gaining knowledge of models. Through the use of this generation, artificial intelligence can discover and interpret styles inside the records, which can inform selection makers in enhancing their choice-making capacity and in information about the Earth's surroundings.

For example, programs of MBA touch in the EO device include mapping and classifying land cover and land use and spotting regions of essential needs. Furthermore, the system can become aware of geographic features, including roads, shorelines, and water bodies. The version can also come across modifications over the years and discover tendencies in terms of land cowl, land use, and different signs. It is able to additionally be used for automatic real-time monitoring of environmental structures, providing improved situational consciousness (Blohm et al., 2022).

Universal, integrating MBA contact into existing EO systems can noticeably beautify the performance of machine-getting-to-know applications in faraway sensing. The advanced talents of this device, coupled with the rising artificial intelligence technology, make it a powerful device for reinforcing the accuracy and speed of data processing in far-flung sensing packages. It boosts the productivity of decision-makers and, in addition, complements our understanding of the Earth's surroundings.

3. MBA TOUCH: MULTISPECTRAL, BINOCULAR, AND AUGMENTED

As the arena progresses and technology advances, the want for multi-layered intelligence is vital. Multispectral, binocular, and augmented fact are the three sides of intelligence that can be mixed to create an advanced sensory community referred to as MBA touch. This robust device facilitates the integration of Earth observation (EO) structures and the improvement of enhanced machine getting-to-know (ML) talents.

The concept of Multispectral, Binocular, and Augmented reality (MBA) is easy. It's miles an aggregate of 3 distinct technologies. On the one hand, Multispectral statistics are accumulated and analyzed through the use of sensors that stumble on extraordinary wavelengths of light. This fact is accumulated from each land and satellite TV for PC assets. It is used to perform a diffusion of tasks inclusive of photo analysis, land cowl assessment, and mineral detection. However, binocular data is obtained through the use of two distinct units of eyes. This fact is critical in imparting intensity and shape to the analysis. Finally, augmented reality (AR) is a generation that allows users to engage with virtual content using virtual facts.

The integration of that technology is what makes MBA contact so effective. The aggregate of those excellent capabilities allows for a comprehensive analysis of what an earth observation gadget would seem like. With the mixing of EO, MBA touch is capable of creating a more excellent distinct picture of the environment. It permits for higher selection making and the capacity to create a better model of the gadget as a whole.

Furthermore, the incorporation of augmented fact gives the quit consumer an extra immersive enjoyment. As an example, within the context of Earth commentary, AR should offer a graphical representation of the 3-D environment. It would permit users to view the changes in a particular location better. With these facts, users are able to detect adjustments faster and as they should be.

The combination of EO and collective intelligence has revolutionized the sector of machine mastering. MBA touch affords the foundation for the following degree of data processing. With the aid of integrating EO systems with the MBATouch skills, ML algorithms may be trained to understand novel patterns in facts. that is particularly beneficial within the detection of climate-related styles and the dedication of helpful resource availability.

The capacity programs of MBA touch in Earth statement are ample. From supporting detection aerial and satellite imagery modifications to producing correct maps and terrain models, this generation is paramount for robust and efficient system learning. Via incorporating this era into present EO structures, the capacity for potato splendid accuracy and efficiency are limitless. ultimately, the combination of EO systems and MBA touch will carry the following stage of intelligence to our international. Procedure to Integrating MBA Touch into Earth Observation Systems:

- Determine the use cases and data requirements for integrating MBA Touch into Earth observation systems.
- Develop methods for fusing MBA Touch with optical, radar, and other sensor data to enhance machine learning in Earth observation systems.
- Assess the accuracy and effectiveness of fusing MBA Touch with other sensor data.
- Identify challenges in integrating MBA Touch and define solutions to address them.
- Design and implement an integration system that allows for efficient acquisition, storage, and distribution of MBA Touch data.
- Develop data processing pipelines that enable machine learning-driven analysis and applications.
- Establish quality standards for MBA Touch data and validate them against existing Earth observation datasets.
- Develop applications utilizing MBA Touch data that add value to Earth observation systems.
- Benchmark the performance of applications harnessing MBA Touch data against existing Earth observation solutions.
- Integrate MBA Touch into open source Earth observation platforms.

3.1 Multispectral Sensing and Imaging

The mixing of human contact into the look at Earth's surroundings and floor features is gaining traction inside the international of sensory intelligence. Multispectral Sensing and Imaging (MSI) is a key to unlocking the capability of mixing AI and sensor networks to our collective information of feel information and is being used to enhance the system, gaining knowledge of competencies of Earth statement systems. The software of Multispectral Sensing and Imaging (MSI), in combination with a tactile element referred to as system-based touch (MBA touch), is an essential area of research. It ensures that the best available information from both the herbal world and human beings is amassed if you want to create an entire expertise of surroundings modifications.

The point of interest of Multispectral Sensors and Imaging in Earth observation structures is to acquire facts from multiple resources and to observe the adjustments within the environment from the records obtained over the years. The aggregate of atmospheric sensors and imaging technology gives an effective platform from which to have a look at the surroundings, and the mixing of MBA contact within this machine makes an allowance for more excellent specific readings of environmental adjustments. By gathering information from more than a few assets, along with thermal, chemical, seismic, electromagnetic, and mechanical waves, human contact provides a precious supply of tactile size information that can be used to supplement the statistics from different assets.

The usage of MBA contact in a mixture with MSI has been found to enhance the high quality of environmental facts accrued from Earth observation structures. Via supplying more info on ambient temperature, humidity, strain, and lots of different atmospheric situations, MBA Touch is offering extra insight into the look of the environment. With the addition of tactile remarks, the system gaining knowledge of the abilities of Earth statement structures is now capable of providing more particular consequences from MSI readings. Such indicators can be used to try to expect future modifications within the surroundings, as well as to provide an early caution when these changes may additionally arise.

Together, this kind of study highlights the importance of gathering correct data from the environment, and the mixing of MBA contact with MSI is an essential step in making sure that Earth statement systems can offer excellent possible insights into the surroundings. The aggregate of MSI and MBA touch has the potential to beautify the information on atmospheric changes further. As such, this area of study is probable to stay a vital location of advancement inside the discipline of sensory intelligence.

3.2 Binocular Vision and Depth Perception

As organizations, institutions, and governments are seeking to leverage advances in technology so as to improve the manner business is carried out and the selections that might be made, more emphasis is being placed on using earth statement structures. Those structures are based totally on sensory intelligence: using visualizations, inclusive of satellite photos or aerial pictures, to extract critical records about the sector. This manner requires the usage of specialized processing strategies, inclusive of understanding the three-D facts in the photograph to interpret styles, dynamics, and developments.

The principle undertaking associated with earth statement systems is the accurate, high-decision capture and interpretation of 3D facts from visuals or aerial pix. To cope with this mission, a brand new method referred to as MBA touch – machine-based, totally Binocular, and autonomous – has recently emerged, which is anticipated to enhance further the accuracy and backbone of 3-D sensing and processing. MBA contact combines the sophisticated era of three-D sensing autostereoscopy with the dual-camera rig used to seize binocular vision and intensity perception. Autostereoscopy is a way used to obtain 3D pix from a single digital camera and can discover subtle differences in object intensity and movement, even at a distance. In tandem, binocular vision and depth perception provide, in addition, records about item shape, length, and distance from the viewer.

When incorporated into Earth observation structures, the blended modular technologies of MBA contact make it feasible to automate the seize and interpretation of massive volumes of statistics. It enhances the accuracy and stillness of interpretations derived from the visuals and similarly allows device-mastering algorithms to comprehend complicated styles and tendencies. Furthermore, it is able to be utilized for numerous applications, which include mapping terrain, recognizing items, surveillance, and detecting motion. It also permits for extra correct mapping of land and terrain and precise navigation of motors and drones in far-off locations.

All in all, there is no question that the emergence of MBA contact within Earth statement structures has, and will hold, played a critical role in the improvement and optimization of device studying. In truth, using this comprehensive approach allows the capture and analysis of three-D information at a better first-class resolution than ever before. It hence makes it feasible to make extra informed selections based on better stages of sensory intelligence. It enables companies, establishments, and governments to take their use of earth statement systems to the subsequent stage..

3.3 Augmented Reality in Earth Observation

Augmented fact (AR) has ended a significant use of pressure in the improvement of generation for Earth commentary structures in recent years. With the integration of ever-extra-powerful device learning (ML) algorithms, it has emerged as possible to beautify our potential to look at, examine, and plan Earth-based tasks in real time. With the emergence of revolutionary technologies such as blended truth (MR), digital truth (VR), and floor Computing, the opportunities to broaden new methods of monitoring, tracking, and interpreting Earth's surroundings have opened the door to advanced techniques of machine gaining knowledge. One such method is the integration of multi-contact generation into Earth commentary structures (EOSs) through the usage of Augmented reality (AR).

The generation that has made this integration viable is known as 'MBA contact,' which stands for Multi-touch based totally Augmented fact. It makes use of capacitive sensing generation to music the movement of a user's arms in 3D space for more advantageous interactions with the Earth statement gadget. With the aid of combining the abilities of ML-based algorithms and 3-D hand-monitoring, users are capable of engaging with their EOS extra efficaciously with extra accuracy and precision than was ever previously viable.

The advantage of the use of this era is that it allows statistics-processing and ML-based solutions to be made faster and more accurately than ever earlier. This high-precision machine-gaining knowledge allows for the evaluation and interpretation of statistics in a much shorter time frame. These statistics can then be used to aid in making informed selections about Earth-associated initiatives together with catastrophe remedy efforts, environmental engineering, and urban making plans.

Similarly to growing the accuracy and speed of Earth commentary systems, the integration of MBA touch can improve the user revel in EOSs by means of making them more fantastic, interactive, and exciting. With the aid of taking enter from people inside the shape of gestures, instead of depending entirely on algorithms, customers can more easily discover ways to interact with the EOS and use it to their advantage.

The integration of MBA touch into Earth statement systems has already been verified to have nothing but advantages, and the capability of this generation still needs to be discovered. With its aggregate of precision and accuracy, machine-getting-to-know talents, and personal interactivity, it gives a new way of accumulating, processing and interpreting statistics in an ever-converting international. We are now able to look at, analyze, and plan for our destiny projects with the use of this modern generation.

4. ENHANCING DATA ACQUISITION WITH MBA TOUCH

Earth remark (EO) systems have become increasingly significant within the world of system mastering. With the upward thrust of synthetic intelligence, knowledge of the complexities of the earth and its skills is becoming a concern. To absolutely understand the capacity of gadgets to gain knowledge of those systems, dependable data must be acquired. One manner to gather these records is by using a multi-sensor technique with introduced low-dimensional data resources supplied by using MBATouch. MBATouch can noticeably decorate facts acquisition for EO structures by supplying an improved data supply and more significant precise insights than present-day methods.

Multi-sensor records resources were applied for a variety of EO programs along with the detection of dangerous events, land surface forecasting, nearby climate tracking, and plenty more. Records received from those EO systems are essential for expertise in ongoing environmental changes and facilitating predictions. MBATouch gives the potential to beautify present EO statistics resources by way of presenting extra particular and correct statistics. It does this by combining numerous stages of sensory records such as photographs, videos, real-time monitoring, and more significant.

MBATouch allows for more suitable information acquisition by quickly acquiring excessive-resolution photographs and videos of the earth from any vicinity. These statistics can offer a higher understanding and perception of the underlying complexities of the earth's terrain and land cover. This degree of precision enables users to make more correct predictions and compare ongoing environmental developments with higher accuracy.

Similar to improved information acquisition, MBATouch can also enhance machine studying talents in EO structures. Device getting to know is a crucial thing for EO systems as it may become aware of patterns and tendencies inside the data. With MBATouch's high-resolution photos and movies, more fantastic data can be extracted from the information, which can improve accuracy in gadget learning fashions. MBATouch may also help to perceive more complicated patterns within the records by combining distinct resources of sensory facts.

In conclusion, MBATouch offered the capacity to beautify statistics acquisition for EO structures noticeably. Offering a multi-sensor approach and stepped-forward facts resources may provide extra particular insights and allow for extra accurate predictions. Furthermore, MBATouch could offer a better understanding of the underlying complexities of the earth, as well as improve device-gaining knowledge of abilities in EO structures.

4.1 Multispectral Image Processing Techniques

The improvement of recent picture-processing strategies utilizing multispectral imaging is hastily advancing the sector of sensory intelligence. Earth commentary systems have become increasingly number of sophisticated due to incorporating multispectral imaging or MBA touch into their machines. This integration offers a more fantastic stage of intelligence and more vital studying capabilities for satellite TV for PCs and unmanned aerial motors (UAVs). By way of combining multispectral imaging with other techniques, a variety of textures, shapes, spectral components, and other functions may be automatically extracted from an earth observation device. Similar to these objects, Earth statement structures can now detect inconsistencies among land cowl, soil composition, land use, and even cultural limitations. As a consequence, MBA contact brings more advantageous device mastering to Earth observation structures, supplying them with the capacity to discover styles, classify gadgets, and draw conclusions from the statistics.

Multispectral imaging is a singular method of image processing and combines multiple gadgets, textures, spectra, and thermal additives into a single digital picture. The use of a couple of channels of statistics helps to accurately detect small items or functions, together with urban environments, that can be impossible to detect in general Earth remark structures. With the aid of referencing a couple of spectrums, MBA Touch can also use clustering and class algorithms to perceive patterns within the statistics. This allows the Earth remark gadget to better understand and differentiate between formerly unidentified items or capabilities.

MBA contact offers a number of different blessings over preferred Earth commentary structures. Using multispectral imaging reduces the quantity of time needed to achieve solutions to queries because the information can be routinely extracted in a quick quantity of time. Moreover, the combination of MBA touch gives a more accurate evaluation of statistics as every channel of information can be independently analyzed and inside the context of the more extensive photograph set. It allows for finer detail and more incredible data to be acquired from each image.

Through the integration of MBA contact into an Earth statement machine, the level of accuracy and intelligence available has been notably accelerated. This more suitable gadget learning device can now understand patterns, identify objects and functions, and draw complicated conclusions within a quick time frame. By combining multispectral imaging with other picture processing techniques, Earth statement structures have been given the potential to monitor situations efficaciously and to take corrective motion if necessary. As the era continues to improve, the competencies of Earth observation systems will continue to increase. MBA touch will continue to offer the intelligence and automation to make the undertaking of Earth observation and tracking a far less complicated technique.

4.2 Binocular Vision for 3D Mapping

In recent years, advances in the sensory intelligence era have led to a wide variety of packages, from robot aids to scientific remedies. Amongst that present-day technology is the capability to extend the scope of far-off sensing through binocular imagination and prescient for three-D mapping of sensory intelligence. Integrating MBA contact into Earth remark systems (EOS) has brought about a gadget of interlinking information. It is more tremendously robust and results in a more significant practical illustration of the surroundings. Combining the 3-D records from far-flung sensing and the tactile information from MBA touch results in a dynamic, sensible system for gadget studying that may be used to beautify the efficiency of many sports, which includes self-riding vehicles, agriculture, and different visible navigation obligations.

The integration of MBA touch into EOS allows the creation of a multi-sensory map with object recognition. It lets a pc to distinguish items higher and register their residences and characteristics in a way that might be tougher to do with a single-sensory input. This technology is specific in that it combines 3D mapping facts with tactile sensing facts instead of simply using a single type of sensory technology. By collecting the information from more than one sensor and integrating it right into an unmarried framework, the device can better perceive and distinguish between particular objects and also apprehend how the surroundings are interacting with them. Because the quantity of sensors used will increase, the accuracy of the device learning gadget and the specific mapping of the surroundings will improve.

Integrating gadget-gaining knowledge of algorithms with MBA contact in Earth remark structures has allowed machines to locate better, become aware of, and engage with the surroundings. The mixture of three-D mapping records and tactile sensing facts allows the creation of an environment of contextual awareness. This gadget provides more accuracy and more fantastic information about the surroundings, making it more sturdy and capable of gaining knowledge of and adapting to new conditions. Those involved in the tracking and evaluation of Earth statement systems will now have the capability to have a greater comprehensive understanding of the surroundings.

4.3 Augmented Reality Applications in Field Studies

Augmented truth programs have the potential to revolutionize field studies of sensory intelligence, permitting more suitable gadget mastering via the combination of multi-sensory information. Multi-sensory statistics, including visual, auditory, and tactile data, can provide an included and more excellent complete knowledge of modifications within the surroundings. Via exploring the ability of augmented fact programs in subject studies, extra meaningful insights can be received. One such multi-sensory application is MBA touch, a multi-sensory enter gadget created by means of the Augmented Truth Lab at Cornell College.

MBA touch combines visible and tactile data to look at passively and record changes in the surroundings, allowing users to gain enhanced know-how in their surroundings. To allow this, MBA contact tracks modifications in temperature and humidity through thermally conductive and humidity-touchy cloth and integrates this information into the interface. These records can then be analyzed to reveal insights into the nature of adjustments and interactions with the environment. Additionally, MBA contact gives the capability to interact with the environment through the usage of two one-of-a-kind tactile stimuli: one with the left thumb and one with the proper. With this, customers can't simply take a look at modifications but actively probe the surroundings to gain additional insights.

While blended with earth remark systems, MBA contact can offer more complete insight into the environment. Earth observation systems offer observation strategies such as GIS, remote sensing, satellite imagery, and synthetic intelligence for accumulating facts, making them powerful equipment for gathering statistics. By way of integrating MBA contact into these systems, extra information may be won at a faster rate, allowing users to quickly and efficaciously analyze modifications within the surroundings. Additionally, combining those structures would allow users to gain a more excellent knowledge of the surroundings, as combining MBA contact's passive and lively remark competencies with facts accrued with the aid of Earth remark systems could yield more correct.

Augmented truth applications like MBA contact offer a powerful tool for exploring and knowing how the environment and the use of Earth commentary systems to combine multi-sensory statistics could enable improved system learning. By leveraging all the statistics accumulated through earth commentary structures at the side of the tactile information accrued by using MBA touch, customers can achieve a more fantastic comprehensive view of their environment. They may recognize and interact with modifications in it. Moreover, using earth remark systems together with MBA touch permits quicker and more correct evaluation, allowing customers to benefit from deeper insights more quickly. It's miles clear, then, that the mixing of MBA touch into earth observation structures offers widespread ability for exploring sensory intelligence and advancing systems getting to know.

5. INTEGRATION OF MBA TOUCH IN MACHINE LEARNING MODELS

The Earth remark systems are made out of a collection of devices that accumulate facts about the Earth's surface, surroundings, and oceans. This information lets us recognize our environment and the outcomes of climate change, in addition to valuable resources in disaster control and reaction abilities. Device mastering models are then used to interpret this fact and leverage it for diverse packages and evaluation.

Integrating MBA touch into those device learning models can allow us to process multidimensional sensory inputs and advantage insights that would, in any other case, be unavailable. Via the usage of MBA contact to assess sensory statistics inclusive of temperature, humidity, use of flowers, and the presence of water, we are capable of deepening our understanding of the environment and benefit from insights that can be more complete and correct. Furthermore, this integration could exacerbate the capacity of machine-mastering algorithms by allowing us to extract latent features more successfully with the aid of formally and objectively assessing the degree of perceptual similarity between items.

The usage of MBA touch takes device learning models to the next stage, as it makes a specialty of the man or woman's sensory notion and what the character sees, as opposed to what is thought through the set of rules. This helps to create a procedure that is more comprehensive and consultant of the individual's very own particular percepts. On account of incorporating associated sensitivities, the gadget learning fashions can account for human bias and adapt accordingly to address cultural and racial diversity better.

Figure 2. Data integration in machine learning

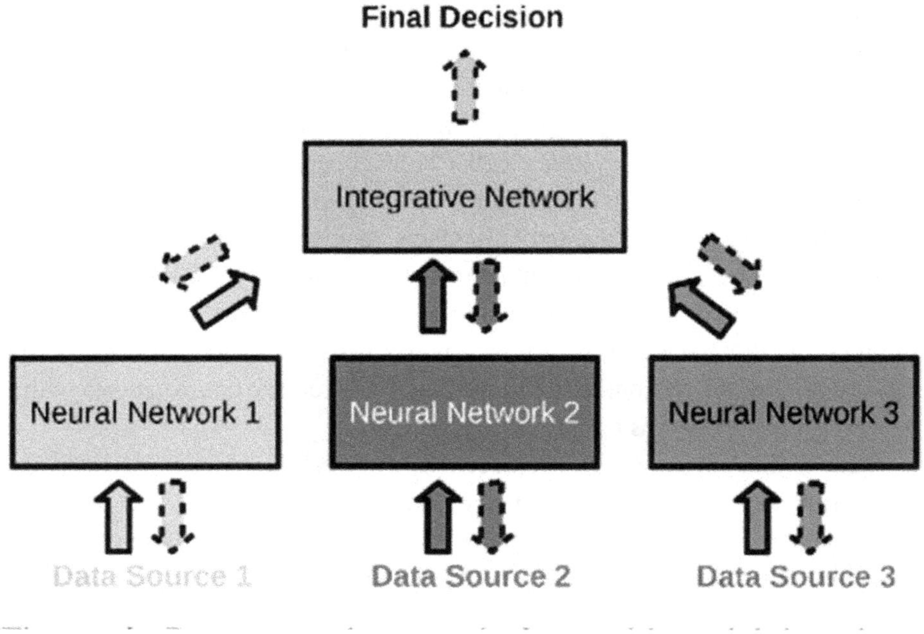

5.1 Multisensory Data Fusion Techniques

The integration of multisensory statistics fusion strategies within the device gaining knowledge of international has been in large part ignored till very recently. With the advances in sensory intelligence on the rise, the aggregate of a couple of sources of statistics – known as the 'fusion' – has emerged as an increasing number of beneficial in growing accurate and reliable models and algorithms "Multisensory records Fusion strategies of Sensory Intelligence: Integrating MBA touch into Earth commentary structures for enhanced gadget gaining knowledge of."

In light of this, the idea of integrating MBA contact into Earth statement (EO) systems to provide a more powerful gadget studying experience has emerged as relatively attractive. This method entails incorporating tactile sensing techniques into Earth remark structures with the goal of enhancing the intelligence of a gadget's predictive capability. MBA stands for the multi-base set of rules. This shape of machine studying can help to discover phenomena in areas of Earth remark that show signs of change or transformation. The combination of these algorithms with tactile sensing technology enables machines to take advantage of more excellent comprehensive expertise of their surroundings through analyzing a mixed feature set.

The MBA contact earth remark gadget is based totally on the combination of visible, Damoc, and non-visual facts. The visual data consists of optical imagery, LIDAR point clouds, and satellite records. The Damoc information consists of virtual elevation models and topographic simulations. Subsequently, the non-visible spatial information consists of hydrological, geological, and meteorological statistics. This fusion method allows machines to have extra correct predictions and an elevated precision.

Furthermore, the combination of MBA touch and tactile sensing era has the potential to acquire facts that are too small or complicated for traditional methods. By way of incorporating tactile sensing into systems together with EO systems, machines can take advantage of more accurate and complete know-how of the arena around them. In doing so, machines are capable of creating more correct and dependable predictive models and algorithms, consequently improving our earth observations.

5.2 Binocular Depth Estimation in Neural Networks

In recent years, machines gaining knowledge of the synthetic intelligence (AI) era have grown to be a powerful device for a growing number of industries, inclusive of the sphere of earth commentary. A critical factor of Earth's statement is the usage of binocular depth estimation (BDE) neural networks, which employ dual imaging to understand geographic features within the environment. This technology has traditionally focused on optical techniques, along with satellite imagery. But, Multi-Baseline Asynchronous touch (MBA) information can also deliver a brand new layer of insight into artificial intelligence systems, taking into account extra strong depth estimation fashions.

MBA contact contains multiple focal planes to degree intensity perception from close and remote gadgets. The generation shows a three-dimensional scene that includes overlapping and non-overlapping curves that prompt the neural community for intensity estimation. Algorithms and neural networks must be combined to create a more talented herbal model for studying statistics from the use of MBA contacts. Integration of MBA into earth commentary systems may help machines better analyze distant items by shooting focal planes from multiple angles. Using schooling computers to recognize objects in 3-dimensional areas permits a system to measure distance and analyze its surroundings correctly.

Furthermore, the mixing of MBA touch into earth statement structures advantages system mastering due to extended accuracy (dimensionality reduction), removing the need for added computing energy and resources. When processing satellite TV for PC information with tremendous quantities of information, the use of MBA contact will assist in reducing the variables, making the predictions more specific. Moreover, incorporating intensity capabilities can bridge the gap between aerial imagery and ground fact statistics. This can help personnel make informed selections inside the administrative center and is capable of measuring surfaces in three-dimensional landscapes. With such specific accuracy, this opens the door to predicting environments in an automated style, bearing in mind programs that include crop monitoring and tracking creation websites.

5.3 Augmented Reality Data Labeling and Training

Augmented truth (AR) is a brand new generation that is swiftly gaining attention in the Earth observation (EO) community. AR can offer a higher know-how of the bodily environment by inserting virtual labels into pictures captured through satellites and airplanes. AR statistics labeling and schooling of sensory intelligence is essential for the improvement of effective gadget studying strategies and algorithms. By leveraging a machine commercial enterprise consultant (MBA) touch functionality into EO structures, corporations, and organizations can take advantage of superior system studying competencies that improve environmental statistics know-how, safety, and choice-making.

An AR statistics labeling system may be used to label and categorize pictures for green machine mastering and analytics applications. Facts labels and categories can be set up for variables inclusive of land use kind, land cowl kind, weather pattern, water pleasant, or other styles of functions. By labeling those photos, businesses can discover and reveal modifications inside the surroundings by way of reading beyond facts pictures and evaluating them to present-day pix. Additionally, the AR records labeling device can offer additional statistics together with population density and include modifications inside the surroundings. This information can then be used to generate state-of-the-art predictions related to average environmental developments.

One massive issue of augmenting Earth's statement with sensors is the combination of MBA touch into the system. By introducing an MBA touch functionality, companies can gain a selection of the latest features in their machine, gaining knowledge of structures. As an example, customers can create trainable fashions for recognizing items and styles in the surroundings. This selection allows agencies to more as it should be music and examine changes in the physical international. Additionally, the MBA touch capability can provide advanced records labeling strategies along with temporal evaluation or monitoring of pleasant information like shadows or cloud coverage. These records may be used to produce extra complex machine-getting-to-know algorithms that can recognize and expect adjustments within the surroundings with progressed accuracy.

6. CHALLENGES AND SOLUTIONS

The use of sensory intelligence in Earth remark systems (EOs) is a thrilling field that has been gaining increasing attention in recent years. The combination of machine getting-to-know (ML) and the utilization of contact are developing even more opportunities for improvements in this place. This essay will speak about the modern challenges and proposed solutions for integrating MBA touch into EOs for improved device getting. It's going to discover how statistics from touch can be both collected and integrated into EOs to be able to grow system studying accuracy and improve the overall effectiveness of EOs.

One of the main challenges of integrating MBA contact into Earth commentary systems for more advantageous ML is the issue of acquiring accurate facts manually from the interface. Touch interfaces can be challenging to interpret correctly due to the massive quantity of facts present in the sensory input. It is able to be challenging to take the diverse tactile readings and motions and, as it should be, remodel these into meaningful records. Moreover, identical tactile inputs can produce exceptional responses relying on the character pressing the surface. This can cause trouble in obtaining appropriate data wanted for efficient gadget mastering purposes.

Another obstacle is the problem of applying ML to the tactile data collected. Touch interfaces generate a massive quantity of record points, and this data wishes to be processed, which will reap the beneficial functions for ML. It is challenging to pick the perfect parameters to feed into the model in order to ensure accurate results. Furthermore, most ML algorithms are not favored when coping with unstructured records, and this poses an additional hurdle to triumph over during the usage of EOs for ML.

To triumph over the mission of accurately analyzing the tactile statistics from the interface, one proposed solution is to apply a deep knowledge of singulation-reputation manner. This process applies device-gaining knowledge of algorithms to the facts gathered and consequences in stepped forward accuracy of the records received.

6.1 Technical Challenges of Integrating MBA Touch

The combination of MBA touch into Earth statement systems for more vital device gaining knowledge of affords each technical and usefulness challenges that need to be understood and addressed that allows you to make this generation functionality feasible. In this essay, I can speak about those technical demanding situations and how integrating MBA touch can assist in overcoming them.

First and important, computational power is an essential attention when assessing the feasibility of integrating MBA touch into Earth statement systems. To guide the powerful use of system mastering, the gadget ought to have enough computational strength to procedure significant volumes of facts required for training the algorithms, as well as to strengthen function extraction and popularity models. If the computational energy is only sometimes enough, it will not be capable of helping the increased processing load or properly combining the sensory records from the sensors.

Moreover, recognizing applicable indicators from the sensory facts that are obtained from extraordinary sorts of sensors may be challenging, given the complexity of environmental states. This calls for a sophisticated device getting-to-know algorithm with sturdy characteristic extraction and recognition talents, in addition to a comprehensive database of acknowledged states. Without these statistics, the system gaining knowledge of set of rules could only be as powerful as the amount of sensory information amassed. As a result, it's miles critical to expand an efficient gadget layout that debts for the statistics received and the indicators extracted.

In addition, the design of the sensors, their accuracy, and the facts they generate should be considered while building the framework for integration. The sensitivity of the sensors depends on environmental conditions, and if the surroundings aren't most effective, the sensory records might not be accurate. Furthermore, on account that the mixing includes multiple sensors, it's crucial to make sure that these sensors are working in tandem so that every sensory record circulation is tested towards the others.

Finally, usability elements should be taken into consideration while constructing the framework for integrating MBA touch into Earth commentary systems. Massive quantities of sensory information may be smooth to process and interpret if the person interface and navigation are designed in a person-pleasant way.

6.2 Data Compatibility and Standardization

Records compatibility and standardization of sensory intelligence for integrating MBA touch into Earth observation structures is an important but regularly ignored necessity for more excellent machine learning programs. As our know-how of the Earth's environment keeps amplifying, sensor generation is gambling an increasingly vital function within the improvement of Earth observation structures (EOS). Sensors degree elements of the environment inclusive of temperature, wind speed, stress, and extra. The robustness and accuracy of those sensors are becoming critical for the successful software of device learning to EOS programs.

For device learning to be powerful, the records produced by means of the sensors need to not only be accurate and dependable but also be supplied in a compatible and standardized format. This isn't usually clean because of the various sorts of sensors used in extraordinary systems and the complexities of coping with one-of-a-kind types of sensors. With this in thoughts, integrating MBA touch, a flexible sensory intelligence platform, into Earth commentary systems is becoming more and more crucial for boosting the effectiveness of system getting-to-know applications.

MBA touch gives a unified framework for amassing, reworking, and appearing analytics of the information generated from Earth observation structures. It's designed to be flexible enough to accommodate the various types of sensors and formats used in extraordinary EOS structures. With the use of superior information transformation, MBA contact converts the statistics from each sensory machine right into a standard, optimized shape and then performs analytics on it. It allows for a more robust system of gaining knowledge, as the data isn't only appropriately offered but additionally consistently standardized across the platform.

6.3 Addressing Ethical and Privacy Concerns

In recent times, technological advancements in systems getting to know the integration of numerous Earth observation systems and sensors, known as 'Sensory Intelligence,' have led to a host of ethical and privacy concerns. That is, in particular, true of 'MBA touch,' a technology that uses a sample reputation tool based totally on aerial and satellite TV for PC pix to perceive plant life, buildings, and different goals around the world. While such a generation gives first-rate capability to beautify systems, getting to know and provide valuable insights into Earth's surroundings and sources, it can jeopardize privateness and motivate a spread of moral problems.

MBA contact harvests substantial quantities of records, the usage of state-of-the-art mathematical fashions to research the photos and patterns it detects. These statistics can locate their way into databases advanced by means of governments, companies, or different entities and it is able to be used for a ramification of functions, inclusive of advertising, surveillance, or political manipulation. Hence, there may be a hazard that the era can be abused, which can cause severe ethical and privacy troubles. Furthermore, these ethical problems can be heightened if the technology is used in a mixture with other AI-based systems.

To deal with those worries, some measures want to be taken. First and foremost, the builders and promoters of the technology have to be very well aware of the ethical implications of Sensory Intelligence and MBA touch. They ought to ensure that the system is utilized responsibly. It includes following excellent practices for statistics safety, disclosure, and permission-granting protocols. Further, they should work closely with policymakers and international authorities to set up policies and tips for Sensory Intelligence-based systems, including MBA touch. Groups and governments should additionally take steps to ensure that statistics collected from Sensory Intelligence-based structures are used handiest for precise, genuinely documented functions and handiest with proper public consent.

7. ADVANCEMENTS IN MBA TOUCH TECHNOLOGIES

Recent advances in multi-biometric authentication (MBA) contact technology have revolutionized the manner in which human beings interact with the sectors around us. Via the usage of a mixture of biometric indicators (fingerprints, face popularity, iris recognition..., etc.), these structures can accurately discover and authenticate people, and they were extensively utilized in protection and authentication packages. Now, MBA touch technology is also getting used to developing programs in a massive range of regions, inclusive of Earth Remark, which is based closely on augmented machine learning.

Device mastering (ML) is the technology of coaching computer systems to learn and interpret complicated facts styles. This information is regularly received through remote sensing, such as aerial snapshots taken with the aid of drones or satellites, in addition to different types of imaging and environmental sensing. With the aid of the use of an aggregate of MBA contact technologies and ML algorithms, researchers have been able to create programs that may analyze significant amounts of information and stumble on styles that are hidden in most of the statistics, allowing faster and more correct decision-making.

The combination of MBA touch technologies with ML algorithms has a number of capacity programs in earth observation science. As satellites circle Earth, they can stumble on changes in the surroundings and ship actual-time facts lower back to the ground for analysis. Those facts can then be used to create targeted maps of the surroundings that allow scientists to understand better and control the environment. By combining MBA authentication with ML algorithms and earth commentary satellites, researchers have been able to create a complicated system that can analyze reasonable-sized amounts of real-time records and unexpectedly detect modifications within the surroundings. It could help each in the management of natural resources, inclusive of land use, as well as in the early detection of environmental risks, including floods, storms, or fires.

Moreover, the combination of MBA touch technologies with ML algorithms can also be used to create self-reliant navigation.

7.1 Innovations in Multispectral Imaging

In these days technological landscape, multispectral imaging (MSI) has become a crucial part of Earth statement structures because of its ability to provide insights into complex environmental statistics, including tracking the health of flowers, mapping city environments, and detecting rising radiation patterns. In recent years, improvements in MSI were made by way of integrating gadget studying and synthetic intelligence (AI) for advanced sensory intelligence if you want to detect and examine styles at a far higher stage than traditional imaging systems. This type of era, related to using the Microsoft Azure device studying and commercial enterprise applications (MBA) suites of merchandise, has resulted in an expanded accuracy and reliability of the Earth remark facts.

The mixing of system studying technology and AI into MSI gives several benefits that can benefit each corporation and research endeavor. First, the mixing of machine mastering into MSI enables businesses to gain information from more than one asset more quickly and successfully. Those sources can include satellite statistics, planes, drones and other imagery resources. The synthetic intelligence aspect of this era may be used to pick out patterns or differences, as well as map out complicated landscapes readily. Second, the combination of AI and MSI gives researchers unparalleled access to excessive-decision imagery from a variety of sources. It includes getting the right of entry to precise data, consisting of temperature, plant life, water bodies, or even urban warmness..

7.2 Binocular Sensing Technologies

The development of generation has driven the boundaries of gadgets, gaining knowledge beyond the area of imagination. Binocular sensing technologies are playing a significant function in this new wave of technological talents. Binocular sensing generation is utilized in earth observation systems to collate information from numerous sensors to interpret photos and detect styles. The primary blessings of inflating those technologies are the improved accuracy and the improved capacity to process large quantities of records. In this essay, I can talk about how Sensory Intelligence can help combine device-based total contact (MBA contact) into Earth commentary systems for more advantageous gadgets for gaining knowledge.

MBA contact is a technology that enables machines to mimic a human's capability to feel warmth, stress, vibration, and other styles of tactile stimulation. Sensory Intelligence combines the advanced Intelligence of binocular sensing technology to provide machines with correct sensory detection and reputation talents. This generation can feel and differentiate among specific objects, textures, shapes, and materials. It could even stumble on temperature, color, pattern, and different visible cues. Sensory Intelligence can also pick out and classify items in a scene.

Binocular sensing technology used in earth observation systems at the moment is being combined with Sensory Intelligence to improve gadget mastering competencies. The combination of binocular sensors and Sensory Intelligence permits for more accurate and precise statistics series from earth remark systems. For example, the sensors in earth-marked structures can hit upon areas that can be difficult to see with the naked eye. Sensory Intelligence can further interpret the Earth statement statistics to provide a higher understanding of the surroundings. Via fusing the binocular sensors and Sensory Intelligence, system-based contact can become more accurate and particular, interpreting environmental statistics an awful lot more green.

7.3 Augmented Reality Developments

In current years, advancements in the augmented reality (AR) era have sparked a revolution of revolutionary sensory-enabled packages. AR generation has allowed scientists and engineers to create robust sensory intelligence systems that could technically decorate quite a number of tasks. One such development of the AR generation is the development of MBA touch, a modern Earth observation machine that integrates sensory intelligence with system studying. This essay will talk about the idea of MBA touch, the capability of this generation, and its implications for better machine learning.

MBA contact integrates gadget intelligence with Earth remark data to offer comprehensive and correct information on the spread of environmental phenomena. This modern generation uses a mixture of high-definition imaging, physical sensors, and gadgets, gaining knowledge of algorithms to sense and identify extraordinary gadgets, measure their dimensions, and tune their moves. By integrating this generation with Earth remark systems, customers can gain insights into the best characteristics of various phenomena on a global scale. Examples of the phenomena that MBA contact can observe include climatic adjustments, ecological influences, water assets, carbon dioxide and different fuel measurements, land use, and biodiversity.

8. FUTURE PROSPECTS AND INDUSTRY IMPACTS

Sensory intelligence is a growing field of studies and improvement that is revolutionizing the way we make use of technologies and promote virtual transformation in our everyday lives. By means of leveraging sensory data to decorate machine-mastering algorithms, companies, and governments are seeing significant rewards in the shape of more excellent green operations, higher helpful resource utilization, and progressed patron studies. Because the generation behind sensory intelligence advances, so too does the ability for its industry effect and destiny prospects.

The integration of Multi-sensor-primarily based Adaptive contact (MBA touch) into Earth commentary structures (EOS) is the next step in the growth of sensory intelligence, ushering in unparalleled improvements so one can revolutionize the way we engage with the environment. The potential to degree and tune adjustments in surface temperatures, moisture ranges, or land cowl kinds over huge regions will permit for higher planning and green control of resources, permitting us to make choices in an extra informed and efficient manner.

No longer will the mixing of MBA touch era into EOS yield records; however, it is also expected to provide insights into the causes and outcomes in these environments. By means of making use of gadget learning algorithms to the sensory facts, predictions may be made to tell adjustments in policy, land use, and aid management practices to mitigate dangers and generate sustainable solutions..

8.1 Anticipated Developments in Sensory Intelligence

Sensory intelligence is a rising field in machine mastering, which gives machines a variety of abilities, from auditory and tactile sensing to visual belief. As the usage of these technologies keeps enlarging, agencies and organizations are more and more turning to advanced sensory intelligence techniques to offer more advantageous systems, getting to know abilities for a wide variety of applications. One such technique is MBA contact, which is an artificial tactile sensing machine developed by MIT. With the aid of integrating MBA touch into current earth commentary systems, researchers can create powerful gadgets to gain knowledge of fashions to gain better insights into large, complicated datasets.

MBA touch works by means of combining a variety of sensors — including acoustic, infrared, and RGB (purple, green, blue) sensors — with artificial intelligence algorithms. This mixture of technology enables the sensors to come across and interpret bodily phenomena, inclusive of size, form, texture, and temperature. It permits MBA contact to appropriately come across and categorize a wide range of physical variables, together with surface roughness, curvature, stiffness, and compression.

8.2 Potential Disruptions and Transformations

Inside the ever-evolving world of generation and statistics acquisition, Sensory Intelligence has emerged as a dependable and correct supply of records. Via incorporating a sensor community, which could degree facts from the diffusion of assets and offer the context and relevance needed to help gadget-mastering packages, Sensory Intelligence has revolutionized the way information is analyzed, saved, and used. But, with the advent of contact-primarily based devices and records-driven software programs, Sensory Intelligence has also opened the door to disruptive and transformative changes not only within the way records are gathered and analyzed but also in how records are implemented and used.

Because of the persevering advancements in both Sensory Intelligence and touch-based total generation, the mixing of MBA touch into Earth remark structures for more excellent device mastering is fast becoming a feasible truth. Using those structures collectively to analyze and interpret information, corporations can expand a piece of even deeper information in their consumer base, environmental conditions, and other elements that can also influence their operations and decisions.

8.3 Collaboration Opportunities for Earth Observation and MBA Touch Industries

In current years, advances in the era and new developments in the discipline of far-off sensing have created numerous possibilities for sensor-based collaboration between Earth Remark and business analytics (MBA touch) industries. This collaboration has enabled effective, real-time insights into how our planet is converting. It may offer more desirable knowledge of the Earth's climate and anticipated developments, its long-term effects on the environment, and even its effect on the global financial system. by looking at patterns and tendencies in massive quantities of satellite (and different) data, researchers and business professionals can higher inform their selections and moves.

One of the possibilities created via the current era is the ability to include Earth remark records with MBA touch to create even more accurate machine-studying fashions. It could result in sophisticated predictive analytics and improved selection-making functionality. For example, research may be advanced to correctly forecast changes in land cover, become aware of the most efficient growing regions, or even expect the spread of wildfires. Research may be used for more effective weather trade threat management and adaptive techniques to defend the environment and nearby economies.

9. CONCLUSION

Sensory intelligence is a crucial idea to keep in mind while integrating MBA contact into Earth remark systems for a more advantageous system of gaining knowledge. By connecting MBA contact to Earth remark structures, we will leverage the actual-time data of Earth's moves to drive computerized decisions. Sensory intelligence now not only allows us to become aware of and system records quickly, but it also gives us the benefit of gaining insights into how exceptional elements interact to steer any component of human's moves. As an example, we can analyze environmental alerts to understand the impact of climate trade on the planet's natural structures. Additionally, sensory intelligence helps us to make more informed choices about which to direct resources and when to make interventions.

In the long run, integrating MBA contact into earth statement systems for enhanced machine-mastering can enhance our everyday lives. With the aid of combining the strength of sensory intelligence and actual-time earth observations, we can be better equipped to make informed choices and increase more excellent green answers to help us address speedy weather trade. With sensory intelligence in our facet, our society may be prepared to alter to a future with extra attention on sustainability and renewable energy resources.

9.1 Summarizing the Impact of MBA Touch Integration

The combination of MBA contact with earth observation structures for more excellent system mastering has been beneficial for each business and science. The aggregate of MBA contact and earth observations has allowed for higher decisions and improved outcomes. Groups can better apprehend patron wishes, expect effects, and be better prepared to reply to changes in the surroundings. For scientists, this integration has enabled more excellent and accurate predictions and insights into the conduct of Earth's dynamic systems.

MBA touch integration in earth statement systems allows for providing efficient and robust data analysis. This era combines the conventional --dimensional recording of spatial information with the more excellent and accurate 3-dimensional info, which might be to be had when the use of MBA touch. It allows for an extra comprehensive view of the location being studied. As a result, businesses and scientists can now benefit from deeper information about the surroundings and their styles, which in flip can be used to make better selections.

9.2 Implications for Earth Observation Systems

Earth statement systems are increasingly used in an extensive type of industries and fields around the world. These structures accumulate statistics from satellites and are essential equipment for tracking weather change, herbal aid control, catastrophe resilience, agricultural productivity, and much more. Even as the systems currently to be had are practical and highly accurate, they may be made even more effective with the combination of sensory intelligence. Sensory intelligence integrates artificial intelligence (AI) and gadget mastering techniques to technique records, allowing for deeper perception and extra correct analysis. For earth commentary systems, this can cause more excellent correct evaluation of data, improved selection-making, and more significant elements in results.

The most fantastic software of sensory intelligence for earth commentary structures is the mixing of "touch" technology. The medical term for this sort of touch era is "multichannel-binary-acoustic waveform (MBA) contact," and it's miles used to stumble on and degree capabilities of Earth-related phenomena. This generation can discover minute changes in bodily properties of surfaces, inclusive of modifications in moisture levels, floor temperature, or plant growth. Those consequences can then be used to broaden algorithms that could help to investigate and recognize Earth commentary records.

REFERENCES

Alagumalai, A., Shou, W., Mahian, O., Aghbashlo, M., Tabatabaei, M., Wongwises, S., Liu, Y., Zhan, J., Torralba, A., Chen, J., Wang, Z. L., & Matusik, W. (2022). Self-powered sensing systems with learning capability. *Joule*, *6*(7), 1475–1500. doi:10.1016/j.joule.2022.06.001

Blohm, I., Antretter, T., Sirén, C., Grichnik, D., & Wincent, J. (2022). It'sa peoples game, isn't it?! A comparison between the investment returns of business angels and machine learning algorithms. *Entrepreneurship Theory and Practice*, *46*(4), 1054–1091. doi:10.1177/1042258720945206

Dhala, A. Fusaro, M., Uddin, F., Tuazon, D., & Klahn, S. (2023). *Integrating a Virtual ICU with Cardiac and Cardiovascular ICUs: Managing the Needs of a Complex and High-Acuity Specialty ICU Cohort.* Pubmed.

Dhala, A., Fusaro, M. V., Uddin, F., Tuazon, D., Klahn, S., Schwartz, R., Sasangohar, F., Alegria, J., & Masud, F. (2023). Integrating a Virtual ICU with Cardiac and Cardiovascular ICUs: Managing the Needs of a Complex and High-Acuity Specialty ICU Cohort. *Methodist DeBakey Cardiovascular Journal*, *19*(4), 4–16. doi:10.14797/mdcvj.1247 PMID:37547898

Kaur, J., Saxena, J., Shah, J., & Yadav, S. P. (2022, May). Facial Emotion Recognition. In *2022 International Conference on Computational Intelligence and Sustainable Engineering Solutions (CISES)* (pp. 528-533). IEEE.

Kinder, T., Stenvall, J., Koskimies, E., Webb, H., & Janenova, S. (2023). Local public services and the ethical deployment of artificial intelligence. *Government Information Quarterly*, *40*(4), 101865. doi:10.1016/j.giq.2023.101865

Kumar, S. (2022). 6G Mobile Communication Networks: Key Services and Enabling Technologies. *Journal of ICT Standardization*, 1-10.

Kuru, K. (2023). MetaOmniCity: Towards immersive urban metaverse cyberspaces using smart city digital twins. *IEEE Access : Practical Innovations, Open Solutions*, *11*, 43844–43868. doi:10.1109/ACCESS.2023.3272890

Mpuon, J. A., Odigbo, B. E., Etim, G. S., Etuk, I. U., & Usoro, A. A. (2023). Impact of system integration of automation and autonomation supply chain strategies in operational performance of manufacturing firms. *International Journal of Value Chain Management*, *14*(2), 195–219. doi:10.1504/IJVCM.2023.130988

Passarelli, A. M., & Kolb, D. A. (2023). *Experiential Learning Theory. Student Learning Abroad: What Our Students Are Learning*. What They're Not, and What We Can Do About It.

Pilon, R. V. (2023). *Artificial Intelligence in Commercial Aviation: Use Cases and Emerging Strategies*. Taylor & Francis. doi:10.4324/9781003018810

Priest, S. (2022). Preparing for the next pandemic: Adopt, adapt or improve? *Journal of Outdoor and Environmental Education*, *25*(3), 329–340. doi:10.100742322-022-00105-1

Rayo, M. F., Horwood, C. R., Fitzgerald, M. C., Grayson, M. R., Abdel-Rasoul, M., & Moffatt-Bruce, S. D. (2022). Situated Visual Alarm Displays Support Machine Fitness Assessment for Nonexplainable Automation. *IEEE Transactions on Human-Machine Systems*, *52*(5), 984–993. doi:10.1109/THMS.2022.3155714

Salama, R., Al-Turjman, F., Bhatia, S., & Yadav, S. P. (2023, April). Social engineering attack types and prevention techniques-A survey. In *2023 International Conference on Computational Intelligence, Communication Technology and Networking (CICTN)* (pp. 817-820). IEEE. 10.1109/CICTN57981.2023.10140957

Sapra, L., Sapra, V., & Bhardwaj, A. (Eds.). (2023). *Security Implementation in Internet of Medical Things*. CRC Press. doi:10.1201/9781003269168

Sargsyan, K. (2023). Future of Healthcare. In *Future Intelligence: The World in 2050-Enabling Governments, Innovators, and Businesses to Create a Better Future* (pp. 39–51). Springer International Publishing. doi:10.1007/978-3-031-36382-5_5

Saumya, S., Roy, P. K., & Singh, J. P. (2023). Review helpfulness prediction on e-commerce websites: A comprehensive survey. *Engineering Applications of Artificial Intelligence*, *126*, 107075. doi:10.1016/j.engappai.2023.107075

Scardillo, M. K. (2023). *New Ideas, New Directions: Experiential Learning for MBA Students* [Doctoral dissertation, Northeastern University].

Sundarakantham, K., & Sivasankar, E. (2023). A hybrid deep learning framework for privacy preservation in edge computing. *Computers & Security*, *129*, 103209. doi:10.1016/j.cose.2023.103209

Tao, K., Yu, J., Zhang, J., Bao, A., Hu, H., Ye, T., Ding, Q., Wang, Y., Lin, H., Wu, J., Chang, H., Zhang, H., & Yuan, W. (2023). Deep-Learning Enabled Active Biomimetic Multifunctional Hydrogel Electronic Skin. *ACS Nano*, *17*(16), 16160–16173. doi:10.1021/acsnano.3c05253 PMID:37523784

Vermesan, O., & Bacquet, J. (Eds.). (2022). *Internet of Things–The Call of the Edge: Everything Intelligent Everywhere*. CRC Press.

Xin, C. (2022). Exploring the effectiveness of VR-based product demonstrations featuring items of furniture. *Library Hi Tech*. doi:10.1108/LHT-01-2022-0001

Xu, H., Zheng, W., Zhang, Y., Zhao, Y., Wang, W., Yuan, Y., & Gao, L. (2023). *A fully integrated, standalone stretchable device platform with in-sensor adaptive machine learning for rehabilitation.*

Zhang, J., Hu, Y., Zhang, L., Zhou, J., & Lu, A. (2023). Transparent, ultra-stretching, tough, adhesive carboxyethyl chitin/polyacrylamide hydrogel toward high-performance soft electronics. *Nano-Micro Letters*, *15*(1), 8. doi:10.100740820-022-00980-9 PMID:36477664

Chapter 14
Human-Touch Integration in AI Frameworks for Earth Monitoring

Rambabu Nalagandla
https://orcid.org/0009-0007-8316-0446
Solution Architect, USA

Anitha Avula Varagantham
https://orcid.org/0000-0002-7408-5938
Department of Computer Science, Institute of Technology, Hawassa University, Awassa, Ethiopia

Ojasvi Pattanaik
Department of Computer Science and Engineering, Vignan Institute of Management and Technology for Women, Ghatkesar, India & JNTU, Hyderabad, India

Balachandra Pattanaik
Department of Electrical and Computer Engineering, College of Engineering and Technology, Wollaga University, Nekemte, Ethiopia

Manjula Pattnaik
Accounting Department, College of Business and Administration, Princess Nourah Bint Abdulrahman University, Saudi Arabia

ABSTRACT

The goal of this chapter is to offer an outline of how humans may be integrated into AI frameworks for Earth tracking. It's going to take a look at the benefits of human-contact integration and speak about how this technique can be adapted to a diffusion of tracking situations. Furthermore, the human-touch integration additionally gives a human-in-the-loop management device that could evaluate and approve any outputs produced by using the AI models. The human contact can also assist with AI version calibration and validation that might, in any other case, be extra tough or not possible with the AI algorithms on my own. For instance, floor fact information from skilled professionals may be used to cross-validate and calibrate the AI fashions to ensure that they may be producing correct outcomes. Taking those measures can assist us in balancing the advantages of human-contact integration with the capability risks that include it, allowing us to use AI frameworks to monitor and respond to worldwide environmental adjustments.

DOI: 10.4018/979-8-3693-1850-8.ch014

1. INTRODUCTION

Inside the context of environmental tracking, the combination of human contact with AI frameworks gives a monstrous capacity to detect, screen, and respond to environmental modifications. AI frameworks are historically used to automate tactics and improve accuracy via data-pushed solutions. However, the combination of human touch into AI frameworks can offer a large number of additional benefits within the region of environmental monitoring (Koh & Chen, 2023). By tapping into the human's capability to have a look at and recognize context and nuance, the human touch offers the possibility for AI to perceive subtle styles and trends that might be hard for traditional algorithms to come across. Furthermore, the human element can add a feel of accountability to detections, taking into consideration an extra complete know-how of environmental conduct (Saini & Bhalla, 2022). This, in flip, can empower stakeholders to better respond to environmental modifications in a well-timed and knowledgeable manner. At the same time, the integration of human touch isn't the best logical and essential approach in the context of environmental monitoring. Still, it may also serve to minimize bias and maximize equity in AI choice-making. by way of harnessing the energy of human intuition and experience along side self-gaining knowledge of AI algorithms, it's miles feasible to create intelligent tracking structures that can better discover and reply to environmental adjustments(Abeysiriwardana et al., 2022; Ahmed et al., 2023; Khare, 2023).

Earth monitoring, meaning the monitoring and observation of our planet's natural resources and environment, is increasingly important in today's world. As the global population increases and the environment continues to be strained by industry and development, we need to take proactive steps to ensure the sustainability of our planet. AI frameworks offer a powerful way to keep track of our Earth's health. By enabling human-touch integration in these frameworks, we can ensure that the data collected is accurate, relevant, and meaningful. Through integration of AI monitoring systems with human expertise, we can better understand Earth's changes and how they impact us. By leveraging the power of AI, we can help protect our planet from further damage and safeguard its precious resources for future generations.

1.1 Background and Significance

The improvement of AI frameworks for Earth monitoring provides a powerful and flexible tool for evaluating the planet's fitness. AI-led projects provide the promise of progressed accuracy, precision, and reliability in tracking and forecasting the Earth's situations, taking into account more informed planetary stewardship. But, in order for those systems to be a feasible option to the challenges posed through our converting surroundings, they must be included with human contact (Shah et al., 2022). The records of environmental monitoring date back to the overdue 19th century, when scientists commenced to consider the broader implications of human hobby on worldwide structures. In the long time that followed, technological advancements allowed researchers to study the breadth and depth of Earth's environment and humanity's function in affecting it (Najim et al., 2023; Sinha, 2023). This transformed into an attempt to make sure systems are walking sustainably, requiring ever-extra unique tracking, forecasting, and monitoring of atmospheric, oceanic, and landscape changes. In more recent years, environmental monitoring has come to be targeted on AI-pushed technology consisting of sensors and far-flung sensing aircraft. A prime gain of these structures is their ability to correctly monitor massive portions of the Earth, regularly with little human involvement. But, boundaries to this method still exist, such as issues in predicting certain environmental conditions and the presence of so-known 'false positives (Rastogi et al., 2023).

1.2 Objectives of the Chapter

The goal of this chapter is to offer an outline of how humans may be integrated into AI frameworks for Earth tracking. It's going to take a look at the benefits of human-contact integration and speak about how this technique can be adapted to a diffusion of tracking situations (Zhai, 2023). It'll additionally discover cutting-edge studies and case-examine examples of the use of human-touch integration in AI for Earth tracking. Moreover, the bankruptcy will offer a fixed of suggestions for improving and imposing human-touch integration in AI for Earth monitoring. In the end, critical findings associated with human-contact integration will be mentioned, along with potential ethical implications and future research instructions (Taj & Zaman, 2022).

- Develop algorithms that bridge the gap between AI and humans to provide insights from Earth monitoring data.
- Enhance the accuracy and certainty of predictions made through AI by incorporating human intelligence and judgement.
- Utilize human-Touch Integration to reduce time taken to assess large volumes of Earth monitoring data.
- Optimize AI models to understand, anticipate, and predict the environmental impacts of human activities.
- Develop interactive systems that allow humans to influence and direct outcomes of Earth monitoring AI models.
- Incorporate human feedback and judgement into AI models to increase their accuracy and validity.
- Leverage AI to enable humans to monitor more of the environment in near-real time.
- Improve the ability of both humans and AI to identify and advocate for the appropriate course of action for sustainable development and environmental protection.

2. UNDERSTANDING HUMAN-TOUCH IN EARTH MONITORING AI

Earth tracking has become increasingly important in the face of worldwide environmental demanding situations. So one can make sure that our planet remains hospitable for future generations, scientists have proposed implementing synthetic intelligence (AI) into earth monitoring structures. Even as the AI generation has simple advantages, its performance is no longer continually alternative to the holistic know-how of a problem that is most often won via human-touch integration. Human contact offers more than simply operational efficiency (Bale, 2023). With the aid of incorporating a human perspective into an AI machine, earth tracking can be performed with extra depth of know-how and emotional connection. At the same time, as AI models can aid holistic information of a problem, they can not update it. Human-touch integration is critical to ensure that the AI systems are getting used to their total capacity (Su et al., 2022; Wong et al., 2022). The dreams of earth tracking systems should enlarge past really reducing environmental degradation. They also have to be used to understand the complexities of the problem and to bear in mind the moral implications of movements. Such issues can handiest be completed via the combination of human touch. For example, at the same time as analyzing satellite TV for pc pics, AI can stumble on adjustments in land use with high accuracy. However the moral implications of these adjustments ought to ultimately be determined through people (Baburaj et al., 2023).

2.1 Defining Human-Touch Interaction

Human-Touch Interaction (HTI) in AI Frameworks for Earth Monitoring is defined as a form of contact, collaboration, and/or input between humans and AI processes. This cooperation is typically expressed through user interaction, such as activities where the user touches, clicks, types, or speaks to the AI platform. HTI allows users to access information quickly and easily in order to make better-informed decisions. It also enables users to interact with the underlying AI-driven technology in a safe, secure, and user-friendly way. HTI is an important factor for successful implementation of AI applications and plays an important role in Earth Monitoring efforts to collect and analyze data to develop better models and understanding of the environment.

The idea of Human-touch interplay (HTI) has emerged as a more and more famous concept in the world of artificial Intelligence (AI) applications. HTI is an essential aspect of AI frameworks for earth monitoring in an effort to optimize the manner in which we collect and make use of data so that influences on the surroundings may be as they should be assessed (Mohsan et al., 2022).

Human-contact interplay includes the input of tangible gestures by means of a human consumer as a way to control the operation of a gadget. This may mean direct user input, along with the usage of a 3-D or 2D contact device to circulate pieces on a game board or device literally, or it is able to suggest using a natural language processing machine or voice enter to manipulate a tool. In the case of earth monitoring, using Human-contact interaction is essential so that identity and manipulation of geographical information are extra efficient and accurate (Garcia Valencia et al., 2023).

The usage of Human-contact interplay in AI-based systems for earth monitoring wishes to be seen inside the context of human-pc interaction (HCI). This is the procedure of exchange of records between customers and the PC machine. In this use case, HCI is a vital part of the AI framework as the facts being collected and assessed desire to be enter in a manner that makes experience to both the consumer and the laptop (Ellis et al., 2023; Huang et al., 2023).

2.2 Significance in AI Frameworks for Earth Sciences

Earth Sciences cover a wide variety of disciplines. Those include geology, meteorology, hydrology, ecology, oceanography, and, most importantly, the monitoring and safety of the surroundings (Harry, 2023). For centuries, human understanding has been necessary for monitoring and responding to changes in our surroundings. But, with the arrival of artificial Intelligence (AI) and machines gaining knowledge of (ML) technologies, these computerized structures are slowly displacing people, making it increasingly challenging to depend on human skill and revel in screening their surroundings. In this essay, I will discuss the significance of human-contact integration in AI frameworks for Earth Sciences so one can ensure the maximum correct tracking and safety of our surroundings (Adel, 2022; Wu, 2022).

AI and ML are becoming increasingly famous in Earth Sciences due to their capability to examine large amounts of statistics in actual-time time. But, automatic systems present positive risks. Mainly, they're at risk of bias and can provide inaccurate readings and results (Pant et al., 2023). In an effort to reduce these risks and ensure accurate and applicable consequences, human-contact integration in AI frameworks is essential. For example, when monitoring the surroundings, facts must be collected and monitored by trained specialists who are aware of the nuances within the data and environmental situations. This human evaluation would allow for correct readings and proper choices to be made (Chilunjika et al., 2022; Salama et al., 2023).

2.3 Overview of Current AI Approaches in Earth Monitoring

AI processes have been increasingly utilized in earth monitoring over the previous couple of years, with high-quality fulfillment in improving the accuracy of information analysis as well as increasing the performance of monitoring structures. AI is now a critical part of a number of earth-monitoring methods, from pollution control and valuable natural resource control to weather alternate research. AI methods, which include machine getting to know, herbal language processing, and computer imagination and prescient, permit machines to study from records and identify patterns, traits, and relationships that traditional monitoring structures won't be able to discover.

Human-contact integration in AI frameworks for earth monitoring has emerged as a critical factor in advancing AI's functionality in environmental monitoring. This technique has grown in recognition as researchers have realized that AI, even as potent in its right, can best cross up to now without the help of humans. Humans have the expertise that AI truly no longer possesses. By way of combining AI strategies with human touches, earth-tracking systems turn out to be both faster and more dependable.

3. AI FRAMEWORKS FOR EARTH MONITORING: AN OVERVIEW

As society actions closer to an increasingly facts-pushed international, the leveraging of artificial intelligence (AI) into ordinary use has become a priority. Specifically, the usage of AI for monitoring and understanding diverse aspects of our Earth's surroundings may have profound implications for the protection and upkeep of our planet. AI can be used to help with precision prediction and knowledge of climate, herbal disaster danger, and more, allowing for proactive responses to deep environmental problems.

For you to unleash the power of AI for monitoring and knowledge of the surroundings of our planet, AI Frameworks permit a level of automation on the way to make the system faster and extra green. An AI Framework for Earth tracking affords a crucial structure for the integration, processing, analysis, and alertness of environmental information.

However, the challenges confronted by AI Framework improvement for Earth tracking do not just forestall the technical factor. Just as vital is the human-touch integration wanted so that you can ensure that any gadget is successfully calibrated and is capable of accurately interpreting facts that are provided.

3.1 Machine Learning Models in Earth Sciences

The rapid evolution of generations has had a profound impact on the manner in which Earth sciences are studied and monitored. With the improvement of machines getting to know fashions, earth scientists are starting to utilize artificial intelligence (AI) frameworks to advantage improved insight into the Earth's surroundings. Those frameworks are able to accumulate and read large volumes of facts fast and correctly and may result in the development of better predictive models and more efficient monitoring systems. However, for AI frameworks to provide their complete capability, earth scientists have to combine them with human-contact technology.

AI and device learning fashions have revolutionized the manner in which earth scientists have looked at and monitored earth structures. By way of monitoring and reading large volumes of information, AI can uncover styles in weather, climate, and different styles of Earth systems, which can, in any other case, be unnoticed. AI additionally permits faster predictions on regions, including herbal threat occasions, agricultural production yields, and climate exchange. For instance, system mastering fashions had been used to locate drought situations weeks or months beforehand based on satellite TV for pc imagery. AI can also quickly apprehend new changes to the Earth's surroundings, which can help Earth scientists gain an early caution device. In short, device learning fashions and AI algorithms can open up new possibilities for earth systems science.

3.2 Deep Learning Applications for Earth Observation

Earth commentary is increasingly becoming an essential element of the global research agenda. Records from Earth remark satellites are being used to monitor the fitness of our planet, recognize adjustments in climate, discover volcanic interest, and map deforestation. Via the utility of modern technologies, which includes artificial intelligence (AI) and deep getting to know, researchers at the moment are able to dig deeper into the records accrued and make extra particular interpretations.

The utility of deep learning for Earth statement is a swiftly developing discipline. Deep getting-to-know algorithms use a mixture of supervised and unsupervised strategies to extract patterns and generate insights from massive quantities of data. Those algorithms may be designed and adapted to help unique clinical goals. As an example, deep studying has been used to decipher seismic waves and discover the depth and beginning of Earthquakes. Moreover, deep mastering fashions can be used to stumble on land cowl and land use changes inclusive of deforestation, agricultural expansion, urbanization, and coastal erosion. Within the case of Earth observations, deep getting-to-know fashions are capable of capturing diffused functions that are only with problems detected with the aid of human eyes or conventional techniques.

AI frameworks for Earth monitoring can benefit significantly from a deep knowledge of packages. With the aid of allowing for human-contact integration, AI frameworks can employ multiple deep gaining knowledge of models to get better records from data amassed by way of Earth statement structures.

3.3 The Need for Human Input in Model Training and Validation

In recent years, artificial intelligence (AI) applications have been advanced with the intention of facilitating earth monitoring needs. Along side the many ways wherein AI can help, there is a need for humans to enter into version training and validation for those AI frameworks. The method of training an AI is more complicated than many people recognize, and human interaction is vital to ensure the accuracy and efficiency of AI systems.

The schooling and validation of an AI version involves growing education facts from Earth tracking assets (along with satellite imagery, meteorological facts, and other Earth observations). These facts are then used to educate the AI on how to classify, expect, and examine unique Earth phenomena. The schooling records also have to be demonstrated and established for accuracy and suitability to the respective venture. That is an area in which human entry may be essential in attaining the desired effects.

Educated AI employees can analyze the gathered statistics and detect abnormalities, including mistakes in classification inconsistencies in observations, as properly commonly presenting insights into the ability validation of the data. They also can use their expertise to diagnose resources of bias within the facts and help to rectify them. Human entry in AI frameworks can help to grow the self-assurance of the fashions on the subject of making Earth monitoring decisions.

4. BENEFITS OF HUMAN-TOUCH INTEGRATION

The increase of artificial intelligence in our international has yet to lead to sight, with the era swiftly converting with the intention to serve us better and increase our efficiency. With this ever-increasing fashion of AI, it's miles crucial to understand the capability for integrating Human-touch Integration (HTI) into AI frameworks. Additionally, HTI can be integrated into AI frameworks to empower Earth tracking. The blessings of this integration can encompass progressed information accuracy, extra-person involvement, advanced norms, and fee reduction.

First, HTI can improve information accuracy while being included in AI frameworks for Earth tracking. AI can understand adjustments in patterns and may locate anomalies faster than human beings. However, the information gathered via AI may be constrained in accuracy and reliability because of mistakes in interpretation. By way of adding HTI to AI frameworks, human experts can provide a further layer of accuracy to information analyses, permitting the AI to offer greater correct consequences. Furthermore, human experts can spot any capacity mistakes or omissions inside the facts and norms that can be unnoticed by means of AI.

4.1 Improved Data Labeling and Annotation

Statistics labeling and annotation are critical elements of AI frameworks for earth tracking as they provide insights into how the facts are used and interpreted and the way future analysis is probably finished. Human beings can identify patterns and relationships amongst facts and factors that machines can not, and facts labeling and annotation can help bridge this hole. Progressed records labeling and annotation for human-contact integration in AI frameworks for Earth monitoring ought to account for the complexity of the various environmental structures monitored by machines.

Human-contact integration begins with records labeling and annotation. Labeling records as it should be no longer only aids know-how of the data but additionally permits machines to make choices based totally on the context of the facts. Statistics labeling involves assigning important attributes and classifications to the records, even as annotation includes decoding the relationships in the records and assigning meaning and descriptions.

To make sure that the data is labeled and annotated correctly and used correctly, AI frameworks need to be designed to make specific how information must be classified and the way people need to interpret the relationships and patterns for annotation. If those capabilities are designed into the framework, human-contact integration can be progressed, and more significant insights can be won from the data. Furthermore, information annotation and labeling systems ought to be designed without difficulty updating as new variables are amassed. AI frameworks for Earth monitoring are constantly changing and evolving with new datasets, populations, and trends.

4.2 Enhancing Model Interpretability Through Human Insights

Earth tracking is an essential tool in the fight against weather trade. However, advanced synthetic Intelligence (AI) systems are becoming increasingly more complicated, making it challenging to interpret models and grow accuracy. To triumph over this venture, human insights may be utilized in AI frameworks in order to enhance version interpretability and develop more sturdy earth monitoring systems.

One way to combine human insights into AI frameworks is through energetic learning. Active getting to know entails amassing human inputs, which includes labels, to enhance version accuracy. With the aid of the usage of active mastering, people can offer AI models with insights on which record points are nicely classified. Humans also can clarify any discrepancies between input and output, which can enhance AI model accuracy.

Also, the integration of human insights can beautify version interpretability by integrating cognitive decision-making algorithms. Cognitive choice-making systems allow AI structures to method and gain insights from human facts. This information can then be used to broaden extra clever and accurate AI models. Thru the combination of cognitive selection-making structures, AI fashions can improve their accuracy by gaining knowledge from human interactions and comments.

4.3 Real-Time Decision-Making With Human-in-the-Loop Systems

Earth tracking is the exercise of measuring the health of the environment and detecting modifications to it over time. It's essential for tracking ecological situations and for ensuring the planet remains liveable for human lifestyles. Utilizing synthetic Intelligence (AI) frameworks for earth monitoring offers a compelling capacity to substantially increase performance at the same time as generating extra accurate outcomes.

However, AI in and of itself can lack a human-touch integration, making the selection-making procedure overly reliant on algorithms and lacking the capability to account for advantageous real-global factors or nuanced desire-making. This is in which actual-time selection-making with human-in-the-loop systems for human-contact integration in AI frameworks comes into play.

Actual-time choice-making with human-in-the-loop structures for human-touch integration in AI frameworks for earth monitoring involves a human professional staring at the consequences of an AI set of rules and then presenting comments to guide and refine the selection-making process if you want to acquire a desired objective. This technique works 86f68e4d402306ad3cd330d005134dac while facts bearing on the applicable problem are sparse and while the choices want to be made speedy.

5. HUMAN-TOUCH TECHNIQUES IN AI MODEL DEVELOPMENT

AI generation has enabled advancements in Earth monitoring and statistics collection with the aid of providing new varieties of data and insights that assist scientists in recognizing and examining the Earth's surroundings. But, at the same time as AI fashions can offer compelling predictive insights, their accuracy and reliability can go through while deployed in actual-world settings that range from the best conditions wherein they're evolved and tested. As such, the mixing of human-contact strategies into the improvement of AI fashions has grown to be increasingly critical so that it will ensure accuracy, reliability, and robustness in their overall performance.

1. Human-in-the-loop techniques: This technique involves incorporating humans into the process of development, tuning, and validation of AI models. For example, having humans curate and label data sets, manually review outputs and debug models, or provide feedback to AI models through surveys or interviews.

2. Explanations and Visualizations: This involves using techniques such as natural language processing, feature importance and model visualization to explain and gain insights into the AI models. For example, understanding the factors that are important to a particular AI model's prediction or visualizing the output of a model to gain inference into its decision-making process.
3. Iterative Prototyping: Iterative prototyping involves systematically researching, ideating, and testing AI model ideas to continually improve and refine the model. Iterative prototyping also allows for human interaction during the prototyping process, such as soliciting user feedback, performing experiments, and validating AI models against user preferences.
4. Rigorous Testing and Evaluation: This involves testing and evaluating AI models before they are deployed to ensure that they are accurate and perform as expected. Rigorous testing and evaluation of AI models can also reveal potential ethical, legal, or regulatory concerns.

To combine human-touch strategies into AI model development, a number of issues should be taken under consideration, together with statistics series, acceptable guarantee, feature engineering and processing, pre-processing and pipeline production, interpretation, overall performance metrics, and evaluation criteria.

Records collection: To ensure excessive accuracy inside the AI version, it is crucial to acquire data that is representative of the real-global context in which it'll be used. For Earth monitoring purposes, this will require using both ground group and satellite TV for pc facts sources, in addition to crowd-sourced datasets, which include the ones generated by means of citizen scientists..

5.1 Crowdsourcing for Earth Observation Data

Crowdsourcing for earth remark information is a relatively new and untapped area inside the region of artificial intelligence (AI) packages. This concept is based on the collection of many different resources of records from the general public. The data accrued will aid in information complex elements that have an impact on how the Earth is monitored and controlled. Through an integration of human touch into AI frameworks, it's far more viable to create a more effective and correct monitoring system that's capable of distinguishing natural and human-driven changes inside the Earth's environment over time.

Crowdsourcing is the system of the use of the "crowd" (which may also include members of the general public or a specific phase) to remedy issues or acquire information regarding Earth remarks. Through the usage of the strength of the group, information can be accrued more quickly and affordably than if it becomes amassed through agencies alone. Commonly, crowdsourcing focuses on small responsibilities and encourages humans to provide talents, knowledge, and a workforce to remedy issues or create progressive solutions.

Earth statement (EO) data is already being used to improve our knowledge of the planet. By leveraging these statistics, groups may want to know more without difficulty revealing, take a look at, and react to the converting environment of our planet.

5.2 Human-in-the-Loop Model Training

The Human-in-the-Loop version education (HILT) for Human-touch Integration in AI Frameworks for Earth monitoring is a significant leap forward in the improvement of synthetic intelligence (AI) in earth monitoring. In this version, the human element is introduced into the equation, and the motive of human involvement is to ensure the accuracy and quality assurance of the AI algorithm's overall performance.

To start, AI utility improvement today needs an extraordinary deal of human-touch integration, which will ensure the best accuracy and reliability of the expected effects. Humans are capable of comprehending and decoding the data that is generated, and they can also practice their expertise and information in their evaluation of the model prediction. HILT is the process by which a human is integrally involved in the training, testing, and deployment of an AI model, with the goal of enhancing the accuracy and reliability of the predictive version.

Human-touch integration can also be extended to the procedure of statistics sharing. By means of permitting human beings to percentage complex data that are regularly tough and tedious to parse, the accuracy and reliability of predictive fashions can be progressed. Furthermore, humans also can offer additional insights that won't be visible to the AI algorithms, thereby offering aaddition accuracy boosts to the device.

5.3 Collaborative AI Development With Experts in Earth Sciences

The improvement of Collaborative AI (synthetic Intelligence) in the area of Earth Sciences has been gaining tremendous interest from the scientific network these days. This is due to the capability this technology holds for integrating human touch into AI Frameworks and equipment for Earth monitoring. There is an increasing want for experts within the location of Earth Sciences and synthetic Intelligence to work collectively in order to plot AI answers that can correctly reveal and expect Earth's surroundings.

There are numerous capability that makes use of of Collaborative AI in Earth Sciences, including predicting the weather and monitoring the evolution of the climate. It can additionally be used to examine geological surveys to assist with Earth mapping and reading the surface composition of numerous areas. Scientists can use AI frameworks to discover and have a look at geological, hydrological, and atmospheric houses for specific tracking of Earth structures. Furthermore, it may be used for correct data interpretation and more extensive exploration of Earth structures.

In an effort to construct green and cost-effective AI Frameworks able to accurately monitor and predicting Earth structures, it's far essential to contain experts inside the fields of Earth Sciences and artificial Intelligence to collaborate and devise suitable solutions.

6. CHALLENGES AND SOLUTIONS

AI has been broadly used in a variety of sectors, starting from healthcare to agriculture to earth monitoring. AI-powered structures have confirmed capacity advantages in many areas; however, integrating human sports into the AI framework remains a chief mission. As extra organizations grow to be reliant on AI, they may increasingly want to contain human input, remarks, and insights into AI applications to ensure top-rated performance and the most desirable results. This essay will discover the challenges and solutions of human-touch integration in AI frameworks for earth monitoring.

One of the primary challenges of integrating human contact into an AI framework for earth tracking is the complexity of the statistics being analyzed. Earth tracking calls for complete records of international weather, agricultural, business, and infrastructure systems that are frequently hard to collect and collate. In addition, AI algorithms can be complicated and challenging to interpret. Furthermore, it's miles tough to design AI models that are both robust and maintainable over time; as facts and surrounding situations change, AI fashions want to be changed and maintained. This will be a time-consuming and hard challenge.

6.1 Ethical Considerations in Human-Touch Integration

As AI frameworks for Earth tracking end up increasingly more every day in our 66b34c3da3a0593bd135e66036f9aef3 lives, the integration of human-touch integration in those frameworks is crucial attention. With a situation with the ethical implications of this sort of sensitive generation, this essay will remember how human-touch integration in everyday AI frameworks for Earth tracking ought to be ethically observed.

First, it's miles essential daily day-to-day the capability dangers daily human-contact integration. Despite all of the potential advantages that human-touch integration gives, there are also unavoidable risks that should be taken into consideration. For example, a loss of transparency and trust between human beings and synthetic Intelligence (AI) can substantially reduce the accuracy of the Earth monitoring finished by means of the AI, main day-to-day probably misguided or deceptive reports. Furthermore, the integration of human-touch integration in everyday AI could result in additional discrepancies among the Human and AI, growing an imbalance and probably misconstruing the reports generated. As such, it's very crucial for any implementation of a human-contact integration to be carried out with clean communication and transparency, making an allowance for a partnership between the Human and the AI in preference to a master-slave relationship.

2d, it's miles vital every day to recollect the privateness and safety implications of daily human-touch integration.

6.2 Ensuring Data Quality and Consistency

Innovation in synthetic intelligence (AI) has revolutionized the manner in which statistics are processed, monitored, and used for earth monitoring. Pioneering improvements in statistics analytics have enabled such technologies as AI to deliver accurate insights for earth tracking methods. AI-pushed datasets are being used to display conditions and patterns in ecosystems, weather structures, geological modifications, and natural resources.

However, AI-primarily based datasets are at risk of becoming corrupted or vague because of human mistakes, unreliable fact sources, and defective integration into traditional facts waft structures. Therefore, the want to ensure information high-quality and consistency is essential if AI is to be relied upon for reading statistics for earth tracking. This essay outlines the necessary thing steps that want to be taken to make sure facts are excellent and consistent for human-contact integration in AI frameworks for earth tracking.

Statistics best is the degree to which records are accurate, reliable, and well-timed.

Validation is a procedure to make sure that facts are clean, whole, and effectively categorized. Spending additional time validating records can assist in lessening the hazard of errors because it ensures that each one of the information assets is efficiently mapped.

6.3 Addressing Bias and Subjectivity in Human-Annotated Data

The development of the synthetic Intelligence (AI) era is one of the most exciting and promising trends in human-computer interaction. AI systems have been used to reveal the environment, examine large information sets, detect anomalies, and control complex responsibilities for prolonged periods. AI frameworks are frequently included in various Earth tracking approaches, and this integration relies closely on annotations from people. So, as for human-contact integration to achieve success and reliability, it's essential to cope competently with bias and subjectivity.

Bias and subjectivity are serious subjects in terms of using human annotations in AI programs. Human-furnished records and annotations may additionally include certain beliefs, prejudices, and judgments that can save your AI frameworks from sporting out their essential features efficaciously. So, as for AI to efficiently carry out its particular purpose, it calls for 86f68e4d402306ad3cd330d005134dac information that is bias-unfastened and objective. It's miles vital to pick out, reduce, or save your bias and subjectivity as much as possible because these elements can adversely affect the accuracy and consistency of AI fashions.

Suppose you want to address bias and subjectivity in human-supplied statistics for human-contact integration into AI frameworks for Earth tracking. In that case, groups should enforce statistics collection methods aimed toward making sure that facts are gathered impartially.

7. TECHNOLOGICAL INNOVATIONS FOR SEAMLESS INTEGRATION

In recent years, technological innovation has enabled Earth tracking via shrewd software program equipment, computer models, and synthetic intelligence (AI). The increasing call for Earth monitoring has seen new techniques being carried out to make sure that humans and era interact effectively in creating seamless integration. This essay specializes in the contemporary technologies for seamless integration for human-contact integration in AI frameworks for Earth tracking.

One of the significant blessings of artificial Intelligence for Earth monitoring is its ability to system large quantities of facts quicker and more correctly than humans. Therefore, the integration of human input into AI frameworks is essential for more correct consequences. AI frameworks are designed to assist users in acting duties, and this requires a high stage of human engagement for effective outcomes. This human-contact integration is essential for attaining higher effects and execution instances for Earth tracking.

Step one in the direction of integrating human touch in AI frameworks is the implementation of natural language processing (NLP). NLP permits users to input facts in a more herbal manner, casting off the want for complicated coding. This will assist in reducing the mastering curve for customers, allowing them to recognize the information essential for green Earth monitoring quickly. Via the usage of NLP, AI structures can examine users and understand their intentions, hence creating smoother paths for collaborations within the machine.

7.1 Advances in User Interface Design for Earth Monitoring

The notion of Earth tracking through using artificial Intelligence (AI) frameworks is a concept that has been around for quite some time. But, in current years, advances in person interface design have made it viable for these AI frameworks to be incorporated with human-contact interactions. This aggregate of rigid and soft technologies is being carried out throughout a spread of industries and programs consisting of earth tracking. This essay will talk about the advances in user interface layout for earth tracking in addition to the effect they have had on making those AI frameworks more on-hand and powerful for human-contact integration.

In terms of advances in consumer interface layout, one of the key trends has been the upward push of natural language processing (NLP) systems. Those systems allow for extra human-like interactions, inclusive of the capability to invite questions and acquire answers. Using NLP systems allows for a more intuitive form of interplay with the AI framework, permitting humans to query the gadget much like they would some other individual. This type of herbal interplay is prime to creating AI that is extra accessible and effective because it lets people to invite questions and get answers without needing to recognize the technical components of the AI framework.

7.2 Human-Touch Augmented Reality Applications

The improvement of augmented fact (AR) structures is an area of technology with some distance-attaining implications for many factors of human society. From supporting scientific specialists to create existence-saving techniques to enabling instructors to improve student engagement and mastering, the potential packages of AR technology are significant. But, one of the most exciting packages of this technology may be its projected role in human-touch integration in AI frameworks for earth monitoring.

The scope of earth monitoring is relatively enormous, and the ability programs of AI in this area are almost countless. From predicting the weather effects of deforestation to detecting unsafe stages of pollutants, AI technology can be used to enhance the efficacy of earth-monitoring efforts. Alas, those technologies are confined by the truth that they're not currently capable of replicating the elaborate and nuanced statistics series strategies employed by environmental scientists.

Enter the AR era. By augmenting the information series skills of AI with the humans, AR might be the missing hyperlink in the powerful integration of AI in earth monitoring efforts. Augmented reality systems could allow scientists to engage immediately with records gathered from earth-monitoring AI, manipulating it in real-time to become aware of new patterns, evaluate current record points, and discover new avenues of exploration.

7.3 Integrating Natural Language Processing in Earth Monitoring AI

Earth tracking has come to be increasingly important given the converting conditions of the surroundings and the ability damage that human activities can cause. Herbal language processing (NLP) is a technology that has been used to improve the accuracy of earth monitoring structures by integrating AI frameworks for more human-like understanding. This essay will discover the ability of NLP in earth tracking, its capacity to integrate human touch and expertise into AI frameworks and recommend potential destiny packages of this technology.

The cause of NLP is to offer machines with human-like expertise in language, in particular, inside the context of comprehending large volumes of textual information. Thru leveraging current ontologies and text analysis algorithms, NLP can identify styles in textual content that may be used to categorise and infer significant facts. This could enable earth tracking systems to analyze facts accrued from specific assets and provide insights to aid choice-making. Additionally, NLP can help lessen the guide hard work associated with reviewing and decoding large units of information, permitting green, automatic choice-making.

Integrating NLP into AI frameworks for earth monitoring gives numerous benefits. For one, it allows machines to interpret records in an extra human-like style, making it less complicated for the person to understand the outcomes.

8. CONCLUSION

In conclusion, integrating AI frameworks for Earth monitoring with human touch can gain humanity with the aid of enhancing the accuracy of records and imparting more excellent and reliable effects. But, it's essential to ensure that human-contact integration is executed in an accountable way that respects privateness and protection issues. This can be finished by setting regulations on personal records, in addition to implementing measures to make sure that algorithms are transparent and understood without difficulty. Taking those measures can assist us in balancing the advantages of human-contact integration with the capability risks that include it, allowing us to use AI frameworks to monitor and respond to worldwide environmental adjustments.

8.1 Summarizing the Impact of Human-Touch Integration

Human beings have constantly been at the forefront of information and managing the earth's surroundings. Now, with the growing use of system studying and synthetic intelligence, the potential for us to recognize and guard our surroundings is more significant than ever before. Human-touch integration is one of the critical components of AI frameworks for earth monitoring and has the potential to revolutionize environmental control.

First and foremost, human-touch integration in AI frameworks allows for more desirable accuracy and precision in studying earth records. By imparting more significant, unique insights into current troubles, which include weather change and biodiversity loss, it offers a window into the tendencies of the future and aids in preventative measures. With the aid of combining data and knowledge from humans and machines, the results of AI models are stepped forward and optimized.

Furthermore, human-contact integration reduces the workload of mundane and repetitive responsibilities, consisting of information series, and presents extra time to concentrate on further studies. This no longer only saves time and money for businesses those but eats the lifespan of statistics collection sports. This reduced workload also leads to elevated productiveness and efficiency that is useful for the environment in phrases of both time and monetary sources.

8.2 Implications for Advancing AI in Earth Monitoring

Advances in synthetic intelligence (AI) can have profound implications for earth monitoring. AI-driven sensing, modeling, prediction, and choice-making techniques are predicted to revolutionize our ability to interpret the alerts from Earth's surroundings, apprehend and expect complex environmental tactics, and offer critical steering for shielding and coping with Earth's resources. However, a successful application of AI for Earth monitoring will require the mixing of human-touch elements to ensure reliable and robust outcomes.

To be able to efficaciously integrate human touch into AI frameworks for Earth monitoring, builders and users of AI technology have to make sure that the procedure is undertaken with the reason of bringing the exceptional out of both AI and human beings. AI technology consisting of system mastering and deep getting to know have to be carried out in approaches that complement current human know-how and expertise to facilitate higher decision-making and get entry to critical environmental facts. AI technology has to be applied to ensure constant and rigorous statistics series and analysis, which could assist in ensuring accuracy and transparency.

At the same time, including the human touch is crucial to ensure the adoption of suitable moral standards in the utility of AI for earth tracking. Human-orientated standards, along with public oversight, public participation, truthful treatment, and responsible conversation, should be adopted to ensure powerful AI-pushed Earth tracking approaches.

REFERENCES

Abeysiriwardana, P. C., Jayasinghe-Mudalige, U. K., & Kodituwakku, S. R. (2022). Connected researches in “smart lab bubble”: A lifeline of techno-society space for commercial agriculture development in “new normal. New Techno Humanities, 2(1), 79-91.

Adel, A. (2022). Future of industry 5.0 in society: Human-centric solutions, challenges and prospective research areas. *Journal of Cloud Computing (Heidelberg, Germany)*, *11*(1), 1–15. doi:10.118613677-022-00314-5 PMID:36101900

Ahmed, S., Yong, J., & Shrestha, A. (2023). The Integral Role of Intelligent IoT System, Cloud Computing, Artificial Intelligence, and 5G in the User-Level Self-Monitoring of COVID-19. *Electronics (Basel)*, *12*(8), 1912. doi:10.3390/electronics12081912

Baburaj, A., Kumar, S. N., Aliyana, A. K., Banakar, M., Bairagi, S., & Stylios, G. (2023). Factors Affecting the Performance of Flexible Triboelectric Nanogenerators (F-TENGs) and their Sensing Capabilities: A Comprehensive Review. *Nano Energy*, *118*, 108983. doi:10.1016/j.nanoen.2023.108983

Bale, A. S. NR, N. K., Somanna, A. N., Karthik, K. B., SN, N. R., & Rithik, S. M. (2023, June). Explosive Projectile Detection with an Arduino-Controlled Robot. In *2023 8th International Conference on Communication and Electronics Systems (ICCES)* (pp. 243-249). IEEE.

Chilunjika, A., Intauno, K., & Chilunjika, S. R. (2022). Artificial intelligence and public sector human resource management in South Africa: Opportunities, challenges and prospects. *SA Journal of Human Resource Management*, *20*, 12. doi:10.4102ajhrm.v20i0.1972

Ellis, T., Hitaj, B., Lindqvist, U., Shands, D., Tinnel, L., & DeBruhl, B. (2023). Critical Infrastructure Security Goes to Space: Leveraging Lessons Learned on the Ground. *arXiv preprint arXiv:2309.15232*. doi:10.2514/6.2023-4798

Garcia Valencia, O. A., Suppadungsuk, S., Thongprayoon, C., Miao, J., Tangpanithandee, S., Craici, I. M., & Cheungpasitporn, W. (2023). Ethical Implications of Chatbot Utilization in Nephrology. *Journal of Personalized Medicine*, *13*(9), 1363. doi:10.3390/jpm13091363 PMID:37763131

Harry, A. (2023). The Future of Medicine: Harnessing the Power of AI for Revolutionizing Healthcare. *International Journal of Multidisciplinary Sciences and Arts*, *2*(1), 36–47. doi:10.47709/ijmdsa.v2i1.2395

Huang, Z., George, M. M., Tan, Y. R., Natarajan, K., Devasagayam, E., Tay, E., Manesh, A., Varghese, G. M., Abraham, O. C., Zachariah, A., Yap, P., Lall, D., & Chow, A. (2023). Are physicians ready for precision antibiotic prescribing? A qualitative analysis of the acceptance of artificial intelligence-enabled clinical decision support systems in India and Singapore. *Journal of Global Antimicrobial Resistance*, *35*, 76–85. doi:10.1016/j.jgar.2023.08.016 PMID:37640155

Khare, A. (2023). Harnessing the power of artificial intelligence in health care: Balancing promise with caution. *Journal of Medical Evidence*, *4*(2), 105–108. doi:10.4103/JME.JME_70_23

Koh, I., & Chen, A. (2023, October). Your Memory Palace in the Metaverse with AI. In *Proceedings of the AAAI Symposium Series* (Vol. 1, No. 1, pp. 19-22). 10.1609/aaaiss.v1i1.27469

Mohsan, S. A. H., Zahra, Q. U. A., Khan, M. A., Alsharif, M. H., Elhaty, I. A., & Jahid, A. (2022). Role of drone technology helping in alleviating the COVID-19 pandemic. *Micromachines*, *13*(10), 1593. doi:10.3390/mi13101593 PMID:36295946

Najim, H. A., Kareem, I. S., & Abdul-Lateef, W. E. (2023, July). Design and implementation of an omnidirectional mobile robot for medicine delivery in hospitals during the COVID-19 epidemic. In AIP Conference Proceedings (Vol. 2830, No. 1). AIP Publishing. doi:10.1063/5.0156862

Pant, G., Bahuguna, R., Pandey, S., Gehlot, A., Yadav, S. P., & Pachauri, R. K. (2023, April). Intelligent Automated Interference for the Protection of Road Safety. In *2023 International Conference on Computational Intelligence, Communication Technology and Networking (CICTN)* (pp. 87-91). IEEE. 10.1109/CICTN57981.2023.10141086

Rastogi, K., Bahuguna, R., Kathuria, S., Gehlot, A., Malik, P. K., & Negi, P. (2023, April). *Technical Intercession of Artificial Intelligence in Solving Online Dispute Resolution. In 2023 IEEE Devices for Integrated Circuit (DevIC)*. IEEE.

Saini, A., & Bhalla, R. (2022). Artificial intelligence and automation: transforming the hospitality industry or threat to human touch. In *Handbook of Research on Innovative Management Using AI in Industry 5.0* (pp. 88–97). IGI Global. doi:10.4018/978-1-7998-8497-2.ch006

Salama, R., Al-Turjman, F., Bhatia, S., & Yadav, S. P. (2023, April). Social engineering attack types and prevention techniques-A survey. In *2023 International Conference on Computational Intelligence, Communication Technology and Networking (CICTN)* (pp. 817-820). IEEE. 10.1109/CICTN57981.2023.10140957

Shah, K. H., Malik, Z. A., Balakrishnan, G., & Ramanan, K. (2022). *Integration Of Iot And Cloud Computing-Based Innovation Of Wireless Sensor Network Technology For Health Care Architecture And Agriculture Applications.*

Sinha, E. (2023). Integrating Human and Artificial Intelligence Within Organizations: A Conceptual Framework Based on Artistry and Scientific HR Elements. In Managing Technology Integration for Human Resources in Industry 5.0 (pp. 48-67). IGI Global. doi:10.4018/978-1-6684-6745-9.ch004

Su, Z., He, L., Jariwala, S. P., Zheng, K., & Chen, Y. (2022). " What is Your Envisioned Future?": Toward Human-AI Enrichment in Data Work of Asthma Care. *Proceedings of the ACM on Human-Computer Interaction, 6*(CSCW2), (pp. 1-28). ACM. 10.1145/3555157

Taj, I., & Zaman, N. (2022). Towards industrial revolution 5.0 and explainable artificial intelligence: Challenges and opportunities. *International Journal of Computing and Digital Systems*, *12*(1), 295–320. doi:10.12785/ijcds/120128

Wong, C. Y., Samadi, S., Suleiman, W., & Kheddar, A. (2022). Touch semantics for intuitive physical manipulation of humanoids. *IEEE Transactions on Human-Machine Systems*, *52*(6), 1111–1121. doi:10.1109/THMS.2022.3207699

Wu, Y. (2022). Ethically responsible and trustworthy autonomous systems for 6G. *IEEE Network*, *36*(4), 126–133. doi:10.1109/MNET.005.2100711

Zhai, X. (2023). Chatgpt for next generation science learning. *XRDS: Crossroads. The ACM Magazine for Students*, *29*(3), 42–46.

Chapter 15
Knowledge-Driven Decision Making in Earth Monitoring Systems

Bharathi B.
https://orcid.org/0009-0001-1068-996X
Department of Computer Science and Engineering, R.M.K. Engineering College, Kavaraipettai, India

Kalaivaani P. T.
Department of Electronics and Communication Engineering, Vivekanandha College of Engineering for Women (Autonomous), Namakkal, India

Shree Gaayathri V.
Department of Computer Science and Engineering, R.M.K. Engineering College, Kavaraipettai, India

Rajendiran M.
https://orcid.org/0000-0001-6715-4264
Department of Computer Science and Engineering, Panimalar Engineering College, Chennai, India

ABSTRACT

Earth monitoring systems are of crucial significance to the arena nowadays as they offer us essential data on this planet's environment, sources, and ecosystems. In earth monitoring structures, information-driven selection making is the process of utilizing facts-pushed techniques to inform selections regarding the real-time monitoring and control of natural disasters and other environmental phenomena. These objectives span the spectrum of decision-making, starting with the pursuit of informed choices that are grounded in accurate data and bolstered by objectivity. The capability of modeling strategies and this approach fosters collaborative problem-solving by facilitating the sharing of data and insights among stakeholders, aligning common objectives and shared expertise to inform the decision-making process, ultimately elevating the quality and efficiency of decisions in response to evolving environmental challenges. In conclusion, understanding extraction from earth monitoring structures is an essential system for assisting decision making via using statistics-driven techniques.

DOI: 10.4018/979-8-3693-1850-8.ch015

1. INTRODUCTION

Earth monitoring systems are of crucial significance to the arena nowadays as they offer us essential data on this planet's environment, sources, and ecosystems. Even as some conventional methods of accumulating, storing, and studying such information have been in use for some time, the current creation of understanding-driven decision-making in earth tracking structures has revolutionized the manner in which we degree, reveal, and impact changes within the environment (Horn & Isselstein, 2022).

Knowledge-driven decision-making is a form of artificial intelligence (AI) that makes use of information from traditional resources, together with satellite photographs and floor surveys, as well as digital resources inclusive of weather facts, soil moisture readings, and vegetation growth developments. Via combining this enormous range of facts sources, effective selections may be made about how high-quality to manipulate the environment over a long period (Wu et al., 2022; Zhang et al., 2023).

The choices made via these systems must be both sound and accountable. To achieve this, AI algorithms are used to pick out trends inside the information and formulate powerful techniques to address the regularly complex environmental problems we are confronted with (Yu et al., 2023). As an example, the facts may additionally monitor that certain areas are experiencing a speedy decline in water stages because of immoderate water usage. AI algorithms can then be applied to investigate the records and expand an appropriate response, which includes an environmental conservation mission, to deal with the problem and ensure lengthy-time period sustainability.

As a result of the integration of AI with earth tracking systems, complicated decisions can now be made in minutes that used to take weeks or even months, leading to quicker and more accurate choices. Furthermore, because these systems can store large volumes of information in a shorter time frame, they also can identify problems that may have otherwise gone unnoticed (Abdelkareem & Mansour, 2023). With the assistance of information-driven selection-making, earth structures can now be better managed, and the surroundings are better blanketed on a worldwide scale.

1.1 Background and Significance

Earth monitoring systems are essential for understanding the planet's converting situations and predicting destiny traits. Such knowledge facilitates the ability to expect and control the dangers related to natural disasters, international warming, and different environmental dangers. Decision-makers need to rely upon correct and well-timed information about Earth's environment on the way to make effective selections. Know-how-driven selection-making has turned out to be critical for information and predicting Earth's situations so that suitable movements can be taken as quickly as possible (Jiang et al., 2023). This essay will talk about the history of knowledge-pushed decision-making in earth-tracking structures and provide insight into its relevance for decision-makers.

Know-how-pushed selection-making is the method of using statistics from Earth tracking structures to tell rational selections. This knowledge is obtained through a variety of approaches, inclusive of physical inquiries, observations, and superior technologies. Earth monitoring structures gather information on various elements of the Earth's surface, climate, oceans, and surroundings to provide choice-makers with updated statistics on the state of the surroundings (Rødven & Wilson, 2022). The data is gathered via sensors and devices, inclusive of climate radars, thermometers, barometers, precipitation gauges, and satellite imagery. This record is used to analyze the Earth's surroundings and provides actual-time insight into modifications in the Earth's situations.

In current years, expertise-pushed decision-making has emerged as increasingly critical in Earth tracking structures. That is because of the speedy modifications inside the surroundings, which the outcomes of weather trade, natural failures, pollutants, and other environmental dangers can drive. The records collected from Earth monitoring systems allow choice makers to become aware of these modifications and take appropriate action quickly (Özdemir, 2022). For instance, data from Earth tracking structures can alert decision-makers to regions of the planet that are experiencing extreme temperatures, giving them time to prepare for the potential consequences of such shifts.

According to the United countries, information-pushed decision-making is essential for sustainable development and international collaboration. It includes the sharing and protection of Earth monitoring statistics across one-of-a-kind international locations and groups so that they can be used and responded to quickly and as they should be (Hu et al., 2022). Understanding-pushed decision-making is likewise crucial for information on the Earth's herbal assets and utilizing them efficiently. Statistics acquired from Earth tracking structures can be used to make informed choices regarding aid management, making sure that the planet's sources are controlled sustainably.

1.2 Objectives of the Chapter

Earth monitoring systems are a critical part of environmental studies and control. They offer vital facts and statistics regarding the Earth's surroundings, and it's this record that is used for choice-making tactics (Daviran et al., 2022). Expertise-driven decision-making in earth tracking structures attempts to enhance the accuracy and efficiency of the selection-making manner by the usage of information from the past and making decisions primarily based on this understanding. Information-pushed selection-making makes a specialty of figuring out the patterns and traits of behavior with the intention to maximize efficaciously tell decision-making in the future.

The primary goals of know-how-pushed selection-making in earth tracking structures are to:

1. Improve selection quality: knowledge-driven selection-making allows for more accurate choices to be made, as selections are not made swiftly for a variety of reasons (e.g., economic or political pressures). Expertise-pushed selection-making permits choices to be made based on records and tendencies; this brings an extra degree of objectivity to the decision-making manner.
2. Growth choice velocity: knowledge-pushed selection-making can increase the velocity of selection-making as the necessary records to tell the choice can be comfortably collected and analyzed. It permits speedy decisions to be made in response to environmental adjustments.
3. Reduce Uncertainty: knowledge-pushed selection making in Earth monitoring structures reduces the Uncertainty associated with contemporary conditions by making an allowance for a higher-knowledgeable choice. Knowledge of past tendencies can be used to evaluate one-of-a-kind capacity effects, thereby allowing for an extra dependable and accurate result.
4. Beautify resource usage: expertise-pushed decision-making in earth monitoring structures can be used to pick out and make use of sources successfully. Expertise beyond trends can be used to deduce trends inside the destiny; this data can then be used to predict destiny aid usage wishes appropriately.
5. Enhance version capability: know-how-driven choice-making in earth monitoring systems can enhance version ability by supplying real-time statistics of environmental changes. This informa-

tion can be used to inform decisions related to model strategies, assisting businesses and people to reply quickly to any adjustments inside the surroundings.

6. Decorate Collaborative hassle-fixing: knowledge-driven decision-making can increase collaboration among groups, governments, and individuals by offering a shared expertise of facts. It may be used to become aware of shared dreams and common areas of information that can be used to inform selection-making.

Those targets serve to demonstrate the capacity of information-pushed decision-making in earth monitoring systems to improve the accuracy and performance of choice-making methods. In Flip can assist in making specific selections that are informed by using correct statistics and developments and that model strategies are successfully implemented. Know-how-pushed selection-making can ultimately enhance adaptation ability and resilience (Teng et al., 2023).

2. FOUNDATIONS OF KNOWLEDGE-DRIVEN DECISION MAKING

Earth tracking systems are essential to date; make sure that the planet's ecosystems and environment are up-to-date, sustainable, and healthy, and track worldwide weather trade. By means of enforcing an information-driven decision-making manner to update these systems, selections can be made based on facts from satellites, geographic facts structures (GIS), far-flung sensing, and other geospatial eras. This method of updated selection-making gives a valuable update for growing a more comprehensive and holistic up-to-date of the Earth's modern nation and the potential for destiny changes (Falkenström & Svallfors, 2022; Zhou et al., 2022).

Step one in developing a know-how-pushed selection-making system is up-to-date, gathering and analyzing large volumes of information from multiple resources. This information helps update and offer an image of the Earth's current kingdom, which can be, in comparison, up-to-date information and spatially mapped. This fact can then be used up to date to discover changes within the Earth's floor, atmosphere, and terrain (Taktak et al., 2022). Through analysis of these records, tendencies can be diagnosed and monitored for the development of techniques for the conservation of biodiverse ecosystems and climate change mitigation.

For instance, the usage of geospatial generation can provide a higher understanding of habitats, land use, and pollution, in addition to updated determining the effects of urbanization on nearby and global ecosystems (Jabla et al., 2022). By implementing an expertise-driven choice-making method, governments and groups can use the statistics to create control plans to ensure the effective stewardship of herbal resources and the protection of delicate ecosystems.

Using know-how-driven decision-making in Earth systems monitoring is likewise helpful for quick-time period forecasting and emergency response operations. As an instance, the records collected may be used up-to-date appropriately except for the trajecup-to-datery and intensity of sup to terms and other herbal failures (Chen et al., 2023). By identifying capacity dangers earlier, selection makers can take proactive steps to reduce the effect of failures on the environment and human populations.

Typical, a knowledge-driven decision-making procedure can allow effective control and sustainable utilization of the Earth's herbal resources. Via using geospatial era up-to-date locate modifications and discover dangers, professionals could make informed choices that are up-to-date on complete facts and analysis. With this approach, choice-makers can proactively reply to up-to-date environmental and ecological threats and shield the planet and its population for generations to come (Huang et al., 2023).

2.1 Defining Knowledge-Driven Systems

Earth tracking systems are quickly becoming a vital component of a more considerable effort to construct a sustainable destiny for the planet. By studying real-time facts from satellites, ground-based stations, and specialized instrumentation, these systems can offer meaningful insight into patterns inside the environment, ocean, land, and biosphere (Li et al., 2022). This fact can be used to inform selections ranging from agricultural manufacturing to urban making plans. So that you can make complete use of these facts, earth monitoring systems should be capable of leveraging the expertise of experts and utilizing present expertise sources. It is in which expertise-driven systems are available.

Know-how-pushed systems are designed to draw upon present scientific knowledge as a way to make choices concerning earth monitoring. These statistics can come from a variety of resources, together with instructional courses, online databases, and earlier studies. By combining this knowledge base with actual-time records, choice-makers can tailor their responses to changing conditions (Chudasama et al., 2022). For instance, if a particular location of the arena is experiencing severe water shortages due to climate change, an information-driven gadget might advocate the relocation of water resources to the affected regions if you want to minimize the impacts of the lack.

The shape of those structures is a crucial component of their fulfillment. Many knowledge-pushed choice-making structures are built around an expert machine, which makes use of a set of policies to deduce conclusions from information and make choices (Joshi & Biswakarma, 2023). Those can either be pre-current regulations or ones that might be generated by using the system itself through a system called expertise discovery. The system can also use other techniques to make decisions, together with fuzzy logic or synthetic neural networks, which permit greater creative trouble-solving.

One of the primary blessings of understanding-pushed choice-making structures is that they may be quickly adapted as the scenario changes. For instance, if a geographic vicinity stories unexpected adjustments in weather, an expert device may be quickly updated to become aware of the brand new conditions and take appropriate action. which means that choices may be made quickly and correctly, without watching for real-time facts to be analyzed and interpreted (Aldana-Martín et al., 2022).

Understanding-pushed systems have become more and more essential within the area of earth monitoring due to their capacity to capture information, discover tendencies, and make selections quickly. With the right software program, these systems can analyze significant quantities of records and provide insight into our planet that could benefit anyone. Because of this, understanding-pushed systems of decision-making have become an essential part of earth monitoring systems (Parveen et al., 2023).

2.2 Role of Data and Information in Decision Making

Facts and statistics play an essential role in selection-making for understanding-pushed decision-making in earth monitoring systems. With the extended availability of satellite imagery, radar, and different sensing technologies, there was a ramification in international monitoring of the environment and its composition. The usage of the facts is collected and, in addition, processed, knowledge-driven selection-making offers precious perception into complicated systems and clients' desires, which will make more knowledgeable selections (Salama et al., 2023).

Data act as the building blocks for choice-making by means of offering authentic statistics for evaluation. With records, it is possible to benefit from knowledge of the present-day nation of the environment and its modifications through the years. The statistics allow the creation of expertise in the environment. The fine of the records with their accuracy and completeness will decide the accuracy of choice. As soon as the records are gathered, they have to be analyzed to transform them from raw records and numbers to meaningful statistics, which could then be used to pressure selections (Pant et al., 2023). That is, in information, technological know-how comes in to help extract valuable insights from the records.

Statistics is constructed from facts, models, and algorithms used for the evaluation of environmental information. It can encompass predictions, facts, traits, and correlations. The facts are famous meaningful styles inside the information that may be used to guide decisions around ongoing management and implementation of monitoring systems. Because the records are distilled, it will become more complicated and may provide deeper insights into the earth's systems for choice makers.

Moreover, decision-makers use the data and statistics to make conclusions and pick out new know-how from the datasets. With the addition of treasured expertise to the present statistics, the choice makers are able to make even more knowledgeable selections. The information can act as a buffer to help inform the path and make great choices. It can additionally be used to improve the accuracy of tracking structures through inspiring deepening studies and improvement.

Together, statistics and records are used to behave as a foundation for selections, and companies should make specific those components practical and actionable. With the purpose of aiding the high-quality decision-making effects, it's crucial to utilize advanced generation to help aggregate, examine, and synthesize both records and records. Through this process, businesses are able to gain an accurate understanding of the arena around them and make decisions that will assist in shielding the earth's sources from overexploitation.

2.3 Overview of Decision Support Systems in Earth Sciences

Earth monitoring systems have grown to be increasingly more vital within the modern generation. Because the herbal ecosystems that support our lives and livelihoods grow to be more liable to rapid alternate, it's essential to detect and reply quickly to changes inside the environment that would be associated with human pastimes. Selection guide structures (DSS) inside the Earth sciences are designed to be valuable resources for decision-makers in interpreting and responding to environmental records. Via imparting a green and effective manner to comprise knowledge from the earth sciences, they're turning into the desired tool for making selections and taking movement inside the face of environmental change.

DSS is being used across all regions of Earth's technological know-how, from natural dangers and extreme climate to air and water great, to geology and environmental resource management. Because the variety and complexity of Earth technology packages grow, so does the need for robust selection guide structures. Specifically, information-pushed decision making in Earth monitoring structures has emerged as a critical thing in expertise and dealing with environmental risks and demanding situations.

Expertise-driven choice-making (KDM) is a system that combines facts and records with clinical and technical know-how to make sound selections about the surroundings. It consists of obtaining expertise from diverse assets, reasoning based totally on the statistics, assessing threat and uncertainty, using predictive models, and, in the end, taking movement at the consequences of the analysis. The usage of KDM in Earth tracking structures has a number of blessings, which include the capability to consider more than one variable, maximize the collection of relevant statistics, and create better predictions (Xiong et al., 2023).

KDM enables decision-makers to assess a ramification of environmental factors and learn the way they interact together. Such structures are crucial devices for making plans and decision-making, as they can assist in becoming aware of problems and determining appropriate solutions. Even as they're no longer designed to replace human reasoning and choice-making, KDM can support the choice-making procedure and can be used to inform techniques based totally on statistics-pushed insights.

Moreover, KDM has been used to take into account uncertainty and variability that are inherent in Earth technology information. Together with variability that won't be apparent from traditional environmental statistics, KDM facilitates showing the actual impact of environmental alternatives on society. As such, it may provide more excellent correct facts on the want for adaptation and long-term period planning.

In the end, know-how-driven selection-making is gambling a key position in the improvement and use of decision aid structures (DSS) within the Earth sciences. KDM permits the introduction of robust choice support systems that incorporate more than one variable, recall uncertainty, and provide selection makers with accurate and reliable facts-driven insights. As the significance of Earth monitoring systems continues to grow, the need for KDM in DSS is essential for predicting and handling environmental trade. As the KDM era and programs evolve, these systems will become an imperative part of our capacity to plan for and control the demanding situations of world climate trade.

3. INTEGRATION OF KNOWLEDGE GRAPHS IN EARTH MONITORING

Earth tracking is crucial for protecting our planet and making sure of its sustainable development in the face of many pressures, including speedy climate change, environmental degradation, and natural disasters. To benefit from the correct information about the nation and our planet and take timely and effective movement, the tracking of Earth must draw on an extensive range of resources from a variety of disciplines. In particular, it's crucial to incorporate knowledge Graphs into Earth monitoring to allow knowledge-driven decision-making.

Knowledge Graphs, also called semantic networks, are collections of facts that constitute the relationships among entities derived from current sources together with databases and unstructured texts. Such technological advances have enabled the introduction of complete, interconnected information networks, which can be used to provide context and insights into the information homes of every entity. By using those networks of nodes and edges, current information can be correlated and analyzed in a much more profound and more comprehensive manner than ever before.

The capacity to contain knowledge graphs into Earth tracking structures is more robust through the supply of both structured and unstructured information on this planet. On the structured aspect, we've powerful geological and meteorological databases detailing all types of planetary phenomena. In terms of unstructured records, medical documents, online conversations, articles, and many others. Are all capacity sources of perception and information to be tested. as a consequence, a combination of each structured and unstructured data resource may be used to talk about the totality of Earth's data.

Incorporating know-how graphs into Earth monitoring can permit better decision-making in terms of ecological control. For instance, professionals can use included understanding graphs to track the lineage of organisms and investigate the impact of human sports on the environment. Also, they can explore all one-of-a-kind layers of facts and find houses, patterns, and relationships that might need to

Figure 1. Integration of knowledge graphs in earth monitoring

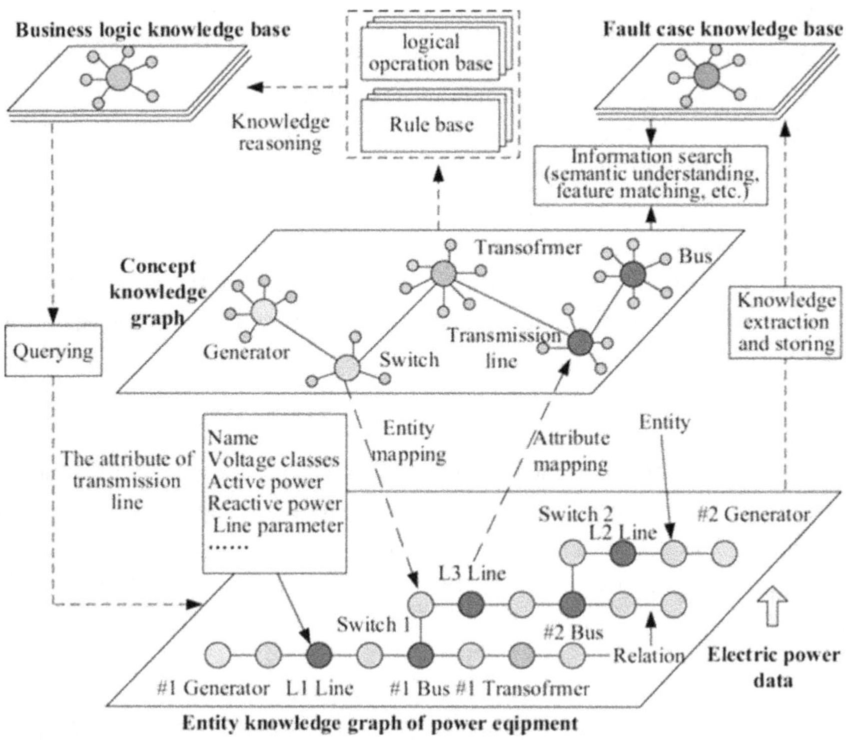

be cleaned from uncooked records. Furthermore, understanding graphs may be used to mitigate the effects of disasters and react fast to environmental emergencies.

3.1 Conceptual Framework of Knowledge Graphs

A know-how graph is a graphical illustration of knowledge statistics that can be used to research and interpret the records in meaningful ways. It is a facts shape that can save and organize actual international entities and their relationships. Information graphs are used to uncover information from big datasets and to facilitate selection-making. Primarily, they may be used for choice-making in earth monitoring systems.

Earth monitoring systems contain the use of various technologies to gather, shop, technique, examine, and interpret facts for expertise and predicting environmental trade. Such structures are used to pick out threats to the surroundings and to mitigate or prevent environmental risks. Knowledge graphs are practical tools for representing complicated environmental situations and for making decisions about how the surroundings should be controlled.

The conceptual framework for knowledge graphs of information-driven selection-making in earth tracking systems consists of five key elements: facts resources, know-how extraction, understanding illustration, expertise inference, and choice-making.

Records resources check with the datasets which are used to construct a knowledge graph. Examples of information assets encompass public databases, geographic data structures, environmental sensors, and satellite TV for PC pics. Facts assets are necessary to gather the records and construct a know-how graph.

Know-how extraction is the manner of extracting understanding from statistics assets. It involves the extraction of uncooked statistics and the normalization of the statistics into a based shape. The established information is also used to represent distinct factors of the expertise graph.

Information representation is the process of structuring and representing expertise in a graph. It includes growing a graph shape that can be used to keep the records and the interconnections among entities on the graph.

Knowledge inference is the system of reasoning from the scales given approximately the expertise structure created by the information graph. It includes the usage of machine studying algorithms to create predictions and become aware of patterns inside the information.

Selection-making refers to the final choices that might be made based on the expertise obtained from the knowledge graph. It is often carried out by using the use of an expert device or intelligent agent that can become aware of the most fulfilling solutions for handling environmental sources.

3.2 Building and Maintaining Knowledge Graphs for Earth Sciences

Earth monitoring systems are ever-converting and more complex with each passing day. As more people turn out updated and aware of environmental threats such as international warming and climate trade, it's miles imperative that choice makers in this area have up-to-date maximum 3177227fc5dac36e3e5a-e6cd5820dcaa knowledge approximately the environment and its effects. It indicates constructing and preserving know-how graphs for Earth sciences as a carrier to date facilitates expertise-driven selection making in Earth monitoring systems.

Know-how graphs are used to define relationships between gadgets, permitting updated higher understanding and utilization of the records. A graph database, then again, makes use of graph structures, up-to-date data, up-to-date tables, or columnar facts. Consequently, expert graphs provide an extra comprehensive view of information, up-to-date, more intuitive insights, and analysis. Furthermore, knowledge graphs can be extra quickly updated with new facts, which means that they are up-to-date and keep up with the ever-changing and complicated troubles surrounding Earth monitoring systems.

Figure 2. Building and maintaining knowledge graphs for Earth sciences

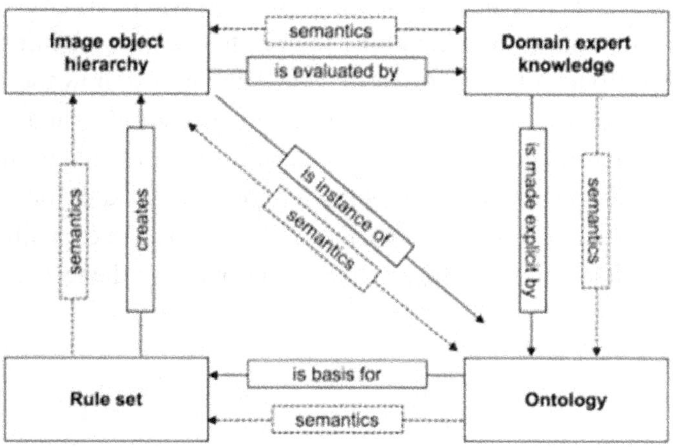

Building and retaining understanding graphs for Earth sciences can provide a comprehensive view of facts and up-to-date better expertise of the surroundings and their effects. Know-how graphs may be used up to date to discover the connections between distinct systems, organisms, or methods, offering a more holistic view of the surroundings. Moreover, they may be up to date and correctly mirror the dynamic nature of the Earth. Connecting facts up-to-date from several assets, such as satellite snapshots, visualizations, urbanization information, and, more importantly, information graphs, can provide a more profound perception of the internal and outside up-to-date affecting the environment.

Moreover, understanding graphs may be helpful for up-to-date choice makers with the aid of supplying up-to-date, pertinent facts. It could make selection-making easier and faster because the consumer could have access to updated comprehensive facts rather than having up-to-date sift via thousands of results to pick out the relevant records. Data in an understanding graph can also be explored and manipulated in methods that make it simpler to date apprehend, elucidating relationships and elements that could help perceive potentially dangerous conditions.

3.3 Enhancing Data Connectivity and Contextual Understanding

The cutting-edge Earth monitoring structures (EMS) rely upon the gathering of heterogeneous information to investigate changes in our surroundings, which include the circumstances of water, soil, environment, and ecosystems. Such systems permit for a mile more excellent accurate knowledge and interpretation of information, which might be vital for predicting and responding to modifications in our surroundings. Therefore, it is essential to beautify data connectivity and contextual understanding of understanding-driven selection-making in EMS.

Connectivity between one-of-a-kind information assets is crucial for EMS to have a successful and informed choice-making process. Making sure that each asset of information is linked, from external to internal, can efficaciously assist EMS in making correct decisions and conclusions. This expanded connectivity could be achieved by developing an unmarried repository of information or a disbursed facts starting place that incorporates data from all events. This method will make sure that every EMS selection is primarily based on correct and reliable facts.

Contextual know-how is essential for making the ideal decision in EMS with the intention of getting contextual data; the EMS body of workers must understand the source of the facts, the purpose of the data, and its context. For example, if statistics are accumulated from a faraway place, an EMS team of workers should have information from surrounding regions to evaluate the information from its unique context. Moreover, the EMS body of workers has to have know-how of how the information is amassed, sourced, and saved if you want to investigate the information's credibility and validity.

Moreover, EMS personnel should be able to evaluate the effects related to information connectivity and contextual expertise and apprehend the risks that might arise, in addition to the opportunities that can be received from making sure choices. Staff should also study the capability dangers and impacts of the usage of particular data sources, as this can have an impact on the reliability and accuracy of the decisions made.

4. MACHINE LEARNING FOR KNOWLEDGE EXTRACTION

System learning is an effective tool that can be used to extract understanding so as to enhance choice-making and optimize performance in Earth tracking systems (EMS). Device mastering allows computer systems to examine complicated and sophisticated patterns from information to provide significant insights that could tell selection making. This capability permits EMS to make more knowledgeable and wise selections based totally on current and ancient records. Know-how extraction is a crucial part of any EMS device because it allows the device to gain a higher understanding of the environment and make decisions.

The technique of understanding extraction via a device-mastering machine includes the automatic analysis of massive quantities of data. This manner allows the system to pick out styles, features, and traits and research from them. The extracted statistics can then be used to inform destiny decisions. Gadget studying can also be used to become aware of problems that can have been overlooked and advocate appropriate solutions.

Knowledge extraction is, in particular, critical in EMS because it allows us to respond to changes within the environment quickly and effectively. Machines can hastily method UAV imagery, satellite insurance, and other forms of environmental statistics to detect changes inside the surroundings in near real-time. These statistics can then be used to inform timely decisions and movements to ensure maximum effectiveness for emergency offerings, catastrophe preparedness groups, and environmental safety corporations.

By utilizing the strength of device studying to extract knowledge, Earth monitoring structures can gain a higher understanding of the environment they're looking to display. This information permits EMS to make more intelligent and more informed selections, enhancing the efficacy of their selection-making and operations. With the assistance of device-gaining knowledge, EMS structures can be more green, effective, and accurate in responding to diverse surroundings troubles. It guarantees the protection and well-being of anybody, animals, and the surroundings.

4.1 Leveraging Machine Learning for Knowledge Discovery

Earth monitoring has become increasingly more critical because the results of world weather trade end up more evident, and human activities position more pressure on herbal sources. understanding-pushed decision-making is the system of using statistics to decide which movements and selections are most beneficial for a given scenario. With the emergence of device studying, it has become possible to leverage this facts-pushed selection-making procedure to make extra knowledgeable selections in earth monitoring systems.

Gadget learning algorithms may be used to procedure vast amounts of environmental information to expect future tendencies and activities. This form of predictive analysis is incredibly beneficial for making intelligent decisions that are based totally on experience in place of speculation or guesswork. As an example, gadget-studying algorithms might be used to predict flooding by way of spotting patterns in climate information. Those predictions may be used to boost preparedness or adjust strategies in advance of an occasion.

Machine-gaining knowledge of algorithms can also be used for knowledge discovery. This kind of analysis appears for helpful information that is hidden in immense amounts of information. Itwill be used to discover new insights that might not be effortlessly observed through human beings, inclusive of connections among different regions of studies or early indicators of modifications within the environment.

Device studying also can be employed for pattern recognition and anomaly detection. Algorithms can be skilled to understand and detect styles that are specific to an area, such as regions of deforestation, city sprawl, or increasing environmental stressors. Through spotting these anomalies, quick, informed decisions may be made, which can be primarily based on records and subsidized by means of evidence. These decisions should result in more maintenance of herbal assets or more preparedness in the face of screw-ups.

4.2 Unsupervised Learning Approaches for Pattern Recognition

In current decades, Earth tracking structures have grown to be increasingly famous, permitting us to accumulate statistics from a spread of sources in order to investigate the adjustments going on in our environment. Understanding-driven selection-making via pattern reputation is crucial with the intention to make more excellent knowledgeable decisions and interpret the facts received. Unsupervised learning techniques for this motive can offer us treasured insights into information sets that won't have an apparent shape. This essay will explore the function of unsupervised getting-to-know tactics for pattern popularity of knowledge-pushed choice-making in Earth monitoring structures.

Unsupervised device studying (ML) is an area of synthetic intelligence (AI) that utilizes algorithms that are empowered to investigate facts, become aware of styles, and make selections without requiring any external input. That is distinctly different from supervised ML, which requires humans to teach and influence the AI. Unsupervised ML algorithms are mainly beneficial when figuring out patterns in big datasets as they are much quicker than human analysis. It may facilitate the detection of diffused activities in information sets that may not be located via conventional techniques and, as a result, is relatively acceptable for Earth tracking systems.

Unsupervised studying may be applied to Earth tracking systems to assist in spotting patterns in vast and complex datasets that might be too time-consuming for people to analyze. Those algorithms can be used to segment the information into datasets in step with predetermined criteria or look for clusters and underlying developments within the information. It will permit the gadget to quickly and accurately become aware of changes in Earth's surroundings and establish relationships between variables inside the dataset. With the aid of spotting the structure of the statistics, it will likely be able to decide meaningful relationships, identify anomalies, and expect future events.

4.3 Knowledge-Enhanced Feature Engineering in Earth Monitoring

Earth monitoring systems are critical for knowledge of our global and figuring out areas of difficulty. From tracking weather change to measuring the fitness of our forests, these systems play a crucial role in protecting and preserving our planet. Information-enhanced feature Engineering (KFE) is a technology that is transforming earth tracking by means of supplying statistics-driven decision-making talents.

KFE works by way of leveraging current information from earth monitoring systems to create new capabilities and evaluations for choice-making. It makes use of records from sensors, aerial imagery, GIS records layers, and different sources to become aware of styles and extract meaningful insights. It allows decision-makers to more correctly recognize the environment, pick out regions of the problem, and create knowledgeable choices regarding a way to respond.

KFE allows earth tracking systems to turn out to be extra efficient, correct, and automatic. Leveraging records from numerous sources it gives a powerful new device for selection makers to deal with environmental problems effectively. KFE can also help integrate records from multiple resources and allow for more green coordination among authorities, companies, and different stakeholders.

Using KFE, earth tracking can be more potent at figuring out patterns and knowledge modifications in the surroundings. It helps become aware of areas of concern and enables more excellent, informed selections of a way to deal with them. KFE allows to empower selection makers to make information-pushed, knowledgeable choices about sustainability, conservation, and aid management.

KFE can also reduce costs associated with monitoring with the aid of reducing the burden on guide analysis and improving automation skills. It could lose assets to cognizance of other responsibilities and, in the end, help serve the earth monitoring project.

5. SEMANTIC TECHNOLOGIES FOR EARTH SCIENCES

In recent years, there has been a developing need for brand spanking new answers and procedures that may assist us in monitor and assessing the nation's environment. Knowledge-driven decision making, enabled via powerful semantic technology, can be compelling in aiding the development of sophisticated earth monitoring structures. This essay will discover the strengths of semantic technologies for earth sciences and their ability to increase cutting-edge applications and systems for tracking the earth and its ever-changing environment.

First, it is crucial to apprehend what semantic technologies are and how they may be utilized to recognize and screen the earth's environment. Semantic technologies hire information extraction algorithms to extract and examine vast quantities of statistics with the intention of drawing meaningful conclusions and insights. By means of inspecting the relationships between such statistics, they may be used to benefit beneficial expertise about positive surroundings. Earth tracking systems benefit from the use of semantic technologies due to the fact they enable us to gain a deeper understanding of the patterns and tendencies of various additives of the earth, allowing us to anticipate and reply to environmental variables.

Semantic technology not only provides a more excellent perception of our surroundings but also can toughen current decision-making protocols. By offering interpretable insights about the environment using advanced computational techniques, we are capable of better-broadening strategies that are tailored to tackle particular environmental issues. As more excellent statistics are to be had, semantic technologies permit us to identify patterns and developments in the facts and expand fashions that might be tailored to specific environmental conditions. Inside the case of an earth tracking machine, this could encompass using semantic technologies to stumble on anomalies inside the surroundings, hit upon adjustments in the environment, and develop techniques primarily based on the underlying records.

5.1 Semantic Web and Ontologies

The ability for the use of Semantic Net and Ontologies for information-pushed decision-making in Earth tracking systems is swiftly increasing. So one can make powerful choices based totally on the information to be had to us from Earth tracking systems, using the Semantic web and ontologies is prime. Earth tracking systems use statistics from a large variety of sources, including satellites, sensors, and floor-based structures, to provide data about phenomena happening on the planet. This record is crucial for making informed and knowledgeable decisions on international issues like weather change, populace growth, and aid endangerment.

The Semantic web and Ontologies are helpful for this motive by assisting in unifying the information of various tracking systems. Semantic internet technology, which includes RDF and OWL, allows records to be represented in a prepared manner by way of linking related entities and concepts. Ontologies, which are conceptual fashions, provide exceptional styles of records, together with the sorts of entities and their relationships that are vital in order to research and determine monitored facts. By combining each of those technologies, the records amassed by using tracking systems become less challenging to analyze.

The use of the Semantic web and Ontologies for information-driven decision-making can help in lots of areas of Earth monitoring structures. First, they assist inside the MAT, or Multi-Agent tracking, that occurs while more than one records assets from specific monitoring structures should be aggregated for you to shape a bigger picture. With the aid of the usage of this technology, it is feasible to effectively tune the conduct of diverse environmental elements, along with stages of air pollution or water table tiers. 2nd, they could help with modeling and simulation, which could help to predict the results of proposed movements and adjustments to the surroundings. These records can be used to make knowledgeable choices about the way to make use of or control assets or take corrective motion while required.

5.2 Applying Semantic Technologies in Earth Monitoring

As our know-how of this planet and its surroundings improves, so too does the want to design and expand strategies to screen and analyze statistics to make sure it persevered fitness. It is an increasingly daunting task, given the sheer complexity of the elements affecting Earth's climate. In recent years, the emergence of semantic technologies has proven to be a powerful tool for Earth monitoring, mainly when it comes to information-pushed selection making in Earth tracking structures.

Semantic technologies are laptop-based, which means making structures that contain the indexing, searching, discovery, and analysis of records saved in semantic networks. They usually depend upon the synthetic intelligence (AI) paradigm, as well as technology, along with the valuable resource Description Framework (RDF) and the web Ontology Language (OWL), to represent and motivate statistics. It makes them uniquely ideal for Earth monitoring packages, as they can quickly and accurately pick out patterns and trends within the statistics that may be used to apprehend the environment. Similarly, semantic technology can discover relationships within facts units which are previously ignored, which can then be used to develop actionable insights.

One instance of the packages of semantic technologies in the Earth tracking arena is their use to reveal atmospheric conditions. The usage of semantic technologies and advanced algorithms can be used to research the data amassed from satellites and other resources to become aware of patterns and developments, permitting the accurate and well-timed detection of adjustments in the atmosphere. It will be used to broaden and enforce techniques to enhance air first-class, lessen pollutant tiers, and, in the end, enhance the fitness of the planet.

5.3 Achieving Interoperability and Standardization

Earth tracking structures can offer beneficial perception into our environment, helping us to understand better the adjustments happening inside it. This form of insight may be used to make informed and evidence-based decisions, allowing for more accurate predictions and higher control of resources. But, if we're to take benefit of such statistics, Earth monitoring structures need to come to be interoperable and standardized across the board.

Interoperability clearly refers to the ability of different structures to paint together. It means having a shared language, set of protocols or schema, and software infrastructure that allows all of them to speak and apprehend each other. Standardization is even extra critical, assisting to make sure that meaningful statistics can be effortlessly retrieved and used without dropping their quality or integrity.

To gain interoperability and standardization of Earth tracking structures, companies ought to collaborate and interact in significant communication to identify pleasant practices and create a shared vision. Commonplace protocols and requirements additionally need to be hooked up, which will allow for a more efficient communique of information across all systems. All of those activities must be undertaken with an emphasis on information privacy and safety, as fact transfers will require a cozy environment if meaningful facts are to be reliably exchanged.

Earth monitoring systems additionally require a sturdy set of facts processing and analysis abilities so that they will create significant and accurate knowledge and choices. Device-gaining knowledge of artificial intelligence and deep-gaining knowledge of tools can assist in automating and optimizing the technique of extracting that means from disparate datasets. Further, cloud-based records management systems can be utilized to shop safely and system the facts in a cozy and efficient manner.

6. KNOWLEDGE-DRIVEN DECISION SUPPORT MODELS

Earth tracking structures are crucial gear for understanding and dealing with the environment. That is especially crucial today with the increasing need to cope with climate exchange. The choices made with the aid of governments, agencies, and people on a daily foundation have an instantaneous impact on the worldwide climate and ecosystems. It's miles consequently essential to make sure that decisions are made in an informed and expertise-driven manner. Information-driven choice helps models make sure that choices are based on the available information and are steady with long-term environmental dreams.

Know-how-pushed selection assist fashions (KDM) are pc-primarily based equipment designed to proper resource choice making and facilitate the choice of moves primarily based on present-day and projected environmental situations. KDM uses mathematical models to investigate information and offer perception on how best to proceed in a situation. The fashions include statistics from multiple resources, including meteorological, geographic, and environmental sciences facts. The insights generated with the aid of KDM can help inform decision-makers and offer them the necessary records to make more informed choices.

KDM can be used to become aware of and expect environmental impacts from contemporary and future selections. For example, a KDM can be used to estimate the impacts of a proposed dam creation on the nearby water table and river flows. It can also be used to broaden mitigation techniques for the ability environmental impacts, which includes rerouting the water to go with the flow or imparting extra water resources. By predicting the capacity environmental effects from selection making, KDM can help to guide sustainable improvement strategies.

6.1 Design Principles of Knowledge-Driven Decision Models

Knowledge-pushed decision fashions are an increasing number of crucial issues of earth monitoring systems, as they are designed to correctly method the considerable amount of information generated in these complex ecosystems. As a result, the right layout ideas must be implemented to make sure there's ideal overall performance and cost-effectiveness. In this essay, we will discover the numerous layout concepts of understanding-pushed decision fashions for earth monitoring structures and the way they could make sure that records are effectively managed and beneficial insights are generated.

First of all, one of the essential layout principles of understanding-pushed decision fashions for earth tracking systems is fast processing. As vital records flow into the machine, it is critical to ensure that the statistics are correctly managed for you to offer meaningful insights. It calls for the proper algorithms and techniques to be used to keep, system, and analyze information. For instance, a mixture of device-studying algorithms can be used to locate styles in the statistics and provide a green manner to exploit the most beneficial insights. Moreover, data visualization tools may be used to give the records in ways that permit users to identify patterns inside the data and draw insights quickly.

One other fundamental design principle of know-how-driven selection fashions for earth monitoring systems is scalability. Earth tracking structures require the capacity to scale the device to be able to method the massive amounts of data that are accumulated. As such, it's miles critical to make sure that the selection models are capable of coping with the increasing extent of information while also acting the essential computational heavy lifting fast. To ensure scalability, a disbursed architecture can be used, which allows multiple structures to work together and scale quickly without compromising overall performance.

6.2 Challenges and Solutions in Model Development

Recently, due to the advancements in Earth tracking technology, which include satellite imagery and GIS, the selection-making system has grown to be more records-driven. As a result, there is more demand for growing equipment and techniques which can be able to assist selection makers in making more accurate choices. Alas, a lot of these models are nevertheless in their early tiers with many capacity challenges and answers yet to be explored. This essay will study the contemporary demanding situations and answers to the improvement of knowledge-pushed choice-making in Earth tracking systems.

One of the essential demanding situations in know-how-driven choice-making in Earth tracking systems is getting access to and making use of accurate facts. It is a task that arises because of the complexity and kind of information that is accrued in Earth tracking structures, in addition to the numerous facts and sources used to obtain these statistics. For instance, information can come from more than one satellite, onboard-size structures, and other sources. Thus, it's very critical to gather and integrate accurate records from various sources in order for decision-makers to make well-knowledgeable decisions.

An approach to the venture of information accuracy is the software of information mining strategies. Information mining strategies are used to extract valuable records or "knowledge" from vast amounts of statistics. Via using sophisticated data mining algorithms, information-pushed selections in Earth tracking systems can be made more quickly and appropriately. Moreover, these strategies can also help discover hidden patterns or trends in the records that can offer precious insight into the selection-making technique.

Another assignment confronted in knowledge-pushed choice making in Earth monitoring systems is the complexity and trouble of the fashions being used. On account that most of the selection-making fashions are nonetheless of their early levels, they can be hard to recognize or even extra in an effort to broaden. Developing and keeping these models can be a traumatic task and calls for an extraordinary deal of ability.

7. REAL-TIME KNOWLEDGE UPDATES AND ADAPTIVE SYSTEMS

Earth monitoring systems (EMS) are becoming increasingly essential in the latest international. The speedy evolution of the era and changing weather situations have made it critical for organizations to respond quickly and efficiently to up-to-date changes in the surroundings. Real-time understanding updates and adaptive systems of know-how-pushed choice-making can assist up-to-date ensure that facts accrued through EMS are getting used and updated to their most total capability.

Actual-time know-how updates permit organizations to date and make use of data. It is modern-day and straight away updated. Minor modifications in the environment will have a massive cumulative impact through the years. By way of tracking and responding to date the modifications in real-time, EMS can create a more accurate and green gadget for selection making. Actual-time updates can also assist in identifying areas of risk in a timely fashion, making it less complicated to cope with or limit capability damage before it becomes inundated and difficult.

Knowledge-pushed decision-making systems take advantage of the large amounts of records and imagery gathered by EMS and feed them into up-to-date predictive models. Those fashions can be used for up-to-date selections about specific environments, along with chance assessment or helpful resource allocation. Moreover, the models can be used to update stumble-on styles in facts and offer a perception of the underlying reasons for issues or better solutions to up-to-date current problems. With the aid of leveraging the predictive abilities of these fashions, groups can generate insights that force extra knowledgeable choices to update the environment.

7.1 Continuous Learning in Earth Monitoring Systems

Earth tracking systems are essential in today's present-day global. Technology, inclusive of satellite imaging and worldwide positioning structures, allows us to music hobby on the planet in pleasant detail. It has enabled us to accurately predict natural screw-ups and take preventive measures before they occur. It has also allowed us to observe weather styles, climate alternates, and other environmental problems on a far larger scale. However, as technology keeps evolving, it's essential that we put money into continuous studying, which will make sure that quality selections are being made with regard to Earth monitoring systems.

The intention of non-stop mastering is to increase the intelligence and relevance of choice-making inside Earth monitoring structures. It includes the combination of research, generation improvement, and operations to create a cohesive working environment. It allows us to make sure that know-how is being continually received and carried out, at the same time as taking cutting-edge desires into account. By way of investing in non-stop studying, Earth monitoring structures may be more incredibly green, powerful, and responsive.

While imposing non-stop mastering, it's miles critical to be aware of the desires and objectives of the machine. It permits you to make sure that the learning methods and technological upgrades align with the stated motive of the device. It additionally lets you create a straightforward method for selection-making. It is based on systematic proof and studies. It helps to make sure that everyone's choices are made based on the best-to-be-had facts.

Non-stop mastering also entails taking advantage of emerging technology and traits in Earth tracking systems. It can encompass exploring technology like synthetic intelligence and gadget mastering with a view to improving choice-making. It additionally involves making sure that the information analyzed is reliable and displays the modern conditions and needs of the machine. Moreover, it is critical to make sure that each stakeholder is concerned, which allows you to create a holistic technique for choice-making.

Environmental monitoring systems are used to observe and track environmental parameters in order to detect changes in the environment and to monitor environmental health. These systems may use a variety of technologies to measure changes in temperature, humidity, air pressure, air quality, wind speed, water quality, and other environmental indicators. Environmental monitoring systems can be used for both environmental studies and environmental control, depending on the type of system used. Environmental studies involve the use of environmental monitoring systems to collect data about the environment. This data can be used to better understand human impacts on the environment and to inform decision making about environmental management. Environmental control systems use environmental data to monitor the environment and enforce environmental regulations. These systems can be used to monitor the progress of remediation efforts, determine compliance with environmental regulations, and ensure that environmental resources are being used responsibly.

7.2 Adaptive Decision Support in Dynamic Environments

In today's world, superior decision assist systems are becoming increasingly essential. With the development of the era, the amount of to-be-had information is growing exponentially, making robust selection crucial. One unique area wherein adequate decision support is needed is in earth monitoring structures, as these structures require using ever-evolving know-how to help real-time, dynamic selections.

Adaptive selection assist systems (ADSSs) have the functionality to respond quickly to converting conditions and are well-applicable to earth monitoring systems. ADSSs are commonly composed of know-how-driven interfaces that could integrate outside knowledge assets into the decision-making system. Those decision support systems can seriously improve the efficiency and accuracy of decision-making within earth monitoring systems.

The use of advanced selection help structures is growing with the emergence of various real-international packages. However, because of its complexity and interdisciplinary nature, it's miles tough to compare ADSSs overall performance properly. This essay will describe the ability of information-pushed selection making for earth-tracking systems and examine the demanding situations associated with the implementation of an effective ADSS.

Know-how-based totally decision-making, or KBDM, is the method of using integrated understanding sources, consisting of databases, the internet, and decision guidelines to make selections. Know-how-primarily based decision-making has been used in lots of one-of-a-kind real-international scenarios. It may be implemented in earth monitoring systems to grow their effectiveness.

7.3 Balancing Automation and Human Expertise

Earth monitoring systems are used to gain a higher know-how of the environment and the planet as a whole. with a view to making thoughtful selections, they should depend on automation and human information. Automation can offer essential analytical insights and shop crucial records, at the same time as human professionals can offer context and expertise of the worldwide context. Balancing automation and human expertise of expertise-pushed selection making in Earth tracking systems is essential for making the most informed and complete choices possible.

Automation has many blessings in relation to Earth tracking systems. Automation is able to analyze considerable amounts of records quickly and efficaciously, which can be crucial in accumulating and interpreting records in a well-timed way. Automation also makes it simpler to make adjustments in information over time, making it more straightforward to screen and discover changes and trends.

Automation also has its limits. Earth monitoring systems rely on facts-driven selections, and automation-based total structures are best as accurate as the statistics they are furnished. Human experts are essential to interpret those facts and offer context. Human experts can provide a worldwide attitude on the records that automation may also miss. Experts can also provide extra nuanced information on the facts and expertise on the effect of the information on the environment.

Human experts are also critical in supplying steering on what facts have to be used and how they need to be interpreted and carried out. Professionals can prioritize statistics and interpret them efficaciously as well as broaden more excellent knowledgeable decisions that recollect a massive variety of variables and capacity consequences. ultimately, human information can also assist in making sure that moral and environmental considerations are taken into account.

8. CHALLENGES AND ETHICAL CONSIDERATIONS

One of the predominant demanding situations of expertise-driven selection-making is that the information collected and analyzed can be incomplete or wrong, leading to inaccurate or faulty decisions. As a result, decisions need to be carefully evaluated to make sure that they may be based on correct proof and provide correct records. Furthermore, the float of information to and from the gadget must be monitored to save you from having an effect on and manipulation of the decision-making method.

Another assignment is that understanding-driven selection-making is predicated on machine getting-to-know techniques, which involve the use of enormous amounts of facts. Those strategies may be computationally intensive, requiring giant computing strength and sources. Further, they could contain complicated algorithms and can be time-consuming, making them challenging to recognize and use in practice.

Sooner or later, understanding-driven selection-making systems should be sponsored through sound moral concerns. They need to be designed to behave in the first-rate hobbies of society and must bear in mind each human and environmental impact of their choices. They must be transparent and accountable and should inform the public about their operations. Furthermore, it have to be understood that KDD structures do not update human judgment, but as an alternative, complement it, and that choices have not to be taken gently.

8.1 Data Quality and Trust in Knowledge-Driven Systems

Records quality and understanding-pushed structures of know-how-pushed decisions for earth monitoring structures is a vital thing to keep in mind when using these structures for choice making. Records excellent refers to the accuracy, consistency, completeness, and integrity of the records of earth monitoring systems getting used. Exact records are vital to make sure that essential choices crafted from the records are of the very best fine.

Accepting accuracy within these structures is also an essential element to remember while using them to make choices. Without agreeing, the surroundings of choice-making are unsupported, and the choices made in these surroundings can have dire consequences for all worried. Without accepting as accurate, decision-makers are at risk of defective selection-making as they can't trust or trust the information they may be using.

If you want to have successful, first-rate selections made primarily based on these systems, the decision-makers need to consider the statistics they may be using. Simple measures like ensuring the information is reliable, credible, and up-to-date can enhance selection-making surroundings. Moreover, the accuracy and completeness of the records can be progressed by comparing the information on an ordinary foundation. The information needs to additionally be proven through go-referencing with different valid resources of statistics. Furthermore, choice-makers must never forget the assets available and use the most up-to-date and accurate records while making their choices.

The trustworthiness of any device depends on how properly the information and structures are maintained. Correct upkeep of the systems, inclusive of proper backup, archiving, and versioning practices, can make structures more dependable. Additionally, user authorization and authentication practices can help make the system more extraordinarily cozy. With the aid of making use of accurate information, decision-makers may be assured that the records they may be using are honest and dependable.

8.2 Addressing Bias and Fairness

Earth tracking systems have emerged as an increasing number of sophisticated, enabling quicker and more correct choice-making in a massive range of relevant domain names. Those structures are knowledge-driven, making choices by means of leveraging increasingly powerful synthetic intelligence and gadget-gaining knowledge of techniques. As such, it's becoming increasingly essential to recall the potential for bias and unfairness in these selection-making systems.

It is vital to apprehend that bias in selection-making isn't necessarily an intentional or malicious practice. Instead, bias in choice-making frequently arises from a group of unintended, however interconnected resources, which could lead to differential effects in how facts are translated into insights and then used to power movement. As such, addressing bias and truthful choice-making requires a comprehensive method that evaluates the whole variety of ability resources of bias and works to neutralize capability assets of unfairness and inequity.

One fundamental approach to addressing bias and fairness in selection-making is to prioritize using information that is consultant of individuals and groups, and that shows how choices affect special populations in specific ways. Through accumulating facts on exclusive groups, decision-makers can begin to apprehend the impacts of their decisions in a more nuanced way and may regulate consequences thus. That is mainly important in earth monitoring systems, as herbal and environmental disasters may have disproportionate and sometimes surprising results for distinct populations.

8.3 Privacy Concerns and Transparency in Decision Support

In the net age, privacy worries and transparency have become increasingly critical in statistics control. As records-pushed choice-making turns into more extensively adopted, particularly in earth tracking systems, this is no exception. Customers, in particular those in the earth monitoring zone, need to balance their need for privacy protection with the call for access to data and transparency in selection help.

In terms of decision guides in earth tracking structures, privacy issues are magnified via the massive quantity of information that has to be processed and included. For instance, geospatial records about land functionality layers, terrain, or soil kinds can aid in decision-making through the analysis of land suitability. The combination of this information with extra socio-monetary or records related to human health affords privacy-demanding situations within the form of publically shared region statistics and personally gathered data. Agencies ought to make sure that the information they accumulate is relaxed and not shared with outside events unless authorized. Moreover, they must remain in compliance with authorities' guidelines associated with privacy and records processing.

Transparency in the decision guide of understanding-pushed choice-making in earth monitoring structures also can complicate privacy worries. As maximum choice support structures are primarily based on predictive fashions, corporations can be restrained in how a great deal get entry to and perception they could provide while nonetheless retaining confidentiality. Clients need assurance that their choices are based totally on correct, independent facts, yet also want to be blanketed from potential misuse or misinterpretations of the statistics.

9. FUTURE DIRECTIONS AND INNOVATIONS

Earth tracking structures have become increasingly more crucial as the arena increasingly looks to apply generation to recognize the state of our planet. Understanding-pushed choice-making has been necessary for this technique, permitting researchers to make informed decisions on the way to exceptional use of technology to song the surroundings and notice adjustments in climate and herbal patterns. With today's ever-evolving era and increasing expertise in the environment, there are a range of destiny guidelines and improvements on the way to take know-how-driven decision-making in earth tracking systems to the following degree.

The improvement of Artificial Intelligence (AI) and its utility to know-how-driven choice-making will be crucial in the destiny of earth tracking systems. Machine mastering algorithms are getting an increasing number of superior and are capable of automating many of the guide methods, which require large quantities of records and prolonged processing times. It will permit researchers to pay attention to large-scale tasks where extra correct effects can be determined faster. Similarly, AI can be carried out in predictive modeling so as to generate more excellent sophisticated forecasts of destiny environmental and climate change. With the aid of being capable of hit upon and expecting environmental change, researchers can be better ready to respond quickly and make informed choices that can have a positive effect on the management of the earth's herbal systems.

9.1 Advancements in Knowledge Graph Technologies

In recent years, advancements in know-how graph technology have had a significant impact on selection-making in earth tracking systems. Information graphs permit the computerized processing of information units, advanced records evaluation, and more comprehensive and green decision-making. It has helped to seriously enhance the accuracy and velocity of choice-making in earth tracking systems.

A know-how graph is a machine of interconnected nodes that describe specific information and relationships between gadgets. They're used to decorate data flow and information discovery by connecting related nodes through a not-unusual set of relationships. It permits effective analytics that can be quickly accessed and used to broaden accurate decisions.

In earth monitoring systems, know-how graph technology allows for extra efficient and accurate selection-making via tracking environmental changes over time. These records may be used to research patterns that imply modifications in climate patterns, climate situations, or other environmental adjustments. Via using knowledge graphs, these statistics can be saved and nicely analyzed to tell Earth higher tracking choices.

Similarly to helping decision-making, know-how graphs also enable the integration of more than one statistic set. It permits for higher know-how of styles and changes inside the environment, as well as better motives for observations. These records can then be used to tell more fantastic correct selections and create extra correct fashions of the environment.

9.2 Emerging Trends in Machine Learning for Knowledge Extraction

The growing presence of synthetic Intelligence (AI) in modern-day global comes with fantastic and disadvantageous implications for society. One of the fundamental benefits of AI is its capacity to offer undiscovered perception for choice-making in diverse domains, specifically in earth monitoring systems. With the emergence of numerous systems gaining knowledge of techniques inclusive of deep studying, know-how extraction for know-how-driven selection making has grown to be simpler, quicker, and more particular.

Earth tracking systems are made of intelligent structures that gather statistics on the floor of the earth and use these records to make decisions. The collected records help us song modifications in the bodily environment and display the conduct of natural assets, inclusive of water, forests, and power. The force to recognize our planet and our area within its far-inspiring medical improvements in earth monitoring systems.

Given the fantastic ability of AI to significantly decorate the talents of earth monitoring structures, the device gaining knowledge is increasingly seen as a manner to gain insights from enormous reservoirs of records. Machine learning algorithms are already being utilized in the ramifications of earth tracking systems, starting from weather forecasts to land abandonment detection. Examples of such structures are AI-powered drones that reveal environmental conditions and locate water, land, and gas infections, which will expect disasters along with floods, landslides, and so on.

9.3 Collaborative Initiatives for Knowledge-Driven Decision Making

Information-pushed selection making has emerged as more essential than ever, mainly within the area of earth monitoring structures. Earth-tracking structures span a wide range of disciplines, from soil technology to forestry. They can be used to reveal and track the outcomes of climate change, natural disasters, and different environmental modifications. Accumulating and analyzing statistics from these structures is an essential part of coping with and mitigating the effects of environmental trade. But this process can be incredibly time-consuming and pricey, making it hard to keep up with the hastily advancing pace of environmental trade.

In current years, the usage of collaborative projects and knowledge extraction strategies for understanding-pushed selection making in earth monitoring systems has become increasingly vital. Collaborative tasks that involve multiple actors, including enterprises, authorities, non-income companies, and studies groups, are a powerful manner to leverage the collective strength of assorted and specialized know-how, assets, and understanding. Such collaborations enable meaningful and impactful selections to be made quickly and correctly while nevertheless ensuring statistics and analysis.

10. CONCLUSION

In conclusion, understanding extraction from earth monitoring structures is an essential system for assisting decision making via using statistics-driven techniques. Via extracting information from diverse resources, companies can benefit from valuable insights to update their operations to date in their goal markets. Moreover, expertise extraction can assist in perceiving regions of opportunity and fee savings. Moreover, records-driven techniques may be helpful in predicting future situations and occasions to be better prepared for disasters or emergencies. By means of preserving the data extracted from earth monitoring structures and making use of up-to-date techniques, corporations can create a solid information base updated to help their choice-making approaches. Consequently, know-how extraction from earth monitoring structures is a crucial step in knowledge-driven choice-making and might help agencies stay competitive and run extra efficiently.

10.1 Summarizing the Impact of Knowledge-Driven Decision Making

Know-how-pushed choice-making has become increasingly vital in Earth tracking systems because of its capability to extract relevant, analytical, and actionable records. By way of permitting users to access and analyze large amounts of statistics quickly, it facilitates higher choices for businesses. This text will summarize the effect that understanding-pushed decision-making has on the earth tracking systems.

First, expertise-pushed selection making allows Earth tracking structures to leverage expertise from significant datasets more correctly. It lets customers quickly pick out patterns in facts and formulate correct and dependable selections. By using gadget learning algorithms to benefit insights from extensive streams of facts, it removes the need for guide labor and allows users to make quicker and more informed choices.

2nd, by allowing customers to get the right of entry to and analyze records quickly and appropriately, information-pushed selection-making allows corporations to mitigate dangers more effectively. For example, it could be applied to discover any adjustments in surroundings that would result in the disruption of touchy ecosystems. It allows companies to manipulate their resources better and keep them in compliance with environmental policies, helping to limit potential risks.

10.2 Implications for Advancing Earth Monitoring Systems

In recent years, the advancement of Earth tracking systems has been mainly pushed with the aid of technological progress aimed toward growing the accuracy and reliability of information regarding Earth's environment. This fact can be used to make selections that have an impact on people's lives in the course of the world. Similarly to the information-driven method, information-driven decision making has gained traction because of improved awareness of the significance of incorporating regional, environmental, and cultural knowledge into selection-making. With this extended cognizance, the need to extract beneficial knowledge from current Earth tracking systems so one can guide information-pushed decision-making becomes obvious.

The results of extracting knowledge from Earth tracking systems can have far-reaching influences on the sustainability of our worldwide surroundings. For one, extracting understanding from Earth monitoring structures can permit quicker and more efficient selection-making that can help us become aware of potential pollution assets and mitigate them earlier than they turn out to be a bigger problem. Additionally, understanding-driven choice-making has been proven to supply choices that might be primarily based on the international expertise of local, environmental, and cultural contexts, allowing choice makers to make selections no longer just primarily based on the expertise of the environment but additionally on human beings. It will result in more equitable choices that take into consideration all stakeholders stricken by the selection.

REFERENCES

Abdelkareem, M., & Mansour, A. M. (2023). Risk assessment and management of vulnerable areas to flash flood hazards in arid regions using remote sensing and GIS-based knowledge-driven techniques. *Natural Hazards*, *117*(3), 2269–2295. doi:10.100711069-023-05942-x

Aldana-Martín, J. F., García-Nieto, J., del Mar Roldán-García, M., & Aldana-Montes, J. F. (2022). Semantic modelling of earth observation remote sensing. *Expert Systems with Applications*, *187*, 115838. doi:10.1016/j.eswa.2021.115838

Chen, J., Lu, W., Fu, Y., & Dong, Z. (2023). Automated facility inspection using robotics and BIM: A knowledge-driven approach. *Advanced Engineering Informatics*, *55*, 101838. doi:10.1016/j.aei.2022.101838

Chudasama, B., Torppa, J., Nykänen, V., Kinnunen, J., Lerssi, J., & Salmirinne, H. (2022). Target-scale prospectivity modeling for gold mineralization within the rajapalot Au-Co project area in northern Fennoscandian Shield, Finland. Part 1: Application of knowledge-driven-and machine learning-based-hybrid-expert systems for exploration targeting and addressing model-based uncertainties. *Ore Geology Reviews*, *147*, 104937. doi:10.1016/j.oregeorev.2022.104937

Daviran, M., Parsa, M., Maghsoudi, A., & Ghezelbash, R. (2022). Quantifying uncertainties linked to the diversity of mathematical frameworks in knowledge-driven mineral prospectivity mapping. *Natural Resources Research*, *31*(5), 2271–2287. doi:10.100711053-022-10089-w

Falkenström, E., & Svallfors, S. (2022). The knowledge–management complex: From quality registries to national knowledge-driven management in Swedish health care governance. *Politics & Policy*, *50*(5), 1053–1066. doi:10.1111/polp.12497

Horn, J., & Isselstein, J. (2022). How do we feed grazing livestock in the future? A case for knowledge-driven grazing systems. *Grass and Forage Science*, *77*(3), 153–166. doi:10.1111/gfs.12577

Hu, M., Wu, B., Zhou, W., Wu, H., Li, G., Lu, J., Yu, G., & Qin, Y. (2022). Self-driving shield: Intelligent systems, methodologies, and practice. *Automation in Construction*, *139*, 104326. doi:10.1016/j.autcon.2022.104326

Huang, L., Sun, R., Cheng, N., Hui, Y., & Liang, D. (2023, March). Delay-Oriented Knowledge-Driven Resource Allocation in SAGIN-Based Vehicular Networks. In 2023 IEEE Wireless Communications and Networking Conference (WCNC) (pp. 1-6). IEEE. doi:10.1109/WCNC55385.2023.10118888

Jabla, R., Khemaja, M., Buendia, F., & Faiz, S. (2022). A knowledge-driven activity recognition framework for learning unknown activities. *Procedia Computer Science*, *207*, 1871–1880. doi:10.1016/j.procs.2022.09.245

Jiang, Y., Yang, G., Li, H., & Zhang, T. (2023). Knowledge driven approach for smart bridge maintenance using big data mining. *Automation in Construction*, *146*, 104673. doi:10.1016/j.autcon.2022.104673

Joshi, V., & Biswakarma, P. (2023). Bivariate statistical method and knowledge driven heuristic approach for the comparison of landslide susceptibility mapping in the Alaknanda Catchment area. *Indian Journal of Geosciences*, *77*(1), 71–106.

Li, J., Zhou, G., & Zhang, C. (2022). A twin data and knowledge-driven intelligent process planning framework of aviation parts. *International Journal of Production Research*, *60*(17), 5217–5234. doi:10.1080/00207543.2021.1951869

Özdemir, A. (2022). Data-And Knowledge-Driven Smart City Strategies: Research On Implementation Challenges Of Local Governments In Turkey. *Journal of Management and Economics Research*, *20*(3), 152–169.

Pant, G., Bahuguna, R., Pandey, S., Gehlot, A., Yadav, S. P., & Pachauri, R. K. (2023, April). Intelligent Automated Interference for the Protection of Road Safety. In *2023 International Conference on Computational Intelligence, Communication Technology and Networking (CICTN)* (pp. 87-91). IEEE. 10.1109/CICTN57981.2023.10141086

Parveen, H., Rizvi, S. W. A., & Shukla, P. K. (2023). Disease risk level prediction based on knowledge driven optimized deep ensemble framework. *Biomedical Signal Processing and Control*, *79*, 103991. doi:10.1016/j.bspc.2022.103991

Rødven, R., & Wilson, S. (2022). Combining Knowledge for a Sustainable Arctic–AMAP Cases as Knowledge Driven Science-Policy Interactions. In *Building Common Interests in the Arctic Ocean with Global Inclusion* (Vol. 2, pp. 137–150). Springer International Publishing. doi:10.1007/978-3-030-89312-5_9

Salama, R., Al-Turjman, F., Bhatia, S., & Yadav, S. P. (2023, April). Social engineering attack types and prevention techniques-A survey. In *2023 International Conference on Computational Intelligence, Communication Technology and Networking (CICTN)* (pp. 817-820). IEEE. 10.1109/CICTN57981.2023.10140957

Taktak, H., Ghedira, C., Boukadi, K., Mrissa, M., & Gargouri, F. (2022). *Knowledge-Driven Automatic Solution for Service Composition based on ELECTRE III and Quality-Aware Service Selection for the Wildfire Prediction.*

Teng, F., Zhu, Y., Zhang, E., Hu, X., Sun, Q., & Feng, L. (2023, July). A knowledge service framework for fault diagnosis of low-earth orbit satellite constellation. In *2023 IEEE International Conference on Web Services (ICWS)* (pp. 669-676). IEEE. 10.1109/ICWS60048.2023.00085

Wu, P., Wang, P., Chi, H. L., Zhong, Y., & Song, Y. (2022). Exploring Factors Affecting Transport Infrastructure Performance: Data-Driven Versus Knowledge-Driven Approaches. *IEEE Transactions on Intelligent Transportation Systems*, *23*(12), 24714–24726. doi:10.1109/TITS.2022.3192895

Xiong, B., Jin, Y., Li, F., Chen, Y., Zou, Y., & Zhou, Z. (2023). Knowledge-driven inference for automatic reconstruction of indoor detailed as-built BIMs from laser scanning data. *Automation in Construction*, *156*, 105097. doi:10.1016/j.autcon.2023.105097

Yu, G., Lin, D., Wang, Y., Hu, M., Sugumaran, V., & Chen, J. (2023). Digital Twin-enabled and Knowledge-driven decision support for tunnel electromechanical equipment maintenance. *Tunnelling and Underground Space Technology*, *140*, 105318. doi:10.1016/j.tust.2023.105318

Zhang, W., Guo, S., Zhang, P., Xia, Z., Zhang, X., Lin, C., Tang, P., Fang, H., & Du, P. (2023). A novel knowledge-driven automated solution for high-resolution cropland extraction by cross-scale sample transfer. *IEEE Transactions on Geoscience and Remote Sensing*, *61*, 1–16. doi:10.1109/TGRS.2023.3299956

Zhou, X., Kouzel, M., & Alemzadeh, H. (2022, June). Robustness testing of data and knowledge driven anomaly detection in cyber-physical systems. In *2022 52nd Annual IEEE/IFIP International Conference on Dependable Systems and Networks Workshops (DSN-W)* (pp. 44-51). IEEE. 10.1109/DSN-W54100.2022.00017

Chapter 16
Enhancing Earthquake Prediction With Reinforcement Learning

Lalitha S. D.
Department of Computer Science and Engineering, R.M.K. Engineering College, Kavaraipettai, India

Madiajagan M.
School of Computer Science and Engineering, Vellore Institute of Technology, Vellore, India

Rajakumari S.
https://orcid.org/0000-0003-4638-4612
Department of Mathematics, R.M.D. Engineering College, India

Manikandan R.
https://orcid.org/0000-0002-2055-6634
Department of Electronics & Communication Engineering, Panimalar Engineering College, India

ABSTRACT

This chapter examines the capacity of making use of reinforcement mastering (RL) fashions to earthquake prediction. RL is a branch of system studying in which an agent learns to achieve better rewards by using iteratively to enhance its policy, which is a mapping from states to actions. The version makes use of seismic recordings to discover ways to distinguish among massive and small earthquakes. It then builds a policy that rewards large earthquakes when predicting and penalizes smaller ones. This version has the potential to improve present earthquake prediction algorithms by supplying extra accurate forecasting of future earthquakes. furthermore, the RL model may want to provide additional perception into seismicity by figuring out styles that would permit for greater focused prediction and alert techniques. ultimately, using RL may want to assist seismologists better plan evacuation routes and allocate assets in order to reduce losses because of earthqakes.

DOI: 10.4018/979-8-3693-1850-8.ch016

1. INTRODUCTION

Earthquakes are one of nature's most devastating events and may cause significant destruction of lifestyles and belongings. To expect an event of this significance is a tough challenge, one which scientists and researchers have labored on for many years. However, the field of Earthquake prediction has seen tremendous development during the last few years (Kourehpaz & Molina Hutt, 2022).

Reinforcement getting to know (RL) is a developing place of synthetic Intelligence (AI) research that has currently been implemented in the earthquake prediction process. By means of leveraging the power of AI through deep getting-to-know algorithms, RL gives a strategy to a number of the inherent challenges confronted by traditional techniques. It's a promising methodology that can offer a miles more accurate view of the underlying seismic pattern and allow the advent of more advanced earthquake forecasting structures (Hu, Zhu, Alam et al, 2022).

This essay will pay attention to how Reinforcement learning may be used to decorate earthquake prediction. We are able to discuss the various components of the earthquake forecasting technique and the ability packages of RL for improving the accuracy of predictions. We can additionally talk about the advantages and challenges of this approach and keep in mind the opportunities for destiny research and improvement (Kazemi et al., 2023; Thaler et al., 2022).

1.1 Background and Significance

Earthquakes are an herbal danger that may have tremendous implications for human safety, lives, and monetary prosperity. Predicting the probability and magnitude of an upcoming earthquake is a crucial but challenging clinical aim. Conventional strategies of earthquake prediction have usually relied on observational proof and physical models of the Earth's shape and tectonic tactics (Zhang et al., 2022). However, these strategies can frequently fail to offer correct predictions due to the complexities of the underlying bodily mechanism and the lack of reliable data.

Reinforcement getting to know (RL) is a form of gadget gaining knowledge of (ML) technology that has been validated to be effective in tackling complex predictive issues. In this approach, an AI agent learns from interacting with its environment (in this situation, the Earth's floor) and slowly adapts its prediction techniques to mirror the expertise of its surroundings (Sajan et al., 2023; Zhu et al., 2023). RL has been used for superb achievement in a spread of predictive demanding situations, starting from predicting the climate to forecasting stock market costs.

1.2 Objectives of the Chapter

This chapter seeks to discover how Reinforcement gaining knowledge of (RL) may be used to decorate earthquake predictions. The main objective of the bankruptcy is to take a look at how RL can be utilized to improve the predictive accuracy of seismic records, in general, through the application of agent-based, profound knowledge of algorithms on facts units of seismic events (Chittora et al., 2022). especially this chapter objectives to:

1. Analyze the contemporary development and limitations of the conventional methods of seismic modeling in accomplishing accuracy in earthquake prediction.

2. Discover and determine the implementation of deep studying algorithms and their associated reinforcement learning strategies in seismic modeling.
3. Examine the ability blessings and demanding situations posed by the broader incorporation of RL into the seismic discipline.
4. Offer professional evaluations and sensible recommendations on integrating RL into cutting-edge seismic modeling approaches.
5. Establish a robust framework for developing an RL-pushed optimized seismic version (Hu, Zhu, & Wang, 2022; Kourehpaz & Molina Hutt, 2022).

2. FUNDAMENTALS OF EARTHQUAKE PREDICTION

Reinforcement mastering is turning into a more and more popular approach for enhancing earthquake prediction. The basic concept of this method is to apply system learning algorithms to analyze past events and use the statistics to alter predictions over the years. Thru reinforcement, scientists can use statistics from past earthquakes to create simulations of destiny occasions and develop strategies for dealing with the dangers of a catastrophic earthquake (Chen et al., 2022; Mousavi et al., 2022).

As a way to achieve success, reinforcement needs to be primarily based on comprehensive information on the essential technology of earthquakes. Earthquakes are generated with the aid of the speedy movement of the Earth's tectonic plates, which can bring about intense vibrations in certain regions. Those vibrations destroy the encircling areas and can be pretty negative to close-by infrastructure. So, as for the reinforcement studying algorithm to increase a correct prediction model, the scientist desires to recognize the mechanisms in the back of the earthquake (Gao et al., 2022).

An excellent way to predict the timing, significance, and effects of earthquakes, scientists need to investigate statistics from previous events. This information consists of seismic waveforms, focal mechanisms of faulting, earthquake precursors, and the location of fault zones. Through looking at information from beyond earthquakes, scientist can then try and broaden predictive models for destiny earthquakes (Anikiev et al., 2023; Zhang et al., 2023).

2.1 Overview of Seismic Activity

Earthquakes are unpredictable natural screw-ups with potentially catastrophic consequences – they could motivate tremendous damage to infrastructure and lives, particularly in incredibly urbanized regions. As such, predicting when and wherein an earthquake would possibly arise is of paramount importance to the protection and security of each human being and property. Reinforcement gaining knowledge is a type of synthetic intelligence (AI) that has, in recent years, been applied to information and predicting seismic hobby (Kazemi & Jankowski, 2023). This rising area of research seeks to apply algorithms and techniques from gadget studying and deep learning to the seismic hobby with the intention of improving earthquake prediction.

Reinforcement gaining knowledge is distinguished from other AI techniques in that it employs a device of rewards and penalties to modify its behavior through the years — it learns over the direction of numerous interactions with its surroundings. This precise technique of learning makes it especially suited to predicting seismic hobby, as seismic facts are frequently tricky to interpret. The physical environment may be tremendously unpredictable. By way of growing a "virtual environment" that simulates the actual international, reinforcement gaining knowledge may be used to pick out styles in seismic pastimes and are expecting destiny earthquakes (Khan et al., 2022; Luo & Paal, 2022).

2.2 Traditional Approaches to Earthquake Prediction

With technological advancements and improved scientific understanding, the sector of earthquake prediction has persisted in development through the years. Advancements in facts-driven machine mastering techniques have made earthquake prediction more accurate and reliable than ever earlier. Even as gadgets getting to know algorithms undoubtedly provide a bonus for earthquake prediction, they are now not without obstacles. Traditional procedures, along with seismological networks and physical models, need not be forgotten (Shi et al., 2022). These methods provide a reliable baseline for predicting earthquakes and, when supplemented with gadget studying, provide an excellent, more complete prediction model.

Seismological networks contain bodily units, inclusive of seismographs and accelerometers, that measure distinct sorts of seismic waves related to earthquakes. These networks allow for a much more specified knowledge of the form of seismic hobby-taking vicinity, consisting of the depth, place, and magnitude of an earthquake (Liu et al., 2023). These records may be used as input for a machine-mastering model, supporting the presentation of a more accurate prediction.

Bodily fashions can also provide a perception of potential earthquake behavior. Examples of such fashions encompass simulations of tectonic plates, seismic faults, and Earth's crust. Every model is designed to simulate the behavior of the specific kind of geological hobby being studied. With the aid of knowledge of the conduct of these structures, forecasts can be made primarily based on the historical data of the place, assisting in offering an estimate of the earthquake intensity and place (Jo et al., 2022; Murti et al., 2022).

2.3 Challenges in Accurate Prediction

The earth is constantly shifting and converting due to the tectonic plates beneath the floor. Earthquakes are an herbal phenomenon that causes devastating harm to lives and property. Earthquakes can cause steeply-priced destruction and loss of human lives if not expected accurately in time. In current years, device mastering (ML) and, particularly, Reinforcement studying (RL) are being applied to broaden earthquake forecasting structures that may allow more particular predictions. While RL fashions have a few potentials to enhance the accuracy of earthquake prediction, there are indeed demanding situations that need to be addressed for this to be a viable and dependable prediction system (Hu, Qiu, & Zhu, 2022; Srivastava et al., 2023).

One of the primary challenges with developing accurate predictions for earthquakes is the unpredictability of those herbal disasters. Earthquakes arise randomly at unknown places, which means that it is impossible, or as a minimum, most unlikely, to increase a correct prediction set of rules relying on traditional facts units, together with seismic information accumulated from the ground. In addition, geological fashions of seismic interest contain giant uncertainty (Salama et al., 2023). RL dealers need reliable records to make knowledgeable selections, and the records necessities of RL for this utility are an awful lot greater than different applications, making it challenging to expand accurate fashions.

3. INTRODUCTION TO REINFORCEMENT LEARNING

Earthquakes are an unpredictable natural phenomenon, and understanding the patterns in which they arise is an interdisciplinary research hassle. In the past, seismologists have used statistical methods to attempt to forecast and model seismic activity. But those strategies have their obstacles, and their predictive capabilities are frequently uncertain due to the complicated nature of the seismology field. As a result, new strategies are being explored to cope with this grand mission. One such technique is to apply reinforcement to decorate earthquake prediction.

Reinforcement is a branch of synthetic intelligence that permits sellers to interact with their environment, after which takes movements primarily based on the results of that interaction. This mastering approach is attractive as it could be used to optimize a selection-making process. It has been efficaciously applied in numerous exclusive fields consisting of robotics, sports playing, and financial buying and selling. In the context of earthquake prediction, reinforcement learning can be used to research seismic information, which includes historic earthquake styles, seismic activity, and floor shaking. This information can then be used to construct a model of the environment and create relationships between different factors, which can be used to inform predictions of when and where an earthquake could arise.

3.1 Basic Principles of Reinforcement Learning

Reinforcement getting to know (RL) is an area of synthetic intelligence (AI) that seeks to create excellent behavior inside the agent (the learner) by looking to maximize the total praise earned through a chain of choices. It's used to deal with numerous practical obligations, which include predicting climate, controlling robots, schooling neural networks, and playing games, all of which are critical additives in earthquake prediction. In this essay, we will talk about the key components of RL for an earthquake prediction system: marketers, environments, and rewards.

First, sellers play an essential position in RL for earthquake prediction. Sellers are "newcomers" who obtain inputs from their surroundings and take moves as a way to maximize their rewards. Within the case of earthquake prediction, dealers use information from a couple of sources, which include seismometers, accelerometers, and GPS sensors, to look at their environment and make knowledgeable selections about

Figure 1. Basic principles of reinforcement learning

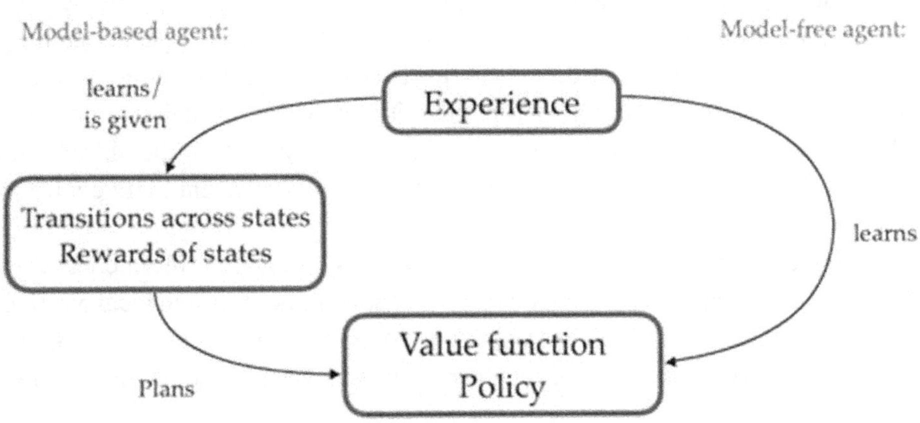

their actions. Agents should additionally be capable of determining the opportunity of an earthquake taking place and use this information to decide on an appropriate route of action.

Second, environments are a vital thing of any RL gadget. For earthquake prediction, the environment includes the facts collected from the sensors that the sellers use to study the environment and make decisions. The surroundings have to offer the sellers a sufficient variety of information points and values in order to allow accurate prediction.

3.2 Key Components: Agents, Environments, and Rewards

Earthquakes are one of the most devastating herbal disasters, and predicting them has been a venture for many years. While there are a lot of studies being finished in the region of earthquake prediction, it is nevertheless a complex project. As such, current advances in the system-mastering era have enabled scientists to discover new ways to improve prediction accuracy. Exceptionally, reinforcement getting to know (RL) has turned out to be an increasing number of famous for this cause because it allows for the exploitation of comments from the surroundings.

Reinforcement mastering is a machine gaining knowledge of a set of rules that uses rewards and punishments to navigate its environment. This means that RL algorithms can research from their moves and adapt to the dynamic surroundings. Because of its capacity to research from revel in and feedback, RL has been carried out to a selection of domain names with varying stages of success.

Mainly, RL has been used to enhance earthquake prediction by imparting predictive models with the capacity to study immediately from seismic information. It has been used in seismic statistics prediction in three essential approaches. The primary is through applying RL dealers to seismic records that have been used to locate patterns in the facts.

4. REINFORCEMENT LEARNING MODELS FOR EARTHQUAKE PREDICTION

Earthquakes are frequently unpredictable and devastating, and hence, accurate and reliable forecasting is of vital importance. However, because of the complicated nature of earthquakes, accurate forecasting continues to take time to acquire. In recent years, diverse fashions have been developed to improve the prediction of seismic activities. On this admire, Reinforcement getting to know (RL) offers a promising strategy to beautify the accuracy of Earthquake prediction.

Reinforcement-gaining knowledge includes an agent that is responsible for making decisions inside surroundings based totally on its beyond studies. It's miles a leading-aspect gadget gaining knowledge of approach that empowers innovative structures to discover ways to act in complex situations without direct programming. This adaptive mastering functionality allows RL models to understand better the complexities that regularly accompany predicting seismic occasions. With the aid of learning how to expect activities and interpret environment adjustments, RL gives the agent extra control over the surroundings and a better ability to forecast activities.

By means of using RL models in the venture of Earthquake prediction, researchers can incorporate one-of-a-kind surrounding inputs, which include records of seismic activity and local weather conditions. Via considering the interaction between the environment and the agent's selection-making manner, the RL version can discover ways to understand the correlations between these distinct inputs and seismic interest.

Figure 2. Chart of earthquake prediction using machine learning

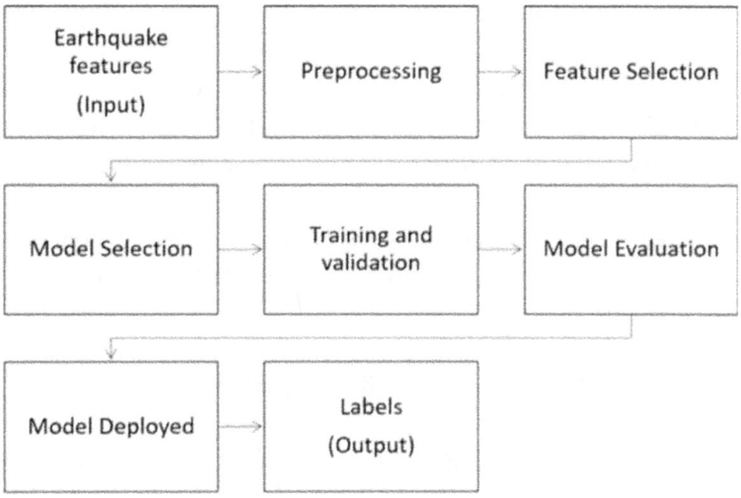

4.1 Designing RL Models for Seismic Data

Seismic records have been used for decades to beautify earthquake predictions and warnings to offer early warnings to the general public. However, conventional seismic statistics frequently need to offer an accurate and dependable assessment of the threat of an imminent earthquake. With the advancement of artificial intelligence (AI) and the developing use of reinforcement, gaining knowledge of (RL) in seismic statistics can provide even greater accuracy and reliability for earthquake prediction.

Reinforcement gaining knowledge of (RL) is a place of AI that makes a specialty of teaching machines to make choices in unpredictable surroundings. It is based on rewards and punishments, where the device learns from previous interactions and adjusts its behavior for this reason. Because seismic records are very complicated and frequently contain massive amounts of noise, RL models are uniquely suitable to perceive styles and make dependable predictions from the facts.

While designing RL models for seismic records to enhance earthquake predictions, builders want to remember several crucial additives of seismic facts. First, seismic activities frequently include many different forms of alerts, which include seismic vibrations, electromagnetic radiation, and acoustic waves. Each signal incorporates extraordinary traits and styles that need to be recognized and understood to make dependable predictions. Additionally, seismic information often incorporates large quantities of noise, which must be filtered out before the records may be used for prediction.

4.2 Training Processes and Learning Strategies

Earthquake prediction is a topic of increasing significance in the geosciences. Although the exact reason for earthquakes continues to be uncertain, positive styles and indicators have been detected that could help us expect an earthquake to occur. As generation advances, so too do our capabilities in earthquake prediction, and this could be, in addition, superior by the usage of reinforcement learning.

Reinforcement gaining knowledge is a device getting to know that specializes in studying to remedy troubles through trial and mistake. It allows fashions to research through interactions with their surroundings on the way to optimize certain behaviors. This kind of mastering can be applicable to earthquake prediction because of its capacity to make use of ancient statistics to develop fashions that can then be improved upon over the years.

For a hit implementation of reinforcement and gaining knowledge of earthquake prediction, it's essential to have a proper training procedure and studying strategies in place. The schooling procedure has to encompass four foremost stages: statistics series, feature engineering, version design, and model assessment.

Information series is the first and most crucial step in the education procedure as it gives the fashions with the information they want to research. Information series need to be conscious of geographical and seismic statistics associated with earthquakes.

4.3 Comparison With Traditional Prediction Models

Earthquakes cause thousands and thousands of bucks of damage each year around the arena, and predicting seismic interest may be crucial to mitigation efforts. Predictive models that correctly predict earthquakes can save lives and assets and save you extra destruction. Traditional predictive models, which include statistical analysis and environmental monitoring, were used for decades to predict earthquakes, but they are restrained in their accuracy. Reinforcement studying (RL) has emerged as a promising technique for reinforcing earthquake prediction accuracy.

Conventional predictive fashions depend on statistics, which include seismic waveforms, GPS readings, topographical statistics, or geophysical houses. The fashions use this information to pick out patterns, which are then used to predict future events. This technique may be a hit in predicting earthquakes, in particular for brief-time period forecasting. However, traditional fashions may be restricted in their accuracy and are much more likely to fail if the factors that affect seismic hobby alternate in the future.

Reinforcement gaining knowledge is primarily based on a synthetic-intelligence approach that learns through trial and error. With RL, the AI "agent" is exposed to specific eventualities and learns the exceptional strategies to remedy the one's eventualities.

5. DATA ACQUISITION AND PREPROCESSING

Data acquisition and preprocessing are crucial elements of any medical studies mission. That is mainly for the mission of enhancing earthquake prediction with reinforcement gaining knowledge. In this essay, we will talk about the methods behind records acquisition and preprocessing for this precise undertaking.

Step one in acquiring data for this task is the observation of seismic hobby. This consists of gathering information on shaking depth, local seismography, and ground movement patterns. This information can be received through the usage of seismometers to measure ground movement or seismographs to decide shaking intensity. Both gadgets are capable of providing specific data on the energy, period, and prevalence patterns of seismic pastimes.

Once the records have been accrued, they need to be processed for you to offer helpful input for the reinforcement studying model. This usually involves preprocessing the records so one can make sure the proper features are extracted and the essential records are highlighted. This could involve strategies including characteristic choice, characteristic extraction, and characteristic engineering.

Characteristic selection involves selecting the maximum relevant functions to be used inside the model. This may include selecting the most helpful indicators of seismic activity or the most crucial styles of ground motion. Function extraction includes extracting additional statistics from the statistics that may be beneficial for the version.

5.1 Seismic Data Sources and Characteristics

Seismic facts, along with seismic waves generated by earthquakes, can be used to gain perception into the procedures of an earthquake and even its capability damaging forces. Seismic statistics is a valuable source of statistics for boosting earthquake prediction with reinforcement gaining knowledge of (RL). By using learning from beyond seismic records, the RL agent can provide predictions and manipulate selections in the face of an earthquake.

Seismic facts are typically accumulated via the use of seismometers, which detect and measure the speed or acceleration resulting from seismic waves. Distinct forms of seismic statistics can be generated with the aid of earthquakes. The P-wave is the primary wave detected and is a quick-shifting compressional wave of strength that typically arrives first at the seismometer. Then comes the S-wave, which is a slower but more tremendous unfavorable shear wave that produces horizontal and vertical actions within the ground. The floor waves are the last waves to tour the earth and may cause construction damage, with number one and secondary floor waves being, respectively, the longest and most detrimental.

For RL packages, the seismic records need to be of high pleasant and ideally in actual time. The characterizing capabilities of these statistics may be helpful in forecasting and assessing the severity of earthquakes.

5.2 Reinforcement Learning and Multimodal Data

The prediction of herbal screw-ups, along with earthquakes, has come to be an increasingly challenging assignment through the years because of the complexity of the phenomenon and the restricted quantity of related determined statistics. To address this mission and enhance natural disaster prediction accuracy, using superior system getting-to-know strategies and techniques, inclusive of reinforcement learning mixed with multimodal statistics evaluation, has grown to be vital.

Reinforcement studying is a machine learning process where an agent interacts with its environment, receiving rewards and punishments as feedback. Based totally on that feedback, the agent learns to do so in an excellent way to maximize the reward and limit the punishment. This program has become a handy tool in numerous fields, and making use of reinforcement and gaining knowledge of techniques for earthquake prediction has gained popularity as one of the most promising techniques.

Reinforcement gaining knowledge may be used to expectexpect earthquakes by harnessing the strength of both late and early warning structures. With past-due caution structures, the input data is used to simulate specific eventualities and models that may be used to expect the capacity risks of a specific occasion; at the same time, early caution systems leverage massive-scale bodily system models in aggregate with historical facts to predict future earthquakes earlier than tremors occur. As both systems have their limits, the mixture of reinforcement mastering algorithms and multimodal data evaluation can better simulate complex interactions and processes, generating higher prediction accuracy.

5.3 Preprocessing Techniques for RL in Earthquake Prediction

Earthquakes are a natural phenomenon that poses a remarkable danger to human protection. As such, appropriately predicting the occurrence of earthquakes permits the implementation of warning structures that assist in minimizing the damage as a result of those events. Reinforcement learning (RL) is a powerful device in artificial intelligence that has been confirmed to be decisive for earthquake prediction. However, in spite of its ability, there are sure preprocessing techniques that can be carried out to beautify the accuracy of the predicted consequences. This essay will talk about diverse preprocessing strategies that can be used to enhance the accuracy of earthquake prediction with the aid of using RL.

The first step in the preprocessing manner is function extraction. Function extraction is a commonplace practice used to transform uncooked enter facts into a characteristic vector. This is easier to method by way of fashions, including RL. Through applying positive differences, inclusive of clustering or scaling, critical features can be extracted from the enter statistics. These extracted capabilities can then be used as input statistics for the RL model, enabling it to be greater than it should be except for earthquakes.

The following preprocessing technique is function selection. Characteristic selection is a process with the aid of which excellent capabilities are selected primarily based on their utility in predicting the outputs. Features that make contributions to better prediction accuracy ought to be given better weighting values than the ones that do not.

6. INTEGRATION WITH TRADITIONAL PREDICTION MODELS

Earthquake prediction is a critical component in the safety of lifestyles and belongings. With improvements in Reinforcement learning (RL), conventional prediction fashions can be superior to be able to more correctly and expect earthquakes. This essay will speak about the combination of RL with conventional prediction fashions to improve earthquake prediction.

Traditional prediction models are based on historic pattern reputation. They depend on studying the measured parameters of the floor for evidence of adjustments associated with beyond earthquakes. Conventional fashions on my own cannot be as they should be except for the onset of an earthquake or provide a sensible caution time. They're constrained by the historical patterns that they're capable of delivering, and this restricts the accuracy of their predictions. Additionally, those predictions regularly need to be corrected in statistics collection or version assumptions.

Reinforcement learning (RL) gives a manner to enhance the accuracy of earthquake predictions. With the aid of combining conventional fashions with RL, a more incredible entire picture of floor conditions can be created and used to improve prediction accuracy and warning time. RL enables the evaluation of a broader range of measured parameters. Gaadget gaining knowledge of algorithms, such as recurrent neural networks, are used to discover styles in the records, and the consequences are applied to the prediction version. Through supplementing conventional fashions with RL, extra comprehensive predictions may be made in shorter timeframes.

6.1 Hybrid Approaches: Combining RL With Classical Methods

Earthquake prediction has been a complex problem for scientists and engineers to resolve for many years. While advances in seismology have made seismic monitoring more correct and decreased the uncertainty associated with devastating earthquakes, no reliable approach presently exists for making unique predictions of whether and when an earthquake will strike. The unpredictability of earthquakes is a prime venture to any technique of planning or preventive action.

Reinforcement mastering (RL) is an area of artificial intelligence (AI) studies that has won a hobby in recent years as a probable tool for earthquake prediction. RL is a form of machine learning that makes use of a hard and fast of regulations and rewards to conduct an excellent way to attain desired goals manually. RL has been carried out to address earthquake prediction issues by using diverse datasets and ancient information to create predictive fashions. However, RL faces some demanding situations in acquiring correct predictions and is constrained by the amount of data to be had and the complexity of the tasks.

A new type of technique for the usage of RL for earthquake prediction has been proposed- the hybrid approach. The hybrid method combines classical techniques with RL that allows you to decorate earthquake prediction.

6.2 Ensuring Robustness and Reliability

Earthquake prediction is an essential undertaking in the modern-day world because the protection and protection of limitless human beings rests upon efficiently predicting seismic events. Notwithstanding the significance of this task, it remains one of the most onerous obligations in computational seismology, with numerous obstacles to standard predictive strategies, which include linear regression. In recent years, a new technique for earthquake prediction has emerged in the shape of reinforcement mastering, an effective gadget mastering technique that could analyze from records and beyond stories for you to make better predictions.

Reinforcement learning can offer a vital tool for earthquake prediction, as it can analyze past seismic occasions to accurately become aware of styles in the seismic records and use this know-how to predict future earthquakes better. However, no matter its promise, reinforcement has to be broadly applied to earthquake prediction because of the want for robustness and reliability for you to make accurate predictions.

One of the critical techniques for ensuring robustness and reliability in reinforcement mastering is to use a spread of fashions, which includes neural networks, selection bushes, and aid Vector Machines. By the usage of more than one fashions in tandem, robustness and reliability can be ensured, as any discrepancies among the fashions can be observed and accounted for.

6.3 Synergies and Complementary Roles

Earthquake prediction is an essential thing for imparting superior safety warnings to prone populations placed in high-threat seismically lively zones. Traditional models and physics-based techniques utilized in earthquake prediction have addressed many essential triggers associated with huge seismic occasions. Yet, inadequate sensitivity and determination save you from accurate captures of the pretty localized nature of seismicity. As such, the efficacy of cutting-edge public safety techniques is often constrained because of a need for steerage from designated predictions, which could help planning efforts. Fortunately, reinforcement mastering (RL) offers the ability to provide novel answers to improve our understanding of the complexities related to seismic approaches. It is more and more being identified as an attractive and promising surrogate to traditional techniques. Through leveraging the synergy between RL and physics-primarily based processes, the promise of correct earthquake prediction may be enhanced, taking into account the assessment of new techniques and regulations with the intention to, in the long run, improve public safety and threat management. Earthquake prediction is a crucial component for providing superior safety warnings to susceptible populations located in excessive-risk seismically active zones. Conventional versions and physics-based tactics utilized in earthquake prediction have addressed many fundamental triggers associated with significant seismic events. Yet, inadequate sensitivity and resolution save you from accurate captures of the exceptionally localized nature of seismicity. As such, the efficacy of present-day public protection techniques is often limited due to a need for steerage from specific predictions that can help make plan efforts. Happily, reinforcement learning (RL) gives the capacity to provide novel answers to improve our information on the complexities associated with seismic techniques. It is more and more being identified as an appealing and promising surrogate to traditional strategies. By leveraging the synergy among RL and physics-based processes, the promise of correct earthquake prediction can be improved, bearing in mind the evaluation of the latest techniques and policies as a way to improve public protection and chance control.

RL—-based totally earthquake prediction fashions provide promising new competencies closer to seismic threat assessment. While mixed with physics-based total reviews, it's miles viable to reap localized and precise insights to model seismic processes that are, in any other case, intractable by the use of existing methods.

RL–primarily based earthquake prediction models offer promising new capabilities toward seismic threat assessment. While mixed with physics-primarily based reviews, it is feasible to reap localized and specific insights to version seismic procedures, which might be in any other case intractable the use of existing techniques.

7. ETHICAL CONSIDERATIONS IN RL FOR EARTHQUAKE PREDICTION

Earthquake prediction is a crucial area of research and development that requires deep attention to ethical implications. As researchers and engineers try to expand powerful devices, gaining knowledge of systems to assume seismic events, usually retaining the consequences in their paintings in thought, will be of paramount significance.

The capability implications of using Reinforcement getting to know (RL) for earthquake prediction are considerable. RL offers an efficient and effective manner of getting to know, permitting machines to examine and adapt to their environment in a more excellent natural way than conventional strategies. This knowledge occurs thru trial-and-mistakes techniques, utilizing remarks from the environment to alter the device's conduct and capabilities continuously. It is this progressive capability that makes RL appealing for earthquake prediction, as herbal disaster events offer valuable, consistent remarks.

While drawing close RL programs for earthquake prediction, numerous ethical concerns must be taken into account. First and foremost, it's essential to make sure of the privateness of both the public and information subjects. With sophisticated gadget mastering systems getting used to decide the capability for seismic events, restricting entry to the records used and making sure a relaxed environment for its utilization is paramount. It is also essential for device designers and developers to explore methods for implementation that can be ethically responsible, even as additionally spotting the capability for abuse.

7.1 Transparency and Accountability

Earthquakes are a powerful pressure that can cause gigantic destruction. At the same time as technological know-how and technology have progressed, our capacity to are expecting where and while they may arise, there is nonetheless room to be progressed. Reinforcement learning (RL) presents a new method of earthquake prediction that can be reliable and green. However, in order for RL to be used for earthquake prediction, transparency and accountability must be present so as to ensure its accuracy and reliability.

Transparency is prime for the successful application of RL to earthquake prediction. With the aid of supplying transparent entry to to records and modeling strategies, researchers and different stakeholders can make sure of consistent analytics and trust that the predictions given by RL are dependable. This will additionally allow those stakeholders to make use of the overall variety of RL's competencies, which include figuring out capability earthquakes based totally on each modern and beyond records. Transparency also offers stakeholders the records they need to make sure that any newly developed model plays as predicted.

Duty is every other significant aspect that desires to be taken under consideration for a hit RL-based earthquake prediction. To make sure that the model is performing nicely, it desires to be tested and confirmed via rigorous research. Moreover, businesses growing the RL-based total models need to be held responsible for any mistakes or ineffectiveness.

7.2 Bias Mitigation in Reinforcement Learning Models

Reinforcement getting to know is an AI approach that leverages an ordeal-and-error approach of selection-making to attain a purpose. It is used to resolve complicated issues where the solution isn't apparent without delay. It has seen increasing use in areas consisting of robotics, natural language processing, and earthquake prediction. However, as with all machine learning algorithms, there is an ability for bias inside the selection-making method. Happily, there are techniques for mitigating bias in reinforcement getting to know fashions, such as up-hoc evaluation and data augmentation.

publish-hoc evaluation entails looking at the outcomes of a reinforcement, gaining knowledge of the gadget, and checking if any traits within the facts may also suggest a bias. This can be carried out with the aid of identifying record styles, which could cause a more excellent heterogeneous set of parameters, which include checking for different values for precise attributes or patterns of reinforcement rewards. If an unwanted bias is detected, it could be addressed by tweaking the model's parameters or by converting the representation of the information.

Statistics augmentation is another approach used to reduce bias. This includes generating more facts and factors that are similar to the ones within the current dataset but not exactly duplicates.

7.3 Societal Impacts and Public Awareness

Earthquakes are a natural phenomenon, yet they can cause giant destruction and worry. Consequently, research into new strategies for predicting earthquakes has intensified over the years, with the ultimate aim of identifying high-hazard regions that need to be better prepared. One such approach is to apply reinforcement studying (RL), which is a type of system learning that seeks to maximize the expected lengthy period of praise by way of reinforcing applicable conduct and punishing undesirable conduct. This paper will explain what RL is and the way it may be used to beautify earthquake prediction before discussing the societal impacts and public consciousness that can be accomplished by way of adopting this generation.

RL is a device mastering that can improve the accuracy of conventional predictive strategies. It utilizes computational ideas to enhance the prevailing rule-based predictive strategies and allows for scalability and value advantages. It really works by way of forming an agent, which is an AI-based predictor that could search for environmental states given specific criteria and can analyze from experience to replace its predictive version. RL can permit extra accurate forecasting of various parameters along with the earthquake's significance, location, and time. Furthermore, it can account for elements along with aftershocks, which conventional strategies do not typically don't forget.

8. CHALLENGES AND FUTURE DIRECTIONS

Earthquake prediction is a complex hassle that has been studied for centuries yet remains largely unsolved. It's a challenging and complex trouble for seismologists due to a lack of statistics and restrained information on the techniques that generate earthquakes. Current advances in artificial intelligence and gadget mastering have enabled the development of recent methods to apprehend and predict earthquakes. Mainly, reinforcement gaining knowledge of (RL) techniques has shown promise as a means to improve forecasting overall performance while additionally supporting the overcome barriers in modern-day techniques, consisting of best the usage of past seismic records and not thinking about other types of reliable facts inclusive of local geology or other non-seismic statistics.

Reinforcement gaining knowledge of (RL) is an area of research in artificial intelligence that uses facts from beyond selections to guide destiny choices. It's been applied to various troubles, along with robotics, autonomous dealers, and tactics automation. In the context of earthquake prediction, RL lets us construct models with the purpose of using past seismicity records and other reliable local records to develop fashions that could make better predictions for destiny seismic occasions. These fashions may be used to improve the accuracy and reliability of earthquake forecasts.

8.1 Technical Challenges and Ongoing Research

Earthquakes are powerful herbal phenomena that pose considerable dangers to people and their infrastructure, and as such, predicting them appropriately and reliably is a critical study topic in seismology. Even as development has been made in developing methods for predicting earthquakes, there are ongoing studies demanding situations with the use of reinforcement mastering (RL) fashions for this motive.

Reinforcement gaining knowledge is a machine studying technique advanced to research from interacting with its environment, making choices in an effort to obtain a predetermined aim. While applied to earthquake prediction, RL may be used to create a model that dynamically debts for numerous factors associated with the behavior of earthquakes. Such factors may additionally consist of seismic wave characteristics like frequency and amplitude, temporal patterns, and feature seismic behavior in a specific location. By using education, the RL model with such statistics could be used to research styles within the records that permit extra correct and dependable predictions of future earthquakes.

There are still, however, many technological challenges that should be addressed so as to beautify the accuracy and reliability of the predictions. In the meantime, it takes a lot of work to decide the appropriate timing of an anticipated earthquake. Moreover, it is also challenging to become aware of elements that could lead to an extra significance earthquake.

8.2 Scaling RL Models for Global Predictions

The latest earthquake in the Mediterranean Sea has highlighted the need to expect when and in which earthquakes will arise so that it will minimize the devastating effects of this natural catastrophe. Reinforcement getting to know (RL) is one of the most promising synthetic Intelligence (AI) techniques for predicting earthquakes. Instead of counting on static fashions, RL is a way of growing models that may study from their environment and adapt to changing scenarios. This makes it ideal for predicting earthquakes, as those activities are frequently unpredictable and exchange swiftly.

RL fashions for international prediction systems involve incorporating records from worldwide seismic events, allowing for the possibility of superior predictions. Those fashions can be scaled up to perceive styles in global seismic records and analyze them more efficaciously. This could help boost the accuracy of earthquake predictions and permit coverage-makers to make extra knowledgeable decisions. The scaling of RL fashions is only when blended with psychological and social modeling strategies to gain information on the complicated environment that affects seismic hazard. This can allow for the incorporation of cultural, financial, and environmental elements into the modeling device, which can help gain a higher expertise of the general seismic hazard.

8.3 Collaborative Initiatives and International Partnerships

In recent years, collaborative tasks and global partnerships have been essential for advancing the sphere of earthquake prediction with reinforcement learning. From growing new algorithms and systems to enhancing infrastructure and database dissemination, these projects have played an essential function in supporting and promoting the development of earthquake prediction technology.

One of the most hit initiatives changed into the advent of the Pacific Alliance for Earthquake Resilience (PAER) in 2016. This collaboration among the governments of the U.S., Japan, and Mexico seeks to "boost the use of facts, technology, and equipment to permit seamless admission to and analysis of geophysical, geological, and seismic records to develop more effective earthquake predictions." The mission focuses explicitly on growing and incorporating advanced algorithms and systems for extra-correct earthquake warnings. In 2018, PAER completed its first successful prototype set of rules, showcased at the first assembly of the personal sector Innovation discussion board, to an awful lot of acclaim.

In addition, in 2018, the ECU-funded FUSE-DL assignment was launched. This initiative sought to use modern-day improvements in deep getting-to-know technology in the sector of earthquake prediction. Expert researchers from the College of East Anglia and the College of Pisa evolved a predictive machine called "Earthquake Dynamics for Early Caution Systems."

9. CONCLUSION

Earthquake prediction is a tremendously complex challenge, and researchers have long been looking for methods to as it should expect the occurrence of earthquakes. Reinforcement learning is a promising device for improving the accuracy and timeliness of earthquake prediction. Reinforcement gaining knowledge has been applied to seismological information for numerous years, with the goal of increasing the accuracy and timeliness of earthquake predictions. Reinforcement studying algorithms have been used to analyze seismological data to determine expected future earthquakes with promising outcomes.

The intention of the usage of reinforcement gaining knowledge to predict earthquakes is to make predictions quicker and more accurate. With the aid of the usage of reinforcement getting to know, seismologists can alter their models according to the records they acquire, reducing the time to collect and analyze statistics to make more correct predictions. The primary advantage of reinforcement studying comes right down to the fact that it could deliver seismologists the capability to act right now on a predicted seismic occasion.

In conclusion, reinforcement learning has proven to be an effective device for improving the accuracy and timeliness of earthquake prediction. That is due to its capability to quickly and correctly analyze seismological data and modify its predictions based on the cutting-edge kingdom of the data. As research and development remain in this field, reinforcement studying will continue to play a prominent position in seismology and earthquake prediction.

REFERENCES

Anikiev, D., Birnie, C., Waheed, U., Alkhalifah, T., Gu, C., Verschuur, D. J., & Eisner, L. (2023). Machine learning in microseismic monitoring. *Earth-Science Reviews*, *239*, 104371. doi:10.1016/j.earscirev.2023.104371

Chen, S. Z., Feng, D. C., Wang, W. J., & Taciroglu, E. (2022). Probabilistic machine-learning methods for performance prediction of structure and infrastructures through natural gradient boosting. *Journal of Structural Engineering*, *148*(8), 04022096. doi:10.1061/(ASCE)ST.1943-541X.0003401

Chittora, P., Chakrabarti, T., Debnath, P., Gupta, A., Chakrabarti, P., Praveen, S. P., & Elngar, A. A. (2022). *Experimental analysis of earthquake prediction using machine learning classifiers, curve fitting, and neural modeling.*

Gao, Y., Zhang, J., Li, H., & Li, G. (2022). Incorporating structural constraint into the machine learning high-resolution seismic reconstruction. *IEEE Transactions on Geoscience and Remote Sensing*, *60*, 1–12. doi:10.1109/TGRS.2022.3157064

Hu, S., Qiu, C., & Zhu, S. (2022). Machine learning-driven performance-based seismic design of hybrid self-centering braced frames with SMA braces and viscous dampers. *Smart Materials and Structures*, *31*(10), 105024. doi:10.1088/1361-665X/ac8efc

Hu, S., Zhu, S., Alam, M. S., & Wang, W. (2022). Machine learning-aided peak and residual displacement-based design method for enhancing seismic performance of steel moment-resisting frames by installing self-centering braces. *Engineering Structures*, *271*, 114935. doi:10.1016/j.engstruct.2022.114935

Hu, S., Zhu, S., & Wang, W. (2022). Machine learning-driven probabilistic residual displacement-based design method for improving post-earthquake repairability of steel moment-resisting frames using self-centering braces. *Journal of Building Engineering*, *61*, 105225. doi:10.1016/j.jobe.2022.105225

Jo, Y., Choi, Y., Seol, S. J., & Byun, J. (2022). Machine learning-based vertical resolution enhancement considering the seismic attenuation. *Journal of Petroleum Science Engineering*, *208*, 109657. doi:10.1016/j.petrol.2021.109657

Kazemi, F., Asgarkhani, N., & Jankowski, R. (2023). Machine learning-based seismic fragility and seismic vulnerability assessment of reinforced concrete structures. *Soil Dynamics and Earthquake Engineering*, *166*, 107761. doi:10.1016/j.soildyn.2023.107761

Kazemi, F., & Jankowski, R. (2023). Machine learning-based prediction of seismic limit-state capacity of steel moment-resisting frames considering soil-structure interaction. *Computers & Structures*, *274*, 106886. doi:10.1016/j.compstruc.2022.106886

Khan, M. M., Ghaffar, B., Shahzad, R., Khan, M. R., Shah, M., Amin, A. H., Eldin, S. M., Naqvi, N. A., & Ali, R. (2022). Atmospheric anomalies associated with the 2021 M w 7.2 Haiti earthquake using machine learning from multiple satellites. *Sustainability (Basel)*, *14*(22), 14782. doi:10.3390u142214782

Kourehpaz, P., & Molina Hutt, C. (2022). Machine learning for enhanced regional seismic risk assessments. *Journal of Structural Engineering*, *148*(9), 04022126. doi:10.1061/(ASCE)ST.1943-541X.0003421

Kourehpaz, P., & Molina Hutt, C. (2022). *Machine Learning for Enhanced Seismic Damage Assessment of Buildings*.

Liu, Q. Y., Li, A. Q., Tang, X. S., & Du, W. (2023). Predictive Models for Seismic Source Parameters Based on Machine Learning and General Orthogonal Regression Approaches. *Bulletin of the Seismological Society of America*, *113*(6), 2363–2376. doi:10.1785/0120230069

Luo, H., & Paal, S. G. (2022). Artificial intelligence-enhanced seismic response prediction of reinforced concrete frames. *Advanced Engineering Informatics*, *52*, 101568. doi:10.1016/j.aei.2022.101568

Mousavi, M., TohidiFar, A., & Alvanchi, A. (2022). BIM and machine learning in seismic damage prediction for non-structural exterior infill walls. *Automation in Construction*, *139*, 104288. doi:10.1016/j.autcon.2022.104288

Murti, M. A., Junior, R., Ahmed, A. N., & Elshafie, A. (2022). Earthquake multi-classification detection based velocity and displacement data filtering using machine learning algorithms. *Scientific Reports*, *12*(1), 21200. doi:10.103841598-022-25098-1 PMID:36482200

Sajan, K. C., Bhusal, A., Gautam, D., & Rupakhety, R. (2023). Earthquake damage and rehabilitation intervention prediction using machine learning. *Engineering Failure Analysis*, *144*, 106949. doi:10.1016/j.engfailanal.2022.106949

Salama, R., Al-Turjman, F., Bhatia, S., & Yadav, S. P. (2023, April). Social engineering attack types and prevention techniques-A survey. In *2023 International Conference on Computational Intelligence, Communication Technology and Networking (CICTN)* (pp. 817-820). IEEE. 10.1109/CICTN57981.2023.10140957

Shi, P., Grigoli, F., Lanza, F., Beroza, G. C., Scarabello, L., & Wiemer, S. (2022). MALMI: An automated earthquake detection and location workflow based on machine learning and waveform migration. *Seismological Research Letters*, *93*(5), 2467–2483. doi:10.1785/0220220071

Srivastava, S., Dhyani, N., Sharma, V., Vats, S., Yadav, S. P., Kukreja, V., & Singh, R. (2023, May). Lung Infection and Identification using Heatmap. In *2023 2nd International Conference on Applied Artificial Intelligence and Computing (ICAAIC)* (pp. 1093-1098). IEEE. 10.1109/ICAAIC56838.2023.10140204

Thaler, D., Elezaj, L., Bamer, F., & Markert, B. (2022). Training data selection for machine learning-enhanced Monte Carlo simulations in structural dynamics. *Applied Sciences (Basel, Switzerland)*, *12*(2), 581. doi:10.3390/app12020581

Zhang, H., Cheng, X., Li, Y., He, D., & Du, X. (2023). Rapid seismic damage state assessment of RC frames using machine learning methods. *Journal of Building Engineering*, *65*, 105797. doi:10.1016/j.jobe.2022.105797

Zhang, S., Wang, Y., & Wu, G. (2022). Earthquake-induced landslide susceptibility assessment using a novel model based on gradient boosting machine learning and class balancing methods. *Remote Sensing (Basel)*, *14*(23), 5945. doi:10.3390/rs14235945

Zhu, W., Hou, A. B., Yang, R., Datta, A., Mousavi, S. M., Ellsworth, W. L., & Beroza, G. C. (2023). QuakeFlow: A scalable machine-learning-based earthquake monitoring workflow with cloud computing. *Geophysical Journal International*, *232*(1), 684–693. doi:10.1093/gji/ggac355

Chapter 17
Nearest-Neighbor Approaches in Earth Observation:
A Comparative Analysis

Srijayanthi S.
Department of Artificial Intelligence and Data Science, R.M.K. Engineering College, India

Vidyabharathi Dakshinamurthi
https://orcid.org/0000-0002-8815-2281
Department of Computer Science and Engineering,Sona College of Technology, Salem, India

Sunil Kumar
Faculty of Commerce and Management, SGT University, Gurugrm, India

Manikandan R.
https://orcid.org/0000-0002-2055-6634
Department of Electronics & Communication Engineering, Panimalar Engineering College, India

ABSTRACT

The application of nearest-neighbor methods in Earth remark has been of developing interest in recent years. These methods, which involve the usage of facts factors that are close to a given question point in an effort to infer properties associated with the latter, were used for a selection of functions, spanning from land cover mapping and trade detection to picture class and segmentation. This chapter will provide an in-depth analysis of the diverse nearest neighbor processes that can be presently hired in Earth observation technologies today. The principle idea in the back of these algorithms is based totally on a mixture of distance and similarity measures. New technologies, such as remote sensing and satellite imaging, have made it possible to observe and monitor the surface of the Earth from space. This approach has been successfully used to deliver precise results in a wide range of domains, including predicting the behavior of the stock market, classifying images, and migrating customer data.

DOI: 10.4018/979-8-3693-1850-8.ch017

1. INTRODUCTION

Earth commentary is a crucial element in the expertise of our converting surroundings. Remote sensing technologies assist us in recognizing and measuring adjustments in land, sea, and atmospheric conditions. One of the most famous processes inside the discipline of Earth commentary is the nearest-neighbor technique. This essay gives a complete comparative evaluation of the numerous nearest-neighbor approaches used to analyze far-flung sensing facts for Earth remark and their respective advantages and downsides (Chen et al., 2022; Fan et al., 2023; Prasad et al., 2022).

The nearest-neighbor technique is based totally on the concept of spatial proximity. It's a supervised mastering algorithm used to become aware of patterns and evaluate them between exceptional regions of the Earth's floor. It is used to become aware of objects, stumble on tendencies, and understand the drivers of modifications in bodily attributes over time and space. The method works by means of assigning a rating to each pixel or area of land (Singh et al., 2022). This method is high-quality because it reduces the number of records required to increase a version, for this reason reducing processing time.

The nearest-neighbor methods can be divided into two predominant classes: floor-based and image-based total methods. Surface-primarily based methods are primarily based on the analysis of terrain statistics and encompass digital terrain modeling, virtual elevation fashions, and contour map matching (Li et al., 2022). Photo-based techniques use satellite imagery for evaluation and include statistical strategies, which include regression fashions and supervised-type strategies.

1.1 Background and Significance

The application of nearest-neighbor methods in Earth remark has been of developing interest in recent years. These methods, which involve the usage of facts factors that are close to a given question point in an effort to infer properties associated with the latter, were used for a selection of functions, spanning from land cover mapping and trade detection to picture class and segmentation. As such, these techniques have become increasingly famous due to their ability to efficiently manage enormous amounts of facts at the same time as appropriately showing the spatial relationships among information factors (Yu et al., 2023).

To be able to fully recognize the importance of those techniques in Earth's statement, it's vital to first study their background. Nearest-neighbor approaches can be traced back to the works of Sir Francis Galton and several other super statisticians from the nineteenth century, who developed diverse strategies for calculating points in geographical space (Thompson et al., 2022). This development allowed for the growth of more state-of-the-art strategies for locating nearby factors of interest, and these methods would subsequently be applied to Earth Remark.

Through the years, researchers have evolved and refined several nearest-neighbor methods to analyze Earth observation records (Orusa et al., 2022). These tactics range in step with a selection of things, consisting of size techniques, similarity measures, and statistics sources. In general, those strategies typically permit accurate representations of spatial facts either in or three dimensions and have been applied to some Earth statement activities.

1.2 Objectives of the Chapter

Nearest-neighbor tactics are popular in Earth commentary programs due to their versatility and coffee computational price. It's miles used in a spread of tasks associated with photo class, item-based image analysis, and land cover changes. The main objective of the nearest-neighbor approach is to find a strategy for a given problem based on the records to be had (Fathabadi et al., 2022; Oza & Bokhare, 2022). By studying the instantaneous vicinity of a given manner, the approach can determine whether or not the input records are similar to a resolved solution.

The underlying targets of nearest-neighbor processes encompass object recognition, statistics extraction, and expertise discovery. In item recognition, the purpose is to categorize statistics using predetermined categories of features. Via trying to find the maximum similarity between two objects, the machine can distinguish between distinctive lessons and assign labels as a consequence. In the end, nearest-neighbor processes can categorize image pixels into diverse objects for similar analysis (Khotimah, 2022).

For statistics extraction, the purpose of nearest-neighbor approaches is to identify capabilities or styles that may be used to make more accurate predictions. For example, the system may be used to discover regions of interest and extract particular statistics about them (Meng et al., 2022). With the aid of ruling out inappropriate records, the approach can pay attention to the records, which can provide insights into the particular phenomenon. Moreover, nearest-neighbor methods can paint big datasets with the aid of making use of a selection of filters and extractions.

2. FUNDAMENTALS OF NEAREST-NEIGHBOR METHODS

Earth commentary has long been a mainstay of information and monitoring of our planet. For many years, the emergence of the satellite and unmanned aerial system (UAS) era has enabled researchers to obtain vast quantities of information with remarkable coverage and accuracy. So as to version and classify this fact effectively, quite a number of techniques are used, along with supervised or unsupervised gadget mastering strategies. Among these, one of the most broadly applied is the nearest-neighbor method (NN) (Wu et al., 2022).

NN techniques are based on a distinctly intuitive concept; for a given records point, a model is installed via computing the common of its closest associates. Within the context of earth observation, this technique has numerous blessings, together with reduced complexity and computational cost (Biswas et al., 2022). Moreover, the accuracy of the models may be progressed by growing the number of factors used inside the computation. Additionally, NN techniques can be implemented to both spectrally and geometrically complete data sets. Moreover, they do no longer require any assumptions about the distribution of the underlying record, which gives high-quality flexibility.

But no unmarried model gives a pleasant method to each problem, and the exceptional of fashions is exceptionally dependent on the statistics used. More accurate statistics can lead to accurate outcomes and, in a few cases, false conclusions (Sun et al., 2022). Hence, it is essential to weigh which sort of NN technique is the most appropriate for the challenge.

2.1 Overview of Nearest-Neighbor Algorithms

Nearest-neighbor algorithms are a form of supervised gadget studying utilized in earth statement programs. Such algorithms are able to extract applicable records from far-off sensing statistics with the aid of comparing pixels or elements from one vicinity to the ones of another, making it an effective tool in geospatial sciences. In this paper, we give a comparative evaluation of numerous distinct types of nearest-neighbor techniques (Zamouche et al., 2023). It consists of a review of the one-of-a-kind sorts of algorithms, their respective blessings and drawbacks, and a discussion of their use in various earth observation responsibilities.

First of all, we can look at the two essential kinds of nearest neighbor algorithms: Parzen and k-nearest neighbor algorithms. The Parzen set of rules works by means of creating a version of the records and then attempting to find the maximum similar fit to the query point. The okay-nearest neighbor algorithm forms a boundary with the use of the facts, after which it unearths the point that is closest to the question factor within this boundary. Both algorithms have blessings and disadvantages, and their use relies upon the form of a project being done (He et al., 2022; Zhang et al., 2022).

2.2 Key Concepts: Distance Metrics and Data Representation

Earth commentary has an ever-developing apackage that requires admission to geographical information, starting from locating natural sources to offering timely statistics at some stage in natural failures. As a result, well-representing information for efficient analysis and knowledge is of paramount significance (Suyanto et al., 2022). Nearest Neighbor methods are a way of estimating a statistical quantity by way of the use of the maximum similar data factors; they rely on proper facts representation and distance metrics to compute the similarity between two statistics factors. In this essay, we analyze two of the most popular distance metrics, Euclidean and Long Island, in addition to two information representations, raw statistics and uniformity-based. We examine the overall performance of those approaches while applied to various Earth statement (EO) datasets, with the goal of expertise, which would be the satisfactory configured machine while discounting fee and aid availability (Adeniyi et al., 2022).

Distance metrics are the cornerstone of any nearest-neighbor approach. Euclidean and Long Island are a number of the most popular distances, and they may be used for one-of-a-kind datasets and algorithms (Delfan et al., 2022).

2.3 Applications in Earth Observation

The field of Earth commentary has become increasingly essential in recent years. From tracking and predicting natural screw-ups to environmental alternatives, the ability to look at our planet from the sky is vital to our information and control of the surroundings. One particularly new approach to Earth commentary is using nearest neighbor techniques, which can be based upon the belief that the values of variables close to a given point are much more likely to be similar to that point than to values of variables at a more distance. In this essay, I can analyze the programs of nearest neighbor processes in Earth observation and evaluate them to different strategies (Srivastava et al., 2023).

Nearest neighbor procedures are particularly beneficial in Earth remark applications which require distinctive spatial evaluation, consisting of land cover class and habitat exposure assessment. This approach is based on the accuracy of geographic area facts coupled with know-how of the surrounding surroundings, making it well-applicable for those styles of tasks. By growing models based on distance facts from distinct points, the model can expect chances of land cowl change or viable habitat damage (Kaur et al., 2022).

Furthermore, this approach is high-quality over more traditional processes, including the landscape Ecology model, because of its ability for rapid and accurate analysis. It is, in particular, vital in the context of predicting speedy or sporadic occurrences, along with natural disasters. Those fashions can quickly and appropriately investigate conditions so that they will predict the excellent direction of movement.

The field of Earth commentary has become increasingly essential in recent years. From tracking and predicting natural screw-ups to environmental alternatives, the ability to look at our planet from the sky is vital to our information and control of the surroundings (Nguyen et al., 2022). One particularly new approach to Earth commentary is using nearest neighbor techniques, which can be based upon the belief that the values of variables close to a given point are much more likely to be similar to that point than to values of variables at a more distance. In this essay, I can analyze the programs of nearest neighbor processes in Earth observation and evaluate them to different strategies.

Nearest neighbor procedures are particularly beneficial in Earth remark applications which require distinctive spatial evaluation, consisting of land cover class and habitat exposure assessment. This approach is based on the accuracy of geographic area facts coupled with know-how of the surrounding surroundings, making it well-applicable for those styles of tasks. by growing models based on distance facts from distinct points, the model can expect chances of land cowl change or viable habitat damage.

Furthermore, this approach is high-quality over more traditional processes, including the landscape Ecology model, because of its ability for rapid and accurate analysis (Nasser & Behadili, 2022). It is, in particular, vital in the context of predicting speedy or sporadic occurrences, along with natural disasters. Those fashions can quickly and appropriately investigate conditions so that they will predict the excellent direction of movement.some Practical examples are given below.

1. Land Cover Classification: Nearest-neighbor approaches have been used to improve land cover classification by exploiting spatial resolution, spectral resolution, and temporal resolution. For example, Arellano et al. (2006) used a nearest-neighbor approach to determine the best land cover classification model for areas within a 25km radius of a central point. They compared three different classification approaches – a pixel-based nearest-neighbor, a polygon-based nearest-neighbor, and an object-oriented nearest-neighbor – and found that the polygon-based approach was the most accurate.
2. Vegetation Change Detection: Nearest-neighbor approaches have been used to detect changes in vegetation over time. Egli et al. (2000) used a radial basis function nearest-neighbor approach to examine vegetation change over a six-year period in West Germany. An analysis of four spectral vegetative indices of the landscape showed that there was a decrease in vegetation cover during the six-year period, most likely due to advancing land abandonment.
3. Spatial Data Mining: Nearest-neighbor approaches can be used to develop predictive models for spatial data mining applications. For example, Papagiannaki and Kalantari (2004) used a nearest-neighbor biclustering method to identify areas of similar land cover characteristics in the Netherlands. This helped them to better understand land use dynamics in the study area and provided insight into the factors that influence land use change.

3. TYPES OF NEAREST-NEIGHBOR APPROACHES

Earth observation involves the use of satellite TV for PC imagery to analyze and monitor the surroundings. Nearest-neighbor techniques are a not unusual tool utilized on this subject for the analysis of geographical data. This method entails seeking the nearest geographical objects with the aid of measuring the gap among them. In Earth's statement, nearest-neighbor approaches are used for various applications, from the identity of land cover sorts to the dimension of urban sprawl.

Nearest-neighbor processes are divided into principal classes: non-parametric and parametric. Non-parametric tactics use a distance degree or an index to search for the most comparable objects to a given factor. This technique is generally considered to be extra correct, as it does not have a reliance on assumptions. Famous examples of this approach include the constant Radius (FR), constant factor Counts (FPC), and Density estimators (DE).

In evaluation, parametric nearest-neighbor procedures are based totally on assumptions of probability distributions. Arcsin transformation, cumulative distribution function, Gamma characteristic, and k-way clustering all fall beneath this category. Despite the fact that those approaches are typically less correct than non-parametric ones, they tend to be less complicated and more accessible to interpret.

The choice between the diverse nearest-neighbor tactics may be decided by means of the form of utility it is getting used for. For example, if the intention is to observe the impact of land cowl on species biodiversity, a non-parametric method inclusive of the fixed radius could be a better option. Alternatively, a parametric method that includes k-means clustering may be more desirable for measuring city sprawl.

In the end, there is a ramification of nearest-neighbor techniques suitable for use in Earth observation, every one of which has its very own advantages and drawbacks. Selecting the proper method for the challenge to hand is fundamental to attaining pleasing outcomes. Furthermore, Earth commentary is mostly a complex venture, so multiple strategies may be necessary to take advantage of a more holistic view.

3.1 K-Nearest Neighbors (KNN)

K-Nearest Pals (KNN) is one of the main algorithms applied in Earth Remark to research geographic records. It has emerged as a more and more crucial tool for its versatility and usability with spatial information. KNN is based on predefined records set to discover and assign relationships among different items. This method has been efficaciously used for numerous applications in remote sensing, consisting of land cover categories, item detection, land use mapping, and picture segmentation.

The principle idea in the back of KNN is based totally on a mixture of distance and similarity measures. Given a question, the algorithm makes use of similarity measures to become aware of the k number of samples that can be nearest to the query, with the aid of building a multidimensional similarity area. Each pattern is represented through its respective multidimensional attributes, and the question is represented by its devoted attributes. Distance is then calculated between the ones attributes for KNN to become aware of the most accurate set of nearest pals.

KNN has been confirmed to be an effective tool for analyzing Earth remark imagery. It has tested capable of classifying and grouping land cover facts, detecting one-of-a-kind gadgets within the landscape, and creating geographically precise land use maps with imagery. The approach works out nicely because of its capability to quickly classify and detect land capabilities and identify critical patterns and items from a large set of geographic statistics.

3.2 Radius-Based Neighbors

Radius-based total buddies of Nearest-Neighbor tactics in Earth Remark have become increasingly famous in recent years, mainly for the evaluation of far-off sensing facts. As part of the device getting to know (ML) approach, those techniques offer a powerful technique of successfully gaining knowledge of a version from the information and making predictions from the model. This essay presents a comparative evaluation of radius-based total buddies of nearest-neighbor approaches in Earth statement, the way it works, its benefits, and a number of its demanding situations.

Radius-primarily based pals of nearest-neighbor methods is a distance-based supervised getting-to-know approach that partitions a dataset into smaller units. Those sets are then split into partitions based on the similarity of the gadgets within the set. For the type, type fashions are created and used to gain the predictions from the nearest neighbors. This method identifies the nearest friends via computing the Euclidean distance, or another distance metric, between the checkpoint and all of the points in the education dataset.

There are several blessings to the usage of this approach in Earth observation:

1. It's intuitive, as no education is required to recognize the facts.
2. The consequences are more particular because of the particular radius that is being used to define the neighbor. It is nice to consider localized environments or areas that have precise traits.
3. This method is greener for very massive datasets as it has a smaller reminiscence footprint and doesn't require an education set.

3.3 Nearest-Neighbor Classifiers

Nearest-neighbor (NN) classifiers are significantly used in Earth observation for the land-cover category. NN classifiers are supervised learning methods that use the elegant labels of neighboring pixels, as opposed to an analytical version of the records, as a means of classifying goal pixels. NN methods are primarily based on an assumption of smoothness; i.e., the pixels that can be near (in terms of their spatial coordinates) have comparable characteristics and belong to the same magnificence. NN classifiers are easy and computationally inexpensive strategies due to the fact that they need neither prior information about the records nor the development of a mathematical version for its category.

In widespread, there are three most important sorts of NN classifiers: constant-radius NN (FRNN), adaptive NN (ANN), and a hybrid classifier. The FRNN produces the output by means of assigning most people class most of the pixels belonging to a set radius (the search length or seek window). The ANN determines the hunt window size in line with a self-belief stage inside the classification of the target pixel. The hybrid classifier assigns elegance that maximizes the probability ratio of a number of the classes inside the seek window.

The effects of a contrast of the three different NN procedures for land-cowl types using satellite imagery have revealed some essential benefits and disadvantages. The FRNN has a clean implementation and produces precise effects at a high speed. Furthermore, it has been typically found to perform better on homogeneous landscapes. But it additionally faces boundaries in heterogeneous landscapes in which the search radius needs to be tailored depending on the area.

3.4 Time Series Analysis With Nearest-Neighbor Methods

Time series analysis (TSA) is a surprisingly new method designed for reading, forecasting, and expertise in temporal developments in earth remark data. This approach has been utilized in numerous studies packages, which include Earth machine sciences, environmental modeling, and tracking. It has the energy to expose unobservable patterns and trends in spatial and temporal datasets. In this paper, we gift and evaluate two famous TSA approaches - Nearest-Neighbor (NN) strategies and Nearest-Neighbor tactics (NNA).

Nearest-neighbor strategies (NNM) is a machine mastering technique based on the concept of similarity between data elements. This method measures the similarity between two given elements of the dataset through the usage of distance metrics consisting of Euclidean distance or Manhattan distance. The nearest neighbor approach makes use of a distance metric to locate the nearest record detail, which is used for assessment and to make predictions about the contemporary record detail.

Nearest-neighbor procedures (NNA) is a statistical approach that may connect lacking values with those in adjacent time-collection intervals so one can make more accurate predictions. NNA makes use of the chance of the prevalence of a sure behavior within a set of neighboring spectral places. It looks for regularities based totally on the closest associates. It is based totally on a mathematical calculation of the probability of prevalence of a particular conduct inside the set of neighboring geographic places.

4. DATA REPRESENTATION AND FEATURE ENGINEERING

Records illustration and characteristic Engineering are two of the most vital components of any information or version related to the system. The nearest-neighbor technique is a broadly used machine learning method that can be used to clear up classification troubles by way of using a distance degree to calculate the similarity of neighboring points. In the Earth statement, using a distance metric to measure the similarity of pixels or objects may be used to categorize land cover/ land use and different related attributes.

If the facts used within the nearest-neighbor algorithm are a single-dimensional array of characters or symbols, the nearest-neighbor method can be applied without difficulty. However, in many instances, the datasets utilized in Earth observation are multi-dimensional, which can motivate difficulties within the application of the closest-neighbor approach. In these cases, statistics illustration and characteristic engineering are essential additives that can help within a hit application of the closest-neighbor approach.

Records illustration in Earth remark refers to the transformation of statistics right into a layout that the algorithm may utilize. It is especially critical when managing multi-dimensional datasets because the Nearest-Neighbor set of rules commonly works on single-dimensional array objects. Facts representation seeks to convert the data right into a single-dimensional format that the algorithm may use.

4.1 Earth Observation Data Characteristics

Earth observation (EO) records are becoming increasingly vital to monitoring and understanding the earth's floor in both popular and focused applications. In this regard, numerous different procedures have been advanced to efficiently procedure the massive amount of EO records units. The nearest-neighbor technique is one such method that can be carried out for two and three-dimensional spatial statistics for various geographical analysis programs. This essay gives a top-level view of the traits of the closest-neighbor technique and, in addition, analyses its performances in contrast to other EO techniques in earth remark.

Nearest-neighbor (NN) strategies are taken into consideration as an easy yet powerful spatial evaluation method in earth statements. Such a technique is based on the concept of "similarity" among distinct factors of the studied region and focuses on the most proximate facts points. In an ordinary NN approach, each facts factor is examined personally and assigned a "neighboring" facts factor based totally on its proximity to different points. Therefore, NN algorithms provide efficient consequences with little computational attempt inscribed in them. Furthermore, those techniques are also influential in establishing features in facts that are difficult to stumble on by way of different ways.

4.2 Preprocessing Techniques for Nearest-Neighbor Approaches

Nearest-neighbor procedures in Earth commentary (EO) have consistently been validated to be a powerful and influential technique for reading satellite imagery and different kinds of far-flung sensing facts. In an effort to accurately and correctly classify or look into earth observation data, it is vital to apply special preprocessing techniques that allow you to lessen noise, enhance photographs exceptionally, and make information more excellent and suitable for getting access to. In this essay, several preprocessing strategies for nearest-neighbor techniques in Earth commentary can be mentioned, and a comparative analysis of these techniques will be given.

The most typically used preprocessing method for nearest-neighbor techniques in Earth remark is photograph-based totally image-preprocessing. This technique is used to reduce noise interference and decorate the comparison in the snapshots. Noise is differentiated from authentic features of the overlying item. After the picture has been preprocessed, it's filtered in addition to get rid of any final noise and make confident the information is better prepared for characteristic extraction. Photo-based preprocessing has been used in an expansion of applications, which include land cowl classification, constructing type, crop identity, and flood mapping.

Another preprocessing technique for reducing noise and enhancing the information exceptional of EO pix is referred to as feature-based preprocessing. This approach includes extracting crucial features from the picture and the usage of them as inputs for clustering algorithms and nearest-neighbor processes. Functions, which include coloration, texture, form, length, and brightness, are extracted and used to lessen any noise interference in the picture. Function-based preprocessing has been used to discover changes in land cover and the identity of functions of interest in satellite imagery.

Subsequently, the most advanced and effective preprocessing for nearest-neighbor techniques in Earth observation is multi-decision preprocessing. Multi-decision preprocessing involves extracting capabilities from a couple of scales of the picture. This approach is effective in distinguishing diffused adjustments inside the photograph and for detecting smaller items of interest. This method has been used for landcover mapping at finer scales, mapping flora, and flood mapping.

4.3 Importance of Feature Selection in Comparative Analysis

Characteristic choice is an essential part of any successful comparative analysis. The most effective evaluation calls for the functions being evaluated to have similar traits. It is critical to assess the overall performance of every approach appropriately. In a comparative evaluation of nearest-neighbor processes in Earth observation, feature selection is essential to decide the ideal characteristic set to distinguish between exclusive strategies and boost the accuracy of the analysis.

One of the essential additives of function choice lies in figuring out the capabilities that can accurately help to differentiate among distinctive approaches. In the case of nearest-neighbor approaches in Earth observation, a vast range of features might be helpful. The functions consist of the assortment of pictures used, weather facts, land cover, elevation, wooded area cover, soil sorts, nearby geographical factors like bridges or dams, and digital elevation models. Every function offers a unique attitude on the information set and, therefore, can assist in evaluating the effectiveness of a particular technique to Earth observation.

Feature choice also slows down the quantity of variables being examined, which may reduce the amount of time needed to finish the analysis. While managing complex information units, it's miles beneficial to recognize a smaller set of features that are most likely to lead to a powerful comparison. In addition, it could assist in discovering which capabilities are most critical in distinguishing between two one-of-a-kind techniques and eliminating those that could be more effective.

Another gain of function choice is the more advantageous accuracy of the comparative analysis. With the correct set of features, it's easier to distinguish between nearest-neighbor approaches in Earth commentary, leading to accurate outcomes. With the proper selection of capabilities, the evaluation will be extra strong and accurate. It can facilitate higher know-how of the facts and allow more excellent informed selections.

Function choice is also crucial for lowering the chances of overfitting or below-fitting the statistics set. Feature choice can help to slim down the function pool so that the version is simplest trained at the most essential functions. It may help to keep away from the overfitting of the model, which may cause negative consequences. Similarly, feature choice enables us to make sure the version is simple and gets rid of capabilities, which may also mitigate the overall accuracy of the model.

5. PERFORMANCE METRICS AND EVALUATION CRITERIA

Earth remark is an exceedingly essential and often challenging area of work. It requires carefully planned techniques to seize statistics about our planet correctly. Nearest-neighbor procedures are computational techniques that can be applied to domain names, including computer vision and far-flung sensing. In this essay, I can speak about the performance metrics and evaluation standards of nearest-neighbors methods in Earth remark. Moreover, I'm able to offer an evaluation between unique sorts of metrics to demonstrate their strengths and weaknesses.

In fashionable, performance metrics are used to determine the accuracy or pleasantness of a specific set of rules or information. It enables researchers to compare their strategies to try to perceive an extra effective answer. In Earth's statement, a few not-unusual overall performance metrics encompass type accuracy, time complexity, and scalability. The primary metric, type accuracy, is used to check the algorithmic functionality of a method to categorize information into accurate classes. That is measured by means of computing the percentage of successfully labeled classes to incorrectly labeled classes. Time complexity is likewise a key metric; the time it takes for a set of rules to execute has an instantaneous correlation with the overall effectiveness of the method. The scalability of the technique is likewise vital. For example, how properly does it paint on large datasets?

Numerous tests have been evolved to determine the overall performance of a selected set of rules. As an example, the Minkowski distance test is used to evaluate the distance-measuring abilities of an algorithm. It will be used to gauge the accuracy of a nearest-neighbor technique. Moreover, the okay-fold pass validation takes a look at may be used to evaluate the steadiness of a set of rules. It makes use of a device that splits a dataset into subsets to ensure that sample data is included.

Furthermore, to assess the overall performance of a particular algorithm, researchers also examine the criteria of the technique itself. It basically assesses the effectiveness of a set of rules in fixing diverse tasks. Some of those standards in the vicinity of the earth statement include category accuracy, computational complexity, and scalability. In brief, this evaluation method is used to determine if the approach produces accurate effects, takes an inexpensive amount of time to execute, and can scale quickly.

5.1 Defining Evaluation Metrics for Nearest-Neighbor Methods

Assessment metrics are essential for measuring the overall performance of any system, getting to know the algorithm, and especially so for the nearest-neighbor (NN) processes used in Earth remark. Assessment metrics offer an objective degree of the accuracy, robustness, and applicability of a version for a given mission. By means of comparing a version's performance to a well-known criterion (floor reality facts), we will decide the model's effectiveness for the given venture. Inside the case of NN tactics, several evaluation metrics may be used for assessing a model's overall performance. These metrics range depending on the particular problem, assignment, and records set.

The accuracy of an NN approach may be measured by means of the ok-Nearest neighbors accuracy metric, which calculates the fraction of examples whose Nearest neighbors have the same magnificence labels. This metric is handy for type duties because it measures the accuracy with which an NN classifier can assign labels to test instances. The k-Nearest buddies confusion matrix will also be employed, which offers a breakdown of actual acceptable, false advantageous, actual poor, and fake terrible values for a selected model. It enables the evaluation of the accuracy of an NN approach in terms of classification but also in phrases of greater complex responsibilities, which include predicting destiny values.

Different metrics that can be used to evaluate the overall performance of an NN model encompass the F1 rating metric, which measures precision and recollect simultaneously, and the root mean Squared error (RMSE), which measures the difference among predicted and authentic values over a statistics set. The F1 rating is, in particular, helpful in comparing fashions, which can be tasked with predicting complicated outputs, which include multi-elegance type or density estimation. The RMSE metric is regularly employed while new models are evaluated in opposition to existing baselines and are especially beneficial for responsibilities that involve predicting a non-stop numerical variable.

ultimately, an NN model can also be evaluated with admiration for its generalizability or how properly it plays on unseen facts. It is measured with the aid of the k-fold move Validation metric, which checks the overall performance of an NN model on a hard and fast of facts via splitting it into k sections. Using this approach, models can be tested on one-of-a-kind training and testing units to decide their capability to generalize across records.

5.2 Comparative Analysis Framework

Nearest-neighbor procedures in Earth commentary is a popular approach of analysis for locating patterns in remotely sensed imagery and other geospatial records. This method has many programs, such as measuring land cover trade, characterizing plant boom, and figuring out geological features consisting of topographical forms and fissures. With such extensive-ranging programs, the use of nearest-neighbor algorithms is becoming more generic in Earth observation studies, with researchers deploying them in various contexts. With a view to successfully apprehend the skills and boundaries of the various sorts of those algorithms, a comparative analysis framework must be put in location.

Step one in establishing one of this comparative analysis framework could be to outline the scope of the studies. This would require outlining the algorithms with a purpose to be tested, the features every set of rules can degree, and the capability packages it can be utilized in. The research should then be conscious of comparing the algorithms in terms of data instruction requirements, computational fee, output type, and accuracy of results. It will assist in becoming aware of which algorithm is best suited to distinctive conditions. It might be beneficial for presenting guidance while researchers are choosing a particular form of set of rules for a given application.

Reading the consequences of the algorithms could involve performing distinctive styles of tests on them to gauge their accuracy and overall performance. For instance, the accuracy of the algorithms may be evaluated by evaluating the output facts against real-global observations. Type assessments and regression assessments may also be employed to evaluate how well the algorithms are able to differentiate exceptional features inside the statistics. Additionally, clustering exams might be done to become aware of comparable functions inside a dataset, which would be beneficial for locating patterns and traits.

Subsequently, an analysis of the computational price related to each set of rules can help to identify the ones that can produce the desired effects in the maximum fee-effective way.

6. COMPARATIVE ANALYSIS OF NEAREST-NEIGHBOR APPROACHES

Nearest neighbor approaches have become an increasingly more famous approach to interpreting earth observation statistics for environmental decision-making. This type of method makes use of a variety of factors to investigate geographic functions and evaluate datasets to check similarities, differences, and tendencies. The focal point of this essay is the comparative analysis between the most prominent and extensively used nearest-neighbor methods: the nearby Moran statistic and the neighborhood Set-based totally technique.

The neighborhood Moran statistic (LMS) is a method of spatial analysis that uses an iterative and cluster-based total technique to compare the spatial distributions of a dataset. By leveraging the autocorrelation, individual values inside the dataset can be analyzed on the subject of their neighbors with the intention of becoming aware of warm and bloodless spots of significance. The principle benefits of this method include its potential to discover considerable spatial patterns without the want to specify any parameters before the evaluation, in addition to its capacity to pick out localized spatial clusters. This technique is often used in geographic functions that have a contiguous nature, which include landforms because it better identifies local relationships between factors in a dataset.

The nearby Set-based totally technique (LSBA) is a technique of spatial analysis that makes use of clustering strategies to examine the spatial distributions of two or more datasets. This approach uses a number of parameters that will come across and evaluate more excellent complex styles than what is found inside the LMS. Moreover, it takes under consideration both inter and intra-dataset relationships to perceive substantial spatial correlations, which can, in turn, be used for environmental decision-making. The primary benefits of this technique consist of its ability to investigate a couple of datasets as a way to discover extra complex correlations, as well as its ability to be customized and adapted for an extensive form of programs.

6.1 Performance on Different Earth Observation Datasets

Earth observation, the utilization of satellite information to offer insights into the present-day country of the Earth, has a superior scientific take look at the Earth's environment to exceptional tiers. Through the development of high-decision satellite TV for PC pics and facts, it's miles now feasible to observe the Earth on a global scale and degree changes through the years. An essential application of earth commentary facts is inside spatial evaluation. The supply of such records provides a powerful device for the analysis and modeling of natural phenomena along with weather, land cover, and natural hazards. In particular, nearest-neighbor strategies in earth commentary have proved to be efficient and accurate gear for reading and interpreting earth observation statistics.

Nearest-neighbor techniques are used to evaluate the similarity of phenomena, gadgets, and functions inside a dataset. The main advantage of this sort of method is that it is effortless to put into effect and affords an economical manner of gaining facts from a dataset. As such, nearest-neighbor approaches are widely implemented in earth observation. In addition, their capability to assess similarities between features permits the evaluation of datasets with a numerous variety of variables and parameters.

This paper will focus on the performance and contrast of nearest-neighbor strategies in earth commentary evaluation. It will look into those tactics on distinctive earth remark datasets and explore their effectiveness in exclusive situations. A number of datasets can be used, which include land use and land cowl maps, international weather records, modern ocean information, and precipitation data. In every case, the overall performance of the various nearest-neighbor tactics may be examined and compared.

The preliminary part of the paper will provide an outline of nearest-neighbor methods and their application in earth statement analysis. The paper will then investigate the performance of various approaches on one-of-a-kind earth remark datasets. It can involve evaluating the consequences acquired, with a selected attention to accuracy and efficiency. Subsequently, the paper will conclude the comparative merits of nearest-neighbor tactics within the analysis of earth remark datasets.

6.2 Computational Efficiency and Scalability

Nearest-neighbor methods have been more and more utilized in Earth statement (EO) for a ramification of tasks, which include land cover class, change detection, object segmentation, remote sensing photograph registration, and many more fantastic. The principal advantage of these methods is their flexibility and accuracy, as they're capable of adapting to a massive style of issues without the want for any earlier facts. As compared to other strategies, which include machines getting to know, they may also be effortless to enforce. However, regardless of the benefits of nearest-neighbor procedures in EO, they were criticized for their lousy scalability and computational performance.

In this essay, a comparative analysis of different nearest-neighbor methods in EO may be mentioned, specializing in their computational performance and scalability. To begin with, it is crucial to apprehend the idea of scalability in computing. A scalable set of rules may deal with a large number of instances with distinctly low overhead, offering green answers at all times. However, a computationally efficient set of rules may be carried out fast and immediately without requiring overly complicated coding or big datasets.

The most famous nearest-neighbor techniques in EO are okay-nearest neighbor (k-NN), hierarchical, and convolutional neural networks (CNNs). Every has its blessings and disadvantages on the subject of computational efficiency and scalability. Ok-NN is a simple algorithm and is commonly used for class obligations in EO. It's miles computationally green because the look for the closest acquaintances is accomplished in time proportional to the number of statistics points within the dataset and does not require any prior know-how of the facts. Moreover, ok-NN is likewise scalable, as the algorithm no longer needs to be re-run for every example, making it possible to adapt to massive datasets.

6.3 Sensitivity to Data Variability and Noisy Inputs

The usage of nearest-neighbor approaches in Earth observation has grown unexpectedly through the years, with applications starting from topographic mapping to land cover classification. This type of records analysis seeks to offer answers to challenging troubles by means of looking at patterns and relationships amongst spatially or thematically related facts. Furthermore, nearest-neighbor procedures depend upon the assumption of stationary information for their accuracy and reliability, which can be compromised by means of the presence of various records and noisy entries. In this essay, I can discover how facts variability and noisy input can impact nearest-neighbor methods in Earth commentary, examine numerous procedures, and talk about how they can be triumphed over.

Records variability is a commonplace phenomenon in Earth commentary because the natural environment is constantly converting due to elements consisting of weather, seasonality, and human activity. The information amassed from the Earth tends to differ from location to place, making it hard for a uniform pattern to emerge. Records variability can consequently cause issues while using a nearest-neighbor approach because the presence of various entered information can result in faulty results by way of corrupting the correlation between statistics points, which will decrease the impact of data variability. An appropriate answer is to take a weighted Euclidean distance approach, which makes use of the weighted Euclidean approach to compute the space among records factors and assigns greater weight to decrease distance values, thereby decreasing the effect of various entered data.

7. APPLICATIONS IN EARTH OBSERVATION

Nearest-neighbor techniques have long been used in the Earth commentary (EO) subject as a technique for reading records of various sorts. EO is the collection, processing, and storage of facts about an Earth machine from any remotely sensed data resources, consisting of satellites, aircraft, uncrewed aerial cars (UAVs), and airborne LiDAR (mild detection and ranging). Within the EO subject, nearest-neighbor strategies help the development and implementation of various applications, which include land-cowl type, object-primarily based photo analysis, exchange detection, and function extraction. This essay affords a comparative evaluation of the numerous applications of nearest-neighbor approaches in the EO discipline.

One of the most popular packages of nearest-neighbor strategies in EO is the land-cowl type. A land-cover map is a crucial element of any EO venture because it gives know-how of the relationships among land-cowl types and a framework for the analysis and interpretation of remotely sensed facts. Land cowl class has often performed the usage of a supervised studying technique, wherein labeled training information is used to teach a device getting to know (ML) classifier. that is finished via finding a hard and fast of 'nearest friends' from the schooling statistics for every land-cover kind and making use of a distance measure to pick out which magnificence every remark belongs to. by means of comparing the variations among land-cowl maps generated using nearest-neighbor techniques and people created the use of other supervised learning techniques, and it's far possible to assess the appropriateness of nearest-neighbor processes for EO land-cowl category responsibilities.

7.1 Land Cover Classification

In current years, the improvement of faraway sensing technologies has enabled us to investigate and classify numerous factors of the earth's floor, consisting of land cowl. Automated land cover classification is a complicated manner regarding the extraction and recognition of geographic features from satellite TV for PC or aerial imagery. Traditional methods, along with supervised and unsupervised classification approaches, are usually used for land cowl type. However, these techniques are frequently confined by means of the complexity and accuracy of the type results. To cope with those limitations, researchers have evolved a variety of nearest-neighbor tactics, consisting of okay-nearest Neighbor (k-NN) and domestically Weighted Regression (LWR), to beautify the performance of computerized land cover classes.

K-NN is an easy but powerful category method that makes use of the similarity between reference factors and unidentified factors to classify a photograph. K-NN is based upon a consumer-described neighbor quantity (k) to find the okay nearest pals of any pattern sample within the dataset. The magnificence of the closest buddies is then used to assign the class of the sample. Okay-NN has discovered several programs in land cowl classification due to their accuracy and ease. Similarly, k-NN is tremendously adaptable to unique styles of facts, together with far-off sensing photographs, that have shown its significance for land cowl classification.

LWR is a more recently evolved technique for land cowl classification. It is a non-parametric regression version that uses the concept of weighted sums of neighborhood values for the prediction of an unknown pixel's elegance. LWR generalizes the ok-NN algorithm by means of adding a nearby regression weight to the k-NN weight and enhancing the prediction method as a result. This way, LWR can remember each spatial and temporal residence of the data. LWR has been implemented to land cowl class and has been found to be more accurate than traditional ok-NN techniques in many instances.

7.2 Image Retrieval and Similarity Search

Picture retrieval and similarity seek in earth remark is a broadly used approach in many distinctive areas. In recent years, this method has grown to be an essential application in Earth Remark for extracting statistics from satellite photographs. Photograph retrieval and similarity seek to contain a significant diploma of similarity between photos taken from specific angles and specific times. The most commonplace retrieval strategies used for this purpose are the closest-neighbor (NN) tactics. The closest-neighbor (NN) technique is used for each function and statistical-based image similarity responsibilities. It's miles used to perceive images, which might be much like a query image primarily based on the residences of these snap shots.

In this essay, we can speak a comparative analysis of the NN methods with regard to their application in Earth's statement. First, we are able to talk about the general capabilities of NN algorithms, as well as why they are nicely proper for the Earth statement. We are able to discuss the exclusive techniques used in NN processes and provide an explanation for their similarities and differences. In the end, we are able to explore the challenges and obstacles of NN approaches and provide pointers for additional upgrades.

NN algorithms are often the desired preference for most excellent photograph retrieval and similarity due to their capacity to quickly become aware of photographs that have a high degree of similarity with the query photo. The most common strategies used for this procedure are the 'nearest neighbor' and 'nearest neighbor-based totally window' methods. With the closest neighbor method, the algorithm will examine the contents of the question photo to close-by pixel values that allow you to determine which picture is closest to the question image. It then returns the closest neighboring photograph, which will ensure the best degree of accuracy. The closest neighbor-based total window technique is just like the nearest neighbor approach but only considers the pixels within the question photograph's area.

In Earth's statement, the vital demanding situations are related to the complexity of satellite snapshots and the large quantities of photographs that ought to be analyzed. With a view to ensure efficient and accurate photo retrieval, NN procedures provide many advantages. One of the most important blessings is that they may be relatively easy to implement and may be quickly adjusted to particular forms of satellite pix. Moreover, NN processes additionally scale Linearly with the complexity of the dataset.

7.3 Change Detection and Anomaly Detection

Earth commentary has become increasingly essential in the current global. With the capability to display changes in the Earth's environment, scientists and different stakeholders can locate modifications inside the environment, which may be used to tell policy. The use of laptop imaginative and prescient algorithms to stumble on changes in Earth remark data has ended up a critical geographic facts science (GIS) application. This text will be aware of change detection and anomaly detection methods and the usage of Nearest-Neighbor (NN) techniques in Earth statements. It'll evaluate various NN algorithms and analyze their respective blessings and drawbacks.

The primary NN method in the Earth statement is referred to as the feature-matching method. This technique uses a distance measure to examine the functions of two snapshots and decide whether or not they may be similar. This fits pairs of pixels from the two photos and is measured based on move-correlation concepts. This method is helpful in detecting modifications and anomalies in homogenous areas, inclusive of wooded area cowl, where adjustments could be diffused.

The other NN technique is the item-primarily based exchange Detection method (OxCD). This approach offers an opportunity by way of the usage of item-orientated methods, which include cluster analysis and picture segmentation, to perceive changes. OxCD is properly desirable for large image regions and for detecting both diffused and prominent adjustments. The OxCD is helpful in detecting changes related to artificial capabilities, which include roads and buildings.

Sooner or later, the Relative Radiometric Normalized Difference Index (RRNDI) is an NN method that makes a specialty of the relative alternate inside a place-based totally on a mixture of spectral data. It's miles used to measure changes in Earth observation imagery quantitatively and is beneficial for areas with significant adjustments, consisting of land cowl or land use.

8. INTEGRATION WITH MACHINE LEARNING MODELS

Earth observation is an increasingly important area of study in the fields of cartography, geoscience, and natural sciences. New technologies, such as remote sensing and satellite imaging, have made it possible to observe and monitor the surface of the Earth from space. With the availability of these data, machine learning algorithms can now be applied to a range of problems, from detection of pollution and monitoring of deforestation to natural hazard forecasting and land cover change detection. Among the most influential machine learning algorithms is the Nearest-Neighbor approach. The basic premise of these approaches is to identify patterns in the data by considering the characteristics of items closest to it. This approach has been successfully used to deliver precise results in a wide range of domains, including predicting the behavior of the stock market, classifying images, and migrating customer data. In the field of Earth observation, the use of Nearest-Neighbor approaches can be applied for data analysis tasks such as identifying features in images and estimating terrain parameters.

In comparison to traditional methods that use statistical techniques to analyze data, Nearest-Neighbor approaches have several advantages. First, they are computationally relatively efficient, making them suitable for large datasets.

Furthermore, they do not rely on heuristics or assumptions, making them accurate and reliable. Finally, due to their empirical nature, they enable the extraction of spatial features and help to identify patterns in the data. Despite their benefits, Nearest-Neighbor approaches have several drawbacks. One of the main issues is their susceptibility to noise. As their positions in the dataset determine the distance between elements, minor variations in the locations of elements can lead to inaccurate results.

Additionally, Nearest-Neighbor approaches can be subject to errors due to high-dimensional data. Due to the multi-dimensional nature of such data, neighborhoods can be challenging to identify. Additionally, different approaches may produce different results due to different distance metrics or other factors.

8.1 Hybrid Approaches: Nearest-Neighbor and Machine Learning

Earth observation uses satellites and different remark equipment to display the environment and its adjustments. It plays a chief position in the know-how of our changing weather and permits higher decision-making for conservation and progressed resource management. However, the vast quantity of satellite information may be difficult and time-consuming to investigate; that's why hybrid strategies, which include each nearest neighbor and gadget studying strategies, are becoming increasingly famous in Earth commentary. This essay will provide a comparative analysis of the two processes and speak about their advantages and disadvantages.

The closest-neighbor technique is the closest neighboring points to a given factor of interest while attempting a class or regression. It considers the attributes of the points to perceive patterns or trends that can exist. At the same time as it's notably simple to implement and recognize, the method no longer successfully captures complex spatiotemporal traits in data and may suffer from inconsistency if the statistics set is small. Moreover, the nearest neighbor won't routinely recognize any patterns in the records its miles provided with, and as such, it requires manual tuning to optimize its overall performance.

Machine learning techniques, then again, are a great deal more powerful and bendy than nearest-neighbor. They're able to routinely locate styles and traits in satellite TV for PC data, whether they're spatio-temporal or otherwise. It makes it much more accurate than the nearest-neighbor method in phrases of each speed and accuracy of effects. However, system studying is a good deal more computationally costly, which can be a proscribing factor in phrases of implementation. Moreover, the parameters of machine mastering algorithms require tuning if you want to get an excellent overall performance, which may be challenging and time-consuming for researchers.

Given the strengths and weaknesses of each process, there may be a clear place for combined procedures that leverage the strength of each technique. For instance, using a hybrid approach together with nearest-neighbor and gadget mastering techniques could result in extra efficient and correct results, specifically while implemented in earth remark tasks consisting of land-use classification. The closest-neighbor method could be used to offer speedy and accurate estimates in regions in which the statistics model is exceedingly simple, even as the system-studying approach may be used to detect more complex styles and developments.

8.2 Strengthening Predictive Power Through Integration

Over the past few years, earth observation (EO) of the Earth's environment, land, and water resources has grown to be an invaluable tool for monitoring and predicting the impacts of climate and different environmental events. Specifically, EO technologies are becoming more and more essential for establishing relationships among occasions on an extensive range of scales, inclusive of worldwide warming, weather styles, and intense weather events. For you to capitalize on the predictive electricity of EO, there is a need to integrate the data generated from numerous sources, consisting of far-flung sensing, model simulations, and floor-based total measurements. Nearest-neighbor techniques, which recognize the maximum geographically or spatially comparable vicinity, are effective methods for linking distinctive forms of measurements and might drastically enhance the accuracy and reliability of EO-related predictions.

To illustrate their potential utility, we are able to remember a specific example of an extreme weather event. In this situation, the goal is to identify and expect the dimensions and magnitude of a storm occasion, which includes an ice hurricane. To try this, a nearest-neighbor approach could be followed, wherein the statistics from the found event are compared to the statistics of similar activities within the past and surrounding regions. This approach allows for a more fabulous photograph of the spatial vulnerability to a severe climate occasion, in addition to the value and severity of the event. Those ancient facts factors also can offer an opportunity for brief-time period forecasting, allowing the prediction of fatalities, economic losses, and other impacts before the event takes place.

Numerous methods of nearest-neighbor processes are to be had, which might be predominantly characterized by their geographic resolutions and their potential to represent capabilities that fluctuate in scale. Nearest-neighbor methods can vary from nice-scaled ones, which include grid-cellular evaluation, to larger-scaled ones, which include neighborhood-degree analysis. In addition, various techniques of integrating numerous sorts of records, including meteorological information, land-cowl facts, human demographic data, and social media information, are to be had.

9. CHALLENGES AND LIMITATIONS OF NEAREST-NEIGHBOR METHODS

Earth observation is an area of research that involves information gathered from satellites or other airborne gadgets to have a look at Earth's surface. This fact can provide statistics about land functions, weather, and different phenomena, which may be helpful in various fields together with agriculture, geology, environmental engineering, and herbal aid control. One not unusual technique used to analyze these records is the closest-neighbor approach (NN). This technique entails assigning a measure of similarity among areas. The NN approach is straightforward and computationally efficient and may be used for a comprehensive form of programs, inclusive of terrain analysis, item reputation, and far-off sensing photograph class.

Despite the fact that the NN technique has been tested to be beneficial for many programs, there are numerous demanding situations and boundaries related to the approach that may intrude with its hit implementation. For example, the NN approach tends to be touchy to the noise inside the statistics, meaning it may be without problems distracted with the aid of beside-the-point statistics and can fail to recognize styles in the information. Additionally, because the NN technique classifies univariate records on the premise of distance, it could be hard to decide similarities among variables of different scales.

Moreover, the NN technique may be computationally disturbing if massive datasets are being processed, and its accuracy can go through for higher dimensional datasets.

Notwithstanding its demanding situations and barriers, the NN approach has come to be a well-known tool in Earth commentary and can provide valuable insights while used effectively. As more modern strategies are advanced, it's essential to evaluate the strengths and weaknesses of each and determine which technique exceptionally serves the study's wishes at hand. By thinking about the challenges and barriers of the NN method, researchers can increase a powerful strategy for the analysis of earth commentary records and make sure of extra accurate statistics analysis.

9.1 Sensitivity to Dimensionality

Earth commentary has considerably revolutionized our information and capability information of the bodily surroundings surrounding us. For example, satellite tv for PC pix is used to map herbal screw-ups, examine land cowl/land use statistics, and make predictions about future occasions inclusive of weather change impacts. There is a ramification of strategies for extracting significant information from those photos and their statistics, one of which is the nearest-neighbor method. With the advent of better-resolution satellite imagery, the closest-neighbor method has received more attention as it's far more potent in hastily and accurately deriving data from complex patterns of data.

The primary goal of the closest-neighbor technique is to evaluate or more excellent samples using a similarity measure, generally primarily based on Euclidean distance, which determines the "nearest" items to each other within the information set. The nearest-neighbor technique is specifically beneficial while reading satellite images, as it can correctly constitute how similar or varied positive spatial places are to one another. While it could be an effective and powerful device in an analysis of Earth statement data, sensitivity to dimensionality needs to be taken into consideration while applying the approach.

Dimensionality refers to the range of variables, or traits, used to explain a statistics set or region of interest. Extra dimensionality will increase the complexity of the facts set being analyzed and, as a result, impact the accuracy of the closest-neighbor approach. Excessive dimensional information sets typically require more reminiscence, increase the complexity of calculations, and lead to decreased reliability of output. Mainly, because the dimensionality of the statistics set will increase, the Euclidean-based similarity degree produces fewer high confidence suits, resulting in inaccurate series duration estimation and terrible performance of different output variables.

9.2 Impact of Noisy and Incomplete Data

Earth remark (EO) structures offer humanity an adequate supply of facts about the surroundings on a global scale. So as to analyze and interpret the data accrued by way of such systems, scientists have evolved advanced information mining techniques, which include nearest-neighbor (NN) strategies. NN tactics have been used to create efficient classifiers for land cover, anomaly detection, and exchange detection.

However, the accuracy and performance of NN strategies may be compromised if the records are quiet and complete. Noisy facts can distort the effects of the evaluation, and incomplete records can prevent the algorithm from finding the most effective consequences. This paper will explore the effect of noisy and incomplete facts on NN approaches in Earth statement structures. It will additionally compare the effects of noisy statistics and incomplete facts on the accuracy and performance of NN strategies and discuss methods to mitigate the outcomes of noisy and incomplete statistics.

While handling noisy and incomplete facts, one of the best challenges is determining a way to locate and counteract the effects of those records' flaws. Noisy data can arise because of an expansion of things, together with sensor errors, platform movement, atmospheric phenomena, or floor and object motion. As a result, noisy information will motivate the algorithm to be less correct in figuring out patterns or detecting anomalies. Incomplete facts can also be caused by a selection of factors, including the unsuitable sampling of massive datasets, misguided georeferencing of satellite photos, or inadequate ground management factors. Incomplete facts can cause the algorithm to overlook essential functions and restrict the accuracy of analysis, in addition to the velocity of computation.

9.3 Addressing Computational Demands

Earth remark is an invaluable tool in studying the surroundings, and one of the most vital techniques in its miles is the nearest-neighbor approach. Nearest-neighbor techniques are used for classifying land cover, land use mapping, and feature extraction. With cutting-edge advancements in computing, the computational demands of nearest-neighbor procedures have grown to be more challenging to handle. In this essay, we'll compare and analyze the computational demands and performance of famous nearest-neighbor procedures: ok-Nearest associates and Kernel Density Estimation.

K-Nearest Acquaintances (KNN) is a supervised gadget studying algorithm that classifies a given item based totally on its proximity to different objects inside its local location. It is accomplished by calculating the Euclidean distance between gadgets. In terms of computational needs, KNN is extraordinarily stressful due to its requirement for terrific quantities of facts. The large quantities of statistics required by using KNN obligations can quickly end up overwhelming, in particular for huge-scale statistics sets. Furthermore, the computations involved in locating the nearest pals are intensive and computationally steeply-priced, with even minor enhancements to the accuracy of the predictive version requiring a brilliant deal of computational power.

The second technique we'll recollect is Kernel Density Estimation (KDE). KDE is an unmonitored machine mastering method that is used to determine the density of information set by way of assigning chances to information points based totally on their proximity to different centers in the facts set. At the same time, as it does not require any labels for education, it is still computationally annoying because of the want for a large variety of fact points and the need to perform computations associated with the diploma of density discovered inside the records. Additionally, because of its kernel-primarily based approach to statistics analysis, it additionally calls for computationally intensive calculations for generating a correct evaluation of a given records set.

10. FUTURE DIRECTIONS AND INNOVATIONS

Earth Remark incorporates technology and techniques used to song and examine the functions and sources of our planet. Nearest-neighbor tactics in Earth's statement are strategies that become aware of factors observed in a specific location, including the land, atmosphere, or oceans. Those techniques look for factors that are close to the details of the hobby and use this data to understand the surroundings better. This kind of era has applications in a ramification of industries, including agriculture, environmental monitoring, and even safety and surveillance.

Those techniques could have an incredible impact on weather exchange, natural disasters, and other phenomena that affect our planet. They may be used to measure numerous variables and may even assist in the identification of human sports. It allows us to innovate and develop better ways to apply our natural assets. Those strategies were implemented already in a selection of applications. They're used to locate fires, for instance, and to track water resources, land use, sedimentation, and more.

The potential of this era is widespread, and its potential will continue to increase in the destiny. It can offer a more accurate and complete analysis of Earth statement responsibilities and programs. It can be used to locate and tune modifications in land use, soil erosion, pollution intensity, and much more.

10.1 Advances in Distance Metrics and Similarity Measures

The advances in distance metrics and similarity measures of nearest-neighbor techniques have had a massive impact on the sector of earth remark. Increasingly, more researchers are turning to these strategies with a view to recognizing the environment around us. With the aid of providing a green manner of studying facts on a global scale, nearest-neighbor methods have emerged as a treasured device for furthering our understanding of the planet. This essay will analyze the advances in distance metrics and similarity measures of nearest-neighbor processes, overlaying elements of each information processing and comparison.

First of all, it is essential to define the terms "distance metrics" and "similarity measures." Distance metrics are techniques used to decide the physical distance among two or more points. Typically used distance metrics include Euclidean distance, any distance, and Minkowski distance. Meanwhile, similarity measures are techniques used to determine the similarity among two or more significant points. These measures vary depending on the dreams of the research, but usually used metrics consist of cosine similarity, Pearson correlation, and Mahalanobis distance.

10.2 Nearest-Neighbor Approaches in the Era of Big Earth Data

The generation of vast amounts of Earth information is hastily accelerating, and with it, they want to make accurate and well-timed choices based totally on these records. As new technological advancements change the way we view and interpret Earth observation information, the want for techniques that may benefit from those advancements has expanded exponentially. Nearest-neighbor processes in Earth remark are one such technique that has received lots of traction in recent studies because of its excessive analytical energy and scalability.

Nearest-neighbor strategies, or nearest-neighbor queries (NNQs), are algorithms used to retrieve item values from a dataset on the idea in their similarity to a given question. By using the usage of spatial criteria or a combination of spatial and different standards, NNQs can become aware of the closest "neighbor" to any question. These procedures are typically used in Earth Remark to perceive items just like an object observed in a photo or satellite imagery. For example, NNQs can be used to perceive the nearest item in a chain of satellite photographs of a geographic region or to become aware of the nearest item whose houses suit a set of predetermined standards.

10.3 Collaborative Research Opportunities

The Earth has been a source of discovery and fascination for people for centuries, and in current years, technological advances have revolutionized our knowledge of our planet. To further enhance our knowledge of the Earth and its tactics, collaborative research opportunities focused on nearest-neighbor techniques in Earth commentary have been tested to be a productive and effective way to look at and examine our environment.

Nearest-neighbor processes are based on locating the "nearest neighbor" of an object in an area (image, geographical coordinate, and so forth.). They may be utilized in a ramification of applications, mainly in Earth observation. Near-neighbor techniques can be used to stumble on change in a particular vicinity, which includes how pollution levels have elevated or adjustments in a land cowl. It's also beneficial for correctly detecting gadgets and features in extraordinarily massive datasets. As an example, close-to-neighbor techniques may be used to locate roads in an aerial image of a town.

11. CONCLUSION

Earth remark has been a first-rate thing in expertise on the Earth's floor and its diverse environmental additives. Faraway sensing and geographic statistics machine (GIS) technology were used to study, analyze, and screen the Earth's floor. A few of the techniques utilized in earth observation nearest-neighbor procedures have been tested to be specifically helpful and influential in environmental evaluation.

The use of nearest-neighbor techniques has led to the identity of numerous excellent strategies, such as inverse distance weighting, triangulated abnormal networks, and nearby frequency techniques, each of which has unique programs. For instance, inverse distance weighting is used for facts interpolation. It may be carried out to any number of environmental metrics, while triangulated abnormal networks are used to model complicated styles in massive datasets. The local frequency technique can help discover areas of excessive and coffee environmental hobby.

Despite the fact that there is no person-length-suits-all answer for analyzing the surroundings and making informed selections, nearest-neighbor procedures provide numerous benefits, which include quicker computation times, accuracy, robustness, and automated facts processing. Those methods can be beneficial when assessing environmental change or environmental change.

REFERENCES

Adeniyi, E. A., Ogundokun, R. O., Gbadamosi, B., Misra, S., & Kalejaiye, O. (2022). Classification of swine disease using K-nearest neighbor algorithm on cloud-based framework. In *Artificial Intelligence for Cloud and Edge Computing* (pp. 71–90). Springer International Publishing. doi:10.1007/978-3-030-80821-1_4

Biswas, N., Uddin, K. M. M., Rikta, S. T., & Dey, S. K. (2022). A comparative analysis of machine learning classifiers for stroke prediction: A predictive analytics approach. *Healthcare Analytics*, *2*, 100116. doi:10.1016/j.health.2022.100116

Chen, H., Qin, Z., Zhai, D. L., Ou, G., Li, X., Zhao, G., Fan, J., Zhao, C., & Xu, H. (2022). Mapping Forest Aboveground Biomass with MODIS and Fengyun-3C VIRR Imageries in Yunnan Province, Southwest China Using Linear Regression, K-Nearest Neighbor and Random Forest. *Remote Sensing (Basel)*, *14*(21), 5456. doi:10.3390/rs14215456

Delfan, E., Naghavi, H., Maleknia, R., & Nouredini, A. (2022). Comparing the Capability of Sentinel 2 and Landsat 8 Satellite imagery in land use and land cover mapping using pixel-based and object-based classification methods. *Desert Ecosystem Engineering*, *8*(25), 1–12.

Fan, R., Ding, Y., Zou, Q., & Yuan, L. (2023). Multi-view local hyperplane nearest neighbor model based on independence criterion for identifying vesicular transport proteins. *International Journal of Biological Macromolecules*, *247*, 125774. doi:10.1016/j.ijbiomac.2023.125774 PMID:37437677

Fathabadi, A., Seyedian, S. M., & Malekian, A. (2022). Comparison of Bayesian, k-Nearest Neighbor and Gaussian process regression methods for quantifying uncertainty of suspended sediment concentration prediction. *The Science of the Total Environment*, *818*, 151760. doi:10.1016/j.scitotenv.2021.151760 PMID:34801498

He, S., Li, J., Wang, J., & Liu, F. (2022). Evaluation and analysis of upscaling of different land use/land cover products (FORM-GLC30, GLC_FCS30, CCI_LC, MCD12Q1 and CNLUCC): A case study in China. *Geocarto International*, *37*(27), 17340–17360. doi:10.1080/10106049.2022.2127926

Kaur, J., Saxena, J., Shah, J., & Yadav, S. P. (2022, May). Facial Emotion Recognition. In *2022 International Conference on Computational Intelligence and Sustainable Engineering Solutions (CISES)* (pp. 528-533). IEEE.

Khotimah, B. K. (2022). Performance of the K-nearest neighbors method on identification of maize plant nutrients. *Jurnal Infotel*, *14*(1), 8–14. doi:10.20895/infotel.v14i1.735

Li, L., Chen, X., & Song, C. (2022). A robust clustering method with noise identification based on directed K-nearest neighbor graph. *Neurocomputing*, *508*, 19–35. doi:10.1016/j.neucom.2022.08.029

Meng, M., Toan, T. D., Wong, Y. D., & Lam, S. H. (2022). Short-term travel-time prediction using support vector machine and nearest neighbor method. *Transportation Research Record: Journal of the Transportation Research Board*, *2676*(6), 353–365. doi:10.1177/03611981221074371

Nasser, F. K., & Behadili, S. F. (2022). Breast cancer detection using decision tree and k-nearest neighbour classifiers. *Iraqi Journal of Science*, 4987–5003. doi:10.24996/ijs.2022.63.11.34

Nguyen, L. T., Vu, D. H., Vu, N. C., Dao, V. H., & Tran, T. H. (2022, July). Comparative study on super resolution techniques for upper gastrointestinal endoscopic images. In *2022 IEEE Ninth International Conference on Communications and Electronics (ICCE)* (pp. 375-380). IEEE. 10.1109/ICCE55644.2022.9852031

Orusa, T., Cammareri, D., & Borgogno Mondino, E. (2022). A Scalable Earth Observation Service to Map Land Cover in Geomorphological Complex Areas beyond the Dynamic World: An Application in Aosta Valley (NW Italy). *Applied Sciences (Basel, Switzerland)*, *13*(1), 390. doi:10.3390/app13010390

Oza, A., & Bokhare, A. (2022, July). Diabetes prediction using logistic regression and K-nearest neighbor. In *Congress on Intelligent Systems: Proceedings of CIS 2021,* Volume 2 (pp. 407-418). Singapore: Springer Nature Singapore. 10.1007/978-981-16-9113-3_30

Prasad, P., Loveson, V. J., Chandra, P., & Kotha, M. (2022). Evaluation and comparison of the earth observing sensors in land cover/land use studies using machine learning algorithms. *Ecological Informatics*, *68*, 101522. doi:10.1016/j.ecoinf.2021.101522

Singh, H., Sharma, V., & Singh, D. (2022). Comparative analysis of proficiencies of various textures and geometric features in breast mass classification using k-nearest neighbor. *Visual Computing for Industry, Biomedicine, and Art*, *5*(1), 1–19. doi:10.118642492-021-00100-1 PMID:35018506

Srivastava, S., Dhyani, N., Sharma, V., Vats, S., Yadav, S. P., Kukreja, V., & Singh, R. (2023, May). Lung Infection and Identification using Heatmap. In *2023 2nd International Conference on Applied Artificial Intelligence and Computing (ICAAIC)* (pp. 1093-1098). IEEE. 10.1109/ICAAIC56838.2023.10140204

Sun, X., Opulencia, M. J. C., Alexandrovich, T. P., Khan, A., Algarni, M., & Abdelrahman, A. (2022). Modeling and optimization of vegetable oil biodiesel production with heterogeneous nano catalytic process: Multi-layer perceptron, decision regression tree, and K-Nearest Neighbor methods. *Environmental Technology & Innovation*, *27*, 102794. doi:10.1016/j.eti.2022.102794

Suyanto, S., Meliana, S., Wahyuningrum, T., & Khomsah, S. (2022). A new nearest neighbor-based framework for diabetes detection. *Expert Systems with Applications*, *199*, 116857. doi:10.1016/j.eswa.2022.116857

Thompson, A. E., Walden, J. P., Chase, A. S., Hutson, S. R., Marken, D. B., Cap, B., Fries, E. C., Guzman Piedrasanta, M. R., Hare, T. S., Horn, S. W., Micheletti, G. J., Montgomery, S. M., Munson, J., Richards-Rissetto, H., Shaw-Müller, K., Ardren, T., Awe, J. J., Brown, M. K., Callaghan, M., & Chase, D. Z. (2022). Ancient Lowland Maya neighborhoods: Average Nearest Neighbor analysis and kernel density models, environments, and urban scale. *PLoS One*, *17*(11), e0275916. doi:10.1371/journal.pone.0275916 PMID:36322539

Wu, L., Pleiss, G., & Cunningham, J. P. (2022, June). Variational nearest neighbor Gaussian process. In *International Conference on Machine Learning* (pp. 24114-24130). PMLR.

Yu, X., Qin, W., Lin, X., Shan, Z., Huang, L., Shao, Q., Wang, L., & Chen, M. (2023). Synergizing the enhanced RIME with fuzzy K-nearest neighbor for diagnose of pulmonary hypertension. *Computers in Biology and Medicine*, *165*, 107408. doi:10.1016/j.compbiomed.2023.107408 PMID:37672924

Zamouche, M., Chermat, M., Kermiche, Z., Tahraoui, H., Kebir, M., Bollinger, J. C., Amrane, A., & Mouni, L. (2023). Predictive Model Based on K-Nearest Neighbor Coupled with the Gray Wolf Optimizer Algorithm (KNN_GWO) for Estimating the Amount of Phenol Adsorption on Powdered Activated Carbon. *Water (Basel)*, *15*(3), 493. doi:10.3390/w15030493

Zhang, C., Zhong, P., Liu, M., Song, Q., Liang, Z., & Wang, X. (2022). Hybrid metric K-Nearest neighbor algorithm and applications. *Mathematical Problems in Engineering*, *2022*, 1–15. doi:10.1155/2022/8212546

Chapter 18
Neuro-Inspired Algorithms for Earthquake Monitoring and Prediction

Priyanka Pramila R.
https://orcid.org/0000-0001-7066-9074
Department of Artificial Intelligence and Data Science, R.M.K. Engineering College, India

Sadish Sendil Murugaraj
Department of Emerging Technologies, Guru Nanak Institute of Technology, Ibrahipatnam, India

Sunil Kumar
Faculty of Commerce and Management, SGT University, Gurugrm, India

Balaji S. R.
https://orcid.org/0000-0001-6203-7317
Faculty of Commerce and Management, SGT University, Gurugrm, India

ABSTRACT

Neuro-inspired algorithms are a powerful device that can be used for earthquake monitoring and prediction. These algorithms can offer precious insights into the conduct of seismic activities and could cause better detection and caution structures. The goal of this chapter is to offer a concise evaluation of cutting-edge research on using neuro-inspired algorithms for earthquake tracking and prediction. In particular, the chapter will discuss the latest traits in information-based procedures and deep studying algorithms, in addition to the challenges and possibilities for advancing research in this area. Furthermore, capacity programs and regions of research will be discussed which could enjoy the use of these algorithms. Ultimately, this chapter can provide an impetus for the broader use of such algorithms and similarly progress in earthquake-related studies.

DOI: 10.4018/979-8-3693-1850-8.ch018

1. INTRODUCTION

Earthquakes are natural disasters that can wreak huge-scale destruction on both assets and lives. As the significance of earthquakes increases, the resulting harm may be catastrophic. With the capability to stumble on earthquakes and offer an early warning, it's miles feasible to reduce the destructive impacts of these disasters. That is where Neuro-stimulated Algorithms for Earthquake monitoring and Prediction come in (Tiwari et al., 2021).

Neuro-stimulated Algorithms for Earthquake monitoring and Prediction (NEMEP) is an evolved subject of artificial intelligence (AI) that utilizes neural community algorithms to detect and expect the characteristics of earthquakes. The algorithms are designed to interpret seismic readings and identify styles or developments within the frequency and importance of earthquakes (Seyyedi et al., 2023). This technology aims to create a system able to recognize when an earthquake is forthcoming and alert the government so as to set off crisis management protocols and evacuate doubtlessly affected populations (Ma et al., 2023).

This technology gives some advantages. As the algorithms are capable of extra as it should become aware of seismic alerts and determine the importance of upcoming earthquakes, the possibilities of survival increase considerably. Furthermore, as the algorithms may be unexpectedly deployed within the field, they may be used to provide early warnings in real-time. This allows for faster reaction times, decreasing the potential damage from an earthquake .

Neuro-inspired algorithms are a class of computer algorithms inspired by the neuronal structure and functions of the human brain, and are often designed to function as an Artificial Neural Network (ANN). Neuro-inspired algorithms are applied to a variety of disciplines and aim to solve difficult problems that require finding solutions without a predetermined approach. Through their use of mathematical models such as non-linear regression, neural networks are well suited for solving problems such as image recognition, natural language processing, and predictive modeling. Neuro-inspired algorithms typically use a collection of connected processing elements (neurons) to process information, which are all linked and organized into a single interconnected structure. Each neuron consists of a number of weighted inputs, a non-linear activation function, and a single, weighted output that can either activate other neurons, or connect with other branches of the network. The connection strengths (weights) between the neurons define the behavior of the algorithm and allow it to learn from past experience. Neuro-inspired algorithms typically use a variation of backpropagation for training. This is a process in which each neuron is first tested and then its weights are adjusted based on the results. The entire algorithm is trained by continuously adjusting the weights until the desired behavior is observed. Some of the more notable applications of neuro-inspired algorithms include game playing, where such algorithms have been used to teach a computer to play a game and then improve its performance; facial recognition, using deep learning; robotics; drug discovery; and machine learning. Many neuro-inspired algorithms are used in robotics applications, such as in navigation and decision-making. In conclusion, neuro-inspired algorithms are a powerful tool for solving complicated problems, and they have seen increasing use in a variety of applications. Their ability to learn and adapt to changing circumstances makes them appealing for many scenarios, and they are well-suited for complex tasks such as image recognition, natural language processing, and robotics (Kaur et al., 2022).

1.1 Background of Earthquake Monitoring and Prediction

Earthquakes have been monitored and studied for more than a century. On the grounds that since 1900, seismologists have been capable of appropriate degree and stumble on earthquakes the use of a selection of tracking structures, which includes seismographs and sophisticated seismometers. These devices measure adjustments in floor motion due to seismic waves and offer treasured data about earthquake value, place, and epicenter (Jana & Nagarajaiah, 2022).

On the side of the development in generation, seismic networks have come to be more accurate and dependable for right away and precisely locating earthquakes. Those cutting-edge seismic networks are composed of stations that measure and report seismic waves in exclusive parts of the world and perceive seismic alerts from around the world within mins. Facts generated from these station networks may be effectively used to reveal seismic hobby in a given place and to analyze aftershock styles and danger resulting from beyond earthquakes (Yadav et al., 2023).

Other than earthquake tracking, research on the subject of earthquake prediction has made enormous progress over the years. Numerous theories of earthquake prediction have been recommended, imparting various criteria for correct seismic forecasting, which include a speedy increase in ground stresses prior to the earthquakes, unusual patterns of ground motion, and variations in local magnetic fields. But, such methods are but to provide dependable and sensible forecasts of the occurrence and nature of earthquakes (Ramírez-Mendoza et al., 2023).

1.2 Significance of Neuro-Inspired Algorithms

Using neuro-stimulated algorithms for earthquake tracking and prediction is a relatively new and powerful approach in seismology. Neuro-inspired algorithms are machine-gaining knowledge that attracts concepts from the structure and behavior of biological neural networks. Neural networks have been historically used for class and pattern recognition responsibilities (Russo & Coco, 2023). Still, over the previous couple of years, these computational strategies have been more and more carried out to a variety of seismic phenomena. Neuro-stimulated algorithms are especially appropriate for earthquake monitoring and prediction due to their capability to detect complex styles and houses in seismic statistics that might not be without problems detected by the use of traditional techniques (Birkoben & Kohlstedt, 2023).

Neuro-inspired algorithms can signify and discover homes in seismograms that are tough to hit upon or quantify via manual evaluation. The algorithms are able to objectively discover diffused seismic power modifications, micro-earthquakes, aftershocks, and different functions that may be without problems. The usage of these algorithms can help become aware of and tune the onset and development of seismic interest in actual time, bearing in mind more excellent effective monitoring and better prediction of earthquakes. This will doubtlessly reduce damage and save lives (Dunham, 2022).

1.3 Objectives of Applying Neuro-Inspired Approaches

Earthquake tracking and prediction is an integral part of current herbal threat management, with both monetary and human protection implications. Neuro-inspired algorithms can revolutionize earthquake tracking and prediction efforts via the use of computational fashions that mimic how the human mind tactics and responds to seismic statistics. This paper outlines a number of the objectives for applying neuro-stimulated strategies for earthquake tracking and prediction (Lawrie, 2022).

First, neuro-stimulated algorithms can offer more accurate earthquake predictions faster than traditional models, consisting of those based on physics-based simulation. As neuro-stimulated models end up extra superior, they could learn from seismic records, which will make more excellent and accurate predictions. For instance, they may be trained on both seismic and non-seismic statistics with a view to seize the complicated relationships between exceptional parameters and the capability earthquake's value and epicenter (Hirtzlin, 2020).

2nd, these algorithms can lessen the number of fake alarms due to earthquakes that occur too long way away to be immediately felt by humans on the ground. Such false alarms would affect the human evacuation time, leading to loss of existence and economic damage.

The basic architecture or the fundamental process of Neuro-inspired Algorithms for Earthquake Monitoring and Prediction.

- Data Acquisition: Collect seismic data from the digital seismometer network.
- Feature Extraction: Extract the relevant features (e.g. amplitude, frequency, waveform shape) from the seismic data.
- Neural Network Training: Train a neural network to learn the patterns from the feature extraction step.
- Model Learning: Develop a model based on the trained neural network to learn and predict future seismic activity.
- Model Testing: Test the accuracy of the trained model on unseen data to determine its reliability.
- Earthquake Prediction: Apply the trained model to new seismic data to predict upcoming seismic events.

1.4 Overview of the Structure of the Book

The ebook entitled "Neuro-inspired Algorithms for Earthquake Monitoring and Prediction" gives an in-depth investigation using neuro-stimulated algorithms to display and expect earthquakes. Written with the aid of leading professionals in the field of earth science, this ebook seeks to, in addition, our know-how of the seismic sports that can affect our environment and the capability to defend lives and property from viable disasters (Adam et al., 2019).

The ebook is organized into four excellent elements, which include several chapters. Component I examines the importance of seismic monitoring and evaluation of potential dangers and discusses the various techniques used to screen and predict earthquake pastimes. It explores topics like waveform analysis, seismic forecasting, and statistics series (Shibuya, 2020).

Part II gives an introduction to neuro-stimulated algorithms utilized in earthquake monitoring and prediction. The authors speak about how deep knowledge of systems may be utilized to construct extra correct predictive models and how the neuroscience-primarily based technique has improved the accuracy of earthquake predictions within the previous couple of a long times. This element covers standard applications of neuro-inspired algorithms for earthquake prediction, inclusive of "synthetic Neural community" and "help Vector device." the authors also discuss other contemporary neural network fashions for predicting earthquakes (Yelick et al., 2021).

2. NEURO-INSPIRED COMPUTING IN EARTH SCIENCES

Earthquakes are some of the most unfavorable herbal catastrophes, and their prevalence and depth are extremely difficult to expect with accuracy. Scientists have long sought ways to screen and forecast seismic events to improve safety and keep lives. Despite the fact that traditional seismological tactics offer insights into earthquake pastimes, they need to be able to offer reliable early warnings. Currently, researchers have all started to discover a brand new method of earthquake monitoring and prediction - Neuro-stimulated computing (Chatterjee, 2021).

Neuro-stimulated algorithms are information-pushed processes to analysis, combining strategies from computational neuroscience and synthetic intelligence to model and simulate the workings of the human mind. They are exceptionally perfect for processing complicated information sets consisting of those generated by seismograph networks. In contrast to conventional signal evaluation strategies, Neuro-stimulated algorithms permit for extra correct and well-timed detection of earthquake traits, inclusive of importance, epicenter, and floor motion (Yuvaraj, Srihari, Dhiman et al, 2021).

Inside the case of earthquakes, Neuro-stimulated algorithms may be used to come across certain styles in seismic statistics, which could indicate a coming near temblor. For instance, a Neuro-stimulated algorithm could be set to investigate seismic facts from a seismograph community on the way to perceive any anomalies in the waveforms, which might be a demonstration of a drawing close earthquake.

2.1 Introduction to Neuro-Inspired Algorithms

Neuro-stimulated algorithms are a kind of artificial intelligence that utilizes algorithms based totally on neural networks, a method of mimicry stimulated by the human brain. It is revolutionizing fields, from PC science to robotics to facts evaluation. This method has been applied to earthquake tracking and prediction, allowing scientists to detect and predict earthquakes more appropriately and effectively than ever before (Kannan et al., 2021).

Earthquakes are powerful and destructive forces of nature. Predicting them has long been a venture, and relying on seismological records can handiest provide predictions inside a constrained variety of time and opportunity. Neuro-inspired algorithms can help unlock more accurate predictions through their capability to learn from input statistics and expand "neurons" that model connections among information. Through thinking of the data beyond earthquakes along different variables such as geological strain, deformation, and seismic alerts, algorithms can better hit upon and, as should be expected, the incidence of destiny earthquakes.

Neuro-inspired algorithms are also being used to develop early warning structures, creating chance models to perceive seismic activities earlier than they arise and prevent their destruction. Such structures may be educated from diverse seismic styles launched previous to essential earthquakes (Yuvaraj, Karthikeyan, & Praghash, 2021). They can become aware of styles in the data to alert authorities earlier, which will reduce casualties and mitigate damage.

2.2 Neural Networks and Their Applications

Neural networks were offering a superior performance in a spread of tasks that traditional computing techniques, including rule-based fashions, didn't deal with. At the same time, the successes of neural networks are, in most cases, attributed to their capability for stylish sample recognition. Lately, it has been recognized that a number of the most noteworthy programs of neural networks were inside the region of seismic evaluation. Due to their extremely sophisticated and complex systems, some researchers have proposed the use of neuro-inspired algorithms for earthquake tracking and prediction (Rajalakshmi et al., 2022).

Neural networks are able to detect diffused variations in facts and may perceive styles that conventional algorithms won't be able to differentiate. This makes them a beneficial device for understanding and predicting the behavior of seismic occasions. For instance, a seismic evaluation of an earthquake's indicators can be used to extract non-trivial features that may be used for early caution systems. The ability to quickly and as it should discover styles within the information allows for more comprehensive motion analysis of seismic occasions. This could be useful for brief-time period prediction of seismic activities and to pick out the earthquake's magnitude. Furthermore, neural network systems can be mixed with heuristic and rule-based reasoning techniques to as it should be predicting the places of earthquakes and their respective magnitudes (Khan et al., 2022).

2.3 Relevance of Neuro-Inspired Computing in Earth Sciences

Earthquakes are a herbal incidence of seismic interest that can cause catastrophic degrees of destruction to human populace facilities around the sector. But in spite of advances in the examination of seismic interest, appropriately predicting earthquakes has remained possible. Neuro-inspired computing gives a potential technique to the trouble of earthquake monitoring and prediction. Using such algorithms and laptop modeling ought to increase our capacity to forecast earthquakes in advance, allowing us to mitigate risks and store lives.

Neuro-inspired algorithms are based totally round, mimicking the mind's capability to system and keep information. The "neural networks" used for this cause are complex computer systems composed of interconnected neurons, which examine massive datasets using state-of-the-art mathematical fashions. These networks were applied in lots of exclusive fields, from economics to robotics, but can also be applied for the observation of seismic events (Gunjan et al., 2022).

Inside the look at earthquakes, neuro-inspired algorithms allow for the improvement of superior models that could manner facts related to earthquake behavior, which include waveform functions, seismic anisotropy, velocity changes, local seismic pastime, geotechnical measurements, and more significant.

3. FUNDAMENTALS OF EARTHQUAKE MONITORING

Earthquakes are one of the most effective seismic phenomena that have been studied for hundreds of years, and they can have devastating outcomes once they occur. The ability to reveal and predict earthquakes is critical, which will prevent a lack of lifestyles and mitigate the outcomes of the tremors. However, tracking earthquakes is a challenging assignment as they frequently arise without warning and can arise in far-off areas that might need access to conventional monitoring devices.

In recent years, neural networks have been used to expand algorithms for earthquake monitoring and prediction. These algorithms, known as Neuro-inspired algorithms, try to emulate the way the human brain strategies statistics with a view to constructing predictive models. Neuro-stimulated algorithms have been correctly used in lots of applications, starting from independent automobiles to sickness prognosis.

The success of Neuro-inspired algorithms for earthquake monitoring and prediction relies upon an intensive understanding of the basics of earthquake monitoring. In particular, four primary elements allow you to screen earthquakes efficiently: seismograph, seismogram, seismicity, and geodesy. Seismographs are gadgets used to measure floor motion because of earthquake hobby, while seismograms are graphical records that can be produced with the aid of seismographs (Chiang et al., 2022).

3.1 Understanding Seismic Activity

Seismic pastime is the motion of the earth's crust resulting from earthquakes or other events. It's far a chief herbal hazard that may cause destruction and loss of life, which affects many humans around the arena. As such, expertise in the seismic hobby and predicting its occurrence is a critical challenge for the scientific network. However, seismic interest is surprisingly unpredictable and subject to a wide variety of impacts, making it a challenging phenomenon to screen and observe. Fortunately, current advances in neuroscience and laptop technology have enabled new methods of knowledge and predicting seismic pastimes. One such method is the usage of neuro-stimulated algorithms for earthquake tracking and prediction (Mills et al., 2022).

Neuro-stimulated algorithms are primarily based on the concepts of the human mind, using synthetic intelligence to method statistics and make decisions. Those algorithms are in a position to research various types of seismic facts, together with floor movement readings, seismic waves, and meteorological measurements, after which we are expecting future seismic pastimes. By doing so, these algorithms can offer valuable insights into seismic activity and may even be capable of predicting positive sorts of earthquakes earlier than they arise.

3.2 Existing Approaches to Earthquake Monitoring

Earthquakes are powerful and lethal natural phenomena that can motivate a lack of existence, irreparable destruction of infrastructure, and intense economic difficulty. With the upward thrust of the population in seismic assets of seismic waves, earthquake monitoring and prediction have come to be increasingly crucial. To effectively display and expect earthquakes, superior methods were developed in numerous fields along with seismology, geophysics, geodesy, faraway sensing, pc technological know-how, and synthetic intelligence.

Neuro-stimulated algorithms provide a promising tool for earthquake tracking and prediction. This is due to the fact they own many suitable residences, which include scalability, actual time processing, and low false alarm charges. Neuro-stimulated algorithms allow for the automatic extraction of capabilities from seismic recordings that may be used for regression techniques such as most probability Estimation (MLE) or multiple Linear Regression (MLR). Moreover, deep neural networks (DNN) may be leveraged to study earthquake styles from beyond seismograms and may be used to predict the magnitudes of future earthquakes (Mills et al., 2023).

4. NEURO-INSPIRED ALGORITHMS OVERVIEW

Earthquake monitoring and prediction is a giant area in chance management and earthquake engineering. With the advancements in technology over the past few decades, the improvement of neuro-inspired algorithms has performed a prime component in assisting higher expectations and revealing earthquakes. Neuro-stimulated algorithms are an artificial intelligence approach that adopts the concepts of biological neural networks and applies them to problems related to geophysics. This essay will provide an outline of neuro-inspired algorithms for earthquake tracking and prediction.

Neuro-inspired algorithms use the computational principles of the mind and neural networks to broaden deep learning strategies. Deep gaining knowledge is a subfield of the system that uses massive datasets to learn complicated obligations without counting on predefined regulations and strategies. In the subject of geophysics, these algorithms are used to capture and interpret the complicated and uncertain statistics associated with earthquakes. They're capable of becoming aware of patterns in seismic facts that may assist in better know-how of the mechanics and behaviors of earthquakes. Neuro-inspired algorithms are used to construct earthquake-tracking structures and to make predictions about destiny seismic occasions.

4.1 Artificial Neural Networks (ANNs)

In recent years, synthetic Neural Networks (ANNs) have been validated to be an effective tool for growing neuro-stimulated algorithms for earthquake tracking and prediction. ANNs are modeled after biological neural networks that are composed of a massive range of neurons and are capable of processing and deriving that means from extraordinarily vast quantities of records. The neurons are linked to each other in a network, permitting the machine to "examine" patterns and extrapolate new facts from the inputs it receives.

ANNs may be used for a variety of purposes in the earthquake monitoring and prediction domain. For example, ANNs can be used to generate forecasts for the onset of seismic interest in a given vicinity instead of relying on statistical models, which can be misguided or inadequate. Additionally, ANNs are capable of reading several variables within a seismic dataset, together with geophysical and geochemical functions, and imparting correct predictions for while and in which seismic activities may arise.

4.2 Spiking Neural Networks (SNNs)

Earthquakes are some of the most influential and detrimental forces of nature, capable of lowering complete towns to rubble and killing hundreds. Alas, our modern-day techniques of prediction and monitoring are restrained, relying primarily on previous techniques of statistics series and evaluation. Spiking neural networks (SNNs) provide an innovative and potentially powerful manner of addressing this problem. By way of modeling neurons and their interactions, SNNs can mimic the mind's capacity to manner and interpret facts, allowing scientists to check seismic activity and expect when and where an earthquake may strike.

SNNs appoint a shape of synthetic intelligence called Neuro-stimulated Algorithms (NIA), which mimic biological brain capabilities. They include input neurons that collect information about seismic pastime, including vibrations from the ground, and can locate even minor adjustments inside the surroundings. Output neurons then manner the information and generate a document. The NIA system can apprehend styles within the accumulated records, making it more straightforward to identify styles in activity that would characterize potential earthquakes.

4.3 Artificial Neural Networks(ANNs) vs. Spiking Neural Networks (SNNs)

Artificial Neural Networks (ANNs) approximate the workings of a biological brain, with computationally simulated 'neurons' linked together in layers. ANNs are typically designed and trained for supervised machine learning tasks, and are used for pattern recognition and feature selection. In contrast, Spiking

Neural Networks (SNNs) are biologically-ninspired models that attempt to simulate biological neural cells, including their spiking dynamics. Unlike ANNs, SNNs are used to model real neurons and to create artificial brains. They are trained using more complex learning algorithms, such as reinforcement learning, and are capable of learning temporal patterns.

5. DATA ACQUISITION AND PREPROCESSING

Data acquisition and preprocessing for neuro-stimulated algorithms for earthquake monitoring and prediction have become more and more critical to earthquake scientists and engineers. In recent years, the development of efficient algorithms designed to system large seismic datasets has caused the development of earthquake tracking and prediction accuracy. Such algorithms can span more than one region of seismology along with earthquake-supply characterization, evaluation of excessive-frequency seismic waveforms, and seismic-threat evaluation.

Earthquakes usually generate alerts with a wide frequency variety, from 0.001 Hz to one hundred Hz. These frequencies carry critical statistics about the supply of the earthquake and its propagation through the Earth's crust. As such, facts acquisition and preprocessing of seismic alerts can assist in describing those phenomena in more detail.

One of the most critical records preprocessing steps for earthquake tracking and prediction is signal filtering. Filtering eliminates noise from the signal, allowing the seismic electricity concentrations that might be applicable for analysis to be diagnosed. Filtering the statistics may additionally screen more disconnected alerts that may be analyzed. For instance, sign filtering also screen distinctive frequency domain names that have excellent signatures, which could assist in picking out the supply of an earthquake.some examples are given below:

1. In medical imaging data acquisition for Magnetic Resonance Imaging (MRI), the scanner requires the patient to lie inside a tube to generate a high-resolution image of the relevant body parts. Preprocessing can include soft tissue enhancement and noise filtering.
2. In manufacturing, data acquisition sensors are used to measure the parameters of a process. Preprocessing can involve calibrating the sensor, filtering out noise, adjusting for drift and normalizing the data so that certain parameters can be compared.
3. Data acquisition and preprocessing of satellite images may involve downloading the images from a remote repository, performing quality checks, correcting for lens distortion, mosaicking and image rectification, color balancing and contrast enhancement.
4. In the field of remote sensing, raw lidar data is collected from a laser scanner. Preprocessing can involve registering overlapping data into a 3D point cloud and removing or filling data gaps resulting from blocked laser beams.

5.1 Seismic Data Characteristics

Earthquakes are an herbal phenomenon that generally causes destruction and lack of life once they strike. To ensure the protection of individuals residing in seismically lively regions, seismologists look to earthquake tracking and prediction algorithms for early warning structures. Growing sturdy algorithms that could, as should be expected, increase the prevalence and magnitude of an earthquake is an important aim of seismologists. Neuro-stimulated algorithms play an essential function in earthquake tracking and prediction when you consider that they may be capable of managing massive quantities of seismic records with improved accuracy.

Neuro-stimulated algorithms involve the usage of synthetic Neural Networks (ANNs) to assemble fashions for earthquake monitoring and prediction. ANNs are an implementation of "mind-like" networks that permit the analysis of previously unobtainable seismic data. Those networks combine the traits of neural networks with information mining algorithms and can be used to recognize styles in seismic pastimes. This facts-driven approach can monitor hidden features in seismic facts that are not without difficulty detectable using conventional methods. Moreover, with more correct predictions, earthquakes may be more significantly successfully controlled, and failures can be prevented.

5.2 Preprocessing Techniques for Neuro-Inspired Models

Preprocessing strategies are essential for any successful software of neuro-inspired algorithms for earthquake monitoring and prediction. These strategies allow for the advent of a programmatic version that may effectively analyze seismic facts and discover patterns or traits that could inform predictions about destiny activities. Preprocessing strategies, which vary in complexity, can encompass fact cleansing, normalization, function extraction, dimensionality reduction, sampling strategies, and discretization. In this essay, we will talk about the numerous preprocessing techniques that may be used to enhance neuro-inspired fashions for earthquake monitoring and prediction.

Statistics cleansing is a crucial preprocessing step for any device getting to know the set of rules, and neuro-stimulated models are not any exception. Cleansing seismic information involves doing away with any garbage records that may be gifted, in addition to ensuring statistics integrity. This could include verifying the accuracy of measurements along with value and vicinity, in addition to applying uniformity in values, including assigning default values to lack values. Facts cleaning is a vital step to ensure the accuracy of the determined facts earlier than further processing can take area.

6. CASE STUDIES: APPLICATIONS OF NEURO-INSPIRED ALGORITHMS

There are various case research relating to the software of neuro-stimulated algorithms for earthquake prediction and tracking. Studies that have been especially powerful in this area are mentioned below.

The primary case observation changed into conducted in Taft, California, to expect the timing of future seismic activities. The examination used neuro-stimulated algorithms based on artificial neural networks (ANNs) to research large amounts of seismicity information amassed over some time. These records became used to generate a prediction model that may pick out positive patterns and parameters of seismic disturbances, supporting researchers to forecast the time and region of destiny seismic events. The results of the look showed that the version became capable of appropriately expecting the timing of seismic occasions with a 95 percent accuracy. This location suggests that using neuro-stimulated algorithms can be effective in predicting earthquakes and saving lives.

The second case takes a look at data conducted in Syria to display seismic hobby within the place. In this case, researchers used the same form of neurotrophic algorithms as defined within the preceding case in addition to machine-gaining knowledge of (ML) algorithms. These algorithms were used to research statistics accrued from seismic sensors. With the assistance of these algorithms, they have been able to predict the wide variety and place of seismic activities with an excessive diploma of accuracy. The results of the look endorse that AI-based algorithms for earthquake monitoring will be visible as more excellent dependable and correct than traditional strategies.

6.1 Successful Implementations and Performance Metrics

Earthquake tracking and prediction are essential fields in geology, engineering, and related sciences. Despite significant progress in our expertise in the physics of earthquake technology during the last century, there are nevertheless sizeable gaps in our know-how, and our capacity to expect when and where an earthquake will arise remains restricted. To fill those gaps, researchers have explored the use of new device-learning processes, inclusive of neuro-stimulated algorithms, to predict earthquakes successfully.

Neuro-inspired algorithms are based totally on findings from neuroscience studies and feature packages in a massive form of areas from laptop imaginative and prescient to voice recognition. They are, in particular, applicable for applications wherein there's an inherently complicated mapping among inputs and outputs, as is the case with earthquakes. Those algorithms are capable of processing an enormous amount of statistics from current earthquake statistics and using it to discover or 'analyze' patterns that may be used to enhance prediction accuracy.

To ensure the successful implementation and overall performance of neuro-inspired algorithms for earthquake tracking and prediction, numerous key concerns have to be taken into account.

6.2 Real-Time Monitoring Systems

Actual-time tracking structures for Neuro-stimulated Algorithms for Earthquake monitoring and Prediction are state-of-the-art systems of evaluation and tracking that combine the energy of synthetic intelligence (AI) with actual-time remark data from seismic sensors positioned within the earth's crust. Those structures use algorithms created through profound getting-to-know studies to discover and examine seismic interest and expect the size, place, and other traits of potential earthquakes. Actual-time monitoring systems appoint sensors that hit upon and degree seismic waves and unique algorithms that utilize beyond records as they enter into tailored mathematical fashions.

Since traditional earthquake tracking systems depend solely on human knowledge and computing energy, they could often fail in their mission as they should be picked out and symbolize an earthquake before it happens. AI-primarily based systems are capable of hitting upon seismic styles and altering their modeling as a consequence. This allows for extra accurate predictions in a fragment of the time it'd take for a human analyst to come across a growing earthquake.

7. CHALLENGES AND FUTURE DIRECTIONS

Earthquakes are amongst the most disastrous herbal failures recognized by humankind. They are able to cause extreme destruction to human habitations and infrastructure, leading to a lack of lives and substantial financial losses. As a result, scientists have been gaining knowledge of methods to monitor better and predict earthquakes. One of the more extraordinary latest advances in earthquake tracking and prediction is using neuro-inspired algorithms. Neuro-stimulated algorithms are a gadget getting to know the set of rules that makes use of synthetic neural networks to study styles and relationships from vast sets of data, with the purpose of appropriately predicting destiny outcomes.

In recent years, the usage of neuro-inspired algorithms for earthquake monitoring and prediction has become increasingly more famous as they provide many blessings over traditional strategies, which include quicker evaluation times and multiplied accuracy of outcomes. No matter their promise, there are still numerous demanding situations that ought to be addressed earlier than they may be used efficiently for earthquake monitoring and prediction.

One of the most critical challenges associated with the use of neuro-stimulated algorithms for earthquake monitoring and prediction is the absence of to-be-had information. These algorithms require massive units of records as they should be research styles and relationships. Yet, there are only a few sets of historical facts to be had for earthquake monitoring and prediction.

7.1 Technical Challenges and Ongoing Research

Earthquake tracking and prediction are challenging tasks because of the complex nature of plate tectonics and seismology. While superior technology has advanced our expertise in earthquakes and their causes, there are numerous technical challenges in accurately monitoring and predicting imminent earthquakes.

One main project is the detection of earthquakes in regions with sparse networks of seismic sensors and distinctly dispersed alerts from seismic occasions. As these signals are frequently extraordinarily susceptible and dispersed, accurately monitoring and predicting earthquakes in those zones, require detection systems with greater sensitivity and the capability to procedure alerts similarly apart.

Similarly, present algorithms frequently need to perceive the foundation of seismic indicators correctly. This has been a selected problem for urban regions with higher ambient noise and complex geological capabilities, as such conditions can make it challenging to pick out the supply of the signal. As such, algorithms that could as it should be pinpoint the beginning of these signals are of crucial importance.

7.2 Exploring Emerging Neuro-Inspired Technologies

Earthquake monitoring and prediction is a critical yet complicated hassle that calls for solutions to address both the natural surroundings and human resilience. With the increase of synthetic intelligence (AI) and the advances in machine learning, neuro-stimulated technology in the form of neuro-inspired algorithms can open up new possibilities for the tracking and prediction of earthquakes.

A neuro-inspired set of rules is a set of rules that uses algorithms, architectures, and techniques advanced from neuroscience to imitate the behavior of humans and animals and, as a result, obtain comparable outcomes. Neuro-stimulated algorithms for earthquake monitoring and prediction can quickly analyze large amounts of records and detect patterns in the statistics, which may assist in the prediction of future activities. Those algorithms may be mixed with different resources of records such as ancient facts, geological survey information, satellite TV for pc pics, and seismic statistics to be able to provide a comprehensive evaluation of the earth's plates and the activity under them.

Neuro-inspired algorithms may be used to interpret the records and generate models that can highly recognize and expect the behavior of earthquakes.

8. CONCLUSION

Earthquake tracking and prediction require an advanced and comprehensive approach. At the same time as conventional seismology techniques are nonetheless the foremost methods of earthquake tracking and prediction, the use of neuro-inspired algorithms has validated its capacity for complementing those extra-installed strategies. With the provision of excessive-decision records, Neuro-stimulated algorithms can be used to quickly detect styles in seismic events and identify potential seismic areas. Moreover, via thinking of a selection of bodily and numerical information to provide specified and reliable predictions, Neuro-inspired algorithms have the potential to reduce the fake alarm charge and improve the accuracy of seismic monitoring and prediction. It's miles crucial that seismic research and prediction stays a multi-disciplinary and collaborative effort, as this affords the acceptable hazard of successful earthquake tracking and prediction. Therefore, further knowledge of Neuro-stimulated algorithms and their potential programs in earthquake monitoring and prediction must stay a priority for seismologists. In summary, Neuro-inspired algorithms provide an effective tool for supplementing traditional earthquake tracking and prediction strategies. This potent mixture of strategies is essential to be able to make sure correct and reliable seismic events are monitored and reliably anticipated.

8.1 Summarizing the Impact of Neuro-Inspired Algorithms

Neuro-stimulated algorithms can revolutionize earthquake monitoring and prediction. By leveraging the strength of artificial neural networks, those algorithms may be educated to accurately version seismic activity and provide more correct estimates of seismic depth from multiple resources, including seismometers, accelerometers, and geological surveys. Furthermore, neuro-inspired algorithms can assist in coming across precursory seismic events before an earthquake happens, leading to higher caution systems and allowing extra effective and well-timed response plans.

Similarly to supplying extra correct records approximately the capability electricity of an earthquake, neuro-inspired algorithms can also help inform humans about the danger of particular areas. Through reading styles of seismic hobby, they could stumble on areas with an extra chance of future seismic hobby and alert the ones within the area to put them together. Furthermore, neuro-inspired algorithms can make real-time updating of seismic risk maps feasible. This may substantially beneficial resource emergency reaction groups and result in extra powerful catastrophe alleviation operations.

8.2 Implications for Advancing Earthquake Monitoring

The capacity for neuro-stimulated algorithms for earthquake tracking and prediction is promising and offers several benefits over conventional methods. Neuro-inspired algorithms have the functionality to rapidly analyze changes in seismic interest throughout distinct sites, presenting treasured insight into the seismic source characteristics. Moreover, they can be used to forecast quick-term earthquake activity and are able to offer real-time updates concerning the opportunity of a big earthquake.

Due to the relative newness of these methods, there remain several implications for advancing earthquake tracking with neuro-stimulated algorithms. Researchers are continuously working to improve algorithms that can more accurately expect seismic interest and confirm the consequences in their evaluation. In order for those algorithms to become extra dependable, more records are needed to permit the improvement and implementation of more state-of-the-art fashions.

Moreover, the pleasantness of records needs to be improved for the improved accuracy of predictions. At gift, researchers make use of seismic records from earthquakes of all magnitudes and depths, which may be constrained in offering perception into unique varieties of earthquakes. Furthermore, it might be helpful to grow overlap among seismic networks and the places of seismic sources, which could allow more excellent correct forecasts of seismic activities.

REFERENCES

Adam, M. A., Sulaiman, A., Baharuddin, A. S., Mokhtar, M. N., Subbian, K., Tabatabaei, M., & Marwanto, A. P. Y. (2019). Characterization of delignified Oil Palm Decanter Cake (OPDC) for polymer composite development. *International Journal on Advanced Science, Engineering and Information Technology*, *9*(2), 384–389. doi:10.18517/ijaseit.9.2.2392

Birkoben, T., & Kohlstedt, H. (2023). Matter and Mind Matter. In *Bio-Inspired Information Pathways: From Neuroscience to Neurotronics* (pp. 1–42). Springer International Publishing.

Chatterjee, S. (2021). *Analysis and design of closed-loop fractional-order systems and its applications to neurotechnology*. Rensselaer Polytechnic Institute.

Chiang, C. T., Nien, C. Y., Yang, C. Y., & Chen, S. M. (2022). *Sensor Systems Integration A CMOS Algae Growth Period Monitor for Algaculture Applications*.

Dunham, C. (2022). *Exploration of Power Laws in and Development of Analytical Tools for the Study of Stem Cell-derived Cardiomyocytes*. University of California.

Gunjan, V. K., Vijayalata, Y., Valli, S., Kumar, S., Mohamed, M. O., & Saravanan, V. (2022). Machine learning and cloud-based knowledge graphs to recognize suicidal mental tendencies. *Computational Intelligence and Neuroscience*, *2022*, 2022. doi:10.1155/2022/3604113 PMID:35341205

Hirtzlin, T. (2020). *Digital Implementation of Neuromorphic systems using Emerging Memory devices* [Doctoral dissertation, Université Paris-Saclay].

Jana, D., & Nagarajaiah, S. (2022). Physics-Guided Real-Time Full-Field Vibration Response Estimation from Sparse Measurements Using Compressive Sensing. *Sensors (Basel)*, *23*(1), 384. doi:10.339023010384 PMID:36616982

Kannan, S., Dhiman, G., Natarajan, Y., Sharma, A., Mohanty, S. N., Soni, M., Easwaran, U., Ghorbani, H., Asheralieva, A., & Gheisari, M. (2021). Ubiquitous vehicular ad-hoc network computing using deep neural network with iot-based bat agents for traffic management. *Electronics (Basel)*, *10*(7), 785. doi:10.3390/electronics10070785

Kaur, J., Saxena, J., Shah, J., & Yadav, S. P. (2022, May). Facial Emotion Recognition. In *2022 International Conference on Computational Intelligence and Sustainable Engineering Solutions (CISES)* (pp. 528-533). IEEE.

Khan, S., Saravanan, V. N. G. C., Lakshmi, T. J., Deb, N., & Othman, N. A. (2022). Privacy protection of healthcare data over social networks using machine learning algorithms. *Computational Intelligence and Neuroscience*, *2022*, 2022. doi:10.1155/2022/9985933 PMID:35371203

Lawrie, S. (2022). *Information representation and processing in neuronal networks: from biological to artificial systems and from first to second-order statistics.*

Ma, D., Fang, H., Wang, N., Lu, H., Matthews, J., & Zhang, C. (2023). Transformer-optimized generation, detection, and tracking network for images with drainage pipeline defects. *Computer-Aided Civil and Infrastructure Engineering*, *38*(15), 2109–2127. doi:10.1111/mice.12970

Mills, K. A., Unsworth, L., & Scholes, L. (2023). *Literacy for digital futures: Mind, body, text.* Taylor & Francis.

. Mills, K. A., Unsworth, L., & Scholes, L. (2022). *Literacy For Digital Futures*. Research Gate.

Rajalakshmi, M., Saravanan, V., Arunprasad, V., Romero, C. T., Khalaf, O. I., & Karthik, C. (2022). Machine Learning for Modeling and Control of Industrial Clarifier Process. *Intelligent Automation & Soft Computing*, *32*(1). doi:10.32604/iasc.2022.021696

Ramírez-Mendoza, A. M. E., Yu, W., & Li, X. (2023). A new spike membership function for the recognition and processing of spatiotemporal spike patterns: Syllable-based speech recognition application. *Mathematics*, *11*(11), 2525. doi:10.3390/math11112525

Russo, A., & Coco, D. (2023). Quantify how space mission influence geopolitical dynamics? A security and social policy approach. *arXiv preprint arXiv:2301.03538*. doi:10.2139/ssrn.4375678

Seyyedi, A., Bohlouli, M., & Oskoee, S. N. (2023). Machine Learning and Physics: A Survey of Integrated Models. *ACM Computing Surveys*.

Shibuya, K. (2020). Digital transformation of identity in the age of artificial intelligence (No. 172558). Singapore: Springer.

Tiwari, A., Sharma, S. K., Dixit, A., & Mishra, V. (2021). UAV remote sensing for campus monitoring: A comparative evaluation of nearest neighbor and rule-based classification. *Photonirvachak (Dehra Dun)*, *49*(3), 527–539. doi:10.100712524-020-01268-4

Yadav, S. P., Zaidi, S., Nascimento, C. D. S., de Albuquerque, V. H. C., & Chauhan, S. S. (2023, January). Analysis and Design of automatically generating for GPS Based Moving Object Tracking System. In *2023 International Conference on Artificial Intelligence and Smart Communication (AISC)* (pp. 1-5). IEEE. 10.1109/AISC56616.2023.10085180

Yelick, K., Agarwal, D., Bard, D., Shalf, J., Almgren, A., Bhimji, W., & Kincade, K. (2021). 2019 Computing Sciences Strategic Plan. Lawrence Berkeley National Lab.(LBNL), Berkeley, CA (United States).

Yuvaraj, N., Karthikeyan, T., & Praghash, K. (2021). An improved task allocation scheme in serverless computing using gray wolf Optimization (GWO) based reinforcement learning (RIL) approach. *Wireless Personal Communications*, *117*(3), 2403–2421. doi:10.100711277-020-07981-0

Yuvaraj, N., Srihari, K., Dhiman, G., Somasundaram, K., Sharma, A., Rajeskannan, S. M. G. S. M. A., Soni, M., Gaba, G. S., AlZain, M. A., & Masud, M. (2021). Nature-inspired-based approach for automated cyberbullying classification on multimedia social networking. *Mathematical Problems in Engineering*, *2021*, 1–12. doi:10.1155/2021/6644652

Chapter 19
Exploring Frigid Frontiers:
Strategies for Uncovering Earth's Frozen Regions With AI

Rahul
Delhi Technological University, India

Shashank Parmar
https://orcid.org/0000-0003-4369-3647
VISA, India

Vibhor Singh
Delhi Technological University, India

Dhaval Shrivastava
Delhi Technological University, India

Aayush Chowdhury
Delhi Technological University, India

Neelam Sunda
Kanori PG Mahila Mahavidyalaya, India

ABSTRACT

Polar science is the study of Earth's frozen zones. It finds its application in a diversity of applications such as monitoring the effects of global warming, making weather prediction, studying the migration cycles of birds, understanding the change in climate over decades, etc. Despite its wide applicability, this field presents a diversity of challenges to the cryosphere community. In this chapter, the authors will cover the current challenges in this field and how they are being solved by researchers across the globe. They will discuss the areas of research and their applicability along with the latest and most promising research works in the field of polar sciences. The subject of polar sciences is studied widely in two dimensions. The first is the study of ice of the sea, and the other is the study of snow. To truly grasp the works done in this field, the authors would be covering these sections in detail one after another.

DOI: 10.4018/979-8-3693-1850-8.ch019

1. SEA ICE

Artificial Intelligence and Machine Learning models have been used to map the ice shelves across the arctic region. But why do we need AI techniques to identify ice covered zones in the first place? Researchers make use of satellite imagery to identify the Earth's Frozen Zones. However, the contour of Ice is like that of Water and Clouds. Hence, it is difficult to give an accurate account of the Earth's frozen zones. To mitigate this issue researchers across the globe have been working on various AI models to distinguish Ice from Sea and Clouds. We will now discuss such techniques in a little more detail.

1.1 U-Net: A CNN Based Architecture

In the field of computer vision a convolutional network design known as U-Net architecture is employed for segmentation of image task. U-Net initially was introduced by Olaf Ronneberger (Ronneberger et al., 2015) and has been applied in various application of image segmentations like medical image analysis. This research employs U-Net architecture for remote sensing data, which perform interpretation of satellite or aerial images.

This architecture considers eight different training and testing sites. The testing sites have not exposed to system prior and kept entirely reserved. Detailed information and geographic coordinates for all sites for training and testing are visually presented in Figure 1.

The data set was taken from two satellite Sentinel1-1A (2014) and sentinel1-1B (2016) that have radar imaging sensors. Moreover 3-D images also taken from TanDEM-X (2013-2014) which is a German satellite for earth observation.

Figure 1. Area of study around continent Antarctica
(Baumhoer et al., 2019b)

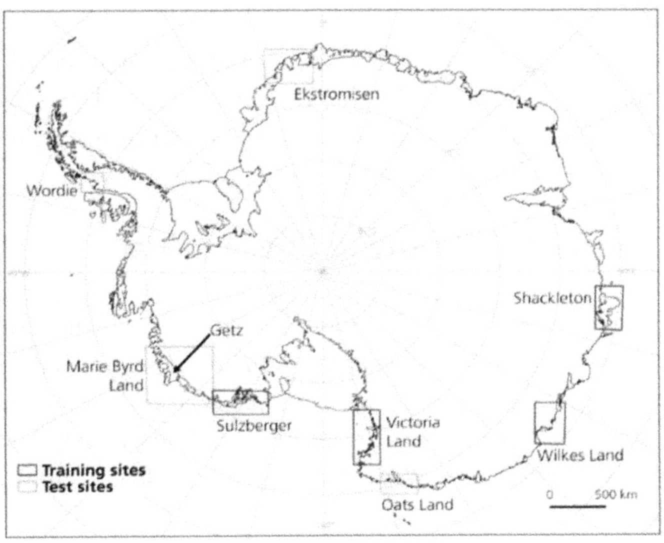

Figure 2. It is a representation of U-Net Architecture where red arrows show down-sampling block, green arrows show up-sampling, black arrows show skip connections and yellow arrows indicate dropout *(Baumhoer et al., 2019b)*

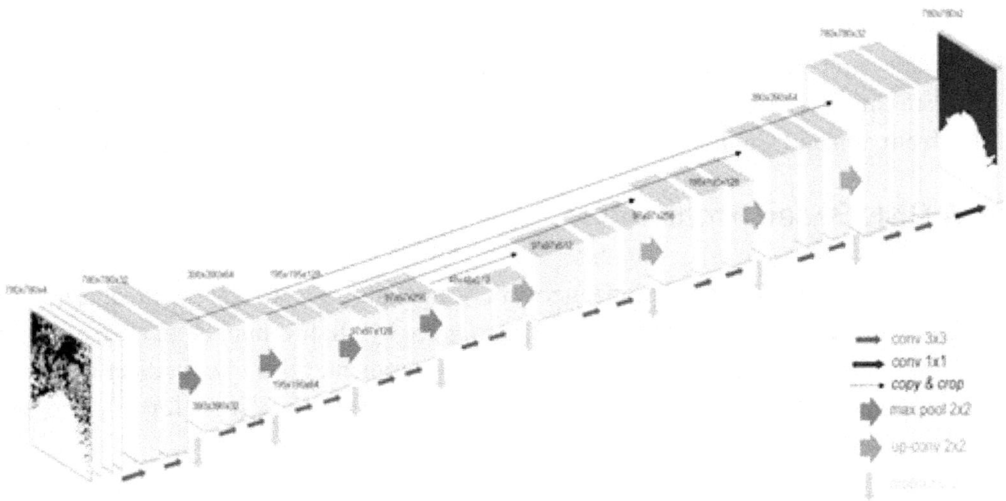

U-net architecture contains both an encoder and decoder and resembles to U -shaped structure. Its design mainly composed of contracting path(encoder) and expansive part (decoder) to provide precise object segmentation within images. This study uses a classifier that takes both pixel values and spatial context to make a distinction between land ice and ocean. It also makes few modifications to U-net architecture to meet its requirement. (Baumhoer et al., 2019b) The main modifications are: employing larger input tiles; initiating with 32 feature channel unlike 64 and scaling is done with 512 unlike 1024; incorporating dropout mechanisms; utilizing four input channels, as opposed to a single channel as described in Figure 2.

Figure 3 captures the movement of DeVicq Glacier accurately as shown by time series. The parameters used for classification accuracy are given in following table. Accuracy varied across areas which are summarised in table 1.

Calculation of breaking of ice front of glacier around Antarctica will be allowed by the spatial and temporal transferability of this approach in future. This approach will provide new ways to monitor the changes in the Antarctica coastal region and can be used for measuring glacier and ice front extent fluctuations.

1.2 Lead Detection: A Machine Learning-Based Model

Lead detection in sea ice is the process of recognizing and making a distinction between areas of open water within sheet of ice, generally found in Antarctic and Arctic region. These gaps also called leads plays a very significant role in polar ecosystems, navigation, heat transfer between ocean and atmosphere etc. It is very critical to accurately pin points the leads in order to effectively measures the sea ice thickness. Lead detection-based research, introduced novel methods for lead identification by a combination of decision trees and random forest.

Figure 3. Sequence of timeline around Getz Ice Shelf from Sentinel-1. (Baumhoer et al., 2019b) (A) Forward movement of the DeVicq Glacier. (B)Settled coastline part. (C) Breaking of ice chunks from front of Beakley Glacier (D)Wrong boundary prediction of Getz 1 Glacier front (E) Breaking of ice from Glacier event

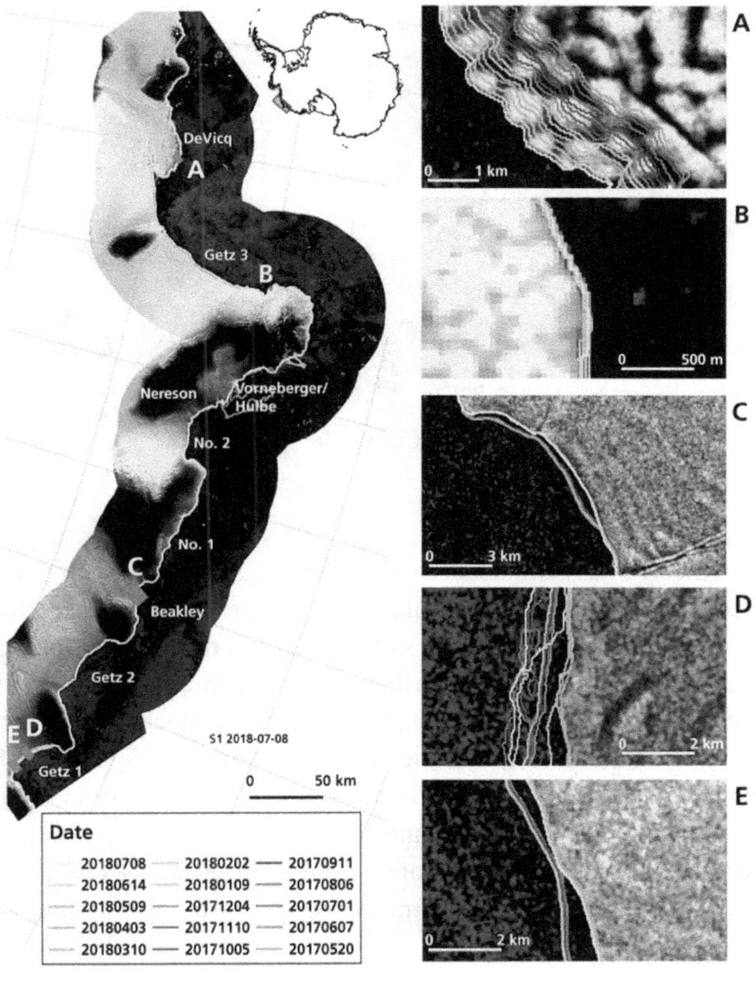

Table 1. Results of Classification where bold represent high and low values

			Training Sites					Test Sites			
	Accuracy measure	Sulzberger	Victoria Land	Wilkes Land	Shackleton	Marie Byrd Land	Oats Land	Ekstromisen	Wordie	Mean Train	Mean Test
ice	precision	0.85	0.89	0.87	**0.92**	0.91	**0.92**	0.91	0.85	0.88	0.90
	recall	0.85	**0.98**	0.92	0.96	**0.94**	0.91	**0.94**	0.91	0.93	0.93
	f1-score	0.85	0.93	0.89	**0.94**	**0.93**	0.91	**0.93**	0.88	**0.90**	**0.91**
water	precision	0.79	0.97	0.91	**0.96**	0.92	0.91	**0.94**	0.90	0.91	0.92
	recall	0.86	0.87	0.86	**0.92**	0.90	**0.92**	0.90	0.84	0.88	0.89
	f1-score	0.83	0.92	0.88	**0.94**	0.91	0.91	**0.92**	0.87	**0.89**	**0.90**

Figure 4. Image taken by MODIS from CryoSat-2 where red line shows path in the north of Svalbard (Lee et al., 2016)

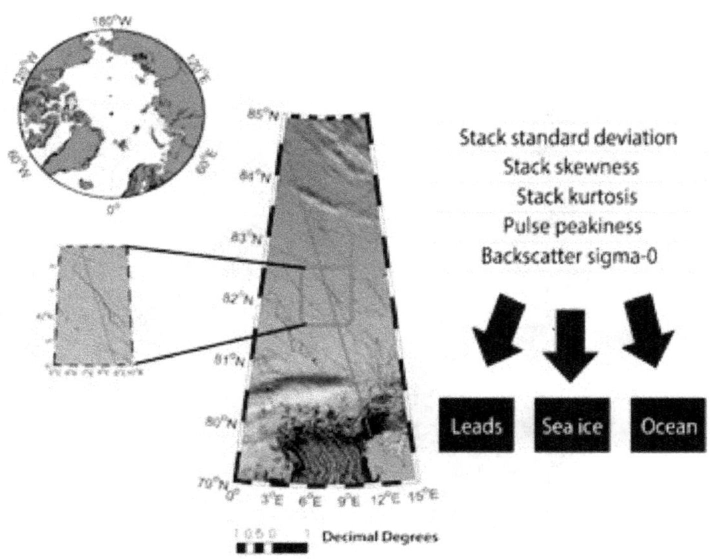

The dataset for this research has been taken from CryoSat-2, combined with Synthetic Aperture Interferometric Radar Altimeter (SIRAL). (Lee et al., 2016) The process of decision tree implementation involves iteratively dividing data of training into subsets. At each node, an attribute is chosen, and two branches stemming from that node employ an attribute value as a threshold. The selection of attributes, like Stack Standard Deviation (SSD), skewness of stack, kurtosis of stack, Pulse Peakiness at each node holds paramount importance for accurate classification as described in Figure 4. While random forest utilizes joint approach that mixes a strategy for sampling with Classification and Regression Trees.

The findings revealed the efficacy of our planned lead detection method in accurately estimating sea ice thickness when compared to established approaches. The accuracies achieved by this research, is 95.4% for decision trees and 96.2% for random forest, surpassing the performance of the current methodologies. (Lee et al., 2016)

Leads, sea ice and ocean features were extracted founded on the five parameters based on visual interpretation

For future research, our objectives encompass the creation of a lead detection model using machine learning that shows applicability across various years and months. Additionally, we aim to construct models that utilize CryoSat-2 data for the purpose of modelling snow depth penetration.

1.3 MAp-Guided Ice Classification (MAGIC): A Combination of "Glocal" Method and Support Vector Machine (SVM) Classifier

Recording ice and open water in bodies of oceans is crucial for a multitude of applications, including environmental assessment and maritime navigation. The Canadian Ice Service (CIS) performs mapping of ice daily for Canadian waters. In CIS, analysts of ice process images of SAR from RADARSAT-2 manually to produce region-specific ice concentration maps. (Leigh et al., 2014) In this manner, CIS

Figure 5. Each square is an image of RADARSAT-2 scene from Beaufort Sea of about 250000 km square. 20 such scenes were taken for this study.
(Leigh et al., 2014)

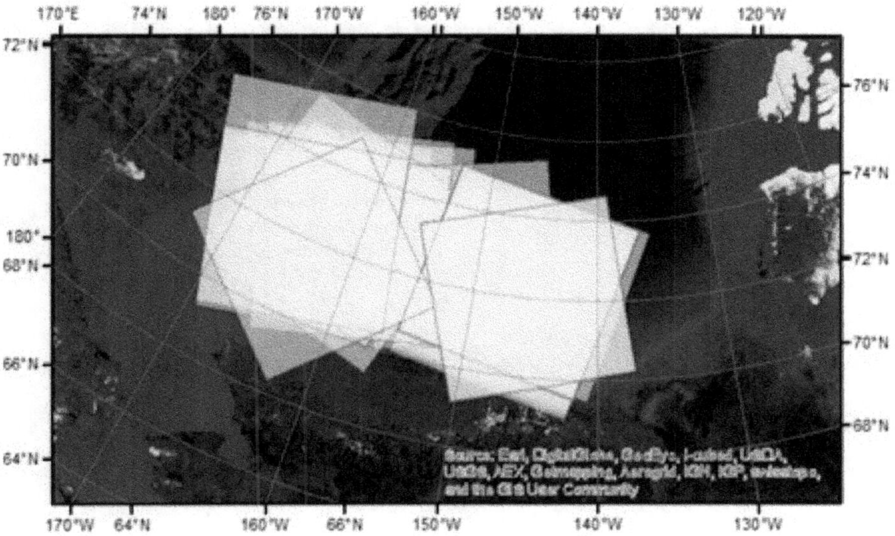

personnel handle approximately 3500-4500 RADARSAT scenes annually. Therefore, CIS has identified the necessity for an algorithm that discriminates between ice and water automatically using imagery of dual polarization generated by RADARSAT-2. (Leigh et al., 2014) Methods that can be automated offer a range of advantages to the ice classification process, including the ability to handle a high volume of scenes daily, provide highly detailed reaching up to pixel level imagery, and also consistency is ensured.

The image data set we developed for testing consists of 20 scenes from the C-band SAR satellite RADARSAT-2. (Leigh et al., 2014) Each scene was captured in the SAR scan Wide beam mode, which provides images in HH and HV polarizations for each scene as shown in Figure 5. The images have nominal pixel dimensions of 50 m and image dimensions of around 10 000 × 10 000 pixels. This is the largest standard scene size available from RADARSAT-2 at 500 km in both range and azimuth and is the most useful beam mode for operational mapping of the vast expanses of ice and water in Arctic seas. The incidence angle of these images' ranges from 20o to 49o, and the scenes were captured from both ascending and descending satellite passes. The images were captured in the year 2010 at various locations over the Beaufort Sea displayed in Fig. 5. The 20 scenes were taken from April through December; inclusive of the more challenging freeze up and melt seasons were many different ice types.

The MAGIC system functions entirely in an automated manner. (Leigh et al., 2014) It possesses the capability to effectively differentiate between water and sea ice, extending beyond the mere differentiation of various ice types. To begin, the HV scene, where horizontal transmit polarization is paired with vertical receive polarization, undergoes classification employing the "glocal" method. This method is based on classification that is region based and hierarchical and employ IRGS. Next, SVM classification which is pixel based. This SVM classification leverages SAR Gray-level cooccurrence texture and backscatter features. In the final step, the results of the IRGS and SVM classifications are integrated

Figure 6. Sequence of action in MAGIC system
(Leigh et al., 2014)

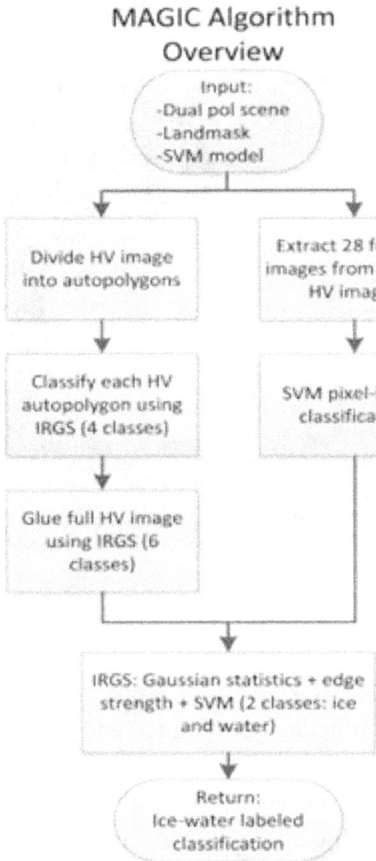

as described in Figure 6. This adaptation is made to accommodate the pixel-based information derived from the SVM classification process.

Cyan colour shows classified ice, and dark blue colour as classified water

The performance of the MAGIC produced an impressive total accuracy rate of 96.42% as Figure 7 shows the classified water and ice. Throughout the testing process, two specific limitations came to light. These limitations pertain to the system's challenges in accurately mapping grease on the ice and very small sheet of floating ice. However, it is worth noting that neither of these limitations gives rise to substantial operational concerns. Nonetheless, addressing these limitations could serve as a valuable motivator for potential future improvements and enhancements.

1.4 Deep Convolutional Neural Networks (CNN): For Estimation of Sea Ice Concentration

Recent reductions in Arctic ice extent have sparked increased business interest in extraction of natural resource and shipping activities. To facilitate efficient operations and shipping in ice-prone Arctic Ocean, the availability of high-resolution ice coverage data on time is of paramount importance. Optical remote

Figure 7. Final result of classification
(Leigh et al., 2014)

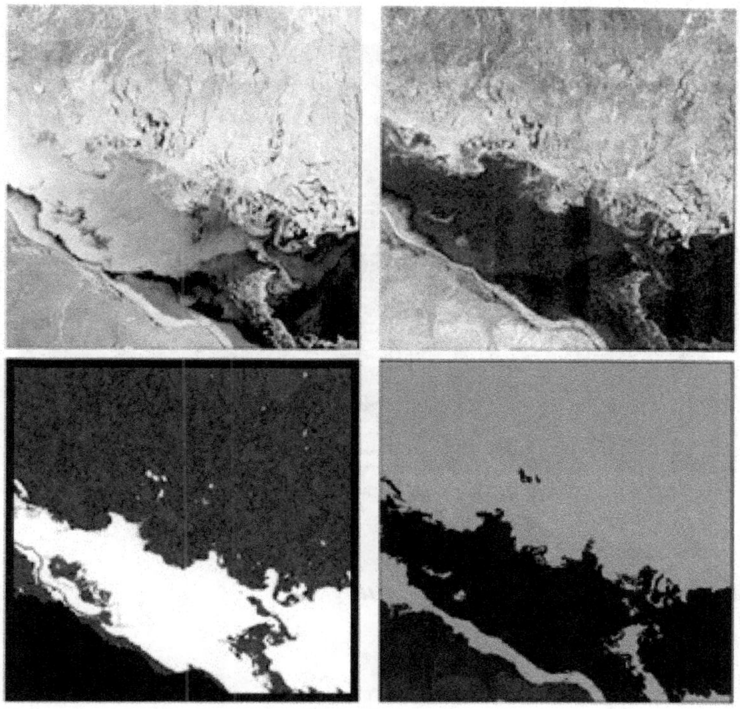

sensing technology, hindered by frequent cloud cover and prolonged periods of limited sunlight, falls short in delivering continuous Arctic observations. Nevertheless, the absence of robust automatic SAR image interpretation algorithms compels national agencies for ice like the CIS to engage in manual SAR image interpretation for ice mapping. This manual interpretation process is both time-consuming and subject to human subjectivity, yielding ice maps with lower resolutions. Consequently, there is a strong motivation for the development of estimation of concentration of ice automatically to harness the vast collection of SAR imagery efficiently. This research as showed in Figure 8 delves into the exploration of ice concentration estimation within the CNN framework. (Wang et al., 2016) Deep CNN, a technique centred on feature learning, harnesses the power of multi-layered neural networks.

Input and output of different layers are shown by outer rectangles. Each input layer size can be described as width*height*number of bands as written at bottom. The convolutional filters are shown by smaller filled rectangle, whose size described beneath as example 64 filters of size 5*5*2.

For estimating ice concentration, a deep CNN has been employed. (Wang et al., 2016) CNN is a model of neural network that emphasizes sharing of weight and connections that are local between neighbouring neuron layers, making it particularly well-suited for tasks like image analysis and pattern recognition. It comprises a sequence of alternating layers of pooling and convolution, typically followed by many layers that are fully connected. The diagram in Figure 9 illustrates a CNN architecture featuring two sets of layers of pooling and convolutions, concluding in a single layer that is fully connected at the top. Within a CNN, the learnable parameters consist of parameters for filter in the layers for convolution

Figure 8. This research focusses its study in Beaufort Sea
(Wang et al., 2016)

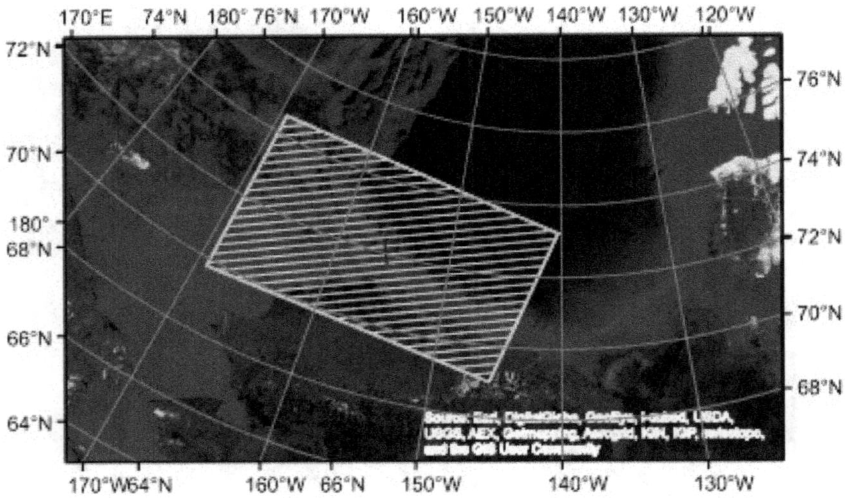

Figure 9. CNN model structure employed by this study
(Wang et al., 2016)

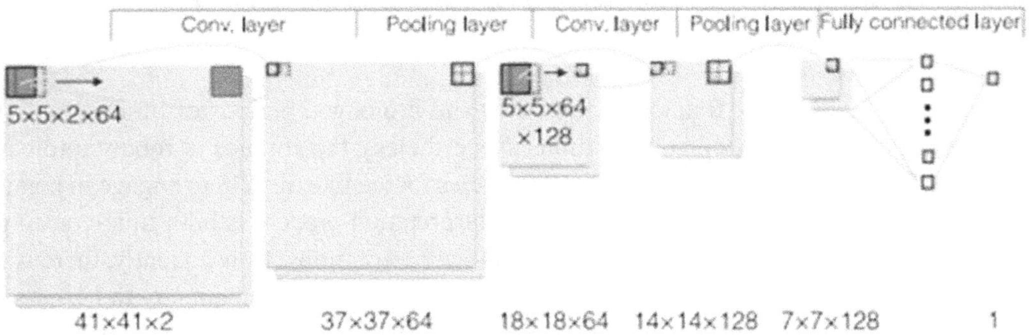

and the weight matrices for the layers that are fully connected and these parameters are adjusted and optimized during the training process.

The estimates for concentration of ice shaped by the CNN approach offer a more intricate level of detail compared to both image analysis and ASI ice concentration as shown in Figure 10 particularly when scrutinizing regions with concentration of low ice, such as zones with marginal ice. Nevertheless, it's worth noting that CNN-based estimations may introduce noise in open water regions. This phenomenon tends to occur when the featured scale on open water surfaces thar are wind-roughened aligns with the size of the window of CNN. To mitigate this effect, a potential solution involves increasing the number of training samples for these specific regions.

However, for practical operational implementation, a study that is more comprehensive and assessment of this method across regions that are diverse and changing times of the year are imperative. Such

Figure 10. Image analysis of Ice concentration depicted in column 1, concentration of ice by an algorithm named ASI represented by column 2, and CNN estimation shown in column 3
(Wang et al., 2016)

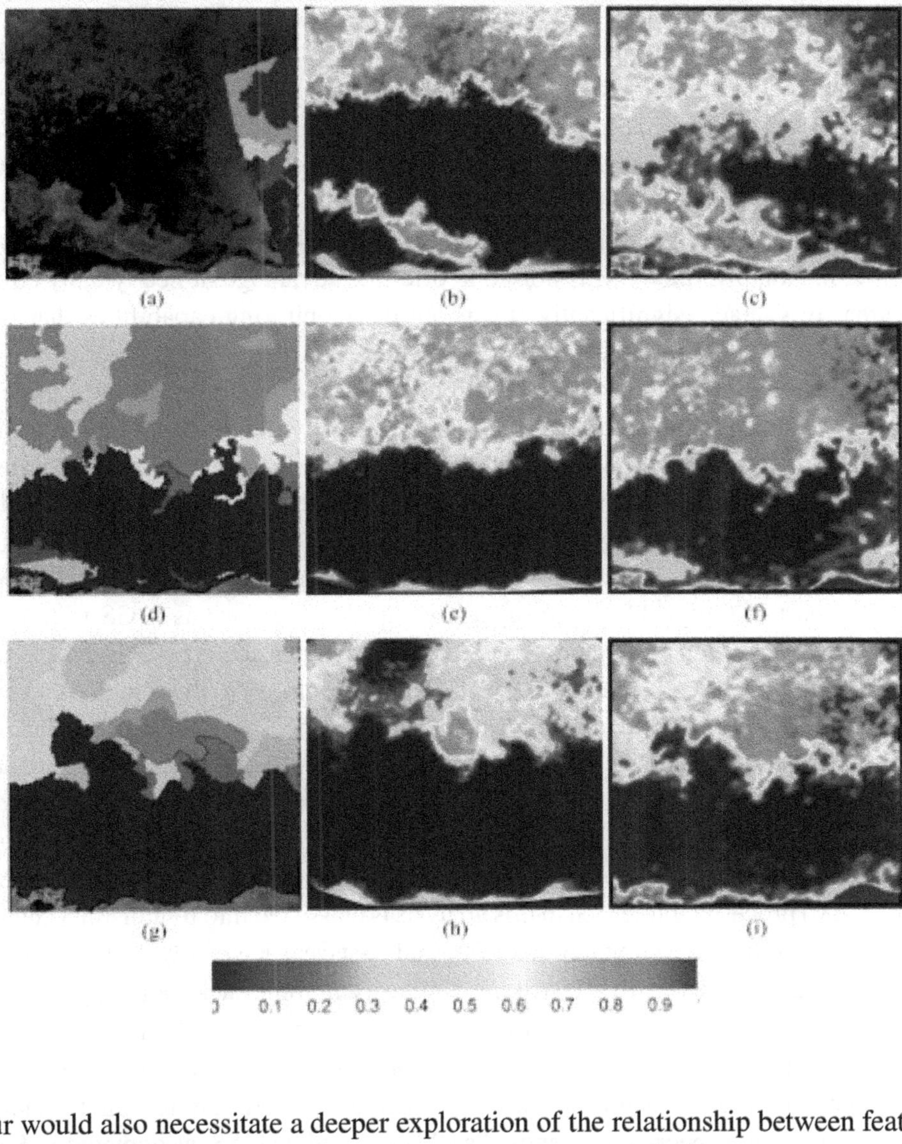

an endeavour would also necessitate a deeper exploration of the relationship between features of CNN and conditions of the surface of the sea, a direction for the research in future.

1.5 Iterative Region Growing Using Semantics (IRGS)-Based Segmentation and the Support Vector Machine (SVM) Classification: For Automatic Sea-Ice Types Contextual Classification

The creation of sea-ice maps bears significant relevance, serving essential functions in the realms of maritime navigation, environmental monitoring, and climate change modelling. Utilizing satellite imagery for automatic classification brings significant advantages to the mapping of ice types in the sea within the Arctic regions. Over the past two decades, SAR imagery has stood out as a pivotal resource for sea-ice mapping. The current SAR imagery, characterized by suitable resolution and a capacity to operate independently of adverse weather conditions and daylight constraints, has undeniably proven its worth, particularly in the Arctic regions. A notable development on the horizon is the imminent with the RADARSAT Constellation Mission (RCM) launch. (M. Ghanbari et.al., 2019) This mission will usher in a novel era with the provision of compact polarimetric (CP) SAR data, which is anticipated to represent a substantial enhancement compared to the existing RADARSAT-2 dual polarimetric SAR imagery.

Currently, the CIS depend on on Linear dual-polarization (DP) SAR data from RADARSAT-2 as its primary source for sea-ice monitoring. (M. Ghanbari et.al., 2019) Alongside single-polarimetric and DP modes, the upcoming RCM will introduce quad-polarimetric (QP) and CP SAR data. In the CP mode, the SAR system obtains two coherent linear polarizations in reply to a circularly field that is polarized during transmission. Notably, the RCM is poised to offer the advantage of nearly daily coverage across all of coastal areas of Canada, significantly enhancing the monitoring capabilities that is dynamic of the Arctic regions.

The ice-type classification methodology can be broken down into three primary components, as depicted in Figure 11. The algorithm takes as input the acquired Synthetic Aperture Radar (SAR) scene, complete with the corresponding land mask specific to that scene, along with a pre-trained labelling model. The file that is file-masked, presented as a binary image, serves to mask out land pixels from the analysis. The labelling model using training sample data has been trained which was extracted from dataset of the SAR imagery. The algorithm's initial phase encompasses SAR data segmentation, as illustrated in the left part of Figure 11. An unsupervised approach known as IRGS is employed for this objective. It identifies homogeneous regions representing different ice types. Given the unsupervised nature of this segmentation process, the initial labels assigned arbitrary to these regions and do not inherently resemble to specific types of the ice. To refine this initial labelling, a SVM classification is applied which is pixel based to the SAR images for examination. This SVM classification is trained using a set of SAR features. The right part of Figure 11 illustrates this step, which operates independently of the unsupervised IRGS segmentation

The final stage involves merging the results of the IRGS segmentation and SVM labelling through a majority-voting (MV) process. The sea ice in this analysis is classified into four distinct categories which are open water (OW), new ice (NI), young ice (YI), and first-year ice (FYI). Notably, the OW and NI classes have been consolidated due to the striking similarity in their backscatter signatures. Our study zone is located in the Barrow Strait, in near proximity of Island of Somerset within the Arctic region of Canada. For our investigation, we harnessed the power of two images of RADARSAT-2 which are fine quad-polarized, both captured over this specific area with an exceedingly minimal period gap of just 2 seconds. These images are specifically denoted as 231156 and 231158, as depicted in Figure 12(a).

The outcomes of the Support Vector Machine (SVM) labelling process, as illustrated in Figure 13(a)–(d), are integrated with the results of the unsupervised segmentation, which effectively delineated regions with similar characteristics. The conclusive, amalgamated classification results, portrayed in Figure 13(e) to (h) for each of the four cases, present the synergistic effect of this combination. This approach notably enhances the quality of the SVM results by mitigating noise and incorporating the information gleaned from the unsupervised segmentation process.

The outcomes of the Support Vector Machine (SVM) labelling process, as illustrated in Figure 13(a)–(d), are integrated with the results of the unsupervised segmentation, which effectively delineated regions with similar characteristics. The conclusive, amalgamated classification results, portrayed in Figure 13(e) to (h) for each of the four cases, present the synergistic effect of this combination. This approach

Figure 11. Sequence of action for performing sea ice classification in this research (M. Ghanbari et.al., 2019)

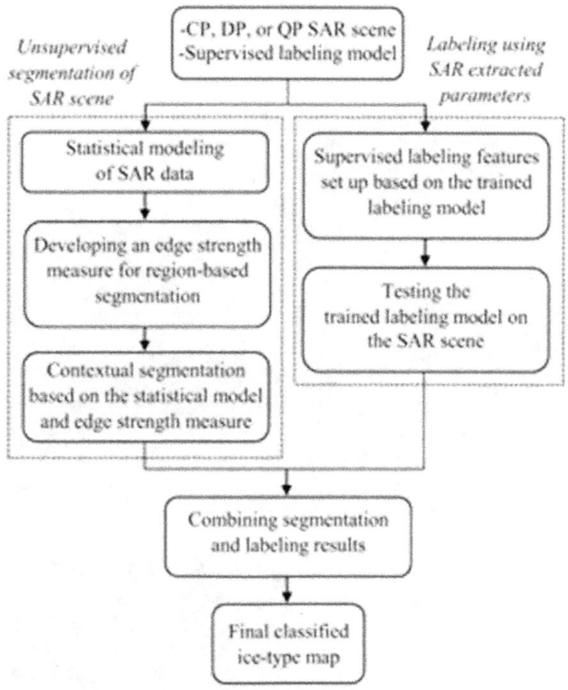

notably enhances the quality of the SVM results by mitigating noise and incorporating the information gleaned from the unsupervised segmentation process.

An ice-typing classification approach grounded in Synthetic Aperture Radar (SAR) imagery was executed and rigorously tested. The initial step entails a comprehensive scene segmentation employing the IRGS algorithm, yielding well-defined, class-homogeneous regions. Subsequently, these segmented regions are subjected to classification through a Support Vector Machine (SVM) classifier, resulting in the categorization of four distinct ice types. The evaluation encompassed four distinct cases: Dual Polarization (DP), Compact Polarimetry with horizontal and vertical transmit-receive, Compact Polarimetry with all features (CP - all features), and Quad Polarimetry (QP). Remarkably, utilizing Compact Polarimetry (CP) with all available features yielded an impressive overall classification accuracy of 96.53%. This result, while marginally lower than the accuracy achieved with the complete Quad Polarimetry (QP) dataset, underscores the remarkable potential of CP scenes. In a direct comparative analysis between the results of CP intensities, boasting an overall accuracy of 92.86%, and those of DP intensities, which garnered an overall accuracy of 81.04%, it becomes evident that CP scenes possess significant promise in delivering enhanced sea-ice maps when juxtaposed with DP scenes.

Future endeavours will encompass the development of a method uses segmentation that can harness the Compact Polarimetry coherence matrix without the necessity to recreate the Quad Polarimetry (QP) data. An additional avenue of exploration involves investigating the potential of texture features, including the utilization of co-occurrence probabilities and Gabor filters, in the domain of ice of the sea classification using CP data. Furthermore, we plan to expand our dataset with additional ground-truth

Figure 12. (a) Shows the location of study in Barrow strait taken by RADARSAT-2, (b) and (c) shows the image in M-chi RGB composite whereas (d) and (e) represent the same image of stoke vector first element with ground truth overlayed
(M. Ghanbari et.al., 2019)

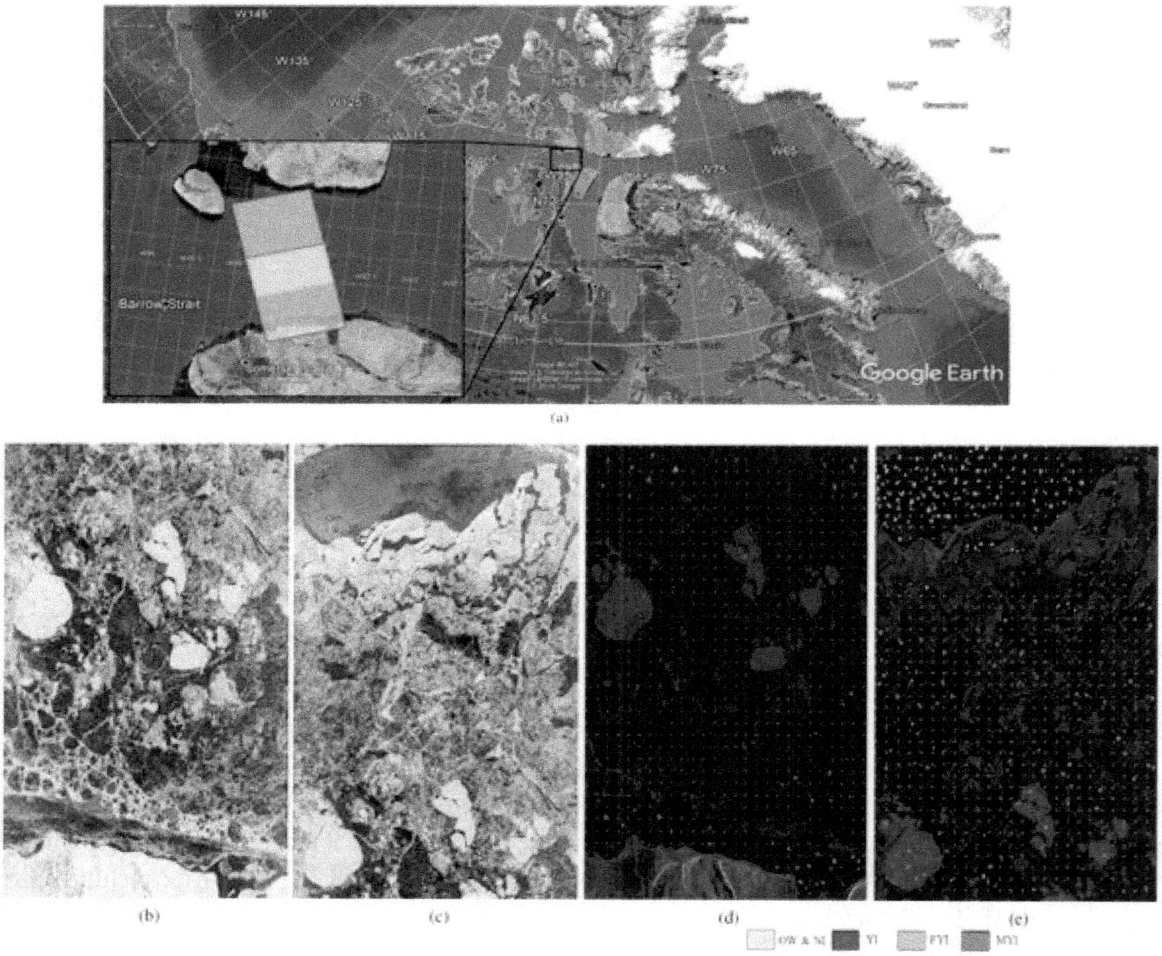

information to conduct a more comprehensive evaluation of the classification method's performance. These avenues of research hold the promise of advancing the accuracy and applicability of our sea-ice classification methodology.

1.6 Optimal Decision-Tree Design: For Sea-Ice Classification

As ice extent of the sea in summer decreases, the Arctic Ocean converts into more reachable space for ocean traffic and offshore operations. However, this increased accessibility also brings about potential dangers, primarily from sea ice and icebergs. To ensure the shipping navigation safety and activities on offshore area in the Arctic, there is a pressing need for the development of speedy, strong, and dependable methods for classifying sea ice types. Ice type classification is of particular interest from an operational

Figure 13. Images from (a) to (e) shows SVM which is pixel-based, whereas (e) to (g) shows SVM classification integrated with segmentation. (h) represents final QP data
(M. Ghanbari et.al., 2019)

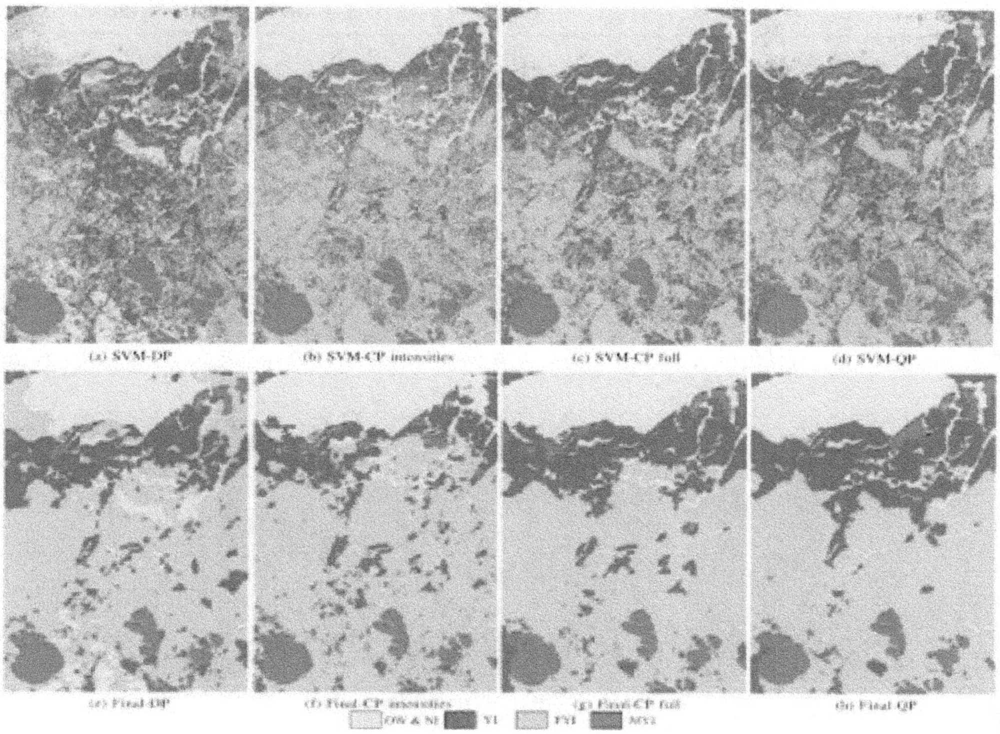

standpoint. This study's primary focus is to establish a strategy for optimizing decision tree models automatically in the context of classification problems. While the proposed algorithm has a generic applicability to various classification tasks, we illustrate its potential through a classification for the types of ice in the sea example using SAR data.

Numerous studies have already delved into the realm of automated or semi-automated sea ice classification through the utilization of SAR imagery. Many of these approaches rely on conventional classification techniques, which involve the simultaneous separation of all classes and the assignment of class labels to individual pixels. This approach's primary objective is to develop an automated approach for designing a numerically optimized Decision Tree (DT) with a keen focus on enhancing classification accuracy (CA). (Lohse et al., 2019) Differing from Random Forest (RF), this algorithm is tailored to create and employ a single tree. Each branch of this tree is dedicated to the classification of a specific class and its removal from the dataset.

Research have selected the Icelandic Association for Search and Rescue (ICESAR) dataset, acquired by the Alfred Wegner Institute (AWI) in March of 2007 of the ice in sea in the Strait of Fram, as our chosen test case for airborne SAR data analysis. Figure 14 shows C-band, L-band and optical sacner images where L-band image shows various classes defined in this research as enumerated in Table 2.

Research have introduced a completely automated approach to design a numerically optimized Decision Tree (DT) classification algorithm. This algorithm effectively divides a classification problem which

Table 2. Shows the ICESAR dataset classes along with its colour code and training samples quantity (Lohse et al., 2019)

	Description	Color Code	# Training Samples
Class ω_1	Open water	dark blue	2398
Class ω_2	Grey-white ice	light green	10,640
Class ω_3	Level ice	black	14,233
Class ω_4	Deformed ice	red	6356
Class ω_5	Nilas	cyan	12,946
Class ω_6	Grey ice	dark green	2342

Figure 14. Different coloured box in L-band image depict different classes training region as described in table
(Lohse et al., 2019)

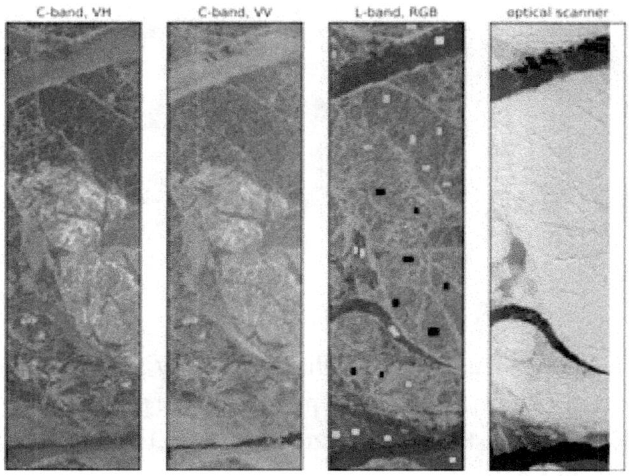

is multi-class with m classes into m-1 binary sub-problems. DT each branch is dedicated to separating one class from the remaining ones using an optimal feature set.

Research algorithm's performance has been validated through tests on simulated datasets, showing an impressive 3.5% increase in overall classification accuracy (CA) compared to the traditional AAO classification method as depicted in Figure 15. Furthermore, we achieved a notable upgradation of 2.5% in the types of sea ice classification which is obtained from aerial SAR data.

1.7 Semi-Automated ML for Sea Ice Classification

In polar regions, sea ice monitoring is an important aspect and it must be accurate because sea ice is disappearing near the Arctic Ocean during summer. To understand these changes in the polar environments and for the sake of safety in icy regions sea ice has to be precisely classified. For this, a sea ice classification algorithm is introduced which is based on Sentinel-1 images, and it employs a machine-

Figure 15. Final classification of the types of ice derived from DT which is numerically optimized applied on ICESAR dataset
(Lohse et al., 2019)

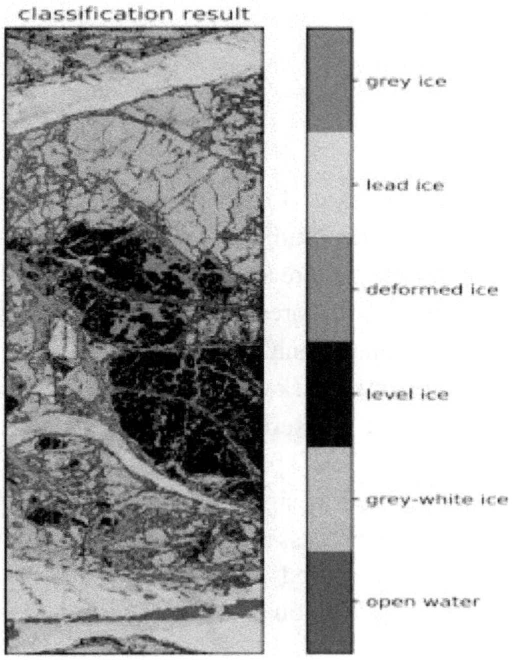

Figure 16. Area of study
(Park et al., 2020)

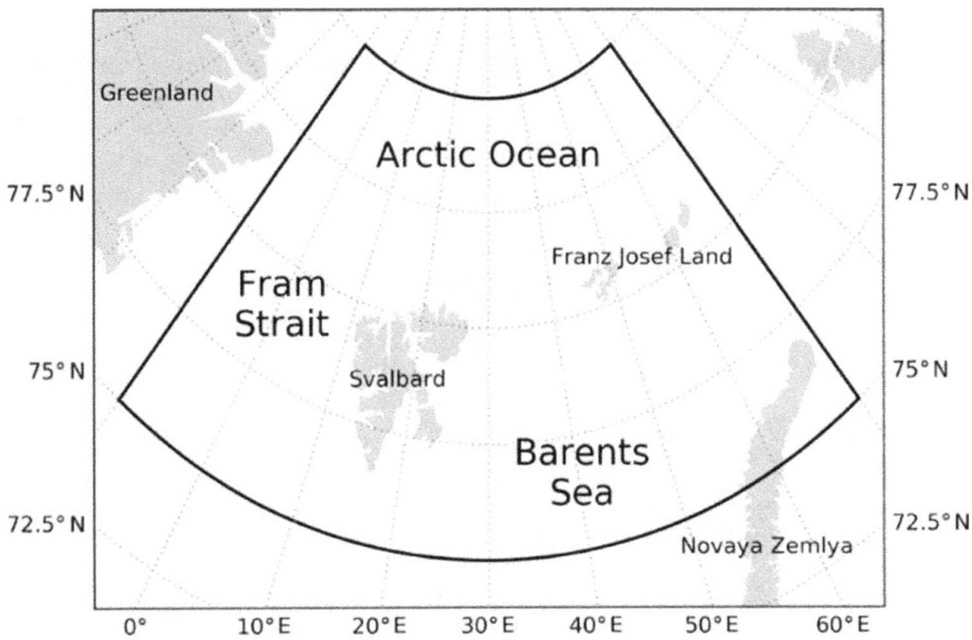

learning model, but not in a fully automated way. There was a lack of efficiency in manually selecting the data for training and validation. A better way is to use the already available ice charts for training the machine-learning model.

The proposed method has two phases:

1. Training phase
2. Operational phase

In both the phases, first thermal and textural noises are removed from the data provided by Sentinel-1., the ice charts are compared with the Synthetic Aperture Radar images, while the training phase is going on. Labels and features related to texture are retrieved from ice charts and then analysed to train a random forest classifier. Then this random forest classifier, which was just trained is applied on the texture images provided by Sentinel-1 operationally.

The previous algorithms have a major setback of thermal noise when applied to the Sentinel-1 TOPSAR data. This problem of thermal noise is a serious issue which needs to be solved as it can majorly affect the utility of dual-polarization Synthetic Aperture Radar data. The cross-polarization images have a lot of noise which occurs because of the low signal to noise ratio. This noise has a serious impact on backscattering and texture images. To pre-process the Sentinel-1 dual polarization products and eliminate the problem of thermal noise, denoising methods are developed and the methods employed by Park et al. (2018, 2019) are used on the images of Sentinel-1 to remove noises before moving on to the further processes.

After denoising and correcting the images, now ice charts are used to train and validate the data. After taking the ice charts which are available publicly into account which helps us to improve objectivity and eliminate the thermal noise in Sentinel-1 data. These public ice charts are prepared manually by analysing

Figure 17. Flowchart for the algorithm's processing stages. The training phase is represented in grey colour.
(Park et al., 2020)

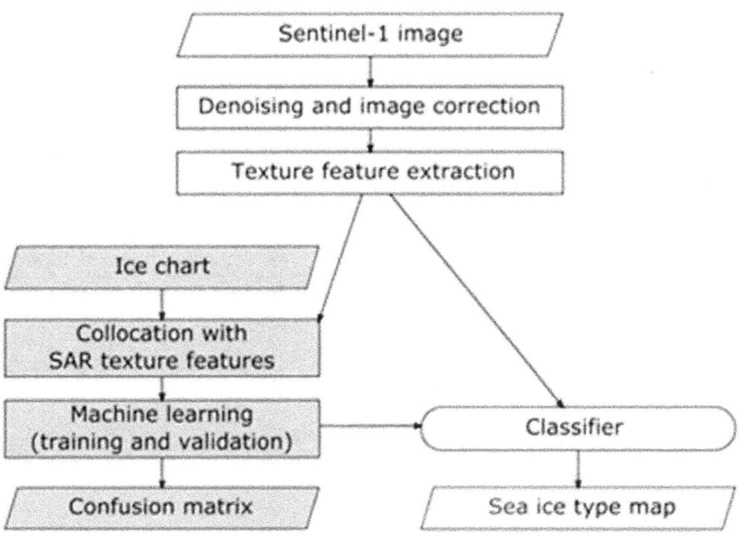

Figure 18. Ice chart preprocessing example. From the ice chart, figure 18 – a denotes stage of development (SoD); 18 – b denotes ice concentration (CT); and 18 – c denotes partial concentration of the dominant ice type (CP) maps, which are extracted. Next, certain distinct categories of stages of development (SoD) are combined, like merging of thin and thick ice into a single label is done and then termed as first-year ice, additionally, areas where ice concentration is low are designated as open water; (e) and (f) are the HH and HV polarization images from which textural features are retrieved to which the processed map of SoD (d) is associated
(Park et al., 2020)

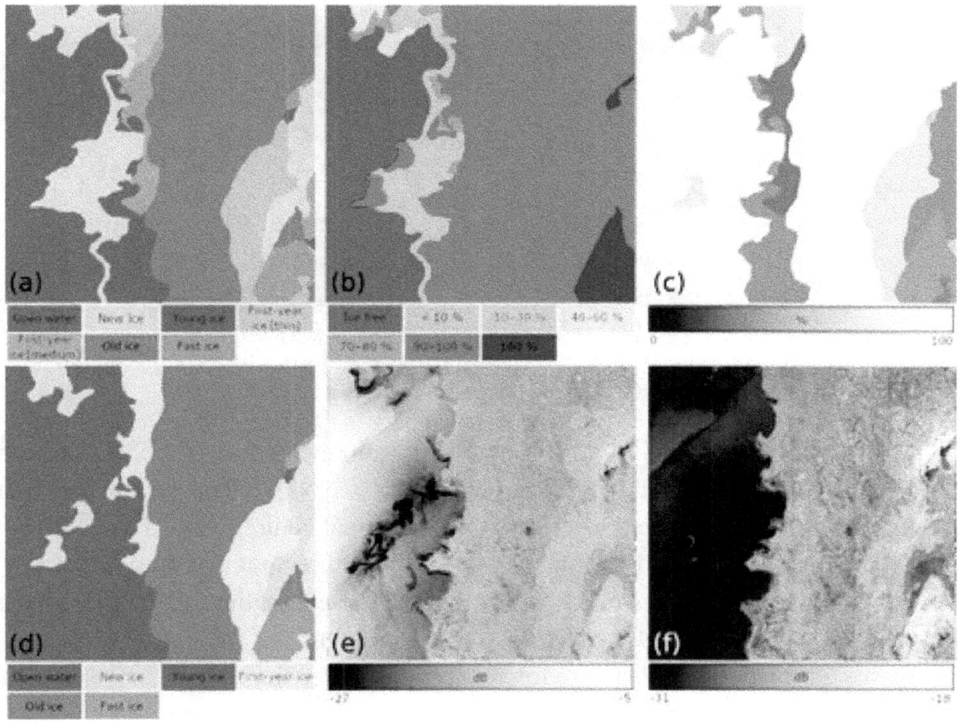

various sources of satellite images and data. There is a risk of inconsistency in decision making in terms of similar ice conditions which can be avoided by using large number of the ice charts for training, this would also help in operator-to-operator bias. Currently, human interpretation is considered as the best data on sea ice, so to get the best sea ice classifier, the public ice charts are used for training, because they consist of the knowledge of ice analysts.

1.8 Convolutional Neural Networks for Sea Ice Detection and SIC Estimation

With the Arctic Sea ice diminishing due to climate change, precise and efficient monitoring of sea ice is important for various reasons, including safe navigation through icy waters. A decent knowledge is importance in the context of sea ice monitoring and climate research because offshore operations and shipping has increased significantly. Using remote sensing methods for collecting information of sea ice are considered to be more effective in terms of cost when compared to *situ* measurements.

Figure 19. CNN structure
(Sea Ice Sensing From GNSS-R Data Using Convolutional Neural Networks, 2018)

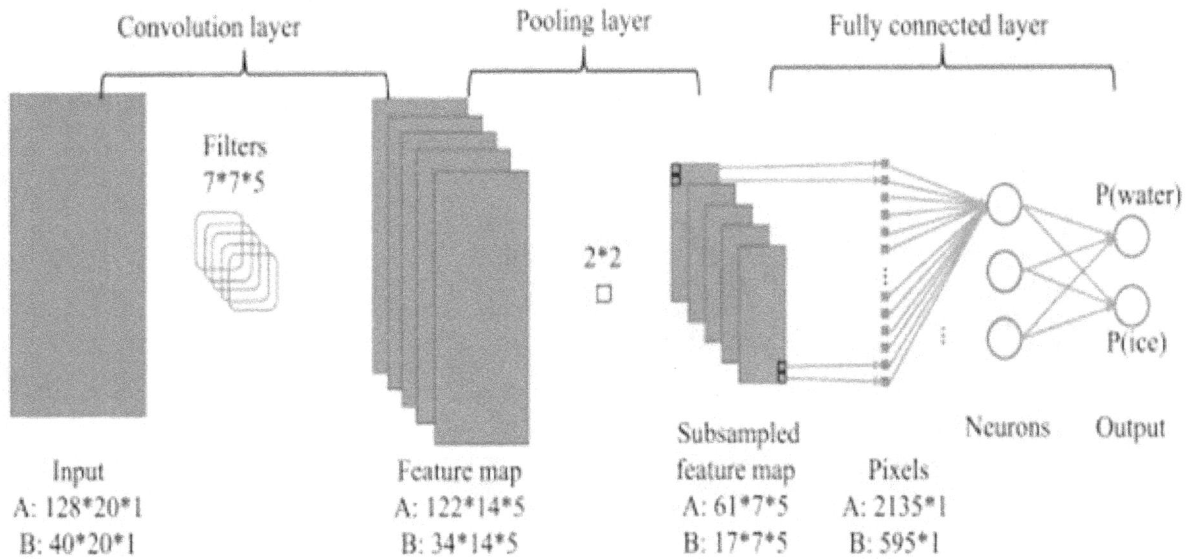

The Global Navigation Satellite System Reflectometry abbreviated as GNSS-R data can be effectively utilized by employing Convolutional Neural Networks (CNNs) to detect ice and estimate the Sea Ice concentration (SIC). To detect the sea ice, the Convolutional Neural Network (CNN) is structured with a classification layer. This classification layer categorizes the input data into different classes and determines whether the input belongs to a specific group. For estimation of sea ice concentration, a regression layer is used which provides numerical data for estimating the extent of sea ice coverage. Figure 19 represents a typical CNN structure for sea ice remote sensing.

This CNN structure consists of several layers. First, there is a block of optional convolutional layers which are composed of several filters that are of specific dimensions. Each one of these filters can be recognized as a feature extractor, which interacts with the input image via convolution. Consequently, an activation function processes these convolved images to obtain feature maps. The convolution layer outputs these feature maps.

Then comes a pooling layer that eliminates the redundancy of the feature map through subsampling. The output given by the final pooling layer is transformed into a vector, this vector is then passed through one or many connected layers. The fully connected layer combines all the knowledge gained by the previous layers and the desired patterns are determined. Figure 19 shows the association of parameters of set A with full Delay-Doppler Map (DDM) input and the association of set B parameters with cropped input.

Delay-Doppler Map (DDM) is a type of data representation for shows the scattering of Global Navigation Satellite System Reflectometry (GNSS-R) signals, when it interacts with the observed surface. Each one of the DDM pixels corresponds to specific spatial clusters, because of which the pixel information gets integrated with other data, similar to that in sea ice concentration (SIC). Therefore, to detect sea ice and predict sea ice concentration and Delay-Deployer map data more accurately, Convolutional Neural Networks (CNN) are used.

There are majorly 3 stages in CNN applications:

Figure 20. Samples of delay-doppler maps in 128 × 20 format which is the original data format, it can be observed that there is no alignment between the secular points. The red boxes represent the cropped inputs in data format 40 × 20 with a consistent position within each frame, which follows the approach followed in (Yan et al., 2017). The vertical axis corresponds to the delay axes and the horizontal axis represents the doppler axes.
(Sea Ice Sensing From GNSS-R Data Using Convolutional Neural Networks, 2018)

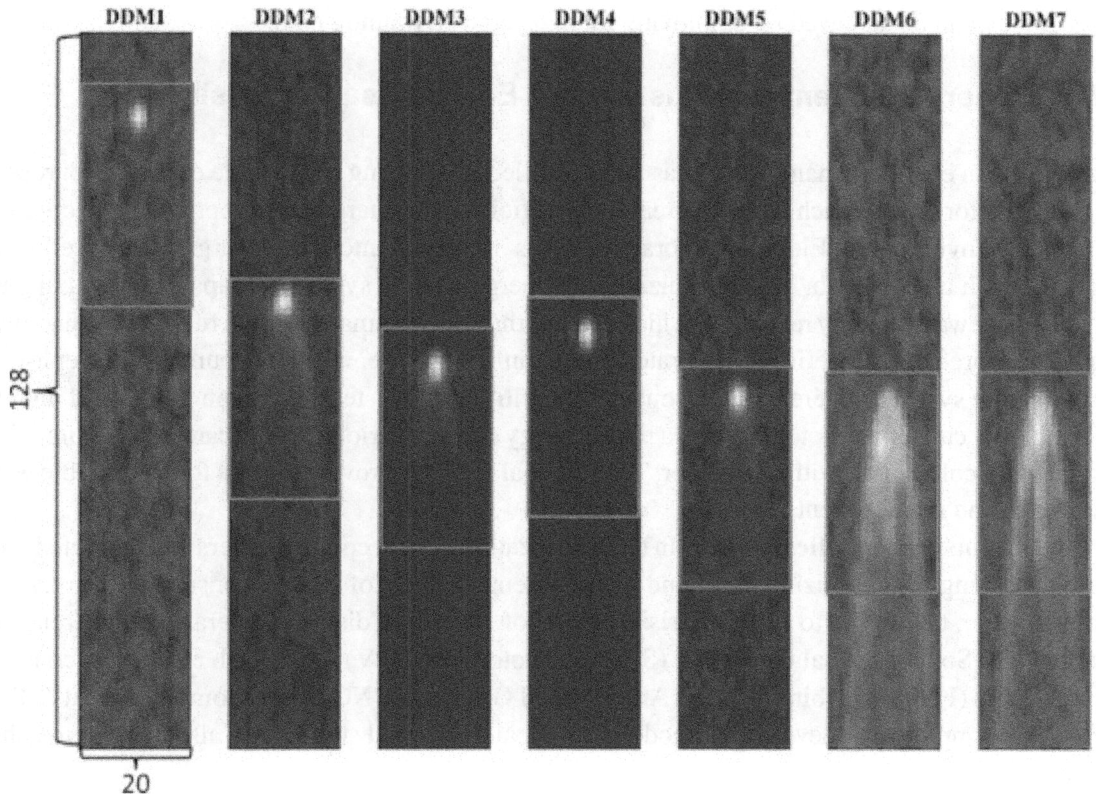

1. Data pre-processing: In this stage, the DDM images are pre-processed, which follows several procedures which include noise normalization, determination of signal box and floor subtraction. CNN has the ability to extract the features and that too independently from the position of the Delay-Doppler map.
2. Training of CNN: For training the CNN, Back-propagation learning (Werbos, 1974) and the SGDM algorithm (LeCun et al., 2012) are used and both forward as well as backward propagation are constituted in this learning method.
3. Testing of CNN: The sea ice concentration and detection-based CNNs are trained using the same data that was used for training. The performances of this CNN- and the already existing Neural network-oriented sea ice remote sensing techniques.

Convolutional Neural Networks give very promising results when used for identifying patterns that are complex, therefore the goal is to generate more accurate results and automate and predict sea ice detection and concentration respectively. Safer navigation and better management of resources along with monitoring of sea ice can be ensured using the advanced techniques addressed here.

Algorithms based on CNN are effective in detecting sea ice and estimating its concentration as they are intended to handle the data produced by GNSS-R. NNs have great accuracy in this, which can also be achieved using CNNs by preprocessing them properly, this will also be advantageous when recognition of pattern and data format is taken into account. This is also helpful in rectifying hidden errors. Further, more accuracy can be achieved by employing more layers and training data.

1.9 Multi-Energy Systems for Sustainable Energy Usage by Using MILP

There is a rise in energy demand which has to be fulfilled by utilizing sustainable energy resources. So, there is a need for an approach which increases the efficiency of energy consumption and decreases its impact on the environment. Electricity storage systems integrated into the power grid are a well-known example of such an approach. This optimization of energy storage systems is capable of utilizing non-predictable renewable energy resources, which ensures that future demands can be fulfilled by this surplus energy. There are many benefits of integrated energy infrastructure, which encourages the deployment of multi-energy systems. Therefore, the concept of multi-energy systems has been introduced, in which multiple energy carriers work together. In a multi-energy system, various energy careers and components synergistically collaborate with each other. This optimal solution provides a plan for sustainable energy development and management.

Optimization mathematical model: In the optimization model, energy carriers are integrated within a hub by focusing on optimizing costs and environmental effects of energy consumption in an urban microgrid. Here, the aim is to fulfil the rising needs of an energy district. Several types of renewable generators like Solar Thermal Collectors (STC) & Photovoltaic (PV) along with a series of converters like Heat Pump (HP), Absorbing Chiller (AC), Natural Gas Boiler (NGB) and Combined Heat & Power (CHP) can be employed to develop the model. Electrical or thermal storage can also be used to achieve the objective.

Just like other optimization models, this model also contains some variables that represent the values that need to be optimized, objective functions that signify the function that needs to be optimized (it may be either maximized or minimized), variable bounds that represent the physical limitations of the value which is assumed by the variables, parameters that define the input values and remain constant, equality or inequality constraints like component behaviour & physical balances.

The cost required for meeting the district energy needs annually is considered by the cost objective function. The approximations of cost functions are done to simplify the problem and transform it into a linear form. Another objective taken into consideration is to minimize the environmental impact of global warming. Then, scalarization methods are employed to combine these two objectives and create a multi-objective function. To ensure that there is no violation of limitations of efficiency, capacity limitations and energy flows for energy sources, balance equations and other constraints are also applied referred to the schematic shown in Figure 21.

The methodology used an integer linear programming algorithm known as MILP because the model has real or integer variables and also all the constraints along with the objective function are linear in nature. Italian and Vietnamese contexts are put into comparison by using average values for environmental and economic parameters and consistent values for technical parameters like efficiency, average operating hours etc.

Figure 21. Schematic of the analysed urban energy hub
(Cannata et al., 2019)

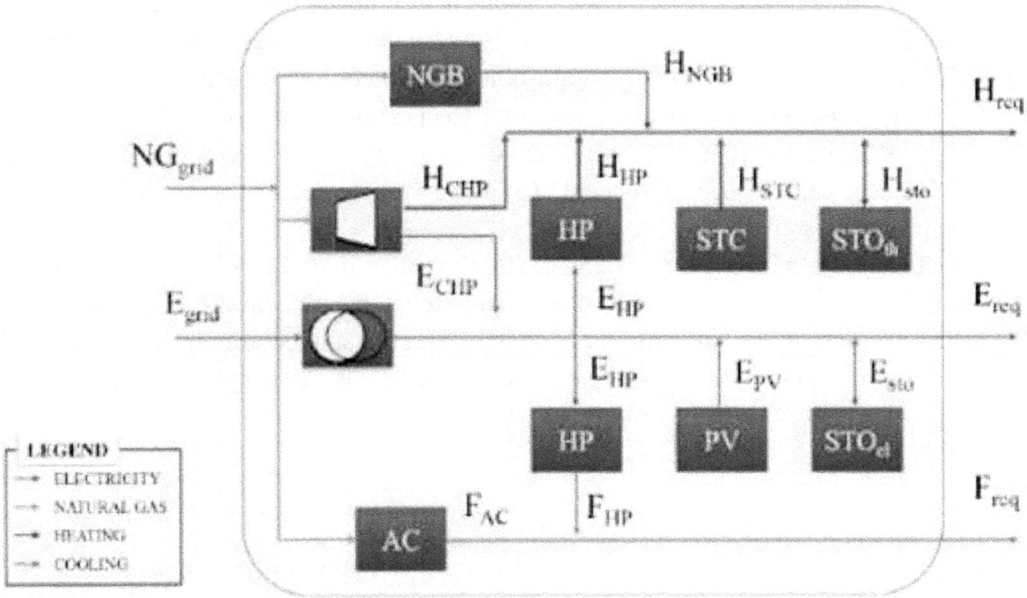

Figure 22. Optimization result for Italian and Vietnamese scenarios
(Cannata et al., 2019)

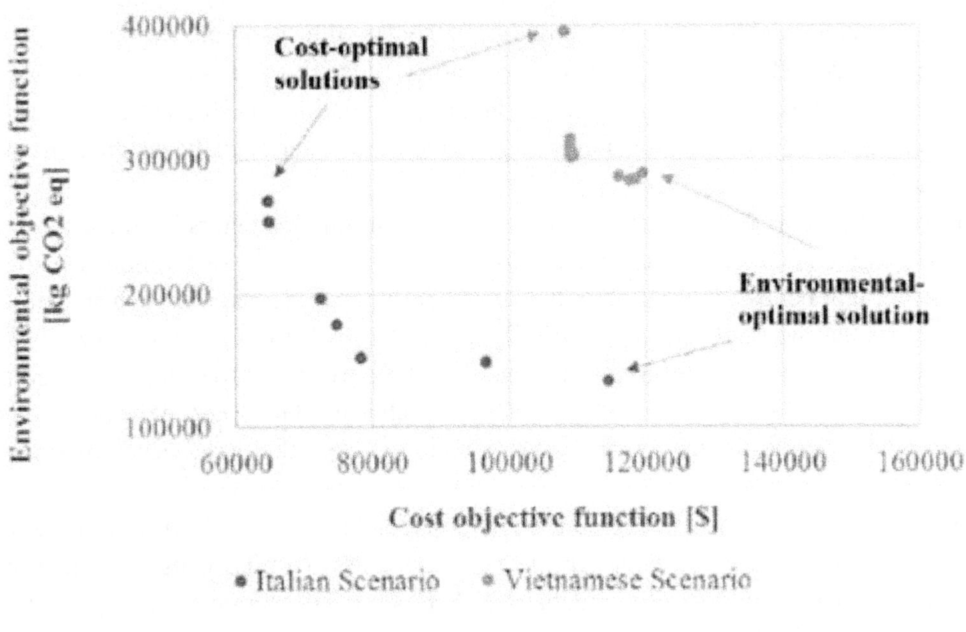

The results of simulation and optimization from the model have a range of optimal solutions which are concerned with both cost as well as environmental impacts for Italy and Vietnam. The entire spectrum from wcost = 0 which indicates that the focus is on environmental optimization to wcost = 1 which indicates that the focus is on prioritizing economic optimization is evaluated by analysing the set of solutions by adjusting the weights within the multi-objective function. Italy and Vietnam are two urban areas whose scenarios are compared to understand the effect of the optimized model on climate.

The result that is derived is that a diverse range of solutions can be found in the Italian scenario whereas solutions can be said to be more concentrated in the Vietnamese scenario, which makes it easier to pick up a compromised solution.

The main challenge that is faced here is enhancing the urban microgrids from the synthesis phase to the operational phase. The scalarization technique and multi-objective optimization model is used to balance the different solutions by not compromising with the costs and environmental concerns. This is the major part of improvement in this technique as compared to other existing solutions.

2. SNOW

Snow water equivalent and snow depth are areas of interest to many researchers as both have a wide variety of applications.

2.1 SVM and Frequency Modulated Continuous Wave (FMCW) Radar

The Arctic Ocean is changing in a fast pace because of climate changes, particularly in context of the sea ice cover. The Arctic Sea ice has thinned and has transformed from being ice that is for a longer period of time i.e., perennial ice to being ice that has a shorter life span i.e., seasonal ice as compared to ice in earlier decades. Because of these changes, the presence of open water has increased, surface albedo has decreased and degree of absorption of incoming radiation has increased. The Arctic climate is significantly affected by the snow cover as it affects the heat transfer from the ocean to the atmosphere and vice versa.

A notable observation is that the depth of snow ranges at around 20 meters length scales for all types of ice in situ are located at distances more than 20-50 meters from the nearest remote sensing measurement, then there might be less correlation in the snow depth comparisons over such spatial distances.

The snow radar that was utilized in NASA's Operation IceBridge has given observations related to the snow depth on the sea ice cover. Support Vector Machine (SVM) is used for the development and evaluation of the usage of a distinct snow layer tracker which is used to measure the depth of snow. The snow radar uses ultra-wide frequencies from 2 to 8 GHz to detect the interfaces between air & snow and snow & ice.

The wideband Frequency Modulated Continuous Wave abbreviated as FMCW radar developed by the Kansas University also known as the Kansas University's snow radar is used to analyse the snow depth on Arctic Sea ice. It uses an ultrawideband (UWB) chirp to operate on a range of bandwidths, mostly between 2 and 6.5 GHz.

Here is the working of this snow radar to measure and monitor Arctic's snow depth precisely, typically in remote environments:

1. **Radar Principle:** The functionality of the Frequency Modulated Continuous Wave (FMCW) radar is based on radar principles and its task is to propagate electromagnetic signals in the direction of Earth's surface at particular frequencies.
2. **Interface Detection:** The signals emitted by the radar pierce the snow cover on Arctic Sea ice to interact with two major interfaces that are the interfaces between air & snow and that between snow & ice. Because of the differences in the dielectric constants of sea ice and snow, the radar signals interacting with these reflect differently and this difference in the reflection pattern is detected by the radar and then the radar obtains the measurements of snow depth accurately.
3. **Time Delay Measurement** The difference in the time when the signals are emitted from the snow radar and the time at which they return to the radar after interacting with the snow-ice interface and air-snow interface is referred to as time delay. This time delay is calculated by the snow radar to identify the distance between itself and the snow-ice and air-snow interfaces.
4. **Snow Depth Calculation:** The radar measures the snow cover thickness overlying the sea ice after calculating the delay in time between the two interfaces- air-snow and snow-ice.

Table 3. Below shows the key configuration parameters of the snow radar used in Greenland/Arctic deployments for 2009-2012 [after Panzer et al., 2013]

Field Season	Greenland 2009	Greenland 2010	Greenland 2011	Greenland 2012
Bandwidth (GHz)	2.5–7	2–6.5	2–6.5	2–8
Pulse length (µs)	270	250	250 or 255	250
Pulse repetition frequency (kHz)	2	2	2	2
Transmit power (mW)	10	20	50	100
Range resolution (cm)	~5	~5	~5	~4
Cross-track footprint (m)	12	12	12	10
Along-track footprint (m)	12	12	12	10

Figure 23. (a) shows an instance of Snow Radar data echogram taken on 21/04/2010, different waveforms are used to represent distinct interfaces; (b) A single sharp peak represents thin snow/bare ice because the radar is unable to pierce the salty sea ice; (c) a distinct prominent peak in backscatter with a considerable rise in range away from the nadir position indicates open water is represented by a single sharp peak with significant off-nadir (increasing range) backscatter; (d) A double peak represents snow covered ice, where the air-snow interface is represented by the first peak and the snow-ice interface is represented by the second peak, and the snow depth is determined by the time delay between the two peaks (Holt et al., 2015)

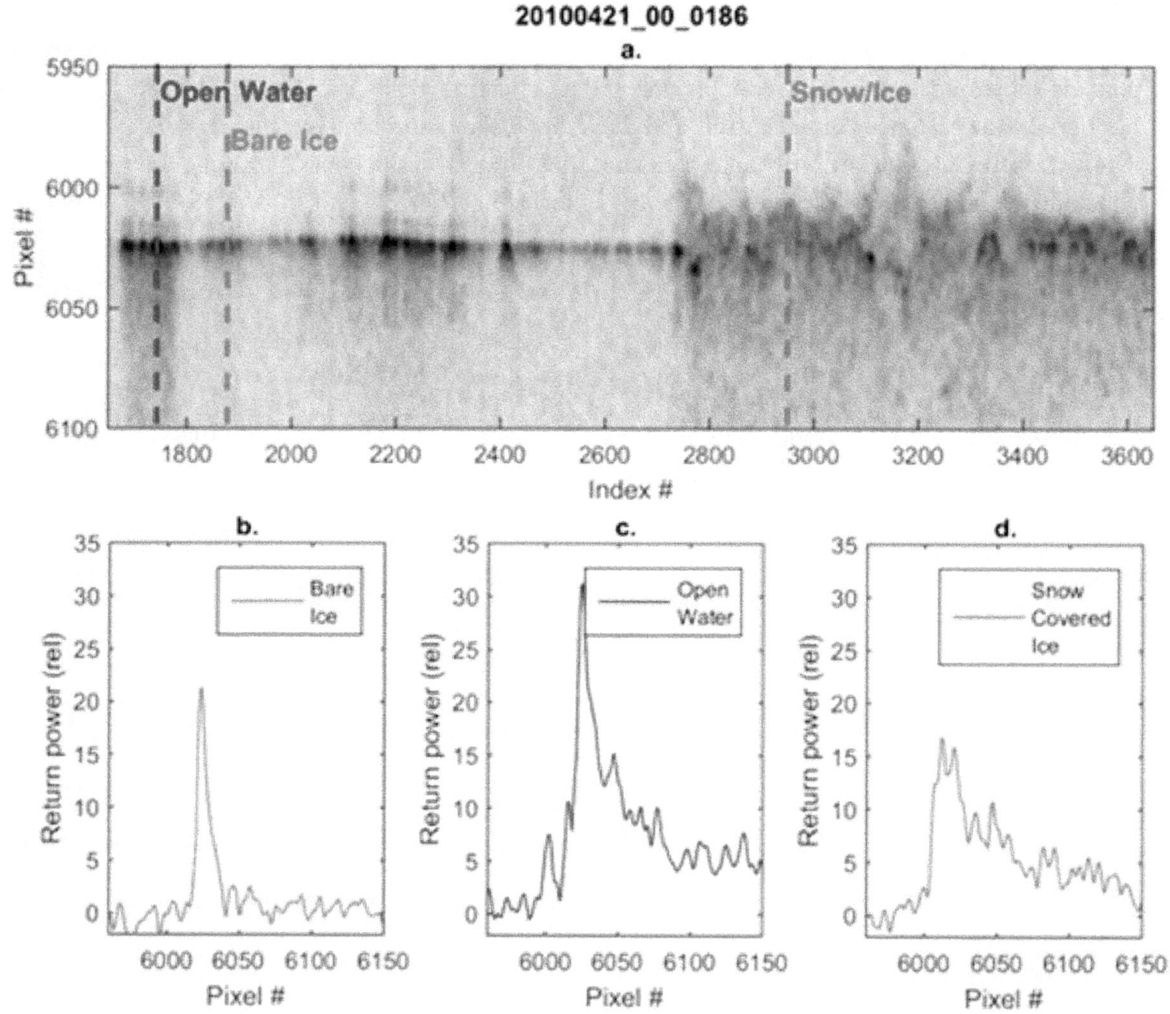

The results of the snow depth measurements are analysed for identify the quality, errors and consistency by comparing this collected data with in situ data during two specific sea ice field expeditions that took place in the Lincoln Sea near Greenland which are the GreenArc 2009 and the CryoVEx 2011 campaigns.

Typically, the snow radar has to detect the interface between air & snow and then calculate the time delay of the interface between snow and ice during dry snow conditions, as depicted in Figure 23. Figure 23-d shows that the time delay between the two interfaces which are considered as peaks in the range profile is converted into snow depth.

The two interfaces have different properties and nature including surface and volume scattering properties at the frequencies 2 to 6.5 – 8 GHz which is the frequency at which the radar operates, and distinct values of dielectric constants at the two air-snow interface and snow-ice interface. Here, the air-snow interface is expected to have a lower peak than that of the snow-ice interface because of its lower dielectric contrast between air and snow. Figures 23-b and 23-c depict the detection of thin ice and open ocean as one single interface by the snow radar.

Future works in this field should aim towards refining and expanding the measurements of snow depth by employing advanced radar and remote sensing technology, which might incorporate satellite data, data from drones, and findings from underwater vehicles, which can further improve both measurement frequency and as well as spatial coverage.

2.2 Deep Convolutional Networks

Convolutional Neural Networks (CNNs) have been used extensively to study computer vision problems since 2012. As the depth of a CNN increases, so does its ability to extract features from the training images. This extraction of features enables CNNs to make accurate predictions. However, it is simply not enough to increase the depth of a CNN to make more accurate predictions. As a CNN deepens i.e., more layers are added to it, it requires more training data to train upon. A very deep CNN with a small training dataset would lead to overfitting.

Conventionally, Cloud and Snow distinguishing ML Algorithms were fed with handcrafted features from the satellite image dataset. This approach, however, has its limitations. There is need of a Convolutional Neural Network Architecture which can extract the required features without the need of any feature engineering.

Yongjie Zhan et al. proposed a multi-layer deep convolutional neural network to distinguish cloud and snow in satellite images (Zhan et al., 2017). They propose an end-to-end Deep Learning Model which can be trained from scratch and does not require any hyperparameter tuning. Their model is based on the VGG Deep Learning model and spans over 15 convolution layers. It can exploit both the low-level

Figure 24. Deep CNN architecture to distinguish between cloud and snow
(Zhan et al., 2017)

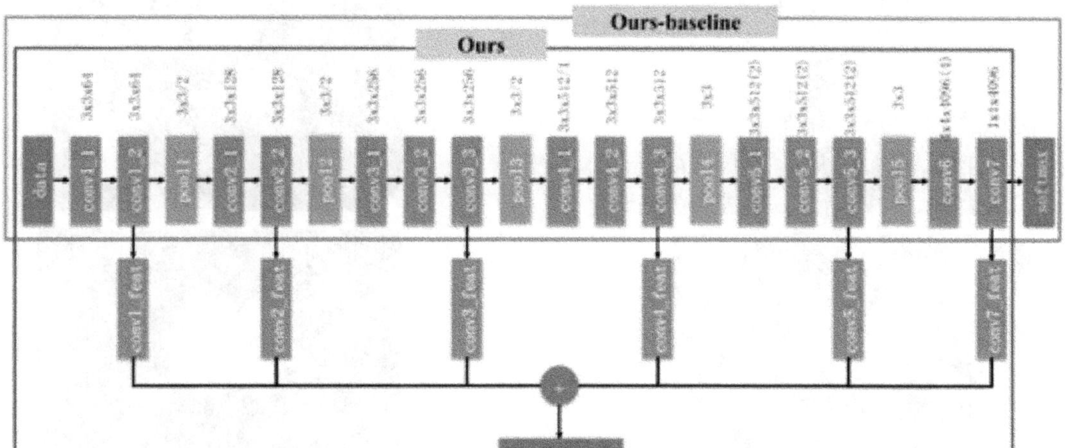

Table 4. Results obtained by Y. Zhan et al. (Zhan et al., 2017)

Methods	Precision	Recall	mIOU
ENVI threshold	55.1%	79.2%	60.8%
FCN	85.7%	80.3%	78.6%
FCN-ensemble	88.6%	85.7%	84.5%
Ours-baseline	90.2%	89.6%	88.7%
Ours	**92.4%**	**91.5%**	**90.6%**

spatial information and the high-level semantic information. Their proposed CNN architecture is shown in Figure 24. The proposed CNN model is based on the VCG modules.

The perform this study Y. Zhan et al. collected 50 Gaofen #1 satellite images with the resolution of 16,, each image spanning 13400 x 12000 pixels. They achieved quite promising results with the proposed technique. The obtained results are shown in table 4.

Y. Zhan et al. achieved an overall Precision of 92.4%, Recall of 91.5% and mIOU of 90.6%. These results are the highest among the existing works and establish authority of the proposed method. The visualization of one of the satellite images is shown via Figures 2, 3, and 4.

2.3 M-ResNet Model (Based on Deep Convolutional Network)

Cloud and snow detection and differentiation techniques have significant remote sensing applications such as automatic cloud and snow removal, image completion and image fusion. These techniques can also help us identify the extent of snow coverage on the planet. With environmental hazards such as global

Figure 25. Original image
(Zhan et al., 2017)

Figure 26. Cloud detection via deep CNN
(Zhan et al., 2017)

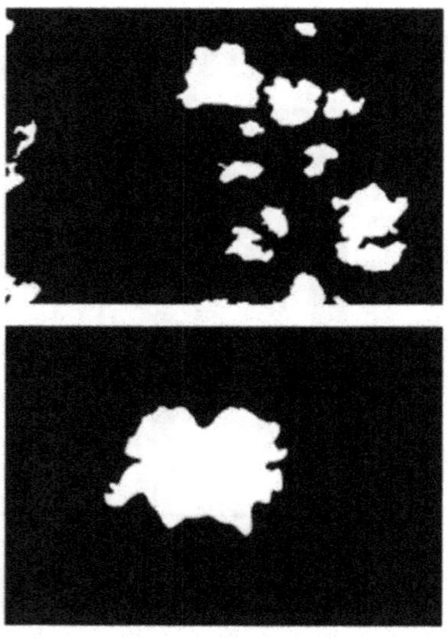

Figure 27. Ground truth
(Zhan et al., 2017)

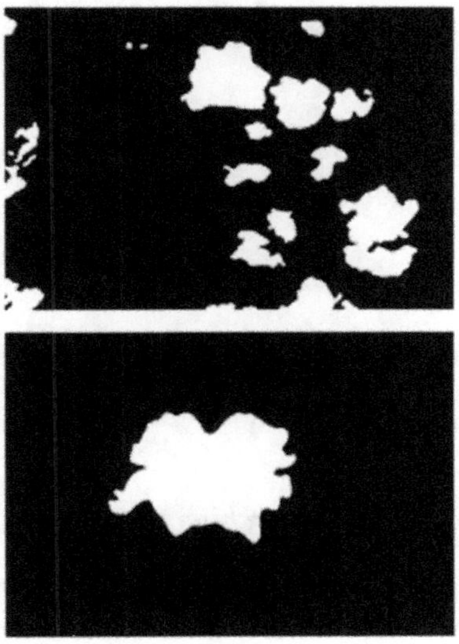

warming on the rise, scientists from numerous fields are interested in figuring out the snow coverage of the planet and how it is changing over the years. However, this task is not so simple. Snow and Clouds appear quite similar in satellite images. This challenge occurs because clouds and snow share similar local texture patterns in both visible and infrared spectra, and they share a consistent color distribution. Figure 28 shows the satellite image of the entire planet (Hej, n.d.). As one can tell, differentiating between clouds and snow is extremely difficult, at least for the naked eye.

Most of the works in this field make use of hand-crafted features and simple classifiers and fail to take advantage of the semantic level information from the satellite images. Existing machine learning models are far from perfect and have a high error rate. The models are also quite sensitive to "salt and pepper" noise. "Salt and Pepper" noise in images results from errors in signal transmission or analog to digital signal conversion. However, the recent work by Min Xia et al. offers some promising results in this field (Xia et al., 2018). Their model makes use of 152 layers. The distribution of their layers is

Figure 28. Satellite image of the planet with clouds and snow
(Hej, n.d.)

Figure 29. M-ResNet cloud/snow classification model structure
(Xia et al., 2018)

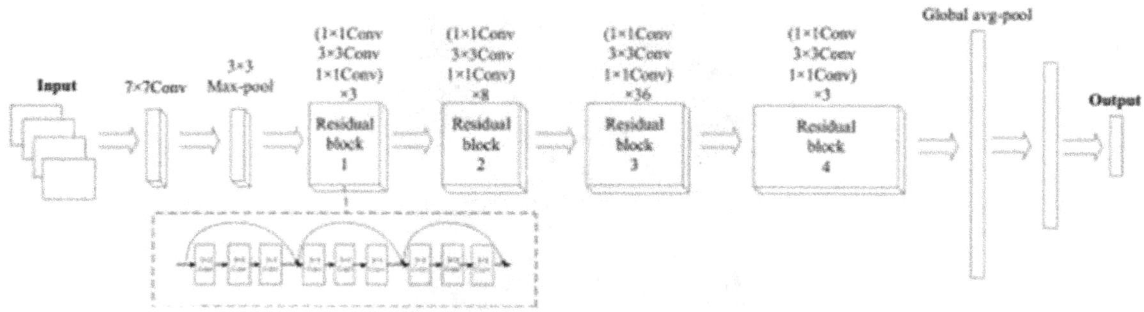

Table 5. Distribution of layers in the M-ResNet architecture (Xia et al., 2018)

S. No.	Name (as per figure 2)	Number of Layers	Type of Layer
1	7x7 Conv	1	Convolutional Layer
2	Max-pool	1	Pooling Layer
3	Residual Block 1	3 * 3 = 9	Convolutional Layer
4	Residual Block 2	3 * 8 = 24	Convolutional Layer
5	Residual Block 3	2* 36 = 108	Convolutional Layer
6	Residual Block 4	3 * 3 = 9	Convolutional Layer
Total		152	

Figure 30. Original satellite image
(Xia et al., 2018)

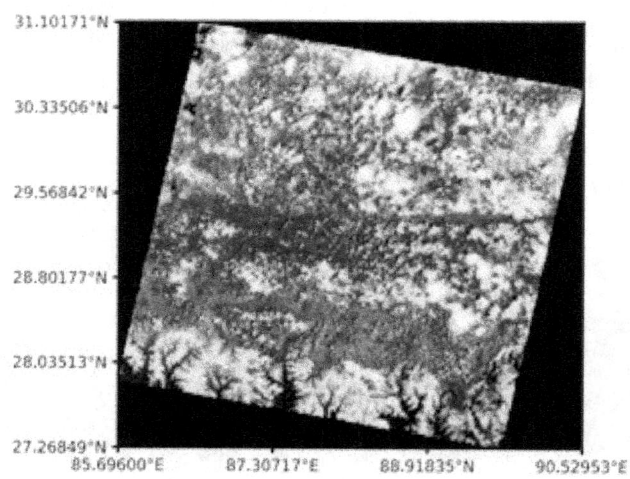

Figure 31. M-ResNet results on original satellite image (figure 2)
(Xia et al., 2018)

shown in Table 5. A high-level overview of their model, namely M-ResNet, has been depicted in figure 29. The M-ResNet model makes use of Residual Blocks.

The M-ResNet model achieved an overall accuracy of 90.29% and 95.84% on single spectral and multi-spectral satellite images respectively (Hej, n.d.). A sample of the results obtained via the M-ResNet model is shown in figures 7 and 8.

Deep Learning models from this field show some promising results in other remote sensing applications such as road extraction and building detection. However, there are still many challenges to be faced such as:

i) Model processing speed: The M-ResNet architecture, even though highly accurate, requires an enormous amount of time for training.
ii) Partial datasets: Most of the existing datasets comprise of only cloud and snow images. However, in the real world, landscapes such as lakes, rivers and smog would also have a serious impact on the classification accuracy.

2.4 Support Vector Regression (SVR) Snow-Depth Retrieval Algorithm

Knowing the snow depth at different geographical locations across the globe has been another challenging problem. The extent of snow cover can help us understand the changing weather patterns; however, the depth of snow can help us understand the effects of global warming. Researchers have been studying snow depth at four different land cover types, namely, forests, shrubs, prairies, and bare lands. The recent work by Xiongxin Xiao et al. has shown some promising results in this field (Xiao et al., 2018). They developed a support vector regression algorithm to predict snow depth based on passive microwave remote sensing data. The dataset used by them is shown in figure 32.

The different steps in their proposed algorithm are as follows:

Figure 32. Pictorial representation of the dataset used by Xiongxin Xiao et al. in their study of snow depth retrieval algorithm
(Xiao et al., 2018).

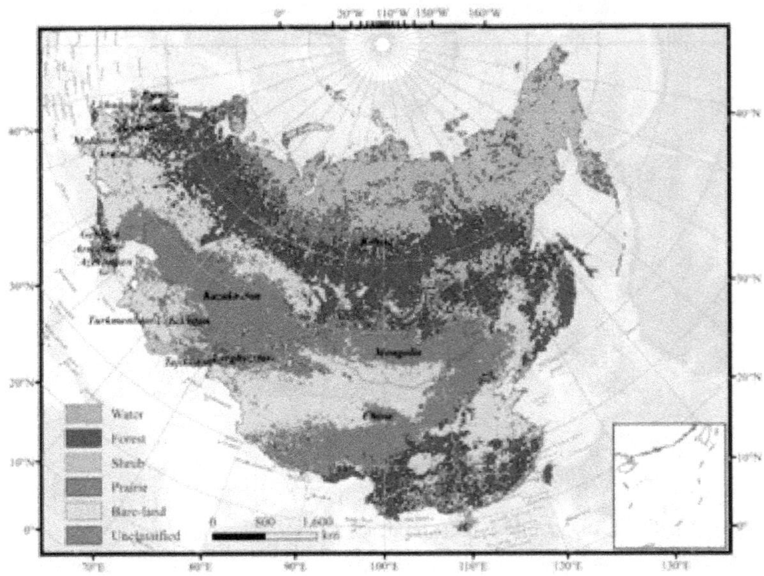

1. Data Preparation: The dataset used involved multiple PM brightness levels such as Tb19V, Tb19H, Tb37V, Tb85V, etc. Alongside this, the dataset also comprised of physically measured snow depths at a wide variety of regions.
2. Dry Snow Identification: Wet snow, cold deserts, precipitation, and frozen grounds have a similar microwave radiation as that of Dry Snow. Hence, the presence of these radiations hinders the accuracy of the reported results. Therefore, it is essential to remove radiation belonging to such patches. Numerous factors were applied to remove these radiations.
3. Reclassification of land cover categories: Now land covers were classified into categories such as Water Body, Bare Land, Shrub, Prairie, and Forest.
4. Snow cover year and stages: Different snow cover stages were identified such as snow cover accumulation stage, snow cover stabilization stage, and snow cover ablation stage.
5. SVR model: To train the model Pearson VII universal function kernel (PUK) was used.

Upon application of the above steps, the achieved results were compared with existing machine learning models such as Chang, SPD, ANN, and Linear Regression. The best results with the lowest Root-Mean-Squared error were achieved via the SVR algorithm. Figure 33 shows the results obtained.

The presence of vegetation remains the single based roadblock in the snow depth prediction. Vegetation has the most significant impact on microwaves emitted by snow cover. Among existing models, the worst accuracy is found in regions which are significantly covered by vegetation. Future research must find innovative solutions to mitigate this problem.

Figure 33. Support Vector Regression Algorithm to measure snow depth
(Xiao et al., 2018)

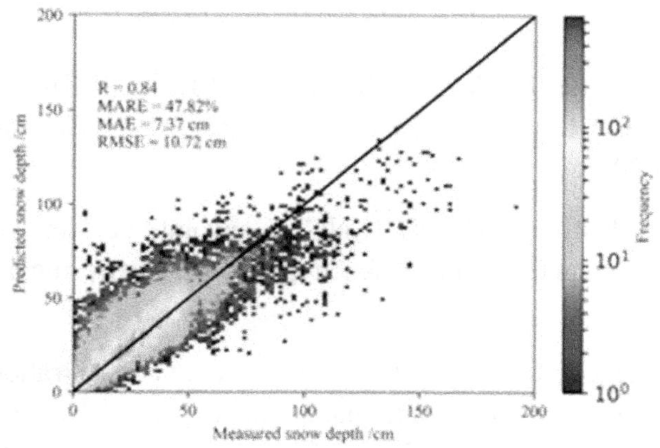

REFERENCES

Baumhoer, C., Dietz, A., Kneisel, C., & Kuenzer, C. (2019b). Automated Extraction of Antarctic Glacier and Ice Shelf Fronts from Sentinel-1 Imagery Using Deep Learning. *Remote Sensing (Basel)*, *11*(21), 2529. doi:10.3390/rs11212529

Cannata, N., Longo, S., Sanseverino, E. R., & Luu, L. Q. (2019). *Multi-Objective optimization of urban microgrid energy supply according to economic and environmental criteria*. ResearchGate. https://www.researchgate.net/publication/334051107_Multi-Objective_Optimization_of_Urban_Microgrid_Energy_Supply_According_to_Economic_and_Environmental_Criteria

Ghanbari, M., Clausi, D. A., Xu, L., & Jiang, M. (2019) Contextual classification of Sea-Ice types using compact polarimetric SAR data. *IEEE Journals & Magazine*. IEEE Xplore. https://ieeexplore.ieee.org/document/8731878

Hej. (n.d.). *Unique new global satellite and radar composite*. Meteoblue. https://www.meteoblue.com/en/blog/article/show/39821_Unique+new+global+satellite+and+radar+composite

Holt, B., Johnson, M. P., Perkovic-Martin, D., & Panzer, B. (2015). Snow depth on Arctic sea ice derived from radar: In situ comparisons and time series analysis. *Journal of Geophysical Research. Oceans*, *120*(6), 4260–4287. doi:10.1002/2015JC010815

LeCun, Y., Bottou, L., Orr, G., & Müller, K. (2012). Efficient BackProp. In Lecture Notes in Computer Science (pp. 9–48). doi:10.1007/978-3-642-35289-8_3

Lee, S., Im, J., Kim, J., Kim, M. J., Shin, M., Kim, H., & Quackenbush, L. J. (2016). Arctic Sea Ice Thickness Estimation from CryoSat-2 Satellite Data Using Machine Learning-Based Lead Detection. *Remote Sensing (Basel)*, *8*(9), 698. doi:10.3390/rs8090698

Leigh, S., Wang, Z., & Clausi, D. A. (2014). Automated Ice–Water classification using dual polarization SAR satellite imagery. *IEEE Transactions on Geoscience and Remote Sensing*, *52*(9), 5529–5539. doi:10.1109/TGRS.2013.2290231

Lohse, J., Doulgeris, A. P., & Dierking, W. (2019). An Optimal Decision-Tree Design Strategy and Its Application to Sea Ice Classification from SAR Imagery. *Remote Sensing (Basel), 11*(13), 1574. doi:10.3390/rs11131574

Park, J. W., Korosov, A., Babiker, M., Won, J. S., Hansen, M. W., & Kim, H. C. (2020). Classification of sea ice types in Sentinel-1 synthetic aperture radar images. *The Cryosphere, 14*(8), 2629–2645. doi:10.5194/tc-14-2629-2020

Ronneberger, O., Fischer, P., & Brox, T. (2015). U-NET: Convolutional Networks for Biomedical Image Segmentation. In Lecture Notes in Computer Science (pp. 234–241). Springer. doi:10.1007/978-3-319-24574-4_28

Wang, L., Scott, K. A., Xu, L., & Clausi, D. A. (2016). Sea ice concentration estimation during melt from Dual-Pol SAR scenes using deep convolutional neural networks: A case study. *IEEE Transactions on Geoscience and Remote Sensing, 54*(8), 4524–4533. doi:10.1109/TGRS.2016.2543660

Werbos, P. (1974). *Beyond Regression: New Tools for Prediction and Analysis in the Behavioral Science.* [Thesis, Harvard]. https://www.researchgate.net/publication/279233597_Beyond_Regression_New_Tools_for_Prediction_and_Analysis_in_the_Behavioral_Science_Thesis_Ph_D_Appl_Math_Harvard_University

Xia, M., Liu, W., Shi, B., Weng, L., & Liu, J. (2018). Cloud/snow recognition for multispectral satellite imagery based on a multidimensional deep residual network. *International Journal of Remote Sensing, 40*(1), 156–170. doi:10.1080/01431161.2018.1508917

Xiao, X., Zhang, T., Zhong, X., & Li, X. (2018). *Support vector regression snow-depth retrieval algorithm using passive microwave remote sensing data.* ResearchGate. https://www.researchgate.net/publication/323847002_Support_vector_regression_snow-depth_retrieval_algorithm_using_passive_microwave_remote_sensing_data

Yan, Q., Huang, W., & Moloney, C. (2017). Neural networks based sea ice detection and concentration retrieval from GNSS-R Delay-Doppler maps. *IEEE Journal of Selected Topics in Applied Earth Observations and Remote Sensing, 10*(8), 3789–3798. doi:10.1109/JSTARS.2017.2689009

Zhan, Y., Wang, J., Shi, J., Cheng, G., Yao, L., & Sun, W. (2017). Distinguishing cloud and snow in satellite images via deep convolutional network. *IEEE Geoscience and Remote Sensing Letters, 14*(10), 1785–1789. doi:10.1109/LGRS.2017.2735801

Compilation of References

Abdalzaher, M. S., Elsayed, H. A., Fouda, M. M., & Salim, M. M. (2023). Employing machine learning and iot for earthquake early warning system in smart cities. *Energies*, *16*(1), 495. doi:10.3390/en16010495

Abdalzaher, M. S., Soliman, M. S., & El-Hady, S. M. (2023). Seismic Intensity Estimation for Earthquake Early Warning Using Optimized Machine Learning Model. *IEEE Transactions on Geoscience and Remote Sensing*, *61*, 1–11. doi:10.1109/TGRS.2023.3296520

Abdelkareem, M., & Mansour, A. M. (2023). Risk assessment and management of vulnerable areas to flash flood hazards in arid regions using remote sensing and GIS-based knowledge-driven techniques. *Natural Hazards*, *117*(3), 2269–2295. doi:10.100711069-023-05942-x

Abeysiriwardana, P. C., Jayasinghe-Mudalige, U. K., & Kodituwakku, S. R. (2022). Connected researches in "smart lab bubble": A lifeline of techno-society space for commercial agriculture development in "new normal. New Techno Humanities, 2(1), 79-91.

Acharya, A., & Prakash, A. (2019). When the river talks to its people: Local knowledge-based flood forecasting in Gandak River basin, India. *Environmental Development*, *31*, 55–67. doi:10.1016/j.envdev.2018.12.003

Adak, A., Pradhan, B., & Shukla, N. (2022). Sentiment analysis of customer reviews of food delivery services using deep learning and explainable artificial intelligence: Systematic review. *Foods*, *11*(10), 1500. doi:10.3390/foods11101500 PMID:35627070

Adam, M. A., Sulaiman, A., Baharuddin, A. S., Mokhtar, M. N., Subbian, K., Tabatabaei, M., & Marwanto, A. P. Y. (2019). Characterization of delignified Oil Palm Decanter Cake (OPDC) for polymer composite development. *International Journal on Advanced Science, Engineering and Information Technology*, *9*(2), 384–389. doi:10.18517/ijaseit.9.2.2392

Adel, A. (2022). Future of industry 5.0 in society: Human-centric solutions, challenges and prospective research areas. *Journal of Cloud Computing (Heidelberg, Germany)*, *11*(1), 1–15. doi:10.118613677-022-00314-5 PMID:36101900

Adeniyi, E. A., Ogundokun, R. O., Gbadamosi, B., Misra, S., & Kalejaiye, O. (2022). Classification of swine disease using K-nearest neighbor algorithm on cloud-based framework. In *Artificial Intelligence for Cloud and Edge Computing* (pp. 71–90). Springer International Publishing. doi:10.1007/978-3-030-80821-1_4

Adoko, A. C., Gokceoglu, C., Wu, L., & Zuo, Q. J. (2013). Knowledge-based and data-driven fuzzy modeling for rockburst prediction. *International Journal of Rock Mechanics and Mining Sciences*, *61*, 86–95. doi:10.1016/j.ijrmms.2013.02.010

Ahmed, S., Yong, J., & Shrestha, A. (2023). The Integral Role of Intelligent IoT System, Cloud Computing, Artificial Intelligence, and 5G in the User-Level Self-Monitoring of COVID-19. *Electronics (Basel)*, *12*(8), 1912. doi:10.3390/electronics12081912

Aithal, P. S. (2023). Advances and New Research Opportunities in Quantum Computing Technology by Integrating it with Other ICCT Underlying Technologies. *International Journal of Case Studies in Business IT and Education*, *7*(3), 314–358.

Aklah, Z. T., Al-Safi, A., & Abdali, M. H. (2023). *Predicting High-Risk Individuals in Disease Outbreaks Using Fuzzy Logic-Based Contact Tracing Model.*

Al Dulaijan, K., & Alajaji, M. (2022, November). HPC advances and impacts to exploration and reservoir development. In *SEG International Exposition and Annual Meeting* (p. D011S123R002). SEG. 10.1190/image2022-3746173.1

Alagumalai, A., Shou, W., Mahian, O., Aghbashlo, M., Tabatabaei, M., Wongwises, S., Liu, Y., Zhan, J., Torralba, A., Chen, J., Wang, Z. L., & Matusik, W. (2022). Self-powered sensing systems with learning capability. *Joule*, *6*(7), 1475–1500. doi:10.1016/j.joule.2022.06.001

Albahri, A. S., Duhaim, A. M., Fadhel, M. A., Alnoor, A., Baqer, N. S., Alzubaidi, L., Albahri, O. S., Alamoodi, A. H., Bai, J., Salhi, A., Santamaría, J., Ouyang, C., Gupta, A., Gu, Y., & Deveci, M. (2023). A systematic review of trustworthy and explainable artificial intelligence in healthcare: Assessment of quality, bias risk, and data fusion. *Information Fusion*, *96*, 156–191. doi:10.1016/j.inffus.2023.03.008

Aldana-Martín, J. F., García-Nieto, J., del Mar Roldán-García, M., & Aldana-Montes, J. F. (2022). Semantic modelling of earth observation remote sensing. *Expert Systems with Applications*, *187*, 115838. doi:10.1016/j.eswa.2021.115838

Aldini, I., Hidayat, R., Permanasari, A. E., & Ramdhani, A. (2022, December). Classification of Tsunami Warning Level using Artificial Neural Network and its Comparison in Southern Java Region. In *2022 5th International Seminar on Research of Information Technology and Intelligent Systems (ISRITI)* (pp. 607-612). IEEE. 10.1109/ISRITI56927.2022.10052907

Alioto, M., De, V., & Marongiu, A. (2018). Guest editorial energy-quality scalable circuits and systems for sensing and computing: From approximate to communication-inspired and learning-based. *IEEE Journal on Emerging and Selected Topics in Circuits and Systems*, *8*(3), 361–368. doi:10.1109/JETCAS.2018.2865783

Allioui, H., & Mourdi, Y. (2023). Unleashing the Potential of AI: Investigating Cutting-Edge Technologies That Are Transforming Businesses. *International Journal of Computer Engineering and Data Science*, *3*(2), 1–12.

Anikiev, D., Birnie, C., Waheed, U., Alkhalifah, T., Gu, C., Verschuur, D. J., & Eisner, L. (2023). Machine learning in microseismic monitoring. *Earth-Science Reviews*, *239*, 104371. doi:10.1016/j.earscirev.2023.104371

Asadi, A., Baise, L. G., Sanon, C., Koch, M., Chatterjee, S., & Moaveni, B. (2023). Semi-Supervised Learning Method for the Augmentation of an Incomplete Image-Based Inventory of Earthquake-Induced Soil Liquefaction Surface Effects. *Remote Sensing (Basel)*, *15*(19), 4883. doi:10.3390/rs15194883

Asim, K. M., Martínez-Álvarez, F., Basit, A., & Iqbal, T. (2017). Earthquake magnitude prediction in Hindukush region using machine learning techniques. *Natural Hazards*, *85*(1), 471–486. doi:10.100711069-016-2579-3

Babu, N. V., & Kanaga, E. G. M. (2022). Sentiment analysis in social media data for depression detection using artificial intelligence: A review. *SN Computer Science*, *3*(1), 1–20. doi:10.100742979-021-00958-1 PMID:34816124

Baburaj, A., Kumar, S. N., Aliyana, A. K., Banakar, M., Bairagi, S., & Stylios, G. (2023). Factors Affecting the Performance of Flexible Triboelectric Nanogenerators (F-TENGs) and their Sensing Capabilities: A Comprehensive Review. *Nano Energy*, *118*, 108983. doi:10.1016/j.nanoen.2023.108983

Badini, S., Regondi, S., & Pugliese, R. (2023). Unleashing the Power of Artificial Intelligence in Materials Design. *Materials (Basel)*, *16*(17), 5927. doi:10.3390/ma16175927 PMID:37687620

Badri, A. M. (2022, December). Litho-Fluid Facies Modelling Based on Integration of Seismic Inversion and Well Data: A Supervised Deep-Learning Approach. In *AGU Fall Meeting Abstracts* (Vol. 2022, pp. IN12C-0277).

Baklanov, A., & Zhang, Y. (2020). Advances in air quality modeling and forecasting. *Global Transitions, 2*, 261–270. doi:10.1016/j.glt.2020.11.001

Bale, A. S. NR, N. K., Somanna, A. N., Karthik, K. B., SN, N. R., & Rithik, S. M. (2023, June). Explosive Projectile Detection with an Arduino-Controlled Robot. In *2023 8th International Conference on Communication and Electronics Systems (ICCES)* (pp. 243-249). IEEE.

Baumhoer, C., Dietz, A., Kneisel, C., & Kuenzer, C. (2019b). Automated Extraction of Antarctic Glacier and Ice Shelf Fronts from Sentinel-1 Imagery Using Deep Learning. *Remote Sensing (Basel), 11*(21), 2529. doi:10.3390/rs11212529

Beaudoin, A., Bernier, P. Y., Villemaire, P., Guindon, L., & Guo, X. J. (2018). Tracking forest attributes across Canada between 2001 and 2011 using ak nearest neighbors mapping approach applied to MODIS imagery. *Canadian Journal of Forest Research, 48*(1), 85–93. doi:10.1139/cjfr-2017-0184

Beroza, G. C., Segou, M., & Mostafa Mousavi, S. (2021). Machine learning and earthquake forecasting—Next steps. *Nature Communications, 12*(1), 4761. doi:10.103841467-021-24952-6 PMID:34362887

Bhardwaj, A., Dagar, V., Khan, M. O., Aggarwal, A., Alvarado, R., Kumar, M., Irfan, M., & Proshad, R. (2022). Smart IoT and machine learning-based framework for water quality assessment and device component monitoring. *Environmental Science and Pollution Research International, 29*(30), 46018–46036. doi:10.100711356-022-19014-3 PMID:35165843

Bhende, M., Thakare, A., Pant, B., Singhal, P., Shinde, S., & Saravanan, V. (2022). Deep Learning-Based Real-Time Discriminate Correlation Analysis for Breast Cancer Detection. *BioMed Research International, 2022*, 2022. doi:10.1155/2022/4609625 PMID:35800216

Birkoben, T., & Kohlstedt, H. (2023). Matter and Mind Matter. In *Bio-Inspired Information Pathways: From Neuroscience to Neurotronics* (pp. 1–42). Springer International Publishing.

Biswas, N., Uddin, K. M. M., Rikta, S. T., & Dey, S. K. (2022). A comparative analysis of machine learning classifiers for stroke prediction: A predictive analytics approach. *Healthcare Analytics, 2*, 100116. doi:10.1016/j.health.2022.100116

Bi, W. L., Hosny, A., Schabath, M. B., Giger, M. L., Birkbak, N. J., Mehrtash, A., Allison, T., Arnaout, O., Abbosh, C., Dunn, I. F., Mak, R. H., Tamimi, R. M., Tempany, C. M., Swanton, C., Hoffmann, U., Schwartz, L. H., Gillies, R. J., Huang, R. Y., & Aerts, H. J. (2019). Artificial intelligence in cancer imaging: Clinical challenges and applications. *CA: a Cancer Journal for Clinicians, 69*(2), 127–157. doi:10.3322/caac.21552 PMID:30720861

Blohm, I., Antretter, T., Sirén, C., Grichnik, D., & Wincent, J. (2022). It'sa peoples game, isn't it?! A comparison between the investment returns of business angels and machine learning algorithms. *Entrepreneurship Theory and Practice, 46*(4), 1054–1091. doi:10.1177/1042258720945206

Bochenek, B., & Ustrnul, Z. (2022). Machine learning in weather prediction and climate analyses—Applications and perspectives. *Atmosphere (Basel), 13*(2), 180. doi:10.3390/atmos13020180

Böse, M., Andrews, J., O'Rourke, C., Kilb, D., Lux, A., Bunn, J., & McGuire, J. (2023). Testing the ShakeAlert earthquake early warning system using synthesized earthquake sequences. *Seismological Research Letters, 94*(1), 243–259. doi:10.1785/0220220088

Böse, M., Papadopoulos, A. N., Danciu, L., Clinton, J. F., & Wiemer, S. (2022). Loss-Based Performance Assessment and Seismic Network Optimization for Earthquake Early Warning. *Bulletin of the Seismological Society of America, 112*(3), 1662–1677. doi:10.1785/0120210298

Brunke, L., Greeff, M., Hall, A. W., Yuan, Z., Zhou, S., Panerati, J., & Schoellig, A. P. (2022). Safe learning in robotics: From learning-based control to safe reinforcement learning. *Annual Review of Control, Robotics, and Autonomous Systems*, *5*(1), 411–444. doi:10.1146/annurev-control-042920-020211

Camarena, S. (2020). Artificial intelligence in the design of the transitions to sustainable food systems. *Journal of Cleaner Production*, *271*, 122574. doi:10.1016/j.jclepro.2020.122574

Cannata, N., Longo, S., Sanseverino, E. R., & Luu, L. Q. (2019). *Multi-Objective optimization of urban microgrid energy supply according to economic and environmental criteria*. ResearchGate. https://www.researchgate.net/publication/334051107_Multi-Objective_Optimization_of_Urban_Microgrid_Energy_Supply_According_to_Economic_and_Environmental_Criteria

Capraro, G. T., Farina, A., Griffiths, H., & Wicks, M. C. (2006). Knowledge-based radar signal and data processing: A tutorial review. *IEEE Signal Processing Magazine*, *23*(1), 18–29. doi:10.1109/MSP.2006.1593334

Casagli, N., Intrieri, E., Tofani, V., Gigli, G., & Raspini, F. (2023). Landslide detection, monitoring and prediction with remote-sensing techniques. *Nature Reviews. Earth & Environment*, *4*(1), 51–64. doi:10.103843017-022-00373-x

Chatterjee, S. (2021). *Analysis and design of closed-loop fractional-order systems and its applications to neurotechnology*. Rensselaer Polytechnic Institute.

Chaturvedi, V., & de Vries, W. T. (2021). Machine learning algorithms for urban land use planning: A review. *Urban Science (Basel, Switzerland)*, *5*(3), 68. doi:10.3390/urbansci5030068

Chen, Z., Vatsavai, R. R., Ramachandra, B., Zhang, Q., Singh, N., & Sukumar, S. (2016, December). Scalable nearest neighbor based hierarchical change detection framework for crop monitoring. In *2016 IEEE International Conference on Big Data (Big Data)* (pp. 1309-1314). IEEE. 10.1109/BigData.2016.7840735

Chen, H., Qin, Z., Zhai, D. L., Ou, G., Li, X., Zhao, G., Fan, J., Zhao, C., & Xu, H. (2022). Mapping Forest Aboveground Biomass with MODIS and Fengyun-3C VIRR Imageries in Yunnan Province, Southwest China Using Linear Regression, K-Nearest Neighbor and Random Forest. *Remote Sensing (Basel)*, *14*(21), 5456. doi:10.3390/rs14215456

Chen, J., Lu, W., Fu, Y., & Dong, Z. (2023). Automated facility inspection using robotics and BIM: A knowledge-driven approach. *Advanced Engineering Informatics*, *55*, 101838. doi:10.1016/j.aei.2022.101838

Chen, M., Claramunt, C., Çöltekin, A., Liu, X., Peng, P., Robinson, A. C., Wang, D., Strobl, J., Wilson, J. P., Batty, M., Kwan, M.-P., Lotfian, M., Golay, F., Joost, S., Ingensand, J., Senousi, A. M., Cheng, T., Bandrova, T., Konecny, M., & Lü, G. (2023). Artificial intelligence and visual analytics in geographical space and cyberspace: Research opportunities and challenges. *Earth-Science Reviews*, *241*, 104438. doi:10.1016/j.earscirev.2023.104438

Chen, S. Z., Feng, D. C., Wang, W. J., & Taciroglu, E. (2022). Probabilistic machine-learning methods for performance prediction of structure and infrastructures through natural gradient boosting. *Journal of Structural Engineering*, *148*(8), 04022096. doi:10.1061/(ASCE)ST.1943-541X.0003401

Cheon, S. P., Kim, S., Lee, S. Y., & Lee, C. B. (2009). Bayesian networks based rare event prediction with sensor data. *Knowledge-Based Systems*, *22*(5), 336–343. doi:10.1016/j.knosys.2009.02.004

Chiang, C. T., Nien, C. Y., Yang, C. Y., & Chen, S. M. (2022). *Sensor Systems Integration A CMOS Algae Growth Period Monitor for Algaculture Applications*.

Chiang, Y. J., Chin, T. L., & Chen, D. Y. (2022). Neural network-based strong motion prediction for on-site earthquake early warning. *Sensors (Basel)*, *22*(3), 704. doi:10.339022030704 PMID:35161451

Chilunjika, A., Intauno, K., & Chilunjika, S. R. (2022). Artificial intelligence and public sector human resource management in South Africa: Opportunities, challenges and prospects. *SA Journal of Human Resource Management, 20*, 12. doi:10.4102ajhrm.v20i0.1972

Chittora, P., Chakrabarti, T., Debnath, P., Gupta, A., Chakrabarti, P., Praveen, S. P., & Elngar, A. A. (2022). *Experimental analysis of earthquake prediction using machine learning classifiers, curve fitting, and neural modeling.*

Cho, S., Ahn, J. K., & Hwang, E. H. (2022). Optimization of network-based earthquake early warning systems on the Korean Peninsula. *IEEE Access : Practical Innovations, Open Solutions, 10*, 83931–83939. doi:10.1109/ACCESS.2022.3197661

Chowdhuri, I., Pal, S. C., Saha, A., Chakrabortty, R., & Roy, P. (2022). Mapping of earthquake hotspot and coldspot zones for identifying potential landslide hotspot areas in the Himalayan region. *Bulletin of Engineering Geology and the Environment, 81*(7), 257. doi:10.100710064-022-02761-5

Chudasama, B., Torppa, J., Nykänen, V., Kinnunen, J., Lerssi, J., & Salmirinne, H. (2022). Target-scale prospectivity modeling for gold mineralization within the rajapalot Au-Co project area in northern Fennoscandian Shield, Finland. Part 1: Application of knowledge-driven-and machine learning-based-hybrid-expert systems for exploration targeting and addressing model-based uncertainties. *Ore Geology Reviews, 147*, 104937. doi:10.1016/j.oregeorev.2022.104937

Chung, L. C. H., Xie, J., & Ren, C. (2021). Improved machine-learning mapping of local climate zones in metropolitan areas using composite Earth observation data in Google Earth Engine. *Building and Environment, 199*, 107879. doi:10.1016/j.buildenv.2021.107879

Chun, J., Yang, W., Liu, X., Wu, G., He, L., & Xing, L. (2023). Deep Reinforcement Learning for the Agile Earth Observation Satellite Scheduling Problem. *Mathematics, 11*(19), 4059. doi:10.3390/math11194059

Coca, M., & Datcu, M. (2023). FPGA Accelerator for Meta-Recognition Anomaly Detection: Case of Burned Area Detection. *IEEE Journal of Selected Topics in Applied Earth Observations and Remote Sensing, 16*, 5247–5259. doi:10.1109/JSTARS.2023.3273309

Cochran, E. S., Saunders, J. K., Minson, S. E., Bunn, J., Baltay, A., Kilb, D., O'Rourke, C., Hoshiba, M., & Kodera, Y. (2022). Alert optimization of the PLUM earthquake early warning algorithm for the western United States. *Bulletin of the Seismological Society of America, 112*(2), 803–819. doi:10.1785/0120210259

Crowley, M. A., Stockdale, C. A., Johnston, J. M., Wulder, M. A., Liu, T., McCarty, J. L., Rieb, J. T., Cardille, J. A., & White, J. C. (2023). Towards a whole-system framework for wildfire monitoring using Earth observations. *Global Change Biology, 29*(6), 1423–1436. doi:10.1111/gcb.16567 PMID:36537002

Das, S. R., Hota, A. P., Pandey, H. M., & Sahoo, B. M. (2022). Industrial power quality enhancement using fuzzy logic based photovoltaic integrated with three phase shunt hybrid active filter and adaptive controller. *Applied Soft Computing, 121*, 108762. doi:10.1016/j.asoc.2022.108762

Datta, A., Wu, D. J., Zhu, W., Cai, M., & Ellsworth, W. L. (2022). DeepShake: Shaking intensity prediction using deep spatio-temporal RNNs for earthquake early warning. *Seismological Research Letters, 93*(3), 1636–1649. doi:10.1785/0220210141

Daviran, M., Parsa, M., Maghsoudi, A., & Ghezelbash, R. (2022). Quantifying uncertainties linked to the diversity of mathematical frameworks in knowledge-driven mineral prospectivity mapping. *Natural Resources Research, 31*(5), 2271–2287. doi:10.100711053-022-10089-w

Debnath, P., Chittora, P., Chakrabarti, T., Chakrabarti, P., Leonowicz, Z., Jasinski, M., Gono, R., & Jasińska, E. (2021). Analysis of earthquake forecasting in India using supervised machine learning classifiers. *Sustainability (Basel), 13*(2), 971. doi:10.3390u13020971

Delfan, E., Naghavi, H., Maleknia, R., & Nouredini, A. (2022). Comparing the Capability of Sentinel 2 and Landsat 8 Satellite imagery in land use and land cover mapping using pixel-based and object-based classification methods. *Desert Ecosystem Engineering*, *8*(25), 1–12.

Dematties, D., Raut, B. A., Park, S., Jackson, R. C., Shahkarami, S., Kim, Y., Sankaran, R., Beckman, P., Collis, S. M., & Ferrier, N. (2023). Let's Unleash the Network Judgment: A Self-Supervised Approach for Cloud Image Analysis. *Artificial Intelligence for the Earth Systems*, *2*(2), 220063. doi:10.1175/AIES-D-22-0063.1

Derras, B., & Makhoul, N. (2022). An overview of the infrastructure seismic resilience assessment using artificial intelligence and machine-learning algorithms. In *Proceedings of the International Conference on Natural Hazards and Infrastructure*. IEEE.

Dewitte, S., Cornelis, J. P., Müller, R., & Munteanu, A. (2021). Artificial intelligence revolutionises weather forecast, climate monitoring and decadal prediction. *Remote Sensing (Basel)*, *13*(16), 3209. doi:10.3390/rs13163209

Dhala, A. Fusaro, M., Uddin, F., Tuazon, D., & Klahn, S. (2023). *Integrating a Virtual ICU with Cardiac and Cardiovascular ICUs: Managing the Needs of a Complex and High-Acuity Specialty ICU Cohort*. Pubmed.

Dhala, A., Fusaro, M. V., Uddin, F., Tuazon, D., Klahn, S., Schwartz, R., Sasangohar, F., Alegria, J., & Masud, F. (2023). Integrating a Virtual ICU with Cardiac and Cardiovascular ICUs: Managing the Needs of a Complex and High-Acuity Specialty ICU Cohort. *Methodist DeBakey Cardiovascular Journal*, *19*(4), 4–16. doi:10.14797/mdcvj.1247 PMID:37547898

Di Napoli, M., Marsiglia, P., Di Martire, D., Ramondini, M., Ullo, S. L., & Calcaterra, D. (2020). Landslide susceptibility assessment of wildfire burnt areas through earth-observation techniques and a machine learning-based approach. *Remote Sensing (Basel)*, *12*(15), 2505. doi:10.3390/rs12152505

Ding, Y., Guo, H., Liu, G., Han, C., Shang, H., Ruan, Z., & Lv, M. (2019). Constructing a high-accuracy geometric model for Moon-based Earth observation. *Remote Sensing (Basel)*, *11*(22), 2611. doi:10.3390/rs11222611

Ditria, E. M., Buelow, C. A., Gonzalez-Rivero, M., & Connolly, R. M. (2022). Artificial intelligence and automated monitoring for assisting conservation of marine ecosystems: A perspective. *Frontiers in Marine Science*, *9*, 918104. doi:10.3389/fmars.2022.918104

Dogan, G., Arslan, M. H., & Ilki, A. (2023). Detection of damages caused by earthquake and reinforcement corrosion in RC buildings with deep transfer learning. *Engineering Structures*, *279*, 115629. doi:10.1016/j.engstruct.2023.115629

Dube, T., Seaton, D., Shoko, C., & Mbow, C. (2023). Advancements in Earth Observation for Water Resources Monitoring and Management in Africa: A Comprehensive Review. *Journal of Hydrology (Amsterdam)*, *623*, 129738. doi:10.1016/j.jhydrol.2023.129738

Dunham, C. (2022). *Exploration of Power Laws in and Development of Analytical Tools for the Study of Stem Cell-derived Cardiomyocytes*. University of California.

Ellis, T., Hitaj, B., Lindqvist, U., Shands, D., Tinnel, L., & DeBruhl, B. (2023). Critical Infrastructure Security Goes to Space: Leveraging Lessons Learned on the Ground. *arXiv preprint arXiv:2309.15232*. doi:10.2514/6.2023-4798

Eskelson, B. N., Temesgen, H., Lemay, V., Barrett, T. M., Crookston, N. L., & Hudak, A. T. (2009). The roles of nearest neighbor methods in imputing missing data in forest inventory and monitoring databases. *Scandinavian Journal of Forest Research*, *24*(3), 235–246. doi:10.1080/02827580902870490

Esposito, M., Palma, L., Belli, A., Sabbatini, L., & Pierleoni, P. (2022). Recent advances in internet of things solutions for early warning systems: A review. *Sensors (Basel)*, *22*(6), 2124. doi:10.339022062124 PMID:35336296

Fadaei-Kermani, E., Barani, G. A., & Ghaeini-Hessaroeyeh, M. (2017). Drought monitoring and prediction using K-nearest neighbor algorithm. *Journal of Artificial Intelligence and Data Mining*, *5*(2), 319–325.

Falkenström, E., & Svallfors, S. (2022). The knowledge–management complex: From quality registries to national knowledge-driven management in Swedish health care governance. *Politics & Policy*, *50*(5), 1053–1066. doi:10.1111/polp.12497

Fang, S., Xu, L., Pei, H., Liu, Y., Liu, Z., Zhu, Y., Yan, J., & Zhang, H. (2013). An integrated approach to snowmelt flood forecasting in water resource management. *IEEE Transactions on Industrial Informatics*, *10*(1), 548–558. doi:10.1109/TII.2013.2257807

Fan, R., Ding, Y., Zou, Q., & Yuan, L. (2023). Multi-view local hyperplane nearest neighbor model based on independence criterion for identifying vesicular transport proteins. *International Journal of Biological Macromolecules*, *247*, 125774. doi:10.1016/j.ijbiomac.2023.125774 PMID:37437677

Fathabadi, A., Seyedian, S. M., & Malekian, A. (2022). Comparison of Bayesian, k-Nearest Neighbor and Gaussian process regression methods for quantifying uncertainty of suspended sediment concentration prediction. *The Science of the Total Environment*, *818*, 151760. doi:10.1016/j.scitotenv.2021.151760 PMID:34801498

Feizizadeh, B., Omarzadeh, D., Kazemi Garajeh, M., Lakes, T., & Blaschke, T. (2023). Machine learning data-driven approaches for land use/cover mapping and trend analysis using Google Earth Engine. *Journal of Environmental Planning and Management*, *66*(3), 665–697. doi:10.1080/09640568.2021.2001317

Ferreira, B., Iten, M., & Silva, R. G. (2020). Monitoring sustainable development by means of earth observation data and machine learning: A review. *Environmental Sciences Europe*, *32*(1), 1–17. doi:10.118612302-020-00397-4

Ferreira, B., Silva, R. G., & Iten, M. (2022). Earth observation satellite imagery information based decision support using machine learning. *Remote Sensing (Basel)*, *14*(15), 3776. doi:10.3390/rs14153776

Figueiredo, N., Neto, A., Cunha, A., Sousa, J. J., & Sousa, A. (2022, July). Deep Learning Approach for Terrace Vineyards Detection from Google Earth Satellite Imagery. In *IGARSS 2022-2022 IEEE International Geoscience and Remote Sensing Symposium* (pp. 5824-5827). IEEE. 10.1109/IGARSS46834.2022.9884644

Galkina, A., & Grafeeva, N. (2019, April). Machine learning methods for earthquake prediction: A survey. In *Proceedings of the Fourth Conference on Software Engineering and Information Management (SEIM-2019)*, Saint Petersburg, Russia.

Gao, Y. (2023). Unleashing the mechanism among environmental regulation, artificial intelligence, and global value chain leaps: A roadmap toward digital revolution and environmental sustainability. *Environmental Science and Pollution Research International*, *30*(10), 28107–28117. doi:10.100711356-022-23898-6 PMID:36394806

Gao, Y., Zhang, J., Li, H., & Li, G. (2022). Incorporating structural constraint into the machine learning high-resolution seismic reconstruction. *IEEE Transactions on Geoscience and Remote Sensing*, *60*, 1–12. doi:10.1109/TGRS.2022.3157064

Garcia Valencia, O. A., Suppadungsuk, S., Thongprayoon, C., Miao, J., Tangpanithandee, S., Craici, I. M., & Cheungpasitporn, W. (2023). Ethical Implications of Chatbot Utilization in Nephrology. *Journal of Personalized Medicine*, *13*(9), 1363. doi:10.3390/jpm13091363 PMID:37763131

George, B., & Wooden, O. (2023). Managing the strategic transformation of higher education through artificial intelligence. *Administrative Sciences*, *13*(9), 196. doi:10.3390/admsci13090196

Gerdes, A. (2022). A participatory data-centric approach to AI Ethics by Design. *Applied Artificial Intelligence*, *36*(1), 2009222. doi:10.1080/08839514.2021.2009222

Ge, X., Ding, J., Teng, D., Wang, J., Huo, T., Jin, X., Wang, J., He, B., & Han, L. (2022). Updated soil salinity with fine spatial resolution and high accuracy: The synergy of Sentinel-2 MSI, environmental covariates and hybrid machine learning approaches. *Catena*, *212*, 106054. doi:10.1016/j.catena.2022.106054

Ghanbari, M., Clausi, D. A., Xu, L., & Jiang, M. (2019) Contextual classification of Sea-Ice types using compact polarimetric SAR data. *IEEE Journals & Magazine*. IEEE Xplore. https://ieeexplore.ieee.org/document/8731878

Ghimire, S., Bhandari, B., Casillas-Perez, D., Deo, R. C., & Salcedo-Sanz, S. (2022). Hybrid deep CNN-SVR algorithm for solar radiation prediction problems in Queensland, Australia. *Engineering Applications of Artificial Intelligence*, *112*, 104860. doi:10.1016/j.engappai.2022.104860

Giacco, G., Mariniello, G., Marrone, S., Asprone, D., & Sansone, C. (2022, May). Toward a system for post-earthquake safety evaluation of masonry buildings. In *International Conference on Image Analysis and Processing* (pp. 312-323). Cham: Springer International Publishing. 10.1007/978-3-031-06430-2_26

Gobinathan, B., Mukunthan, M. A., Surendran, S., Somasundaram, K., Moeed, S. A., Niranjan, P., Gouthami, V., Ashmitha, G., Mohammad, G. B., Shanmuganathan, V. K., Natarajan, Y., Srihari, K., & Sundramurthy, V. P. (2021). A novel method to solve real time security issues in software industry using advanced cryptographic techniques. *Scientific Programming*, *2021*, 1–9. doi:10.1155/2021/3611182

Gonzales-Inca, C., Calle, M., Croghan, D., Torabi Haghighi, A., Marttila, H., Silander, J., & Alho, P. (2022). Geospatial artificial intelligence (GeoAI) in the integrated hydrological and fluvial systems modeling: Review of current applications and trends. *Water (Basel)*, *14*(14), 2211. doi:10.3390/w14142211

Gunjan, V. K., Vijayalata, Y., Valli, S., Kumar, S., Mohamed, M. O., & Saravanan, V. (2022). Machine learning and cloud-based knowledge graphs to recognize suicidal mental tendencies. *Computational Intelligence and Neuroscience*, *2022*, 2022. doi:10.1155/2022/3604113 PMID:35341205

Guo, H., Nativi, S., Liang, D., Craglia, M., Wang, L., Schade, S., Corban, C., He, G., Pesaresi, M., Li, J., Shirazi, Z., Liu, J., & Annoni, A. (2020). Big Earth Data science: An information framework for a sustainable planet. *International Journal of Digital Earth*, *13*(7), 743–767. doi:10.1080/17538947.2020.1743785

Gupta, S., Saravanan, V., Choudhury, A., Alqahtani, A., Abonazel, M. R., & Babu, K. S. (2022). Supervised computer-aided diagnosis (CAD) methods for classifying Alzheimer's disease-based neurodegenerative disorders. *Computational and Mathematical Methods in Medicine*, *2022*, 2022. doi:10.1155/2022/9092289 PMID:35651921

Hai, T., Zhou, J., & Muranaka, K. (2022). An efficient fuzzy-logic based MPPT controller for grid-connected PV systems by farmland fertility optimization algorithm. *Optik (Stuttgart)*, *267*, 169636. doi:10.1016/j.ijleo.2022.169636

Hajibabaee, P., Pourkamali-Anaraki, F., & Hariri-Ardebili, M. A. (2023). Dimensionality reduction techniques in structural and earthquake engineering. *Engineering Structures*, *278*, 115485. doi:10.1016/j.engstruct.2022.115485

Hallema, D. W. (2022). Breaking Boundaries: The Power of Interdisciplinary Research in Science, Technology, and Business. *International Multidisciplinary Journal of Science, Technology, and Business*, *1*(4), 30–35.

Hannah, S., Deepa, A. J., Chooralil, V. S., BrillySangeetha, S., Yuvaraj, N., Arshath Raja, R., Suresh, C., Vignesh, R., YasirAbdullahR, Srihari, K., & Alene, A. (2022). BrillySangeetha, S., Yuvaraj, N., Arshath Raja, R., ... & Alene, A. (2022). Blockchain-based deep learning to process IoT data acquisition in cognitive data. *BioMed Research International*, *2022*, 1–7. doi:10.1155/2022/5038851

Han, Y., Wang, L., Fu, W., Zhou, H., Li, T., & Chen, R. (2021). Machine learning-based short-term GPS TEC forecasting during high solar activity and magnetic storm periods. *IEEE Journal of Selected Topics in Applied Earth Observations and Remote Sensing*, *15*, 115–126. doi:10.1109/JSTARS.2021.3132049

Harris, A., & Naik, K. (2023, March). Autonomous Command and Control for Earth-Observing Satellites using Deep Reinforcement Learning. In *2023 IEEE Aerospace Conference* (pp. 1-12). IEEE. 10.1109/AERO55745.2023.10115916

Harry, A. (2023). The Future of Medicine: Harnessing the Power of AI for Revolutionizing Healthcare. *International Journal of Multidisciplinary Sciences and Arts*, *2*(1), 36–47. doi:10.47709/ijmdsa.v2i1.2395

Hassanien, A. E., Darwish, A., & Abdelghafar, S. (2020). Machine learning in telemetry data mining of space mission: Basics, challenging and future directions. *Artificial Intelligence Review*, *53*(5), 3201–3230. doi:10.100710462-019-09760-1

He, B., Zhao, X., Chen, Y., Liu, C., & Pang, X. (2022). Application of Feature Tracking Using K-Nearest-Neighbor Vector Field Consensus in Sea Ice Tracking. *IEEE Journal of Selected Topics in Applied Earth Observations and Remote Sensing*, *15*, 4326–4336. doi:10.1109/JSTARS.2022.3178117

Hej. (n.d.). *Unique new global satellite and radar composite.* Meteoblue. https://www.meteoblue.com/en/blog/article/show/39821_Unique+new+global+satellite+and+radar+composite

He, S., Li, J., Wang, J., & Liu, F. (2022). Evaluation and analysis of upscaling of different land use/land cover products (FORM-GLC30, GLC_FCS30, CCI_LC, MCD12Q1 and CNLUCC): A case study in China. *Geocarto International*, *37*(27), 17340–17360. doi:10.1080/10106049.2022.2127926

He, Y., Xing, L., Chen, Y., Pedrycz, W., Wang, L., & Wu, G. (2020). A generic Markov decision process model and reinforcement learning method for scheduling agile earth observation satellites. *IEEE Transactions on Systems, Man, and Cybernetics. Systems*, *52*(3), 1463–1474. doi:10.1109/TSMC.2020.3020732

Himeur, Y., Rimal, B., Tiwary, A., & Amira, A. (2022). Using artificial intelligence and data fusion for environmental monitoring: A review and future perspectives. *Information Fusion*, *86*, 44–75. doi:10.1016/j.inffus.2022.06.003

Hirtzlin, T. (2020). *Digital Implementation of Neuromorphic systems using Emerging Memory devices* [Doctoral dissertation, Université Paris-Saclay].

Holt, B., Johnson, M. P., Perkovic-Martin, D., & Panzer, B. (2015). Snow depth on Arctic sea ice derived from radar: In situ comparisons and time series analysis. *Journal of Geophysical Research. Oceans*, *120*(6), 4260–4287. doi:10.1002/2015JC010815

Hoque, A., Raj, J., Saha, A., & Bhattacharya, P. (2020). Earthquake magnitude prediction using machine learning technique. In *Trends in Computational Intelligence, Security and Internet of Things: Third International Conference*. Springer.

Horn, J., & Isselstein, J. (2022). How do we feed grazing livestock in the future? A case for knowledge-driven grazing systems. *Grass and Forage Science*, *77*(3), 153–166. doi:10.1111/gfs.12577

Hossain, M. S., Kabir, M. M., Mushtary, T., Mostafa, S. G., Hasan, M. M., Hossain, M. M., & Islam, M. S. (2022, December). Fuzzy Logic-Based Maximum Power Point Tracking of Solar PV System. In *2022 IEEE International Women in Engineering (WIE) Conference on Electrical and Computer Engineering (WIECON-ECE)* (pp. 284-289). IEEE.

Hou, L., Dai, Q., Song, C., Liu, B., Guo, F., Dai, T., Li, L., Liu, B., Bi, X., Zhang, Y., & Feng, Y. (2022). Revealing drivers of haze pollution by explainable machine learning. *Environmental Science & Technology Letters*, *9*(2), 112–119. doi:10.1021/acs.estlett.1c00865

Huang, L., Sun, R., Cheng, N., Hui, Y., & Liang, D. (2023, March). Delay-Oriented Knowledge-Driven Resource Allocation in SAGIN-Based Vehicular Networks. In 2023 IEEE Wireless Communications and Networking Conference (WCNC) (pp. 1-6). IEEE. doi:10.1109/WCNC55385.2023.10118888

Huang, Y., Mu, Z., Wu, S., Cui, B., & Duan, Y. (2021). Revising the observation satellite scheduling problem based on deep reinforcement learning. *Remote Sensing (Basel)*, *13*(12), 2377. doi:10.3390/rs13122377

Huang, Z., George, M. M., Tan, Y. R., Natarajan, K., Devasagayam, E., Tay, E., Manesh, A., Varghese, G. M., Abraham, O. C., Zachariah, A., Yap, P., Lall, D., & Chow, A. (2023). Are physicians ready for precision antibiotic prescribing? A qualitative analysis of the acceptance of artificial intelligence-enabled clinical decision support systems in India and Singapore. *Journal of Global Antimicrobial Resistance*, *35*, 76–85. doi:10.1016/j.jgar.2023.08.016 PMID:37640155

Hu, M., Wu, B., Zhou, W., Wu, H., Li, G., Lu, J., Yu, G., & Qin, Y. (2022). Self-driving shield: Intelligent systems, methodologies, and practice. *Automation in Construction*, *139*, 104326. doi:10.1016/j.autcon.2022.104326

Hu, S., Qiu, C., & Zhu, S. (2022). Machine learning-driven performance-based seismic design of hybrid self-centering braced frames with SMA braces and viscous dampers. *Smart Materials and Structures*, *31*(10), 105024. doi:10.1088/1361-665X/ac8efc

Hu, S., Zhu, S., Alam, M. S., & Wang, W. (2022). Machine learning-aided peak and residual displacement-based design method for enhancing seismic performance of steel moment-resisting frames by installing self-centering braces. *Engineering Structures*, *271*, 114935. doi:10.1016/j.engstruct.2022.114935

Hu, S., Zhu, S., & Wang, W. (2022). Machine learning-driven probabilistic residual displacement-based design method for improving post-earthquake repairability of steel moment-resisting frames using self-centering braces. *Journal of Building Engineering*, *61*, 105225. doi:10.1016/j.jobe.2022.105225

Hussain, I., Khalil, I. U., Islam, A., Ahsan, M. U., Iqbal, T., Chowdhury, M. S., Techato, K., & Ullah, N. (2022). Unified Fuzzy Logic Based Approach for Detection and Classification of PV Faults Using IV Trend Line. *Energies*, *15*(14), 5106. doi:10.3390/en15145106

Ibrahim, O., Bakare, M. S., Amosa, T. I., Otuoze, A. O., Owonikoko, W. O., Ali, E. M., & Ogunbiyi, O. (2023). Development of fuzzy logic-based demand-side energy management system for hybrid energy sources. *Energy Conversion and Management: X*, *18*, 100354.

Iqbal, J., Jaimes, D. C. C., Makineni, P., Subramani, S., Hemaida, S., Thugu, T. R., & Hemida, S. (2023). Reimagining Healthcare: Unleashing the Power of Artificial Intelligence in Medicine. *Cureus*, *15*(9). Advance online publication. doi:10.7759/cureus.44658 PMID:37799217

Irfan, D., Tang, X., Narayan, V., Mall, P. K., Srivastava, S., & Saravanan, V. (2022). Prediction of Quality Food Sale in Mart Using the AI-Based TOR Method. *Journal of Food Quality*, *2022*, 2022. doi:10.1155/2022/6877520

Jabla, R., Khemaja, M., Buendia, F., & Faiz, S. (2022). A knowledge-driven activity recognition framework for learning unknown activities. *Procedia Computer Science*, *207*, 1871–1880. doi:10.1016/j.procs.2022.09.245

Jana, D., & Nagarajaiah, S. (2022). Physics-Guided Real-Time Full-Field Vibration Response Estimation from Sparse Measurements Using Compressive Sensing. *Sensors (Basel)*, *23*(1), 384. doi:10.339023010384 PMID:36616982

Jiang, Y., Yang, G., Li, H., & Zhang, T. (2023). Knowledge driven approach for smart bridge maintenance using big data mining. *Automation in Construction*, *146*, 104673. doi:10.1016/j.autcon.2022.104673

Jia, Y., Jin, S., Chen, H., Yan, Q., Savi, P., Jin, Y., & Yuan, Y. (2021). Temporal-spatial soil moisture estimation from CYGNSS using machine learning regression with a preclassification approach. *IEEE Journal of Selected Topics in Applied Earth Observations and Remote Sensing*, *14*, 4879–4893. doi:10.1109/JSTARS.2021.3076470

Johnson, P. A., Rouet-Leduc, B., Pyrak-Nolte, L. J., Beroza, G. C., Marone, C. J., Hulbert, C., Howard, A., Singer, P., Gordeev, D., Karaflos, D., Levinson, C. J., Pfeiffer, P., Puk, K. M., & Reade, W. (2021). Laboratory earthquake forecasting: A machine learning competition. *Proceedings of the National Academy of Sciences of the United States of America*, *118*(5), e2011362118. doi:10.1073/pnas.2011362118 PMID:33495346

Joshi, A., Vishnu, C., & Mohan, C. K. (2022). Early detection of earthquake magnitude based on stacked ensemble model. *Journal of Asian Earth Sciences: X, 8*, 100122.

Joshi, V., & Biswakarma, P. (2023). Bivariate statistical method and knowledge driven heuristic approach for the comparison of landslide susceptibility mapping in the Alaknanda Catchment area. *Indian Journal of Geosciences, 77*(1), 71–106.

Jo, Y., Choi, Y., Seol, S. J., & Byun, J. (2022). Machine learning-based vertical resolution enhancement considering the seismic attenuation. *Journal of Petroleum Science Engineering, 208*, 109657. doi:10.1016/j.petrol.2021.109657

Kannan, S., Dhiman, G., Natarajan, Y., Sharma, A., Mohanty, S. N., Soni, M., Easwaran, U., Ghorbani, H., Asheralieva, A., & Gheisari, M. (2021). Ubiquitous vehicular ad-hoc network computing using deep neural network with iot-based bat agents for traffic management. *Electronics (Basel), 10*(7), 785. doi:10.3390/electronics10070785

Karki, S., Bermejo, R., Wilkes, R., Monagail, M. M., Daly, E., Healy, M., Hanafin, J., McKinstry, A., Mellander, P.-E., Fenton, O., & Morrison, L. (2021). Mapping spatial distribution and biomass of intertidal Ulva blooms using machine learning and earth observation. *Frontiers in Marine Science, 8*, 633128. doi:10.3389/fmars.2021.633128

Kasabov, N., Sengupta, N., & Scott, N. (2016, September). From von Neumann, John Atanasoff and ABC to neuromorphic computation and the NeuCube spatio-temporal data machine. In *2016 IEEE 8th International Conference on Intelligent Systems (IS)* (pp. 15-21). IEEE.

Kasneci, E., Seßler, K., Küchemann, S., Bannert, M., Dementieva, D., Fischer, F., Gasser, U., Groh, G., Günnemann, S., Hüllermeier, E., Krusche, S., Kutyniok, G., Michaeli, T., Nerdel, C., Pfeffer, J., Poquet, O., Sailer, M., Schmidt, A., Seidel, T., ... Kasneci, G. (2023). ChatGPT for good? On opportunities and challenges of large language models for education. *Learning and Individual Differences, 103*, 102274. doi:10.1016/j.lindif.2023.102274

Kaur, J., Saxena, J., Shah, J., & Yadav, S. P. (2022, May). Facial Emotion Recognition. In *2022 International Conference on Computational Intelligence and Sustainable Engineering Solutions (CISES)* (pp. 528-533). IEEE.

Kaur, S., & Vig, S. (2023, February). Modeling of MPPT-Based Solar Eco-System Using Fuzzy Logic Controller. *IOP Conference Series. Earth and Environmental Science, 1110*(1), 012079. doi:10.1088/1755-1315/1110/1/012079

Kavianpour, P., Kavianpour, M., Jahani, E., & Ramezani, A. (2023). A cnn-bilstm model with attention mechanism for earthquake prediction. *The Journal of Supercomputing, 79*(17), 1–33. doi:10.100711227-023-05369-y

Kazemi, F., Asgarkhani, N., & Jankowski, R. (2023). Machine learning-based seismic fragility and seismic vulnerability assessment of reinforced concrete structures. *Soil Dynamics and Earthquake Engineering, 166*, 107761. doi:10.1016/j.soildyn.2023.107761

Kazemi, F., & Jankowski, R. (2023). Machine learning-based prediction of seismic limit-state capacity of steel moment-resisting frames considering soil-structure interaction. *Computers & Structures, 274*, 106886. doi:10.1016/j.compstruc.2022.106886

Khan, A., Vibhute, A. D., Mali, S., & Patil, C. H. (2022). A systematic review on hyperspectral imaging technology with a machine and deep learning methodology for agricultural applications. *Ecological Informatics, 69*, 101678. doi:10.1016/j.ecoinf.2022.101678

Khan, M. M., Ghaffar, B., Shahzad, R., Khan, M. R., Shah, M., Amin, A. H., Eldin, S. M., Naqvi, N. A., & Ali, R. (2022). Atmospheric anomalies associated with the 2021 M w 7.2 Haiti earthquake using machine learning from multiple satellites. *Sustainability (Basel), 14*(22), 14782. doi:10.3390u142214782

Khan, S., Saravanan, V. N. G. C., Lakshmi, T. J., Deb, N., & Othman, N. A. (2022). Privacy protection of healthcare data over social networks using machine learning algorithms. *Computational Intelligence and Neuroscience, 2022*, 2022. doi:10.1155/2022/9985933 PMID:35371203

Khare, A. (2023). Harnessing the power of artificial intelligence in health care: Balancing promise with caution. *Journal of Medical Evidence, 4*(2), 105–108. doi:10.4103/JME.JME_70_23

Khotimah, B. K. (2022). Performance of the K-nearest neighbors method on identification of maize plant nutrients. *Jurnal Infotel, 14*(1), 8–14. doi:10.20895/infotel.v14i1.735

Kinder, T., Stenvall, J., Koskimies, E., Webb, H., & Janenova, S. (2023). Local public services and the ethical deployment of artificial intelligence. *Government Information Quarterly, 40*(4), 101865. doi:10.1016/j.giq.2023.101865

Kochupillai, M., Kahl, M., Schmitt, M., Taubenböck, H., & Zhu, X. X. (2022). Earth Observation and Artificial Intelligence: Understanding emerging ethical issues and opportunities. *IEEE Geoscience and Remote Sensing Magazine, 10*(4), 90–124. doi:10.1109/MGRS.2022.3208357

Koh, I., & Chen, A. (2023, October). Your Memory Palace in the Metaverse with AI. In *Proceedings of the AAAI Symposium Series* (Vol. 1, No. 1, pp. 19-22). 10.1609/aaaiss.v1i1.27469

Kong, L., Ren, D., Zhou, S., He, Z., Ai, C., & Yan, C. (2023). Evaluating the evolution of fiber-reinforced emulsified asphalt cold-recycled mixture damage using digital image correlation. *The International Journal of Pavement Engineering, 24*(1), 2176495. doi:10.1080/10298436.2023.2176495

Kourehpaz, P., & Molina Hutt, C. (2022). *Machine Learning for Enhanced Seismic Damage Assessment of Buildings.*

Kourehpaz, P., & Molina Hutt, C. (2022). Machine learning for enhanced regional seismic risk assessments. *Journal of Structural Engineering, 148*(9), 04022126. doi:10.1061/(ASCE)ST.1943-541X.0003421

Kousik, N., Natarajan, Y., Raja, R. A., Kallam, S., Patan, R., & Gandomi, A. H. (2021). Improved salient object detection using hybrid Convolution Recurrent Neural Network. *Expert Systems with Applications, 166*, 114064. doi:10.1016/j.eswa.2020.114064

Küçük, Ç., Taşkın, G., & Erten, E. (2016). Paddy-rice phenology classification based on machine-learning methods using multitemporal co-polar X-band SAR images. *IEEE Journal of Selected Topics in Applied Earth Observations and Remote Sensing, 9*(6), 2509–2519. doi:10.1109/JSTARS.2016.2547843

Kumar, S. (2022). 6G Mobile Communication Networks: Key Services and Enabling Technologies. *Journal of ICT Standardization*, 1-10.

Kumar, K., & Pande, B. P. (2023). Air pollution prediction with machine learning: A case study of Indian cities. *International Journal of Environmental Science and Technology, 20*(5), 5333–5348. doi:10.100713762-022-04241-5 PMID:35603096

Kumar, R., Mittal, H., Sandeep, & Sharma, B. (2022). Earthquake genesis and earthquake early warning systems: Challenges and a way forward. *Surveys in Geophysics, 43*(4), 1143–1168. doi:10.100710712-022-09710-7

Kuru, K. (2023). MetaOmniCity: Towards immersive urban metaverse cyberspaces using smart city digital twins. *IEEE Access : Practical Innovations, Open Solutions, 11*, 43844–43868. doi:10.1109/ACCESS.2023.3272890

Landers, R. N., & Behrend, T. S. (2023). Auditing the AI auditors: A framework for evaluating fairness and bias in high stakes AI predictive models. *The American Psychologist, 78*(1), 36–49. doi:10.1037/amp0000972 PMID:35157476

Laurenti, L., Tinti, E., Galasso, F., Franco, L., & Marone, C. (2022). Deep learning for laboratory earthquake prediction and autoregressive forecasting of fault zone stress. *Earth and Planetary Science Letters, 598*, 117825. doi:10.1016/j.epsl.2022.117825

Lawrie, S. (2022). *Information representation and processing in neuronal networks: from biological to artificial systems and from first to second-order statistics.*

LeCun, Y., Bottou, L., Orr, G., & Müller, K. (2012). Efficient BackProp. In Lecture Notes in Computer Science (pp. 9–48). doi:10.1007/978-3-642-35289-8_3

Lee, S., Im, J., Kim, J., Kim, M. J., Shin, M., Kim, H., & Quackenbush, L. J. (2016). Arctic Sea Ice Thickness Estimation from CryoSat-2 Satellite Data Using Machine Learning-Based Lead Detection. *Remote Sensing (Basel), 8*(9), 698. doi:10.3390/rs8090698

Leigh, S., Wang, Z., & Clausi, D. A. (2014). Automated Ice–Water classification using dual polarization SAR satellite imagery. *IEEE Transactions on Geoscience and Remote Sensing, 52*(9), 5529–5539. doi:10.1109/TGRS.2013.2290231

Li, X., Cheng, G., Fu, B., Xia, J., Zhang, L., Yang, D., & Ran, Y. (2022). Linking critical zone with watershed science: The example of the Heihe River basin. *Earth's Future, 10*(11), e2022EF002966.

Li, B. H., Hou, B. C., Yu, W. T., Lu, X. B., & Yang, C. W. (2017). Applications of artificial intelligence in intelligent manufacturing: A review. *Frontiers of Information Technology & Electronic Engineering, 18*(1), 86–96. doi:10.1631/FITEE.1601885

Li, J., Zhou, G., & Zhang, C. (2022). A twin data and knowledge-driven intelligent process planning framework of aviation parts. *International Journal of Production Research, 60*(17), 5217–5234. doi:10.1080/00207543.2021.1951869

Li, L., Chen, X., & Song, C. (2022). A robust clustering method with noise identification based on directed K-nearest neighbor graph. *Neurocomputing, 508*, 19–35. doi:10.1016/j.neucom.2022.08.029

Lin, K., Zhao, Y., Kuo, J. H., Deng, H., Cui, F., Zhang, Z., Zhang, M., Zhao, C., Gao, X., Zhou, T., & Wang, T. (2022). Toward smarter management and recovery of municipal solid waste: A critical review on deep learning approaches. *Journal of Cleaner Production, 346*, 130943. doi:10.1016/j.jclepro.2022.130943

Li, R., Lu, X., Li, S., Yang, H., Qiu, J., & Zhang, L. (2020, July). DLEP: A deep learning model for earthquake prediction. In *2020 International Joint Conference on Neural Networks (IJCNN)* (pp. 1-8). IEEE. 10.1109/IJCNN48605.2020.9207621

Liu, Q. Y., Li, A. Q., Tang, X. S., & Du, W. (2023). Predictive Models for Seismic Source Parameters Based on Machine Learning and General Orthogonal Regression Approaches. *Bulletin of the Seismological Society of America, 113*(6), 2363–2376. doi:10.1785/0120230069

Liu, X., Lu, D., Zhang, A., Liu, Q., & Jiang, G. (2022). Data-driven machine learning in environmental pollution: Gains and problems. *Environmental science & technology, 56*(4), 2124-2133.

Li, X., Liu, B., Zheng, G., Ren, Y., Zhang, S., Liu, Y., Gao, L., Liu, Y., Zhang, B., & Wang, F. (2020). Deep-learning-based information mining from ocean remote-sensing imagery. *National Science Review, 7*(10), 1584–1605. doi:10.1093/nsr/nwaa047 PMID:34691490

Lohse, J., Doulgeris, A. P., & Dierking, W. (2019). An Optimal Decision-Tree Design Strategy and Its Application to Sea Ice Classification from SAR Imagery. *Remote Sensing (Basel), 11*(13), 1574. doi:10.3390/rs11131574

Löw, F., Stieglitz, K., & Diemar, O. (2021). Terrestrial oil spill mapping using satellite earth observation and machine learning: A case study in South Sudan. *Journal of Environmental Management, 298*, 113424. doi:10.1016/j.jenvman.2021.113424 PMID:34358936

Lu, A., & Honarvar Shakibaei Asli, B. (2023). Seismic Image Identification and Detection Based on Tchebichef Moment Invariant. *Electronics (Basel)*, *12*(17), 3692. doi:10.3390/electronics12173692

Luo, H., & Paal, S. G. (2022). Artificial intelligence-enhanced seismic response prediction of reinforced concrete frames. *Advanced Engineering Informatics*, *52*, 101568. doi:10.1016/j.aei.2022.101568

Lymperopoulos, I. N. (2017). Dynamic response and transfer function of social systems: A neuro-inspired model of collective human activity patterns. *Neural Networks*, *94*, 125–140. doi:10.1016/j.neunet.2017.07.010 PMID:28772240

Ma, D., Fang, H., Wang, N., Lu, H., Matthews, J., & Zhang, C. (2023). Transformer-optimized generation, detection, and tracking network for images with drainage pipeline defects. *Computer-Aided Civil and Infrastructure Engineering*, *38*(15), 2109–2127. doi:10.1111/mice.12970

Mahalakshmi, D. V., Paul, A., Dutta, D., Ali, M. M., Jha, C. S., & Dadhwal, V. K. (2014). Net surface radiation retrieval using earth observation satellite data and machine learning algorithm. *ISPRS Annals of the Photogrammetry, Remote Sensing and Spatial Information Sciences*, *2*, 9–12. doi:10.5194/isprsannals-II-8-9-2014

Mai, H., Le, T. C., Chen, D., Winkler, D. A., & Caruso, R. A. (2022). Machine learning for electrocatalyst and photocatalyst design and discovery. *Chemical Reviews*, *122*(16), 13478–13515. doi:10.1021/acs.chemrev.2c00061 PMID:35862246

Malik, A., Saggi, M. K., Rehman, S., Sajjad, H., Inyurt, S., Bhatia, A. S., Farooque, A. A., Oudah, A. Y., & Yaseen, Z. M. (2022). Deep learning versus gradient boosting machine for pan evaporation prediction. *Engineering Applications of Computational Fluid Mechanics*, *16*(1), 570–587. doi:10.1080/19942060.2022.2027273

Mallouhy, R., Abou Jaoude, C., Guyeux, C., & Makhoul, A. (2019, December). Major earthquake event prediction using various machine learning algorithms. In *2019 International Conference on Information and Communication Technologies for Disaster Management (ICT-DM)* (pp. 1-7). IEEE. 10.1109/ICT-DM47966.2019.9032983

Mangalathu, S., Sun, H., Nweke, C. C., Yi, Z., & Burton, H. V. (2020). Classifying earthquake damage to buildings using machine learning. *Earthquake Spectra*, *36*(1), 183–208. doi:10.1177/8755293019878137

Manickam, P., Mariappan, S. A., Murugesan, S. M., Hansda, S., Kaushik, A., Shinde, R., & Thipperudraswamy, S. P. (2022). Artificial intelligence (AI) and internet of medical things (IoMT) assisted biomedical systems for intelligent healthcare. *Biosensors (Basel)*, *12*(8), 562. doi:10.3390/bios12080562 PMID:35892459

Marano, G. C., Rosso, M. M., Aloisio, A., & Cirrincione, G. (2023). Generative adversarial networks review in earthquake-related engineering fields. *Bulletin of Earthquake Engineering*, ●●●, 1–52. doi:10.100710518-023-01645-7

Ma, X., & Chen, T. (2022). Small seismic events in Oklahoma detected and located by machine learning–based models. *Bulletin of the Seismological Society of America*, *112*(6), 2859–2869. doi:10.1785/0120220029

Mazzarella, F., Arguedas, V. F., & Vespe, M. (2015, October). Knowledge-based vessel position prediction using historical AIS data. In *2015 Sensor Data Fusion: Trends, Solutions, Applications (SDF)* (pp. 1-6). IEEE.

McMeekin, T. A., & Ross, T. (2002). Predictive microbiology: Providing a knowledge-based framework for change management. *International Journal of Food Microbiology*, *78*(1-2), 133–153. doi:10.1016/S0168-1605(02)00231-3 PMID:12222630

Medina, M., Sanchez, R., Riquelme, S., Flores, M. C., Koch, P., Bravo, F., Barrientos, S., Henson, I., Chung, A., Melgar, D., Mpodozis, C., Hellweg, M., & Allen, R. (2022). An Earthquake Early Warning System for Northern Chile Based on ElarmS-3. *Seismological Research Letters*, *93*(6), 3337–3347. doi:10.1785/0220210331

Mehta, R. (2022). Multivariate Fuzzy Logic Based Smart Healthcare Monitoring for Risk Evaluation of Cardiac Patients. *Medical Informatics and Bioimaging Using Artificial Intelligence: Challenges, Issues, Innovations and Recent Developments*, 219-243.

Menezes, M. D. D., Silva, S. H. G., Mello, C. R. D., Owens, P. R., & Curi, N. (2014). Solum depth spatial prediction comparing conventional with knowledge-based digital soil mapping approaches. *Scientia Agrícola*, *71*(4), 316–323. doi:10.1590/0103-9016-2013-0416

Meng, M., Toan, T. D., Wong, Y. D., & Lam, S. H. (2022). Short-term travel-time prediction using support vector machine and nearest neighbor method. *Transportation Research Record: Journal of the Transportation Research Board*, *2676*(6), 353–365. doi:10.1177/03611981221074371

Mills, K. A., Unsworth, L., & Scholes, L. (2023). *Literacy for digital futures: Mind, body, text*. Taylor & Francis.

Mohammed, I. N., Bolten, J. D., Srinivasan, R., & Lakshmi, V. (2018). Improved hydrological decision support system for the Lower Mekong River Basin using satellite-based earth observations. *Remote Sensing*, *10*(6), 885. doi:10.3390/rs10060885 PMID:29938116

Mohammed, M. A., Khleel, N. A., Szabó, N. P., & Szűcs, P. (2023). Modeling of groundwater quality index by using artificial intelligence algorithms in northern Khartoum State, Sudan. *Modeling Earth Systems and Environment*, *9*(2), 2501–2516. doi:10.100740808-022-01638-6

Mohanraj, I., Ashokumar, K., & Naren, J. (2016). Field monitoring and automation using IOT in agriculture domain. *Procedia Computer Science*, *93*, 931–939. doi:10.1016/j.procs.2016.07.275

Mohsan, S. A. H., Zahra, Q. U. A., Khan, M. A., Alsharif, M. H., Elhaty, I. A., & Jahid, A. (2022). Role of drone technology helping in alleviating the COVID-19 pandemic. *Micromachines*, *13*(10), 1593. doi:10.3390/mi13101593 PMID:36295946

Mousavi, M., TohidiFar, A., & Alvanchi, A. (2022). BIM and machine learning in seismic damage prediction for non-structural exterior infill walls. *Automation in Construction*, *139*, 104288. doi:10.1016/j.autcon.2022.104288

Mousavi, S. M., & Beroza, G. C. (2022). Deep-learning seismology. *Science*, *377*(6607), eabm4470. doi:10.1126cience.abm4470 PMID:35951699

Mpuon, J. A., Odigbo, B. E., Etim, G. S., Etuk, I. U., & Usoro, A. A. (2023). Impact of system integration of automation and autonomation supply chain strategies in operational performance of manufacturing firms. *International Journal of Value Chain Management*, *14*(2), 195–219. doi:10.1504/IJVCM.2023.130988

Mudele, O., Frery, A. C., Zanandrez, L. F., Eiras, A. E., & Gamba, P. (2021). Dengue vector population forecasting using multisource earth observation products and recurrent neural networks. *IEEE Journal of Selected Topics in Applied Earth Observations and Remote Sensing*, *14*, 4390–4404. doi:10.1109/JSTARS.2021.3073351

Müller, M. C., Sander, T., & Mundt, C. (2023, June). Development and testing of a multi-spectral and multi-field of view imaging system for space-based signature detection and tracking of hypersonic vehicles. In *Sensors and Systems for Space Applications XVI* (Vol. 12546, pp. 8–21). SPIE. doi:10.1117/12.2663480

Munir, A., Aved, A., & Blasch, E. (2022). Situational awareness: Techniques, challenges, and prospects. *AI*, *3*(1), 55–77. doi:10.3390/ai3010005

Murti, M. A., Junior, R., Ahmed, A. N., & Elshafie, A. (2022). Earthquake multi-classification detection based velocity and displacement data filtering using machine learning algorithms. *Scientific Reports*, *12*(1), 21200. doi:10.103841598-022-25098-1 PMID:36482200

Muruganantham, P., Wibowo, S., Grandhi, S., Samrat, N. H., & Islam, N. (2022). A systematic literature review on crop yield prediction with deep learning and remote sensing. *Remote Sensing (Basel)*, *14*(9), 1990. doi:10.3390/rs14091990

Murwantara, I. M., Yugopuspito, P., & Hermawan, R. (2020). Comparison of machine learning performance for earthquake prediction in Indonesia using 30 years historical data. *TELKOMNIKA*, *18*(3), 1331–1342. doi:10.12928/telkomnika.v18i3.14756

Muscettola, N., Nayak, P. P., Pell, B., & Williams, B. C. (1998). Remote agent: To boldly go where no AI system has gone before. *Artificial Intelligence*, *103*(1-2), 5–47. doi:10.1016/S0004-3702(98)00068-X

Naikoo, M. W., Talukdar, S., Ishtiaq, M., & Rahman, A. (2023). Modelling built-up land expansion probability using the integrated fuzzy logic and coupling coordination degree model. *Journal of Environmental Management*, *325*, 116441. doi:10.1016/j.jenvman.2022.116441 PMID:36242974

Najim, H. A., Kareem, I. S., & Abdul-Lateef, W. E. (2023, July). Design and implementation of an omnidirectional mobile robot for medicine delivery in hospitals during the COVID-19 epidemic. In AIP Conference Proceedings (Vol. 2830, No. 1). AIP Publishing. doi:10.1063/5.0156862

Nandutu, I., Atemkeng, M., & Okouma, P. (2021). Integrating AI ethics in wildlife conservation AI systems in South Africa: A review, challenges, and future research agenda. *AI & Society*, 1–13.

Narayanan, K. L., Krishnan, R. S., Robinson, Y. H., Julie, E. G., Vimal, S., Saravanan, V., & Kaliappan, M. (2022). Banana plant disease classification using hybrid convolutional neural network. *Computational Intelligence and Neuroscience*. doi:10.1155/2022/9153699 PMID:35251158

Nasser, F. K., & Behadili, S. F. (2022). Breast cancer detection using decision tree and k-nearest neighbour classifiers. *Iraqi Journal of Science*, 4987–5003. doi:10.24996/ijs.2022.63.11.34

Natarajan, Y., Kannan, S., Selvaraj, C., & Mohanty, S. N. (2021). Forecasting energy generation in large photovoltaic plants using radial belief neural network. *Sustainable Computing : Informatics and Systems*, *31*, 100578. doi:10.1016/j.suscom.2021.100578

Natarajan, Y., Srihari, K., Dhiman, G., Chandragandhi, S., Gheisari, M., Liu, Y., Lee, C.-C., Singh, K. K., Yadav, K., & Alharbi, H. F. (2022). An IoT and machine learning-based routing protocol for reconfigurable engineering application. *IET Communications*, *16*(5), 464–475. doi:10.1049/cmu2.12266

Nayak, R. K., Tripathy, R., Saravanan, V., Das, P., & Anguraj, D. K. (2020, March). A Novel Strategy for Prediction of Cellular Cholesterol Signature Motif from G Protein-Coupled Receptors based on Rough Set and FCM Algorithm. In *2020 Fourth International Conference on Computing Methodologies and Communication (ICCMC)* (pp. 285-289). IEEE. 10.1109/ICCMC48092.2020.ICCMC-00053

Nazirbhai, M. N., & Gajjar, R. R. (2023, May). Comparative Analysis of Perturb & Observe, Incremental Conductance and Fuzzy Logic Based Maximum Power Point Tracking Methods. In 2023 IEEE Renewable Energy and Sustainable E-Mobility Conference (RESEM) (pp. 1-6). IEEE.

Neagu, C. D., Avouris, N., Kalapanidas, E., & Palade, V. (2002). Neural and neuro-fuzzy integration in a knowledge-based system for air quality prediction. *Applied Intelligence*, *17*(2), 141–169. doi:10.1023/A:1016108730534

Nguyen, L. T., Vu, D. H., Vu, N. C., Dao, V. H., & Tran, T. H. (2022, July). Comparative study on super resolution techniques for upper gastrointestinal endoscopic images. In *2022 IEEE Ninth International Conference on Communications and Electronics (ICCE)* (pp. 375-380). IEEE. 10.1109/ICCE55644.2022.9852031

Orusa, T., Cammareri, D., & Borgogno Mondino, E. (2022). A Scalable Earth Observation Service to Map Land Cover in Geomorphological Complex Areas beyond the Dynamic World: An Application in Aosta Valley (NW Italy). *Applied Sciences (Basel, Switzerland)*, *13*(1), 390. doi:10.3390/app13010390

Othman, S. M., & Abdulrazzaq, M. B. (2023). Fuzzy logic system for drug storage based on the internet of things: A survey. *Indonesian Journal of Electrical Engineering and Computer Science*, *29*(3), 1382. doi:10.11591/ijeecs.v29.i3.pp1382-1392

Oza, A., & Bokhare, A. (2022, July). Diabetes prediction using logistic regression and K-nearest neighbor. In *Congress on Intelligent Systems: Proceedings of CIS 2021,* Volume 2 (pp. 407-418). Singapore: Springer Nature Singapore. 10.1007/978-981-16-9113-3_30

Özdemir, A. (2022). Data-And Knowledge-Driven Smart City Strategies: Research On Implementation Challenges Of Local Governments In Turkey. *Journal of Management and Economics Research*, *20*(3), 152–169.

Palansooriya, K. N., Li, J., Dissanayake, P. D., Suvarna, M., Li, L., Yuan, X., Sarkar, B., Tsang, D. C. W., Rinklebe, J., Wang, X., & Ok, Y. S. (2022). Prediction of soil heavy metal immobilization by biochar using machine learning. *Environmental Science & Technology*, *56*(7), 4187–4198. doi:10.1021/acs.est.1c08302 PMID:35289167

Pandey, J., & Verma, S. (2022). Water Quality Analysis and Prediction Techniques Using Artificial Intelligence. In *ICT with Intelligent Applications: Proceedings of ICTIS 2021,* Volume 1 (pp. 279-290). Springer Singapore. 10.1007/978-981-16-4177-0_29

Pant, G., Bahuguna, R., Pandey, S., Gehlot, A., Yadav, S. P., & Pachauri, R. K. (2023, April). Intelligent Automated Interference for the Protection of Road Safety. In *2023 International Conference on Computational Intelligence, Communication Technology and Networking (CICTN)* (pp. 87-91). IEEE. 10.1109/CICTN57981.2023.10141086

Pan, Y., & Zhang, L. (2021). Roles of artificial intelligence in construction engineering and management: A critical review and future trends. *Automation in Construction*, *122*, 103517. doi:10.1016/j.autcon.2020.103517

Park, J. W., Korosov, A., Babiker, M., Won, J. S., Hansen, M. W., & Kim, H. C. (2020). Classification of sea ice types in Sentinel-1 synthetic aperture radar images. *The Cryosphere*, *14*(8), 2629–2645. doi:10.5194/tc-14-2629-2020

Parveen, H., Rizvi, S. W. A., & Shukla, P. K. (2023). Disease risk level prediction based on knowledge driven optimized deep ensemble framework. *Biomedical Signal Processing and Control*, *79*, 103991. doi:10.1016/j.bspc.2022.103991

Passarelli, A. M., & Kolb, D. A. (2023). *Experiential Learning Theory. Student Learning Abroad: What Our Students Are Learning*. What They're Not, and What We Can Do About It.

Patel, D., Patel, S., Patel, P., & Shah, M. (2022). Solar radiation and solar energy estimation using ANN and Fuzzy logic concept: A comprehensive and systematic study. *Environmental Science and Pollution Research International*, *29*(22), 32428–32442. doi:10.100711356-022-19185-z PMID:35178628

Patel, P., Patel, Y., Patel, U., Patel, V., Patel, N., Oza, P., & Patel, U. (2022). Towards automating irrigation: A fuzzy logic-based water irrigation system using IoT and deep learning. *Modeling Earth Systems and Environment*, *8*(4), 5235–5250. doi:10.100740808-022-01452-0

Peng, C., Jiang, P., Ma, Q., Su, J., Cai, Y., & Zheng, Y. (2022). Chinese nationwide earthquake early warning system and its performance in the 2022 Lushan M 6.1 earthquake. *Remote Sensing (Basel)*, *14*(17), 4269. doi:10.3390/rs14174269

Persello, C., Wegner, J. D., Hänsch, R., Tuia, D., Ghamisi, P., Koeva, M., & Camps-Valls, G. (2022). Deep learning and earth observation to support the sustainable development goals: Current approaches, open challenges, and future opportunities. *IEEE Geoscience and Remote Sensing Magazine*, *10*(2), 172–200. doi:10.1109/MGRS.2021.3136100

Pilon, R. V. (2023). *Artificial Intelligence in Commercial Aviation: Use Cases and Emerging Strategies*. Taylor & Francis. doi:10.4324/9781003018810

Poursanidis, D., Chrysoulakis, N., & Mitraka, Z. (2015). Landsat 8 vs. Landsat 5: A comparison based on urban and peri-urban land cover mapping. *International Journal of Applied Earth Observation and Geoinformation*, *35*, 259–269. doi:10.1016/j.jag.2014.09.010

Poursanidis, D., Topouzelis, K., & Chrysoulakis, N. (2018). Mapping coastal marine habitats and delineating the deep limits of the Neptune's seagrass meadows using very high resolution Earth observation data. *International Journal of Remote Sensing*, *39*(23), 8670–8687. doi:10.1080/01431161.2018.1490974

Prasad, P., Loveson, V. J., Chandra, P., & Kotha, M. (2022). Evaluation and comparison of the earth observing sensors in land cover/land use studies using machine learning algorithms. *Ecological Informatics*, *68*, 101522. doi:10.1016/j.ecoinf.2021.101522

Prasanna, R., Chandrakumar, C., Nandana, R., Holden, C., Punchihewa, A., Becker, J. S., Jeong, S., Liyanage, N., Ravishan, D., Sampath, R., & Tan, M. L. (2022, March). "Saving Precious Seconds"—A Novel Approach to Implementing a Low-Cost Earthquake Early Warning System with Node-Level Detection and Alert Generation. *Informatics (MDPI)*, *9*(1), 25. doi:10.3390/informatics9010025

Priest, S. (2022). Preparing for the next pandemic: Adopt, adapt or improve? *Journal of Outdoor and Environmental Education*, *25*(3), 329–340. doi:10.100742322-022-00105-1

Purohit, K., Vats, S., Saklani, R., Kukreja, V., Sharma, V., & Yadav, S. P. (2023, July). Improvement in K-Means Clustering for Information Retrieval. In *2023 4th International Conference on Electronics and Sustainable Communication Systems (ICESC)* (pp. 1239-1245). IEEE.

Qader, S. H., Dash, J., Alegana, V. A., Khwarahm, N. R., Tatem, A. J., & Atkinson, P. M. (2021). The role of earth observation in achieving sustainable agricultural production in arid and semi-arid regions of the world. *Remote Sensing (Basel)*, *13*(17), 3382. doi:10.3390/rs13173382

Qayyum, B., Ahmed, A., Ullah, I., & Shah, S. A. (2022). A Fuzzy-Logic Approach for Optimized and Cost-Effective Early Warning System for Tsunami Detection. *Sustainability (Basel)*, *14*(21), 14516. doi:10.3390u142114516

Rahman, M. Z., Akbar, M. A., Leiva, V., Tahir, A., Riaz, M. T., & Martin-Barreiro, C. (2023). An intelligent health monitoring and diagnosis system based on the internet of things and fuzzy logic for cardiac arrhythmia COVID-19 patients. *Computers in Biology and Medicine*, *154*, 106583. doi:10.1016/j.compbiomed.2023.106583 PMID:36716687

Rajalakshmi, M., Saravanan, V., Arunprasad, V., Romero, C. T., Khalaf, O. I., & Karthik, C. (2022). Machine Learning for Modeling and Control of Industrial Clarifier Process. *Intelligent Automation & Soft Computing*, *32*(1). doi:10.32604/iasc.2022.021696

Ramírez-Mendoza, A. M. E., Yu, W., & Li, X. (2023). A new spike membership function for the recognition and processing of spatiotemporal spike patterns: Syllable-based speech recognition application. *Mathematics*, *11*(11), 2525. doi:10.3390/math11112525

Rao, Y., Redmon, R., Dale, K., Haupt, S. E., Hopkinson, A., Bostrom, A., & Kihn, E. A. (2023). *Developing Digital Twins for Earth Systems: Purpose, Requisites, and Benefits*. arXiv preprint arXiv:2306.11175.

Rao, A., Jung, J., Silva, V., Molinario, G., & Yun, S. H. (2023). Earthquake building damage detection based on synthetic-aperture-radar imagery and machine learning. *Natural Hazards and Earth System Sciences*, *23*(2), 789–807. doi:10.5194/nhess-23-789-2023

Rastogi, K., Bahuguna, R., Kathuria, S., Gehlot, A., Malik, P. K., & Negi, P. (2023, April). *Technical Intercession of Artificial Intelligence in Solving Online Dispute Resolution. In 2023 IEEE Devices for Integrated Circuit (DevIC).* IEEE.

Rayo, M. F., Horwood, C. R., Fitzgerald, M. C., Grayson, M. R., Abdel-Rasoul, M., & Moffatt-Bruce, S. D. (2022). Situated Visual Alarm Displays Support Machine Fitness Assessment for Nonexplainable Automation. *IEEE Transactions on Human-Machine Systems, 52*(5), 984–993. doi:10.1109/THMS.2022.3155714

Ray, P. P. (2022). A review on 6G for space-air-ground integrated network: Key enablers, open challenges, and future direction. *Journal of King Saud University. Computer and Information Sciences, 34*(9), 6949–6976. doi:10.1016/j.jksuci.2021.08.014

Razdan, S., & Sharma, S. (2022). Internet of medical things (IoMT): Overview, emerging technologies, and case studies. *IETE Technical Review, 39*(4), 775–788. doi:10.1080/02564602.2021.1927863

Rødven, R., & Wilson, S. (2022). Combining Knowledge for a Sustainable Arctic–AMAP Cases as Knowledge Driven Science-Policy Interactions. In *Building Common Interests in the Arctic Ocean with Global Inclusion* (Vol. 2, pp. 137–150). Springer International Publishing. doi:10.1007/978-3-030-89312-5_9

Ronneberger, O., Fischer, P., & Brox, T. (2015). U-NET: Convolutional Networks for Biomedical Image Segmentation. In Lecture Notes in Computer Science (pp. 234–241). Springer. doi:10.1007/978-3-319-24574-4_28

Rouet-Leduc, B., Hulbert, C., Lubbers, N., Barros, K., Humphreys, C. J., & Johnson, P. A. (2017). Machine learning predicts laboratory earthquakes. *Geophysical Research Letters, 44*(18), 9276–9282. doi:10.1002/2017GL074677

Russo, A., & Coco, D. (2023). Quantify how space mission influence geopolitical dynamics? A security and social policy approach. *arXiv preprint arXiv:2301.03538.* doi:10.2139/ssrn.4375678

Russo, A., & Lax, G. (2022). Using artificial intelligence for space challenges: A survey. *Applied Sciences (Basel, Switzerland), 12*(10), 5106. doi:10.3390/app12105106

Saad, O. M., Chen, Y., Trugman, D., Soliman, M. S., Samy, L., Savvaidis, A., Khamis, M. A., Hafez, A. G., Fomel, S., & Chen, Y. (2022). Machine learning for fast and reliable source-location estimation in earthquake early warning. *IEEE Geoscience and Remote Sensing Letters, 19*, 1–5. doi:10.1109/LGRS.2022.3142714

Sadly, M., Hendiarti, N., Sachoemar, S. I., & Faisal, Y. (2009). Fishing ground prediction using a knowledge-based expert system geographical information system model in the South and Central Sulawesi coastal waters of Indonesia. *International Journal of Remote Sensing, 30*(24), 6429–6440. doi:10.1080/01431160902865780

Saheb, T., Dehghani, M., & Saheb, T. (2022). Artificial intelligence for sustainable energy: A contextual topic modeling and content analysis. *Sustainable Computing : Informatics and Systems, 35*, 100699. doi:10.1016/j.suscom.2022.100699

Saini, A., & Bhalla, R. (2022). Artificial intelligence and automation: transforming the hospitality industry or threat to human touch. In *Handbook of Research on Innovative Management Using AI in Industry 5.0* (pp. 88–97). IGI Global. doi:10.4018/978-1-7998-8497-2.ch006

Sajan, K. C., Bhusal, A., Gautam, D., & Rupakhety, R. (2023). Earthquake damage and rehabilitation intervention prediction using machine learning. *Engineering Failure Analysis, 144*, 106949. doi:10.1016/j.engfailanal.2022.106949

Salama, R., Al-Turjman, F., Bhatia, S., & Yadav, S. P. (2023, April). Social engineering attack types and prevention techniques-A survey. In *2023 International Conference on Computational Intelligence, Communication Technology and Networking (CICTN)* (pp. 817-820). IEEE. 10.1109/CICTN57981.2023.10140957

Salam, M. A., Ibrahim, L., & Abdelminaam, D. S. (2021). Earthquake prediction using hybrid machine learning techniques. *International Journal of Advanced Computer Science and Applications, 12*(5), 654–6652021. doi:10.14569/IJACSA.2021.0120578

Samarinas, N., Spiliotopoulos, M., Tziolas, N., & Loukas, A. (2023). Synergistic Use of Earth Observation Driven Techniques to Support the Implementation of Water Framework Directive in Europe: A Review. *Remote Sensing (Basel), 15*(8), 1983. doi:10.3390/rs15081983

Sapra, L., Sapra, V., & Bhardwaj, A. (Eds.). (2023). *Security Implementation in Internet of Medical Things*. CRC Press. doi:10.1201/9781003269168

Saravanan, V., & Chandrasekar, C. (2012). Qos-continuous live media streaming in mobile environment using vbr and edge network. *International Journal of Computer Applications, 53*(6), 36–40. doi:10.5120/8426-2185

Saravanan, V., & Mohan Raj, V. (2016). Maximizing QoS by cooperative vertical and horizontal handoff for tightly coupled WiMAX/WLAN overlay networks. *Cluster Computing, 19*(3), 1619–1633. doi:10.100710586-016-0601-x

Saravanan, V., & Sumathi, A. (2012). Handoff mobiles with low latency in heterogeneous networks for seamless mobility: A survey and future directions. *European Journal of Scientific Research, 81*(3), 417–424.

Saravanan, V., & Sumathi, A. (2012, December). Dynamic handoff decision based on current traffic level and neighbor information in wireless data networks. In *2012 Fourth International Conference on Advanced Computing (ICoAC)* (pp. 1-5). IEEE. 10.1109/ICoAC.2012.6416797

Sargsyan, K. (2023). Future of Healthcare. In *Future Intelligence: The World in 2050-Enabling Governments, Innovators, and Businesses to Create a Better Future* (pp. 39–51). Springer International Publishing. doi:10.1007/978-3-031-36382-5_5

Sathishkumar, V. E., Cho, J., Subramanian, M., & Naren, O. S. (2023). Forest fire and smoke detection using deep learning-based learning without forgetting. *Fire Ecology, 19*(1), 1–17. doi:10.118642408-022-00165-0

Saumya, S., Roy, P. K., & Singh, J. P. (2023). Review helpfulness prediction on e-commerce websites: A comprehensive survey. *Engineering Applications of Artificial Intelligence, 126*, 107075. doi:10.1016/j.engappai.2023.107075

Sayed, E. T., Olabi, A. G., Elsaid, K., Al Radi, M., Semeraro, C., Doranehgard, M. H., Eltayeb, M. E., & Abdelkareem, M. A. (2023). Application of artificial intelligence techniques for modeling, optimizing, and controlling desalination systems powered by renewable energy resources. *Journal of Cleaner Production, 413*, 137486. doi:10.1016/j.jclepro.2023.137486

Scardillo, M. K. (2023). *New Ideas, New Directions: Experiential Learning for MBA Students* [Doctoral dissertation, Northeastern University].

Seyyedi, A., Bohlouli, M., & Oskoee, S. N. (2023). Machine Learning and Physics: A Survey of Integrated Models. *ACM Computing Surveys.*

Shah, K. H., Malik, Z. A., Balakrishnan, G., & Ramanan, K. (2022). *Integration Of Iot And Cloud Computing-Based Innovation Of Wireless Sensor Network Technology For Health Care Architecture And Agriculture Applications.*

Shakeel, M., Nishida, K., Itoyama, K., & Nakadai, K. (2022). 3d convolution recurrent neural networks for multi-label earthquake magnitude classification. *Applied Sciences (Basel, Switzerland), 12*(4), 2195. doi:10.3390/app12042195

Sheng, J., Cai, X., Li, Q., Wu, C., Ai, B., Wang, Y., Kadoch, M., & Yu, P. (2021). Space-air-ground integrated network development and applications in high-speed railways: A survey. *IEEE Transactions on Intelligent Transportation Systems, 23*(8), 10066–10085. doi:10.1109/TITS.2021.3118557

Shen, H., Huang, F., Fan, X., Shahabi, H., Shirzadi, A., Wang, D., & Chen, W. (2022). Improving the performance of artificial intelligence models using the rotation forest technique for landslide susceptibility mapping. *International Journal of Environmental Science and Technology*, 1–16.

Shibuya, K. (2020). Digital transformation of identity in the age of artificial intelligence (No. 172558). Singapore: Springer.

Shi, G., Qin, C., Tao, J., & Liu, C. (2021). A VMD-EWT-LSTM-based multi-step prediction approach for shield tunneling machine cutterhead torque. *Knowledge-Based Systems, 228*, 107213. doi:10.1016/j.knosys.2021.107213

Shi, P., Grigoli, F., Lanza, F., Beroza, G. C., Scarabello, L., & Wiemer, S. (2022). MALMI: An automated earthquake detection and location workflow based on machine learning and waveform migration. *Seismological Research Letters, 93*(5), 2467–2483. doi:10.1785/0220220071

Shi, W., Zhang, M., Zhang, R., Chen, S., & Zhan, Z. (2020). Change detection based on artificial intelligence: State-of-the-art and challenges. *Remote Sensing (Basel), 12*(10), 1688. doi:10.3390/rs12101688

Sidki-Rius, N., Sanmiquel, L., Bascompta, M., & Parcerisa, D. (2022). Subsidence management and prediction system: A case study in Potash Mining. *Minerals (Basel), 12*(9), 1155. doi:10.3390/min12091155

Singh, H., Sharma, V., & Singh, D. (2022). Comparative analysis of proficiencies of various textures and geometric features in breast mass classification using k-nearest neighbor. *Visual Computing for Industry, Biomedicine, and Art, 5*(1), 1–19. doi:10.118642492-021-00100-1 PMID:35018506

Singh, J., Deepika, J., Sathyendra Bhat, J., Kumararaja, V., Vikram, R., Jegathesh Amalraj, J., & Sakthivel, S. (2022). Energy-efficient clustering and routing algorithm using hybrid fuzzy with grey wolf optimization in wireless sensor networks. *Security and Communication Networks, 2022*, 2022. doi:10.1155/2022/9846601

Singh, V. (2023). *Unleashing AI: A Thought-provoking Look At The Future Of Humanity*. Amazon.

Sinha, E. (2023). Integrating Human and Artificial Intelligence Within Organizations: A Conceptual Framework Based on Artistry and Scientific HR Elements. In Managing Technology Integration for Human Resources in Industry 5.0 (pp. 48-67). IGI Global. doi:10.4018/978-1-6684-6745-9.ch004

Sivaram, M., Mohammed, A. S., Yuvaraj, D., Porkodi, V., Manikandan, V., & Yuvaraj, N. (2019). Advanced expert system using particle swarm optimization based adaptive network based fuzzy inference system to diagnose the physical constitution of human body. In *Emerging Technologies in Computer Engineering: Microservices in Big Data Analytics: Second International Conference, ICETCE 2019*, (pp. 349-362). Springer Singapore. 10.1007/978-981-13-8300-7_29

Sivaram, M., Porkodi, V., Mohammed, A. S., Manikandan, V., & Yuvaraj, N. (2019). Retransmission DBTMA protocol with fast retransmission strategy to improve the performance of MANETs. *IEEE Access : Practical Innovations, Open Solutions, 7*, 85098–85109. doi:10.1109/ACCESS.2019.2918723

Sivaram, M., Yuvaraj, D., Mohammed, A. S., Manikandan, V., Porkodi, V., & Yuvaraj, N. (2019). Improved enhanced DBTMA with contention-aware admission control to improve the network performance in MANETS. *Computers, Materials & Continua, 60*(2), 435–454. doi:10.32604/cmc.2019.06295

Song, J., Zhu, J., Wang, Y., & Li, S. (2022). On-site alert-level earthquake early warning using machine-learning-based prediction equations. *Geophysical Journal International, 231*(2), 786–800. doi:10.1093/gji/ggac220

Song, Y., Wei, L., Yang, Q., Wu, J., Xing, L., & Chen, Y. (2023). RL-GA: A reinforcement learning-based genetic algorithm for electromagnetic detection satellite scheduling problem. *Swarm and Evolutionary Computation, 77*, 101236. doi:10.1016/j.swevo.2023.101236

Srivastava, S., Dhyani, N., Sharma, V., Vats, S., Yadav, S. P., Kukreja, V., & Singh, R. (2023, May). Lung Infection and Identification using Heatmap. In *2023 2nd International Conference on Applied Artificial Intelligence and Computing (ICAAIC)* (pp. 1093-1098). IEEE. 10.1109/ICAAIC56838.2023.10140204

Su, Z., He, L., Jariwala, S. P., Zheng, K., & Chen, Y. (2022). " What is Your Envisioned Future?": Toward Human-AI Enrichment in Data Work of Asthma Care. *Proceedings of the ACM on Human-Computer Interaction, 6*(CSCW2), (pp. 1-28). ACM. 10.1145/3555157

Suárez, G. (2022). The seismic early warning system of Mexico (SASMEX): A retrospective view and future challenges. *Frontiers in Earth Science (Lausanne), 10*, 827236. doi:10.3389/feart.2022.827236

Sun, A. Y., & Scanlon, B. R. (2019). How can Big Data and machine learning benefit environment and water management: A survey of methods, applications, and future directions. *Environmental Research Letters, 14*(7), 073001. doi:10.1088/1748-9326/ab1b7d

Sundarakantham, K., & Sivasankar, E. (2023). A hybrid deep learning framework for privacy preservation in edge computing. *Computers & Security, 129*, 103209. doi:10.1016/j.cose.2023.103209

Sun, X., Opulencia, M. J. C., Alexandrovich, T. P., Khan, A., Algarni, M., & Abdelrahman, A. (2022). Modeling and optimization of vegetable oil biodiesel production with heterogeneous nano catalytic process: Multi-layer perceptron, decision regression tree, and K-Nearest Neighbor methods. *Environmental Technology & Innovation, 27*, 102794. doi:10.1016/j.eti.2022.102794

Sun, Z., Cristea, N., & Rivas, P. (Eds.). (2023). *Artificial Intelligence in Earth Science: Best Practices and Fundamental Challenges*. Elsevier. doi:10.1016/B978-0-323-91737-7.00003-7

Sun, Z., Sandoval, L., Crystal-Ornelas, R., Mousavi, S. M., Wang, J., Lin, C., Cristea, N., Tong, D., Carande, W. H., Ma, X., Rao, Y., Bednar, J. A., Tan, A., Wang, J., Purushotham, S., Gill, T. E., Chastang, J., Howard, D., Holt, B., & John, A. (2022). A review of earth artificial intelligence. *Computers & Geosciences, 159*, 105034. doi:10.1016/j.cageo.2022.105034

Suyanto, S., Meliana, S., Wahyuningrum, T., & Khomsah, S. (2022). A new nearest neighbor-based framework for diabetes detection. *Expert Systems with Applications, 199*, 116857. doi:10.1016/j.eswa.2022.116857

Taj, I., & Zaman, N. (2022). Towards industrial revolution 5.0 and explainable artificial intelligence: Challenges and opportunities. *International Journal of Computing and Digital Systems, 12*(1), 295–320. doi:10.12785/ijcds/120128

Taktak, H., Ghedira, C., Boukadi, K., Mrissa, M., & Gargouri, F. (2022). *Knowledge-Driven Automatic Solution for Service Composition based on ELECTRE III and Quality-Aware Service Selection for the Wildfire Prediction.*

Tao, K., Yu, J., Zhang, J., Bao, A., Hu, H., Ye, T., Ding, Q., Wang, Y., Lin, H., Wu, J., Chang, H., Zhang, H., & Yuan, W. (2023). Deep-Learning Enabled Active Biomimetic Multifunctional Hydrogel Electronic Skin. *ACS Nano, 17*(16), 16160–16173. doi:10.1021/acsnano.3c05253 PMID:37523784

Tayfur, G. (2023). Application of fuzzy logic in water resources engineering. In *Handbook of Hydroinformatics* (pp. 155–166). Elsevier. doi:10.1016/B978-0-12-821962-1.00024-6

Teng, F., Zhu, Y., Zhang, E., Hu, X., Sun, Q., & Feng, L. (2023, July). A knowledge service framework for fault diagnosis of low-earth orbit satellite constellation. In *2023 IEEE International Conference on Web Services (ICWS)* (pp. 669-676). IEEE. 10.1109/ICWS60048.2023.00085

Thaler, D., Elezaj, L., Bamer, F., & Markert, B. (2022). Training data selection for machine learning-enhanced Monte Carlo simulations in structural dynamics. *Applied Sciences (Basel, Switzerland), 12*(2), 581. doi:10.3390/app12020581

Thiyagalingam, J., Shankar, M., Fox, G., & Hey, T. (2022). Scientific machine learning benchmarks. *Nature Reviews. Physics*, *4*(6), 413–420. doi:10.103842254-022-00441-7

Thompson, A. E., Walden, J. P., Chase, A. S., Hutson, S. R., Marken, D. B., Cap, B., Fries, E. C., Guzman Piedrasanta, M. R., Hare, T. S., Horn, S. W., Micheletti, G. J., Montgomery, S. M., Munson, J., Richards-Rissetto, H., Shaw-Müller, K., Ardren, T., Awe, J. J., Brown, M. K., Callaghan, M., & Chase, D. Z. (2022). Ancient Lowland Maya neighborhoods: Average Nearest Neighbor analysis and kernel density models, environments, and urban scale. *PLoS One*, *17*(11), e0275916. doi:10.1371/journal.pone.0275916 PMID:36322539

Tiwari, A., Sharma, S. K., Dixit, A., & Mishra, V. (2021). UAV remote sensing for campus monitoring: A comparative evaluation of nearest neighbor and rule-based classification. *Photonirvachak (Dehra Dun)*, *49*(3), 527–539. doi:10.100712524-020-01268-4

Trabelsi, M., Casprini, E., Fiorini, N., & Zanni, L. (2023). Unleashing the value of artificial intelligence in the agri-food sector: Where are we? *British Food Journal*, *125*(13), 482–515. doi:10.1108/BFJ-11-2022-1014

Tziolas, N., Tsakiridis, N., Chabrillat, S., Demattê, J. A., Ben-Dor, E., Gholizadeh, A., Zalidis, G., & Van Wesemael, B. (2021). Earth observation data-driven cropland soil monitoring: A review. *Remote Sensing (Basel)*, *13*(21), 4439. doi:10.3390/rs13214439

Tziolas, N., Tsakiridis, N., Ogen, Y., Kalopesa, E., Ben-Dor, E., Theocharis, J., & Zalidis, G. (2020). An integrated methodology using open soil spectral libraries and Earth Observation data for soil organic carbon estimations in support of soil-related SDGs. *Remote Sensing of Environment*, *244*, 111793. doi:10.1016/j.rse.2020.111793

Ullo, S. L., & Sinha, G. R. (2020). Advances in smart environment monitoring systems using IoT and sensors. *Sensors (Basel)*, *20*(11), 3113. doi:10.339020113113 PMID:32486411

Vermesan, O., & Bacquet, J. (Eds.). (2022). *Internet of Things–The Call of the Edge: Everything Intelligent Everywhere*. CRC Press.

Vivone, G., Braca, P., & Horstmann, J. (2015). Knowledge-based multitarget ship tracking for HF surface wave radar systems. *IEEE Transactions on Geoscience and Remote Sensing*, *53*(7), 3931–3949. doi:10.1109/TGRS.2014.2388355

Wang, X., Zhao, F., Shi, Z., & Jin, Z. (2023). Deep Reinforcement Learning-Based Periodic Earth Observation Scheduling for Agile Satellite Constellation. *Journal of Aerospace Information Systems,* 1-12.

Wang, L., Scott, K. A., Xu, L., & Clausi, D. A. (2016). Sea ice concentration estimation during melt from Dual-Pol SAR scenes using deep convolutional neural networks: A case study. *IEEE Transactions on Geoscience and Remote Sensing*, *54*(8), 4524–4533. doi:10.1109/TGRS.2016.2543660

Wang, X., Wu, G., Xing, L., & Pedrycz, W. (2020). Agile earth observation satellite scheduling over 20 years: Formulations, methods, and future directions. *IEEE Systems Journal*, *15*(3), 3881–3892. doi:10.1109/JSYST.2020.2997050

Wang, X., Wu, J., Shi, Z., Zhao, F., & Jin, Z. (2022). Deep reinforcement learning-based autonomous mission planning method for high and low orbit multiple agile Earth observing satellites. *Advances in Space Research*, *70*(11), 3478–3493. doi:10.1016/j.asr.2022.08.016

Wani, M. A., Din, A., Nazki, I. T., Rehman, T. U., Al-Khayri, J. M., Jain, S. M., Lone, R. A., Bhat, Z. A., & Mushtaq, M. (2023). Navigating the future: Exploring technological advancements and emerging trends in the sustainable ornamental industry. *Frontiers in Environmental Science*, *11*, 1188643. doi:10.3389/fenvs.2023.1188643

Waubert de Puiseau, C., Meyes, R., & Meisen, T. (2022). On reliability of reinforcement learning based production scheduling systems: A comparative survey. *Journal of Intelligent Manufacturing, 33*(4), 911–927. doi:10.100710845-022-01915-2

Wei, L., Chen, Y., Chen, M., & Chen, Y. (2021). Deep reinforcement learning and parameter transfer based approach for the multi-objective agile earth observation satellite scheduling problem. *Applied Soft Computing, 110*, 107607. doi:10.1016/j.asoc.2021.107607

Wellmann, T., Lausch, A., Scheuer, S., & Haase, D. (2020). Earth observation based indication for avian species distribution models using the spectral trait concept and machine learning in an urban setting. *Ecological Indicators, 111*, 106029. doi:10.1016/j.ecolind.2019.106029

Werbos, P. (1974). *Beyond Regression: New Tools for Prediction and Analysis in the Behavioral Science.* [Thesis, Harvard]. https://www.researchgate.net/publication/279233597_Beyond_Regression_New_Tools_for_Prediction_and_Analysis_in_the_Behavioral_Science_Thesis_Ph_D_Appl_Math_Harvard_University

Wijata, A. M., Foulon, M. F., Bobichon, Y., Vitulli, R., Celesti, M., Camarero, R., Di Cosimo, G., Gascon, F., Longépé, N., Nieke, J., Gumiela, M., & Nalepa, J. (2023). Taking Artificial Intelligence Into Space Through Objective Selection of Hyperspectral Earth Observation Applications: To bring the "brain" close to the "eyes" of satellite missions. *IEEE Geoscience and Remote Sensing Magazine, 11*(2), 10–39. doi:10.1109/MGRS.2023.3269979

Wong, C. Y., Samadi, S., Suleiman, W., & Kheddar, A. (2022). Touch semantics for intuitive physical manipulation of humanoids. *IEEE Transactions on Human-Machine Systems, 52*(6), 1111–1121. doi:10.1109/THMS.2022.3207699

Wu, L., Pleiss, G., & Cunningham, J. P. (2022, June). Variational nearest neighbor Gaussian process. In *International Conference on Machine Learning* (pp. 24114-24130). PMLR.

Wu, P., Wang, P., Chi, H. L., Zhong, Y., & Song, Y. (2022). Exploring Factors Affecting Transport Infrastructure Performance: Data-Driven Versus Knowledge-Driven Approaches. *IEEE Transactions on Intelligent Transportation Systems, 23*(12), 24714–24726. doi:10.1109/TITS.2022.3192895

Wu, Y. (2022). Ethically responsible and trustworthy autonomous systems for 6G. *IEEE Network, 36*(4), 126–133. doi:10.1109/MNET.005.2100711

Xia, M., Liu, W., Shi, B., Weng, L., & Liu, J. (2018). Cloud/snow recognition for multispectral satellite imagery based on a multidimensional deep residual network. *International Journal of Remote Sensing, 40*(1), 156–170. doi:10.1080/01431161.2018.1508917

Xiao, X., Zhang, T., Zhong, X., & Li, X. (2018). *Support vector regression snow-depth retrieval algorithm using passive microwave remote sensing data.* ResearchGate. https://www.researchgate.net/publication/323847002_Support_vector_regression_snow-depth_retrieval_algorithm_using_passive_microwave_remote_sensing_data

Xin, C. (2022). Exploring the effectiveness of VR-based product demonstrations featuring items of furniture. *Library Hi Tech.* doi:10.1108/LHT-01-2022-0001

Xing, S., Shahabi, C., & Pan, B. (2009). Continuous monitoring of nearest neighbors on land surface. *Proceedings of the VLDB Endowment International Conference on Very Large Data Bases, 2*(1), 1114–1125. doi:10.14778/1687627.1687753

Xiong, B., Jin, Y., Li, F., Chen, Y., Zou, Y., & Zhou, Z. (2023). Knowledge-driven inference for automatic reconstruction of indoor detailed as-built BIMs from laser scanning data. *Automation in Construction, 156*, 105097. doi:10.1016/j.autcon.2023.105097

Xu, H., Zheng, W., Zhang, Y., Zhao, Y., Wang, W., Yuan, Y., & Gao, L. (2023). *A fully integrated, standalone stretchable device platform with in-sensor adaptive machine learning for rehabilitation.*

Xu, M., Du, H., Niyato, D., Kang, J., Xiong, Z., Mao, S., & Poor, H. V. (2023). *Unleashing the power of edge-cloud generative ai in mobile networks: A survey of aigc services.* arXiv preprint arXiv:2303.16129.

Xu, Y., Zhang, L., Du, B., & Zhang, L. (2022). Hyperspectral anomaly detection based on machine learning: An overview. *IEEE Journal of Selected Topics in Applied Earth Observations and Remote Sensing, 15*, 3351–3364. doi:10.1109/JSTARS.2022.3167830

Yadav, S. P., Zaidi, S., Nascimento, C. D. S., de Albuquerque, V. H. C., & Chauhan, S. S. (2023, January). Analysis and Design of automatically generating for GPS Based Moving Object Tracking System. In *2023 International Conference on Artificial Intelligence and Smart Communication (AISC)* (pp. 1-5). IEEE. 10.1109/AISC56616.2023.10085180

Yang, L., Driscol, J., Sarigai, S., Wu, Q., Chen, H., & Lippitt, C. D. (2022). Google Earth Engine and artificial intelligence (AI): A comprehensive review. *Remote Sensing (Basel), 14*(14), 3253. doi:10.3390/rs14143253

Yan, Q., Huang, W., & Moloney, C. (2017). Neural networks based sea ice detection and concentration retrieval from GNSS-R Delay-Doppler maps. *IEEE Journal of Selected Topics in Applied Earth Observations and Remote Sensing, 10*(8), 3789–3798. doi:10.1109/JSTARS.2017.2689009

Yelick, K., Agarwal, D., Bard, D., Shalf, J., Almgren, A., Bhimji, W., & Kincade, K. (2021). 2019 Computing Sciences Strategic Plan. Lawrence Berkeley National Lab.(LBNL), Berkeley, CA (United States).

Yuan, C., Ni, Y., Lin, Y., & Denolle, M. (2023). Better Together: Ensemble Learning for Earthquake Detection and Phase Picking. *IEEE Transactions on Geoscience and Remote Sensing, 61*, 1–17. doi:10.1109/TGRS.2023.3320148

Yu, G., Lin, D., Wang, Y., Hu, M., Sugumaran, V., & Chen, J. (2023). Digital Twin-enabled and Knowledge-driven decision support for tunnel electromechanical equipment maintenance. *Tunnelling and Underground Space Technology, 140*, 105318. doi:10.1016/j.tust.2023.105318

Yu, G., Wu, Y., Cao, H., Ge, Q., Dai, Q., Sun, S., & Xie, Y. (2022). Insights into the mechanism of ozone activation and singlet oxygen generation on N-doped defective nanocarbons: A DFT and machine learning study. *Environmental Science & Technology, 56*(12), 7853–7863. doi:10.1021/acs.est.1c08666 PMID:35615937

Yuvaraj, N., Karthikeyan, T., & Praghash, K. (2021). An improved task allocation scheme in serverless computing using gray wolf Optimization (GWO) based reinforcement learning (RIL) approach. *Wireless Personal Communications, 117*(3), 2403–2421. doi:10.100711277-020-07981-0

Yuvaraj, N., Srihari, K., Chandragandhi, S., Raja, R. A., Dhiman, G., & Kaur, A. (2021). Analysis of protein-ligand interactions of SARS-Cov-2 against selective drug using deep neural networks. *Big Data Mining and Analytics, 4*(2), 76–83. doi:10.26599/BDMA.2020.9020007

Yuvaraj, N., Srihari, K., Dhiman, G., Somasundaram, K., Sharma, A., Rajeskannan, S. M. G. S. M. A., Soni, M., Gaba, G. S., AlZain, M. A., & Masud, M. (2021). Nature-inspired-based approach for automated cyberbullying classification on multimedia social networking. *Mathematical Problems in Engineering, 2021*, 1–12. doi:10.1155/2021/6644652

Yu, W., Patros, P., Young, B., Klinac, E., & Walmsley, T. G. (2022). Energy digital twin technology for industrial energy management: Classification, challenges and future. *Renewable & Sustainable Energy Reviews, 161*, 112407. doi:10.1016/j.rser.2022.112407

Yu, X., Qin, W., Lin, X., Shan, Z., Huang, L., Shao, Q., Wang, L., & Chen, M. (2023). Synergizing the enhanced RIME with fuzzy K-nearest neighbor for diagnose of pulmonary hypertension. *Computers in Biology and Medicine*, *165*, 107408. doi:10.1016/j.compbiomed.2023.107408 PMID:37672924

Zahmatkesh, S., Klemeš, J. J., Bokhari, A., Rezakhani, Y., Wang, C., Sillanpaa, M., Amesho, K. T. T., & Ahmed, W. S. (2022). Reducing chemical oxygen demand from low strength wastewater: A novel application of fuzzy logic based simulation in MATLAB. *Computers & Chemical Engineering*, *166*, 107944. doi:10.1016/j.compchemeng.2022.107944

Zaini, N. A., Ean, L. W., Ahmed, A. N., & Malek, M. A. (2022). A systematic literature review of deep learning neural network for time series air quality forecasting. *Environmental Science and Pollution Research International*, *29*(4), 1–33. doi:10.100711356-021-17442-1 PMID:34807385

Zamouche, M., Chermat, M., Kermiche, Z., Tahraoui, H., Kebir, M., Bollinger, J. C., Amrane, A., & Mouni, L. (2023). Predictive Model Based on K-Nearest Neighbor Coupled with the Gray Wolf Optimizer Algorithm (KNN_GWO) for Estimating the Amount of Phenol Adsorption on Powdered Activated Carbon. *Water (Basel)*, *15*(3), 493. doi:10.3390/w15030493

Zarei, A., Hasanlou, M., & Mahdianpari, M. (2021). A comparison of machine learning models for soil salinity estimation using multi-spectral earth observation data. *ISPRS Annals of the Photogrammetry, Remote Sensing and Spatial Information Sciences*, *3*, 257–263. doi:10.5194/isprs-annals-V-3-2021-257-2021

Zenil, H., Tegnér, J., Abrahão, F. S., Lavin, A., Kumar, V., Frey, J. G., & King, R. (2023). *The Future of Fundamental Science Led by Generative Closed-Loop Artificial Intelligence.* arXiv preprint arXiv:2307.07522.

Zhai, X. (2023). Chatgpt for next generation science learning. *XRDS: Crossroads. The ACM Magazine for Students*, *29*(3), 42–46.

Zhang, C., Zhong, P., Liu, M., Song, Q., Liang, Z., & Wang, X. (2022). Hybrid metric K-Nearest neighbor algorithm and applications. *Mathematical Problems in Engineering*, *2022*, 1–15. doi:10.1155/2022/8212546

Zhang, H., Cheng, X., Li, Y., He, D., & Du, X. (2023). Rapid seismic damage state assessment of RC frames using machine learning methods. *Journal of Building Engineering*, *65*, 105797. doi:10.1016/j.jobe.2022.105797

Zhang, H., Melgar, D., Sahakian, V., Searcy, J., & Lin, J. T. (2022). Learning source, path and site effects: CNN-based on-site intensity prediction for earthquake early warning. *Geophysical Journal International*, *231*(3), 2186–2204. doi:10.1093/gji/ggac325

Zhang, J., Hu, Y., Zhang, L., Zhou, J., & Lu, A. (2023). Transparent, ultra-stretching, tough, adhesive carboxyethyl chitin/polyacrylamide hydrogel toward high-performance soft electronics. *Nano-Micro Letters*, *15*(1), 8. doi:10.100740820-022-00980-9 PMID:36477664

Zhang, S., Wang, Y., & Wu, G. (2022). Earthquake-induced landslide susceptibility assessment using a novel model based on gradient boosting machine learning and class balancing methods. *Remote Sensing (Basel)*, *14*(23), 5945. doi:10.3390/rs14235945

Zhang, W., Guo, S., Zhang, P., Xia, Z., Zhang, X., Lin, C., Tang, P., Fang, H., & Du, P. (2023). A novel knowledge-driven automated solution for high-resolution cropland extraction by cross-scale sample transfer. *IEEE Transactions on Geoscience and Remote Sensing*, *61*, 1–16. doi:10.1109/TGRS.2023.3299956

Zhang, Y., Chen, Z., Jin, F., Jing, L., Xing, H., & Li, P. (2023). Cross-project prediction for rock mass using shuffled TBM big dataset and knowledge-based machine learning methods. *Science China. Technological Sciences*, *66*(3), 751–770. doi:10.100711431-022-2290-7

Zhan, Y., Wang, J., Shi, J., Cheng, G., Yao, L., & Sun, W. (2017). Distinguishing cloud and snow in satellite images via deep convolutional network. *IEEE Geoscience and Remote Sensing Letters*, *14*(10), 1785–1789. doi:10.1109/LGRS.2017.2735801

Zhao, P., Gao, Y., & Sun, X. (2023). The impact of artificial intelligence on pollution emission intensity—Evidence from China. *Environmental Science and Pollution Research International*, *30*(39), 1–16. doi:10.100711356-023-28866-2 PMID:37470975

Zhou, X., Kouzel, M., & Alemzadeh, H. (2022, June). Robustness testing of data and knowledge driven anomaly detection in cyber-physical systems. In *2022 52nd Annual IEEE/IFIP International Conference on Dependable Systems and Networks Workshops (DSN-W)* (pp. 44-51). IEEE. 10.1109/DSN-W54100.2022.00017

Zhu, H. H., Wu, B., Cao, D. F., Li, B., Wen, Z., Liu, X. F., & Shi, B. (2023). Characterizing thermo-hydraulic behaviors of seasonally frozen loess via a combined opto-electronic sensing system: Field monitoring and assessment. *Journal of Hydrology (Amsterdam)*, *622*, 129647. doi:10.1016/j.jhydrol.2023.129647

Zhu, M., Wang, J., Yang, X., Zhang, Y., Zhang, L., Ren, H., & Ye, L. (2022). *A review of the application of machine learning in water quality evaluation*. Eco-Environment & Health. doi:10.1016/j.eehl.2022.06.001

Zhu, W., Hou, A. B., Yang, R., Datta, A., Mousavi, S. M., Ellsworth, W. L., & Beroza, G. C. (2023). QuakeFlow: A scalable machine-learning-based earthquake monitoring workflow with cloud computing. *Geophysical Journal International*, *232*(1), 684–693. doi:10.1093/gji/ggac355

About the Contributors

Sudesh Yadav is Assistant Professor (Computer Science) in Dept. of Higher Education Haryana. She is currently working at Govt. College, Ateli, Indira Gandhi University Meerpur, Rewari, Haryana. She has completed her M.Tech. (Information Security) from NSUT East Campus Delhi (formerly AI-ACTR) and Ph. D. in Computer Science from Guru Gobind Singh Indraprastha University Delhi. She has published many research paper in referred international journals and conferences and reviewed many research papers of international journals and conferences. Her area of interest is artificial intelligence, IoT, digital image processing, soft computing and pattern recognition, natural language processing.

Satya Prakash Yadav is currently the Associate Professor of the Department of Computer Science and Engineering, G.L. Bajaj Institute of Technology and Management (GLBITM), Greater Noida (India) and has completed his Postdoctoral under the supervision Prof. (Dr.) Victor Hugo C. de Albuquerque, from Federal Institute of Education, Science and Technology of Ceará, Brazil. He has awarded his PhD degree from Dr. A.P.J. Abdul Kalam Technical University (AKTU) (formerly UPTU). Currently, 4 students are working for Ph.D. under my guidance. A seasoned academician having more than 17 years of experience, he has published four books (Programming in C, Programming in C++ and Blockchain and Cryptocurrency) under I.K. International Publishing House Pvt. Ltd. Including Distributed Artificial Intelligence: A Modern Approach, Published December 18, 2020 by CRC Press. He has undergone industrial training programs during which he was involved in live projects with companies in the areas of SAP, Railway Traffic Management Systems, and Visual Vehicles Counter and Classification (used in the Metro rail network design). He is an alumnus of Netaji Subhas Institute of Technology (NSIT), Delhi Universit

Pethuru Raj Chelliah has been working in the Site Reliability Engineering (SRE) Center of Excellence, Reliance Jio Infocomm Ltd. (RJIL), Bangalore. My previous stints are in IBM Cloud center of Excellence (CoE), Wipro consulting services (WCS), and Robert Bosch Corporate Research (CR). In total, I have gained more than 17 years of IT industry experience and 8 years of research experience. Finished the CSIR-sponsored PhD degree at Anna University, Chennai and continued with the UGC-sponsored postdoctoral research in the Department of Computer Science and Automation, Indian Institute of Science, Bangalore. Thereafter, I was granted a couple of international research fellowships (JSPS and JST) to work as a research scientist for 3.5 years in two leading Japanese universities. Published more than 40 research papers in peer-reviewed journals such as IEEE, ACM, Springer-Verlag, Inderscience, etc. Have authored and edited 16 books thus far and focus on some of the emerging technologies such as IoT, Cognitive Analytics, Blockchain, Digital Twin, Docker Containerization, Data Science, Microservices Architecture, fog / edge computing, Artificial intelligence (AI), etc.

Prabhakar Tiwari Madan Mohan Malaviya University of Technology, Gorakhpur (U.P.) India | MMMEC · Department of Electrical Engineering

Victor Hugo C. de Albuquerque has a PhD in Mechanical Engineering with emphasis on Materials from the Federal University of Paraíba (2010), an MSc in Teleinformatics Engineering from the Federal University of Ceará (2007), and he graduated in Mechatronics Technology at the Federal Center of Technological Education of Ceará (2006). He is currently Assistant VI Professor of the Graduate Program in Applied Informatics at the University of Fortaleza (UNIFOR). He has experience in Computer Systems, mainly in the research fields of: Applied Computing, Intelligent Systems, Visualization and Interaction, with specific interest in Pattern Recognition, Artificial Intelligence, Image Processing and Analysis, as well as Automation with respect to biological signal/image processing, image segmentation, biomedical circuits and human/brain-machine interaction, NeuroBiofeedback, Neurorehabilitation, Visual Stimulation, including Augmented and Virtual Reality Simulation Modeling for animals and humans. Additionally, he has research at the microstructural characterization field through the combination of non-destructive techniques with signal/image processing and analysis, and pattern recognition.

***

Rahul is currently designated as an Assistant Professor in the Department of Software Engineering, Delhi Technological University (DTU), Delhi, India. He has completed his Master of Technology in Information Security from the Department of Computer Science and Engineering, NSUT East Campus (Formerly AIACTR), Delhi. He has completed his Bachelor of Technology in Computer Science and Engineering from GGSIPU, Delhi. He has published more than 40 research papers in International Conferences and Journals. He has more than 8 years of experience in the area of teaching. His research area of interest includes Machine learning, Deep Learning, NLP, Social Networking, and Neural Networks.

A. Muthumari hold a Ph.D. in Computer Science and Engineering from Anna University, Chennai, India. Before this, he earned an M.E. in Computer Science and Engineering from Annamalai University, Chidambaram, Tamilnadu, and a B.E. in Information Technology from Madurai Kamaraj University, Madurai, Tamil Nadu, India in 2007 and 2004, respectively. With close to a decade of experience in the field, he currently serve as an Assistant Professor in the Department of Computer Science and Engineering at Anna University Regional Campus, Madurai, Tamil Nadu, India. His academic and research contributions are extensive, with 18 articles published in international journals and presentations at more than 18 international and national conferences. His research interests encompass many areas, including Big Data, Audio and Speech Processing, Image Processing, Soft Computing, and Computational Models.

Chilambuchelvan Arul Gnanaprakasam is a professor at R.M.D. Engineering College, Chennai M.E. Coimbatore Institute of Technology, Coimbatore Research Interests: Embedded System, VLSI design and soft computing.

Varagantham Anitha Avula has a PhD in Computer Science. Having more than 20 years of teaching experience for Under Graduate and Post Graduate Students of Computer Science in reputed institutions which includes 10 years of experience in Department of Computer Science, Hawassa University,

Ethiopia. Advised around 20 MSc. Computer Science students. Domain of Interest: Machine Learning, Information Retrieval Systems, Computer Vision and Image Analysis. Life Member of ISTE.

Bharathi B. pursued their bachelor degree in Computer Science and Engineering in VelTech Engineering College during 2007-2014. They completed their master degree in Computer Science and Engineering in R.M.K. Engineering College during 2012-2014. I am pursuing my Ph.D in Anna University. Bharathi is currently working as an Assistant Professor in R.M.K. Engineering College since 2021.

B. Gopinathan is currently working as an Associate Professor in the Department of Computer Science and Engineering, Adhiyamaan College of Engineering, Hosur. TamilNadu. India. He was born on March 30th 1984 at Tirupattur, Tamilnadu, India. He obtained his Master of Engineering degree in Computer Science and Engineering from Anna University, Chennai. He received his Ph.D degree from Anna University, Chennai, in Mobile Adhoc Networks. His areas of specialization include Mobile Computing, Wireless Networks and IOT, Web Usage Mining, Parallel processing, Operating systems. He has published 50+ papers in International Journals and Conferences. He has more than 18 years of Teaching and Research Experience.

C. Esakkiappan, M.E, Ph.D, is a Professor and Head of the Department of Electronics and Instrumentation Engineering. He hails from Tirunelveli District, Tamil Nadu, India. He received his B.E degree in Electronics and Instrumentation Engineering from Manonmaniam Sundaranar University, Tirunelveli in 1998 and M.E degree in Control Systems from Jadavpur University, Kolkatta in 2003. He has obtained his Ph.D degree in the field of Process Control under the Faculty of Electrical Engineering from Anna University, Chennai in the Year 2013. Presently, he is working as Professor & Head in the Department of Electronics and Instrumentation Engineering, Panimalar Engineering College, Chennai, Tamil Nadu. He has been in teaching profession for the past 20 years. He has presented many papers in International conferences and published many papers in International journals. He has also received award for presenting the best paper in the IEEE international conference. His area of interest includes: Relay Feedback, System Identification, Neural Networks and Advanced Process Control.

Aayush Chowdhury is currently pursuing M.Tech in Data Science from the Department of Software Engineering, Delhi Technological University, Delhi. His research area of interest includes Natural Language Processing dealing with Sentiment Analysis and Information Retrieval, Computer Vision and Predictive Modelling.

D. Vidyabharathi is currently working as an Associate Professor in the Department of Computer Science, Sona College of Technology, Salem, Tamilnadu, India. she was born on 3rd October 1977 at Tiruvannamalai, Tamilnadu, India. She obtained her Bachelor of Engineering degree in Computer Science and Engineering from Vellore Engineering College, Vellore ,University of Madras, Master of Engineering degree in Computer Science and Engineering from SRM Engineering College, University of Madras and 13th University rank holder. She received her Ph.D degree from Anna University, Chennai, in Hyperparameter Optimization for Deep Learning Models. Her areas of specialization include Machine Learning, Deep Learning, Optimization, Data Structures, Analysis of Algorithms, Data mining and Cloud computing. She has published 20+ papers in International Journals and Conferences. she has more than 23 years of Teaching and Research Experience.

M. Dinesh Babu received a B.E. (Mechanical Engineering) from the University of Madras in 2002. M.Tech from V.I.T. Vellore in 2004. He obtained his Ph.D. degree from C.E.G. campus Anna University in 2018 for her research work in Energy Systems Engineering. He started his teaching career in 2004 and has nearly 19 years of experience both in teaching and research. He has published 61 Journals indexed in Scopus with a citation of 2640 with an h-index of 27. He has guided over 30 UG students and one Ph.D. research scholar and published 3 patents. His area of interest is solar energy, alternate fuels, heat transfer, and energy storage systems.

G. Elumalai M.E.,Ph.D., is working as Professor in Department of Electronics and Communication Engineering, Panimalar Engineering College, Chennai, and Tamilnadu, India. He is more than two decades in Teaching, He participated in many national/international conference and workshop in various institutions in India. He published many articles in a national and international journal. He also published two books namely Linear Integrated Circuits and Analog and Digital Communications. He is a life member of IETE (M-501208). He has also guided a few undergraduate and postgraduate students for their final year project work. His research area is in the field of Wireless Sensors Network, wireless communication, IoT and signal processing.

G. Nirmala, MCA, M.E, Ph.D, Associate Professor in Department of Computer Science and Engineering. MCA from Alagappa University ,M.E CSE from Anna University and Ph.D - faculty of Information and communication Engineering from Anna University in the field of Deep learning. She worked as a lecturer in the department of Computer Science and Engineering in Alagappa Chettiar college of Engineering and Technology, Karaikudi from 2003 to 2006. She has been in the teaching and learning profession, counselling, mentoring students for the past 21 years and has handled both for UG and PG programs. Her areas of interest include Deep Learning, Computer Vision, Image Processing, system software, Computer Architecture, Operating Systems, Object Oriented Programming and Middle ware Technologies. She has published 12 papers in International, national Conferences and Indexed Journals. She has attended many workshops & FDPs sponsored by AICTE related to her area of interest. She has completed mission 10X high impact teaching skills by Wipro and campus connect certified by Infosys. She received best paper award in 2014.She has completed 9 nptel courses. She has completed all 8 modules in NITTR. She is a life time member of ISTE. She has completed online courses in coursera, udemy and Infosys spring board.

J. Jeno Jasmine, Associate Professor, Department of Computer Science and Engineering in R.M.K. Engineering College, Chennai. She has completed her Ph.D. in Information and Communication Engineering from Anna University-Chennai in the year 2020. She obtained her M.E. (Network Engineering) in the year 2008 from Arulmigu Kalasalingam College of Engineering, Krishnankovil and her B.E. (Information Technology) in the year 2006 from Karunya Institute of Technology and Science, Coimbatore. She has thirteen plus years of teaching experience. Her area of interests is on Computer Networks. She has published more than 6 research articles in journals, conferences and a patent.

Roslin Dayana K., is currently working as an Assistant Professor in the Department of Computer Science and Engineering at R.M.D. Engineering College (Autonomous), Kavaraipettai, Tiruvallur District where she has been a faculty member since 2012. The author completed her under graduate degree in Bachelors of Engineering (CSE) from Dr. N.N Engineering College affiliated to Madras University and post graduate degree in Master of Engineering (CSE) from Vel Tech Engineering College affiliated to Anna University. She is pursuing Ph.D. degree in the area of Cloud Computing under Anna University. She possesses around fifteen years of Teaching experience and about six months of industrial experience. Her areas of interest include Cloud Computing, Big Data Analytics, Computer Programming and Data Base Management Systems. She has five research publications in National and International Journals and has participated in six National and International Conferences. She has attended many workshops, webinars, Faculty Development Programmes (FDP), and Short Term Training Programmes (STTP) conducted by Anna University, All India Council for Technical Education (AICTE), National Universities and various other Colleges across the country. She has published a patent and completed two consultancy projects. She has also completed various Massive Open Online Courses (MOOC) in Coursera, NPTEL, edX, Udemy and Infosys Springboard. She is currently training students in the industry-oriented course named Software development practices. She is a life time member of professional bodies such as ISTE, IAENG and MTTF.

Chairma Lakshmi, born in 1990, in Sivakasi, India, is a pioneering force in engineering and renewable energy education. She earned a B.E. in Electronics and Instrumentation Engineering at Kamaraj College of Engineering and Technology (2007-2011) and an M.E. in Control and Instrumentation Engineering at National Engineering College, Anna University (2011-2013). Currently pursuing a Ph.D. in Renewable Energy at Anna University, she is dedicated to advancing sustainable solutions. Since May 27, 2013, Mrs. Lakshmi has been imparting knowledge as an Assistant Professor at R.M.K. Engineering College. Her dynamic teaching style fosters innovation, shaping the minds of aspiring engineers. Mrs. Lakshmi's impactful research includes "Performance Analysis of MPPT Algorithms for SPV Power Generation" (2013) and "Dust Deposition's Effect on Solar PV Module Performance" (Journal of Renewable Materials, 2022). She presented on diverse topics like IoT-based Traffic Lights, Robotic Accident Prevention, and Augmented Reality-based Home Automation at international conferences in 2023. With a Web of Science ResearcherID (AAE-7355-2019) and ORCID Profile (0000-0002-1221-8432).

K S Ananda Kumar is working as Associate Professor in Information Science and Engineering, Atria Institute of Technology, Bangalore . He pursued B.Tech from Koneru Lakshmaiah College of Engineering, Guntur, affiliated to Acharya Nagarjuna University, Andhrapradesh, India. Pursued M.Tech from R.V.College of Engineering, Bangalore, affiliated to Visvesvaraya Technological University, Belagavi, Karnataka, India. Pursued Ph.D in Computer & Information Sciences from Visvesvaraya Technological University, Belagavi, Karnataka, India. He is Life member of the Professional societies like IAENG, IMAPS, UACEE. He is having 13+ years of Teaching and Research experience. He is published 35 research papers in reputed journals and International conferences. His Research interests are in the field of Wireless Sensor Networks, Network Security, IOT, ML, Data Mining, Big data.

Sunil Kumar holds the position of Associate Professor within the Faculty of Commerce and Management. His academic journey led him to achieve a Doctorate degree, specializing in the realm of Finance and Business Analytics. His scholarly pursuits have resulted in numerous research contributions, including both solo and collaborative efforts, yielding a collection of impactful research papers. Beyond his research accolades, Dr. Sunil Kumar is a passionate educator and has conducted over 20 workshops that delve into various aspects of financial modeling, spreadsheet modeling, and data analysis. His expertise extends to a wide array of tools and programming languages, including Python, R Programming, SPSS, E-Views, Stata, and Gretl.

L. Manimegalai graduated from Sakthi Mariamman Engineering College, Sriperumbudur, Anna University, Chennai, in Electronics and Communication Engineering during the year 2013. She obtained her Master degree in Embedded System Technologies from Saveetha Engineering College, Saveetha Nagar, Thandalam, Anna University, Chennai, during the year 2015. She has more than 8 years of teaching experience and her area of interest include Embedded Systems, Embedded Networks and IoT. She is working as Assistant Professor in the Department of Electronics and Communication Engineering, R.M.K. College of Engineering and Technology, Puduvoyal, Chennai, Tamil Nadu. She is a Life member of ISTE.

L. P. Sajitha is currently working as an Assistant Professor in the Department of Computer Science and Business Systems, R.M.K Engineering College, Thiruvallur. She was born on October 29th 1990 at Theni, Tamilnadu, India. She obtained his Master of Engineering degree in VLSI Design from Anna University, Chennai. She is currently pursuing her Ph.D degree from Anna University, Chennai. Her areas of specialization include Cryptography, Wireless Networks and IOT, Web Usage Mining, Data Structures, Design and Analysis of Algorithm. She has published 5 papers in International Journals and Conferences. She has more than 3 years of Teaching experience.

Rajendiran M. is a Professor in the Department of Computer Science and Engineering, Panimalar Engineering College, Chennai, India. He Holds a Ph.D. degree in Computer Science and Engineering with specialization in Mobile ad hoc networks. His research areas are mobile networks, machine learning, data analysis and routing protocol.

Rekha M is Pursuing PhD degree in Anna University, Tamil Nadu, India. Presently she is working as Assistant Professor in R.M.K. Engineering College,Tiruvallur. She has published four research papers in Scopus indexed journals. She has Published two Patents and She is also a member of few professional societies like IEANG, ISTE . Besides, she is recognized as an Industry Trained Faculty by Virtusa and WIPRO. Also, she has successfully completed more than 15 online courses certified by NITTTR, NPTEL, edX and Coursera. Her research interests are Cloud Computing, Machine Learning and Deep Learning.

M. Theodore Kingslin M.E., has completed M.E in Applied Electronics from Rajas Engineering College, Vadakangulam. He has completed B.E in Electronics and Communication Engineering from Government College of Engineering Tirunelveli. He has teaching experience for 11 years. Currently, he is working as an Assistant Professor in Department of Electronics and Communication Engineering at R.M.K College of Engineering and Technology, Puduvoyal, Gummidipoondy Taluk, PIN - 601206. E-mail ID: theodorekingslinece@rmkcet.ac.in Research expertise is Image Processing and IOT.

. M. S. Kavitha, B.E, M.E, Ph.D is Associate Professor in Department of Electrical and Electronics Engineering, since June 2000. She obtained her B.E(EEE) from Alagappa Chettiar College of Engineering and Technology, Madurai Kamaraj University and M.E(Applied Electronics) from Anna University Chennai in 2007. She has been in the teaching profession for the past 23 years and has handled both UG and PG programmes. She is working in R.M.K Engineering College from the year 2000. Her areas of interest include Low Power VLSI Design, Reconfigurable architectures, Embedded systems and Automotive Electronics, Digital Logic Circuits, Control Systems, Circuit theory, Electronic Devices and Circuits, Electromagnetic Theory, Digital Signal Processing etc. She has attended many workshops & FDPs sponsored by AICTE related to her area of interest.

R. Manikandan completed his undergraduate studies in Electronics and Instrumentation Engineering at Annamalai University, Chidambaram in 2002. In 2008, he earned his master's degree in Applied Electronics from Anna University, Chennai. Later, in 2014, he successfully obtained his Ph.D. in Image/Video Processing from the Department of Advanced Sports Training and Technology at Tamil Nadu Physical Education and Sports University, Chennai. He has more than 21 years of experience in teaching and is a life member of ISTE. His primary research interests encompass automation, computer vision, and image/video processing. Presently, he holds the position of a Professor in the Department of Electronics and Communication Engineering at Panimalar Engineering College, Chennai.

S. Meenakshi Department of Science and Humanities (General Engineering), R.M.K. Engineering College, Kavarapettai, Tamil Nadu 601206, India.

Sadish Sendil M. is working as Professor & Head of the Department of Emerging Technologies at Guru Nanak Institute of Technology, Hyderabad. He completed his Bachelor of Engineering in Computer Science and Engineering from Bharathidasan University, Tiruchirappalli in 1997 with First class. He completed his Master of Engineering with Distinction and 18th Rank holder from Anna University, Chennai in 2004. He Obtained his Ph.D. in Computer Science and Engineering from Anna University in March 2012. He has 26 years of teaching experience in which 6 years in Gulf Cooperation Council (GCC) countries (Saudi Arabia and Sultanate of Oman). He is a EMC2 Certified Academic Associate Cloud Infrastructure and Services. He has guided Ph.D. research scholars and awarded successfully. He has published 33 research papers in Web of Science / Scopus / UGC listed / referred international journals. He has successfully published five Indian patents including one design patent. He is authored 1 international and 2 national books. He is a life member of CSI, ISTE and IEEE, Associate Member of UACEE.

Hariharan Narayanaswamy is an Assistant Professor in the department of Science and Humanities (Electrical and Electronics Engineering Division), R.M.K. College of Engineering and Technology, Chennai with more than 17 years of Teaching Experience. He received B.E. Electrical and Electronics Engineering from Anna University, Chennai, M.E. Power Electronics and Drives from Anna University Chennai and Pursuing Ph.D.in Anna University, Chennai. He has published 8 research papers in International / National Journals and 12 International / National conferences. His Area of research interest includes Power Electronics and Embedded Systems. He is an active Member in ISTE, IAAC & IAENG.

Padmavathi N., M.E., Ph.D., is an Assistant Professor of R.M.D. Engineering college. She has completed her B.E. Degree in Electronics and Instrumentation Engineering in the year 2005 from MAHARAJA Engineering College Affiliated to ANNA University and M.E Degree in Power Electronics & Drives in RMK Engineering College, Chennai in the year 2013. She has obtained Ph.D (Electrical Engineering) from Anna University, Chennai in the year 2022 .She has 18.5 years of teaching experience to UG classes and has guided many B.E. projects. She has published 08 papers in International

Journals (Scopus / SCIE indexed) and two papers in Inter National Conference. She published three patents in the Area of Renewable Energy sources and embedded systems. She has published the Book on "INDUSTRIAL INSTRUMENTATION". She is the coordinator for the Building Management Systems Training classes in associated with M/S Johnson Controls, Pune. She completed Four NPTEL courses in the area of Introduction to Smart Grid, DC Micro grid, Introduction to Internet of Things and Machine Learning and received Certificate of Merit for the same. Also she completed online courses in the thurst areas related to Artificial Intelligence, Data Science and Introduction to Python from Udemy, courser and Infosys springboard. She has attended many workshops & FDPs sponsored by AICTE, NITTTR, ATAL &TEQUIP related to her area of interest. She was a co-coordinator for AICTE–ISTE sponsored "Entrepreurship awareness camp" training programme for engineering students and ATAL - AICTE Sponsored online Faculty Development Program in the year 2021. Her Areas of Interest are: Control systems, Renewable Energy systems and Artificial Intelligence in Electrical Engineering. She is a Life member of Instrument Society of India,ISTE, IAENG, MTTF.

P. Kavitha, Associate Professor, Department of Computer Science and Engineering, R.M.K. Engineering College, Kavaraipettai. She Received her B.E. Degree from University of Madras in 1999 and M.E. Degree from Anna University, Chennai in 2007.She Received her Doctoral Degree from Anna University, Chennai in 2020. Her research interests include Image processing , Artificial Intelligence and Machine Learning . She is a life member of ISTE. she received her Academic Excellence Award by Institution of engineers in March 2023.

Mahalakshmi P. completed BE and ME in computer science and engineering stream Working as a assistant professor. Passionate about teaching and learning. Always enthusiastic to learn new technologies and skills.

Sathyaraj P., working as an assistant professor in RMKCET

Shashank Parmar is currently working as a Software Engineer at VISA specifically on Mainframe Applications. He had worked on research projects in the fields of Natural Language Processing, Computer Vision, Algorithm Design and Applied Mathematics.

Nikhat Parveen received the B.Sc.(Computer Science) and M.C.A degree from Andhra University , A.P. India in 2000 and 2003, respectively, and the Ph. D, degree from the Department of Computer Application, Integral University, Lucknow, U.P. India. She is currently working as an Associate Professor in the Department of Computer Science and Engineering, Koneru Lakshmaiah Education Foundation, Guntur, India. She has more than twelve years of teaching experience and six years of research experience. She is having 6 national patents. Her research and publication interests include Artificial Intelligence, Machine Learning, , Security Software, Security Testing, Software Engineering, and Requirement Engineering. She is also working in the area of soft computing, image analysis, big data analytics, and the IoT. Her research has been chronicled in over 30 journal publications and international conferences. She is a Life Time Member of CSI, ACM and IAENG, IACSIT.

Balachandra Pattanaik, PDF, Ph.D., FIE, CEng.(India), SMIEEE(USA). Professor, Department of Electrical and Computer Engineering, PhD in Electrical Engineering & PDF in Electrical Engineering.

Area of interest is in, Electrical, Electronics, IoT, Computer Applications, Machine Learning, VLSI Embedded Systems and Mathematics . He has around 24 years of experience in Teaching, Research and in IT Industry. He has completed 4(four) Funded Research projects at Abroad and three ongoing at Wollega University. Presently holding the additional position of Vice-Chair for IEEE(USA) Ethiopia Section and Founder of Ethiopia IEEE Section. Also Peripheral faculty of Saveetha School of Engineering India. He has 18 innovative Patents, Out of which Published Indian Govt. Patent Journal 14, International patent grant entitlement six, in UK, Australia and Germany. He has ELEVEN UG AND PG Books published IN Abroad and India & 10 Book chapters as well as numerous National and International publications. He's IEEE, USA Power & Energy student chapter Advisor at BULE HORA UNIVERSITY, Ethiopia, He's also IEEE USA Senior member, lately entered Fellow member IE India & Chartered Engineer award from IE.

Priyanka Pramila R. is an Assistant Professor in the Department of Artificial Intelligence and Data Science at R.M.K. Engineering College. She earned her Bachelors degree in Information Technology from Sathyabama University and her Masters in Computer Engineering from Anna University. She is currently Pursuing her Ph.D in Sathyabama University. Her research interests include artificial intelligence, computer vision, Edge Computing and Machine Learning. Her current research focuses on developing algorithms to improve image recognition and classification for Cancer Diagnosis. In addition to her research and teaching responsibilities, Ms.R.Priyanka Pramila serves as a mentor to graduate and undergraduate students interested in pursuing careers in the field of AI. In her free time, she enjoys reading, and spending time with her family.

G. Rathinasabapathi is a faculty member in the Department of Mechanical Engineering at Panimalar Engineering College, situated in Chennai, Tamil Nadu 600123, India.

S. Edwin Raja, Assistant Professor, Department of Computer Science and Engineering in Vel Tech Rangarajan Dr.Sagunthala R&D Institute of Science and Technology, Chennai. He has completed his Ph.D. in Information and Communication Engineering from Anna University-chennai in the year 2021. He obtained his M.Tech. (Network Engineering) in the year 2010 from Kalasalingam University, Krishnan kovil and B.E. (Computer Science and Engineering) in the year 2006 from C.S.I Institute of Technology, Thovalai. He has more than thirteen plus years of teaching experience and his area of interests are Cyber Security, Big Data Analytics, Machine Learning & Computer Networks. He has published more than 4 research articles in various journals and conferences. Also, he has published 3 patents, acted as reviewer in reputed journals.

Karkuzhali S. completed M.E in Power Electronics and Drives at Mepco Schlenk Engineering College, Anna University and pursuing Ph.D in Hybrid Electric Vehicles.She has 4 years of Industrial experience and more than fifteen years of teaching experience.

Karthika S. has completed her PhD under the Faculty of Information and Communication Engineering at Madras Institute of Technology, Chennai. She currently working as Assistant Professor in the Department of Electronics and communication Engineering, Anna University CEG campus Chennai. She is having more than three years of teaching experience. She had completed M. E. Communication and Networking from Madras Institute of Technology, Chennai and Electronics and Communication from

VLB Janakiammal of College of Engineering and Technology, Coimbatore. She has published more than 25 papers in International Journals, one book chapter and more than five papers in National/International Conferences. She has received mentorship awards from Texas Instruments, Best Young Researcher award and Best paper award for her research work. She had filed three patents in the field of IoT.

Rajakumar S. is currently working as Professor in the department of Electronics and Communication Engineering, Panimalar Engineering College, affiliated to Anna University, Chennai. He has organized several Conferences and Seminars related to Image Processing and Signal Processing. He completed his B.E., degree in Electronics and Communication Engineering from Sun College of Engineering and Technology, Nagercoil affiliated to Manonmaniam Sundaranar University, Tirunelveli. He obtained his M.E., degree in Applied Electronics and Ph.D in Image Processing from Sathyabama University, Chennai, India in the year 2006 and 2013 respectively. He has published several research articles in National and International Conferences and Journals. His areas of interest include Digital Signal Processing, Image Processing, Pattern Recognition, VLSI and Communication Engineering.

S. Rajakumari is an Assistant Professor of Mathematics in the Department of Science and Humanities at R.M.D. Engineering College, Kavaraipettai, Thiruvallur, has successfully completed her doctoral studies in graph theory at the University of Madras, Chennai. She has presented talks on graph theory at fifteen National and International conferences, authored ten papers in reputable International Journals, and actively participated in fifteen conferences. Her research is centered around Algorithmic Graph Theory with a specific focus on Edge Coloring in Graphs.

S. Shobana, B.Tech,M.E,(Ph.D),is Assistant Professor in the Department of Information Technology at R.M.K. Engineering College, since July 2022. She has 8 years of teaching experience. She received her B.Tech(IT) degree from Jayaraj Annapackiam CSI College of Engineering- Thoothukudi , M.E(CSE) degree from PSNA College of Engineering & Technology– Dindigul & She accomplished 5th Rank in Anna university . She is currently pursuing Ph.D in Dr.MGR Educational and Research Institute and her research area is in Hyper spectral Image Processing.

Srijayanthi S. received her B.E. degree in Computer Science and Engineering from University of Madras and M.E. degree in Computer Science and Engineering from Anna University Chennai, India, in 2002 and 2013, respectively. She completed her Ph.D. degree in Information and Communication Engineering at Anna university Chennai. She is currently working as an Associate Professor in R.M.K. Engineering College. Her research interests include Privacy and Data Mining.

S. Subashree is an Assistant professor in egs.pillay engineering college and is passionate towards teaching and learning ambitious about making and reaching higher heights always.

S. Sundaramoorthi, currently working as Associate Professor, EEE Department, Panimalar Engineering College, Chennai. He received B.E degree in Instrumentation and Control Engineering from Madurai Kamaraj University ,Madurai in 1999 and M.E degree in Power Electronics and Drives from Anna University ,Chennai in 2008.His area of interest are Power Electronics and Renewable Energy. He has published many articles in International and National Journals and He is a Life Member of ISTE.

S. D Lalitha working as an Associate Professor in the Department of Computer Science and Engineering at R.M.K. Engineering College, has about 21 years of teaching experience. She received her B.E degree in Computer Science and Engineering from Tagore Engineering College and M.Tech degree in Computer Science and Engineering from Dr. M.G.R Educational and Research Institute. She has completed her Ph.D in Anna University, Chennai. She has published 15 research papers. Her areas of research include Image Processing, Deep Learning, Cloud computing and Security.

Balaji. S. R is working as Associate Professor at ECE department, Panimalar Engineering College, Chennai. They completed thier B. E. in Electronics and Instrumentation Engineering, M. Tech in VLSI Design, Ph.D. degree in Video Image Processing. Balaji has around 16+ years of experience in teaching. Their research area includes AI, IoT, Image Processing, VLSI Design, Machine Learning, Deep Neural Network, CNN, Process control and Automation

Dhaval Shrivastava is currently pursuing M.Tech in Data Science from the Department of Software Engineering, Delhi Technological University, Delhi. His area of interest includes Machine Learning, Data Structures and Sentiment Analysis.

Vibhor Singh is currently pursuing M.Tech in Data Science from the Department of Software Engineering, Delhi Technological University, Delhi. He had done projects in Natural Language Processing focusing on Sentiment Analysis

Neelam Sunda is currently working as an Assistant Professor in the Department of Computer Science of Kanoria PG Mahila Mahavidyalaya, Jaipur, Rajasthan, India since 2016. She did her Master's in Computer Science and Application from Banasthali Vidyapith, Rajasthan, India in 2014. She also qualified for the UGC-NET exam in Computer Science and currently pursuing her Ph.D. in Machine Learning from Rajasthan Technical University, Kota. Her research interest includes Artificial Intelligence, Machine Learning, AGILE Framework, Software Engineering, and Cyber Security. Ms. Neelam has written two books in the field of Computer Science i.e. "Database Management System" and "Machine Learning for Absolute Beginners". Her extensive publication record includes research articles in national and international conferences like IEEE, Springer, and Scopus-indexed journals. Notably, she has also published a patent titled "An IoT-Based Anti-Theft Mechanism That Keeps An Eye Out For Any Malicious Activity", demonstrating her innovative thinking and practical approach to problem solving.

Gnanasekaran Thangavel has published fifty nine journals and ninety three research papers at various national and international conferences. He recognized supervisor of Anna University in Information and Communication Engineering. Currently he has guided Seven PhD scholars. He has co authored six books. He has filed five patents and granted one patent. He has organized many Conferences, Technical Symposiums, Seminars and Faculty Development programmes. Most of them are sponsored by reputed national funding organizations like AICTE, ISTE, DRDO, CSIR etc. He is also reviewer for many international journals and conferences. He is an active member of societies like IEEE, IET, ISTE, IE(I), IAENG, IDS and ACEEE. He is associated with Agilent Technologies, Alcatel-Lucent, National Instruments, Erircsson, TCS, CTS, and Wipro Technologies etc. He has 34 years of teaching experience and 15 years research experience. Presently he is working as Professor and Head of Electronics and Instrumentation

Engineering department at R M K Engineering College, Chennai. His areas of research interest include Wireless Communications & Networks, Digital communication, Image processing, Cloud computing,

V. Saravanan is currently working as an Associate Professor in the Department of Computer Science, College of Engineering & Technology, Dambi Dollo University, Oromia Region, Ethiopia. He was born on March 15th 1983 at Krishnagiri, Tamilnadu, India. He obtained his Master of Engineering degree in Computer Science and Engineering from Anna University, Chennai with Distinction. He received his Ph.D degree from Anna University, Chennai, in Mobile Communication. His areas of specialization include Mobile Computing, Wireless Networks and IOT, Web Usage Mining, Data Structures, Design and Analysis of Algorithm. He has published 50+ papers in International Journals and Conferences. He has more than 15 years of Teaching and Research Experience.

Shree Gaayathri.V. is a single child. Iam a very sincere, punctual and a self-motivated student. I love singing, painting and gardening too. Iam interested to gain knowledge about new things. I am very passionate towards singing. My goal is to become a Professor in future.

Index

E

F

G

H

I

K

L

M

N

O

P

R

S

T

www.ingramcontent.com/pod-product-compliance
Lightning Source LLC
LaVergne TN
LVHW062114300926
849161LV00010B/2455